INSULIN—THE CROOKED TIMBER

Insulin—The Crooked Timber

A History From Thick Brown Muck to Wall Street Gold

Kersten T. Hall

OXFORD
UNIVERSITY PRESS

UNIVERSITY PRESS

Great Clarendon Street, Oxford, OX2 6DP,
United Kingdom

Oxford University Press is a department of the University of Oxford.
It furthers the University's objective of excellence in research, scholarship,
and education by publishing worldwide. Oxford is a registered trade mark of
Oxford University Press in the UK and in certain other countries

Published in the United States of America by Oxford University Press
198 Madison Avenue, New York, NY 10016, United States of America

British Library Cataloguing in Publication Data
Data available

Library of Congress Control Number: 2021943457

ISBN 978–0–19–285538–1

DOI: 10.1093/oso/9780192855381.001.0001

Printed and bound by
CPI Group (UK) Ltd, Croydon, CR0 4YY

To Michelle, Matthew, and Edward, for all your endless patience and support. And to everyone in the NHS diabetes team who have looked after me these past ten years and kept the insulin flowing.

Out of the crooked timber of humanity, no straight thing was ever made.

Immanuel Kant, German philosopher 1724–1804

Contents

List of Figures ix

Preface—On the Shoulders of Giants . . . ? xv

Acknowledgements xviii

Introduction—**Taming the Tiger** 1

1 **The Pissing Evil**—*a colourful description of diabetes by 17th-century English physician Thomas Willis* 4

2 **Thick Brown Muck**—*Canadian scientist Fred Banting wins the Nobel Prize for the discovery of insulin …and is furious* 16

3 **The Vision of Ezekiel**—*clinicians are stunned at the power of insulin to save lives, but it proves to be a double-edged sword* 52

4 **A Greek Tragedy**—*German clinician Georg Zuelzer snatches defeat from the jaws of victory* 89

5 **The Wasp's Nest**—*insulin proves to be a poisoned chalice for its discoverers* 101

6 **Be Careful What You Wish For**—*the case of Romanian scientist Nicolai Paulescu underlines the truth of an old proverb* 125

7 **'In Praise of Wool'**—*the humble wool fibre sets in motion a revolution in biochemistry* 137

8 **A Boastful Undertaking**—*a discovery made in a fume-filled stable offers the key to unlocking insulin* 166

9 **The Blobs That Won a Nobel Prize**—*or two, all thanks to some coloured spots on a piece of filter paper* 191

10 **The Prophet in the Labyrinth**—*biochemist Erwin Chargaff helps unlock the secrets of DNA, but fears where this may lead* ... 219

11 **The Clone 'Wars'**—*a conflict in which insulin proves to be a decisive weapon* 252

12 **Wall Street Gold**—*in an act of modern-day alchemy, insulin makes stock market history* 299

13 **'Don't You Want Cheap Insulin?'**—*what is it exactly that we want from science? And does the story of insulin have any lessons for us today?*   328

Bibliography   351
Notes   380
Index   445

List of Figures

1 Boston diabetes clinician Elliott P. Joslin (1869–1962). Insulin Collection, F. G. Banting (Frederick Grant, Sir) Papers, Thomas Fisher Rare Book Library, University of Toronto; MS COLL 76 (Banting), Box 63, Folder 3A. Reproduced courtesy of the Thomas Fisher Rare Book Library, University of Toronto. Online at https://insulin.library.utoronto.ca/ islandora/object/insulin%3AP10132.   13

2 Photograph of Fred Banting (1891–1941) taken on December 27, 1922. Insulin Collection, F. G. Banting (Frederick Grant, Sir) Papers, Thomas Fisher Rare Book Library, University of Toronto; MS COLL 76 (Banting), Box 63, Folder 3A.
Reproduced courtesy of the Thomas Fisher Rare Book Library, University of Toronto. Online at https://insulin.library.utoronto.ca/islandora/object/ insulin%3AP10042.   14

3 Photograph of Professor John J. R. Macleod (1876–1935) [ca. 1928]. Insulin Collection, University of Toronto Archives. B1995-0034.
Reproduced courtesy of the Thomas Fisher Rare Book Library, University of Toronto. Online at https://insulin.library.utoronto.ca/islandora/object/ insulin%3AP10134.   17

4 Banting's proposed plan of research. Diagram by K. T. Hall.   25

5 Charles Best (1899–1978). Insulin Collection, C. H. Best (Charles Herbert) Papers, Thomas Fisher Rare Book Library, University of Toronto MS COLL 241 (Best) Box 47, Folder 3.
Reproduced courtesy of the Thomas Fisher Rare Book Library, University of Toronto. Online at https://insulin.library.utoronto.ca/islandora/object/ insulin%3AP10070.   26

6 Photograph of Frederick Banting and Charles Best (1899–1978) with a dog on the roof of the Medical Building (August 1921). Insulin Collection, Best (Charles Herbert) Papers, Thomas Fisher Rare Book Library, University of Toronto; MS COLL 241 (Best) Box 109, Folder 4.
Reproduced courtesy of the Thomas Fisher Rare Book Library, University of Toronto. Online at https://insulin.library.utoronto.ca/islandora/object/ insulin%3AP10077.   27

7 Photograph of laboratory 221. Insulin Collection, University of Toronto Archives, Thomas Fisher Rare Book Library, University of Toronto. Original glass negative in U of T Archives: A65-0004/204 MS COLL 76 (Banting), Box 63, Folder 3B.
Reproduced courtesy of the Thomas Fisher Rare Book Library, University of Toronto. Online at https://insulin.library.utoronto.ca/islandora/object/ insulin%3AP10043.   31

8   Israel S. Kleiner (1885–1966). Insulin Collection, Best (Charles Herbert) Papers, Thomas Fisher Rare Book Library, University of Toronto MS COLL 241 (Best) Box 109, Folder 34.
    Reproduced courtesy of the Thomas Fisher Rare Book Library, University of Toronto.   33

9   a) 'Chart showing sugar levels in blood and urine for dogs 92 and 409,' August 11–31, 1921. Insulin Collection, F. G. Banting (Frederick Grant, Sir) Papers, Thomas Fisher Rare Book Library, University of Toronto; MS COLL 76 (Banting) Mapcase. b) Chart 2 from Banting and Best 1922b, 257. Insulin Collection, F. G. Banting (Frederick Grant, Sir) Papers, Thomas Fisher Rare Book Library, University of Toronto; MS COLL 76 (Banting), Box 62, Folder 13. Offprint.
    a) Reproduced courtesy of the Thomas Fisher Rare Book Library, University of Toronto. Online at https://insulin.library.utoronto.ca/islandora/object/ insulin%3AM10002; (b) Reproduced courtesy of Elsevier and the Thomas Fisher Rare Book Library, University of Toronto. Online at https://insulin.library.utoronto.ca/islandora/object/ insulin%3AP10080.   37

10 Formal photograph of Leonard Thompson. Insulin Collection, F. G. Banting (Frederick Grant, Sir) Papers, Thomas Fisher Rare Book Library, University of Toronto; MS COLL 76 (Banting), Box 12, Folder 1.
    Reproduced courtesy of the Thomas Fisher Rare Book Library, University of Toronto. Online at https://insulin.library.utoronto.ca/islandora/object/ insulin%3AP10046.   45

11 Photograph of James Bertram Collip (1892–1965) [ca. 1920]. Insulin Collection, Collip (James Bertram) Papers, Thomas Fisher Rare Book Library, University of Toronto MS COLL 269 (Collip), Item 1. This photograph is reproduced from the original held by Barbara Collip-Wyatt. Reproduced courtesy of the Thomas Fisher Rare Book Library, University of Toronto. Online at https://insulin.library.utoronto.ca/islandora/object/ insulin%3AP10005.   46

12 Newspaper headline: 'Discovery of Extract That Has Power to Restore Capacity Lost in Diabetes is Made Public by Dr. John R. Murlin'. Insulin Collection, University of Toronto, Board of Governors, Thomas Fisher Rare Book Library, University of Toronto Archives. A1982-0001, Box 24, Folder 11.
    Reproduced courtesy of the Thomas Fisher Rare Book Library, University of Toronto. Online at https://insulin.library.utoronto.ca/islandora/object/ insulin%3AC10031.   62

13 'Toronto Doctors Honored: Nobel Prize to Banting. Dr. J. J. R. Macleod also.' October 1923. MS COLL 76 (Banting) Scrapbook 1, Box 2, Page 81; Insulin Collection, F. G. Banting (Frederick Grant, Sir) Papers.
    Reproduced courtesy of the Thomas Fisher Rare Book Library, University of Toronto. Online at https://insulin.library.utoronto.ca/islandora/object/ insulin%3AC10111.   88

14 Georg Zuelzer (1870–1949). Photographer: Suse Byk. Published by: 'Zeitbilder' 15/1930. Vintage property of Ullstein Bild. © Getty Images. Reproduced under license from Getty Images.   90

15 Photo spread from the *Toronto Daily Star* covering Banting's funeral, March 5, 1941. MS COLL 76 (Banting) Scrapbook 2, Box 2, 116. Title based on contents of item.
Reproduced courtesy of the Thomas Fisher Rare Book Library, University of Toronto.   109

16 a) Charles Best and Norman B. Taylor, *The Physiological Basis of Medical Practice: A University of Toronto Text in Applied Physiology*, 1943, fig, 232, 575. Pencheon Collection/BES, Special Collections, Brotherton Library, University of Leeds. b) 'Chart showing sugar levels in blood and urine for dogs 92 and 409,' August 11–31, 1921. Insulin Collection, F. G. Banting (Frederick Grant, Sir) Papers, Thomas Fisher Rare Book Library, University of Toronto; MS COLL 76 (Banting) Mapcase.
a) Reproduced with permission of Special Collections, Brotherton Library, University of Leeds. b) Reproduced with Courtesy of the Thomas Fisher Rare Book Library, University of Toronto. Online at https://insulin.library. utoronto.ca/islandora/object/insulin%3AM10002.   115

17 Nicolai Paulescu (1869–1931). Insulin Collection, Best (Charles Herbert) Papers, Thomas Fisher Rare Book Library, University of Toronto MS COLL 241 (Best) Box 109, Folder 39.
Reproduced with Courtesy of the Thomas Fisher Rare Book Library, University of Toronto. Online at https://insulin.library.utoronto.ca/islandora/ object/insulin%3AP10081.   126

18 Glycine and the other nineteen amino acids found in proteins. Diagram by K. T. Hall.   138

19 What is a protein? Diagram by K. T. Hall.   144

20 William Astbury (1898–1961).
Reproduced with permission of Special Collections, Brotherton Library, University of Leeds.   145

21 Richard L. M. Synge (1914–1994). Credit: Elliott & Fry, 1952. © National Portrait Gallery, London.
Reproduced under license by National Portrait Gallery, London.   147

22 Archer John Porter Martin (1910–2002).
Credit: Elliott & Fry, 1950. © National Portrait Gallery, London.
Reproduced under license by National Portrait Gallery.   150

23 Longfield Stables, Headingley Lane, Leeds. a) With kind thanks to Mr. Matthew Synge. (b) and (c) by K. T. Hall.   156

24 'A most faithful servant'. (a) by K. T. Hall; (b) and (c) Diagram of chromatography apparatus and amino acid chromatogram taken from Fig. 3 and Plate 1, respectively. In Consden, Gordon, and Martin, 'Qualitative Analysis of Proteins: A Partition Chromatographic Method Using Paper', *The Biochemical Journal* 38 (1944): 224–232.
Reproduced with permission of Portland Press.   160

25 Archer Martin vs William Astbury: dipeptides land the knockout punch.
Diagram by K. T. Hall. 163

26 Dorothy Crowfoot Hodgkin (1910–1994). Ramsey & Muspratt bromide
print, ca. 1937. © National Portrait Gallery, London.
Reproduced under license by National Portrait Gallery. 169

27 Dorothy Hodgkin by Maggi Hambling. Oil on canvas, 1985. © National
Portrait Gallery, London. Reproduced under license by National Portrait
Gallery. 177

28 Dorothy Wrinch (1894–1976). Courtesy of Special Collections, Smith
College. 180

29 Dorothy Wrinch's cyclol hypothesis. Diagram by K. T. Hall. 182

30 a) 'Blobs Win Nobel Prize,' 'News Chronicle', November 7, 1952. b) 1977
Stamp Commemorating Development of Partition Chromatography. a) The
papers and correspondence of Richard Laurence Millington Synge. Trinity
College Library, Cambridge. GBR/0016/SYNG. b) © Royal Mail Group
Limited 198

31 Fred Sanger (1918–2013).
Reproduced with kind permission of MRC Laboratory of Molecular Biology. 200

32 Sanger's strategy for an attack on the amino acid sequence of insulin.
Diagram by K. T. Hall. 204

33 'A Boastful Undertaking'—The complete amino acid sequence of gramicidin
S. Diagram by K. T. Hall. 205

34 a) Sanger's strategy for determining the complete amino acid sequence of
the A and B chains of insulin. b) The complete amino acid sequence of the
A and B chains of bovine insulin as determined by Sanger. a) Diagram by
K.T. Hall. Figure of the chromatogram is for peptide B1gamma, taken from
Fig. 6, p. 469, Sanger, F., and H.Tuppy. 'The Amino-acid Sequence in the
Phenylalanyl Chain of Insulin.' *The Biochemical Journal* 49 (1951): 463–80.
Reproduced with permission of Portland Press. b) Based on sequence
shown in Sanger, F., E. O. P. Thompson, and R. Kitai. 'The Amide Groups
of Insulin,' *The Biochemical Journal* 59 (1954): 509–18. Reproduced with
permission of Portland Press. 208

35 Dorothy Hodgkin photographed in 1989 with her model of the 3D structure
of insulin. Credit: Corbin O'Grady Studio/Science Photo Library. 211

36 Commemorative plaque in The Eagle pub. Photograph by K. T. Hall. 219

37 Oswald T. Avery (1877–1955) celebrating Christmas, December 1940. ©
Oswald T. Avery Collection, US National Library of Medicine. 223

38 Florence Bell (1913–2000).
Reproduced with kind thanks of Mr. Chris Sawyer. 226

39 'Woman Scientist Explains,' News clipping from MS419 A.1, Astbury
Papers, Special Collections, Brotherton Library, University of Leeds.
Reproduced with permission of Special Collections, Brotherton Library,
University of Leeds. 227

40 DNA replication and base-pairing. Diagram by K. T. Hall. 229

41 Portrait of biochemist Erwin Chargaff (1905–2002). Credit: Science Photo Library.
Reproduced with kind permission of the American Philosophical Society.   230

42 George Gamow's model for how the genetic code might work. Diagram by K. T. Hall. Based on Gamow, G. 'Possible Relation between Deoxyribonucleic Acid and Protein Structure,' *Nature* 173 (1954): 318.   237

43 DNA makes RNA makes protein. Diagram by K. T. Hall.   244

44 a) A specific RNA triplet could bind to only one particular type of amino acid. b) 'UUU are great Marshall' c) Marshall Nirenberg (1927–2010) on telephone receiving congratulations for winning the Nobel Prize. a) Diagram by K. T. Hall; b) Courtesy of Marshall W. Nirenberg. © US National Library of Medicine; c) © US National Library of Medicine.   246

45 Herb Boyer sitting in the auditorium at the Asilomar conference. US National Library of Medicine. Courtesy of US National Library of Medicine.   252

46 Stanley N. Cohen. US National Library of Medicine. Courtesy of US National Library of Medicine.   253

47 Paul Berg. Paul Berg Papers, US National Library of Medicine. Courtesy of US National Library of Medicine.   255

48 Berg's proposed experiment. Diagram by K. T. Hall.   259

49 'The Servant with the Scissors.' Diagram by K. T. Hall.   260

50 The Asilomar Meeting. © Maxine Singer Papers. US National Library of Medicine. Courtesy of US National Library of Medicine.   270

51 Walter Gilbert. US National Library of Medicine. Courtesy of US National Library of Medicine.   279

52 Walter Gilbert's strategy to clone and express human insulin in bacteria. Diagram by K. T. Hall.   282

53 Protestors campaigning against recombinant DNA research hold a sign quoting Adolf Hitler, 'We will create the perfect race' in front of a panel (which included Maxine Singer, Paul Berg, and NIH Director Donald Fredrickson) at a meeting of the National Academy of Sciences held in Washington D.C. in 1977. Courtesy of US National Library of Medicine.   291

54 Statue of Herb Boyer and Bob Swanson on the Genentech campus. Reproduced with permission of Genentech Corporate Relations.   301

55 Genentech strategy to clone and express human insulin. Diagram by K. T. Hall.   303

56 The team involved in the successful cloning and expression of somatostatin. Photograph kindly provided by Professor Art Riggs and reproduced with permission of the City of Hope National Medical Center, Duarte, California.   306

57 The team involved in the successful cloning and expression of human insulin, ca. 1978. Photograph kindly provided by Professor Art Riggs and reproduced with permission of the City of Hope National Medical Center, Duarte, California.   310

58 Rosalyn Yalow (1921–2011). Credit: National Library of Medicine/Science
   Photo Library.   313
59 Insulin analogues. Diagram by K. T. Hall.   342

While every effort has been made to contact copyright holders of illustrations, the author and publishers would be grateful for information about any illustrations where they have been unable to trace them and would be glad to make amendments to further editions.

Preface – On the Shoulders of Giants . . . ?

When Theresa May first received the shock diagnosis of type 1 diabetes while serving as Home Secretary in 2013, her response was one of admirable stoicism: 'It's just part of life—so it's a case of head down and getting on with it'.[1] Echoing a piece of advice given to her by athlete Sir Steve Redgrave, she resolved that 'Diabetes must learn to live with me rather than me live with diabetes. That's the attitude'.

Her grit in dealing with such a life-changing event was impressive, but had she been born just over 100 years ago, no amount of stoicism, however heroic, would have saved her from a grim and cruel fate. For before the discovery of insulin, a diagnosis of type 1 diabetes was an inevitable death sentence.

Thankfully, that is no longer the case. Thanks to daily injections of insulin, Theresa May went on to lead the nation; England and Exeter rugby centre Henry Slade has won 26 caps; and actor James Norton may well yet take up 007's license to kill.[2,3]

Like them, I owe my life to insulin. But why go to the trouble of writing a book about it? Why not simply take the daily injections and be grateful? What more needs to be said? Why worry about who discovered it, who unravelled its chemical structure, and who first grasped that ordinary gut bacteria could be coaxed into making it? And if you don't even have diabetes, well, why waste precious hours of your life reading a book about the history behind insulin? My motivation here could be expressed as 'in for a penny, in for a pound'.

History matters. The common ground between the protestors who toppled the statue of Edward Colston in Bristol in June 2020 and those who objected to their actions is that both groups felt passionately that the past was too important to be forgotten. For me, the history of the discovery of insulin is worth recording here.

Admittedly, an intimate knowledge of the history of science is unlikely to bring about social and political justice, or atonement for past wrongs in the way that those who toppled the statue of Colston hoped to achieve. And while the products of science can save lives, the medical benefits of knowing its history are yet to be demonstrated.

But before casting this book aside as just another dry science book, it might be worth reflecting on the words of the molecular biologist Marshall Nirenberg, who was awarded the 1968 Nobel Prize in Physiology or Medicine for helping to decipher the genetic code. A year earlier, Nirenberg had given an address in which he set out the challenge that this technology would bring:

> When man becomes capable of instructing his own cells, he must refrain from doing so until he has sufficient wisdom to use this knowledge for the benefit of

mankind. The reason I state this problem well in advance of the need to do so is because the decisions concerning the application of this knowledge ultimately must be made by society, and only an informed society can make such decisions wisely.[4]

In 1968, the idea that we could instruct our own cells still belonged to the realms of science fiction, but in 2020, just over half a century after Nirenberg gave this address, the Nobel Prize in Chemistry was awarded to Jennifer Doudna and Emmanuelle Charpentier for developing CRISPR-Cas 9—in effect, a pair of molecular scissors that allows precise alteration of the genetic code. CRISPR-Cas 9 may revolutionize what it means to be human, but while it raises the very real possibility of being able to treat debilitating genetic diseases, it also allows permanent alterations to the human genome that could be passed on down the generations. With the announcement in 2018 of the birth of two babies whose DNA had been altered using CRISPR whilst they were at the embryonic stage of development by Chinese scientist He Jiankui, this technology will bring formidable ethical, social, and political challenges.[5] But long before genetic engineering had become a reality, Nirenberg had already recognized that the challenges it raised could not be left to scientists alone to solve. The debates around this new technology were too important to remain confined to the laboratory and needed the engagement of what Nirenberg called an 'informed society'.

For a scientifically informed society, we need first to understand how science works. Surely, the best way of understanding science is to study its history? For, as the philosopher and mathematician Alfred North Whitehead warned in an address to the British Association for the Advancement of Science in 1916, 'A science which hesitates to forget its founders is lost'.[6] But as Nirenberg's fellow Nobel laureate Sydney Brenner once lamented, scientists are sometimes the worst offenders as regards ignorance about their own history:

> For most young molecular biologists, the history of their subject is divided into two epochs: the last two years and everything else before that. The present and the very recent past are perceived in sharp detail but the rest is swathed in a legendary mist where Crick, Watson, Mendel, Darwin—perhaps even Aristotle—coexist as uneasy contemporaries.[7]

Brenner's observation was more than just the typical jeremiad grumblings of a middle-aged man about today's youth and their failings. Having once been one of those young molecular biologists of whom Brenner despaired, I think I can appreciate his point: why should we care about who made a discovery, or how it was made? Surely all that matters is that people in white coats keep on making breakthroughs and finding miracle cures, so that a younger generation can build on these earlier successes.

Alongside ignorance of the history of science, the particular way in which it is remembered can also be problematic. It is tempting to lay the blame for this on Sir Isaac Newton. In a letter written to Robert Hooke in 1675, Newton famously said that 'If I have seen further it is by standing on the shoulders of Giants'. Actually, it turns out that Newton was not being original but rather was most likely quoting earlier sources, such as the twelfth-century philosopher Bernard of Chartres.[8] But regardless of its origin, it is a misattributed quote that is firmly established in popular consciousness. It conjures up an image that scientific progress is achieved by a succession of either lone geniuses or unsung pioneers far ahead of their time, whose work marks out a smooth, steady triumphant ascent to the lofty pinnacles of present-day knowledge.

Neither of these two approaches are helpful. Both obscure, rather than illuminate, the real workings of science. However, a far more compelling—and honest—description of how science really works comes from Immanuel Kant (1724–1804), the German Enlightenment philosopher. Known as the 'Sage of Königsberg', Kant once observed that 'Out of the crooked timber of humanity no straight thing can ever be made'. Kant's phrase was a favourite of the philosopher Isaiah Berlin (1909–1997) who used it as the title for his exploration of the roots of fascism and totalitarianism, *The Crooked Timber of Humanity: Chapters in the History of Ideas*, first published in 1990. According to Nicholas Lezard,[9] Berlin used Kant's phrase 'many times as a stick with which to beat those who would try to build heaven on earth or fit humanity into a straitjacket of their own design'. However, it also lends itself wonderfully to the tortuous process of scientific discovery. It works particularly well when describing the discovery of insulin—which is not a tale of bold, lone geniuses or saints who set to work on improving the lot of humanity. Instead, it is a story of monstrous egos, toxic insecurities, and bitter career rivalry that at times resembles 'Game of Thrones' but enacted with lab coats and pipettes, rather than chain mail and poisoned daggers. Some of the cast involved deserve our sympathy, while others are just plain odious. They regularly fought, sometimes wept, very often made mistakes, and in at least one case, were partial to nude sun-bathing.

Since I began writing this book, this story has taken on a powerful new relevance for us all, whether or not we happen to have diabetes. For as experts appeared on our television news reports each evening talking of PCR, antibody tests and mRNA vaccines, or presenting graphs of infection rates at daily press briefings during the first lockdowns of the Covid-19 pandemic, science—and its importance—came crashing into all of our lives.

In years to come, as we look back on the COVID pandemic, there will once again be a temptation to tell a story of lone geniuses who boldly pioneered miraculous vaccines to bring the disease under control. But, like the discovery and development of insulin as told here, our success in bringing COVID-19 under control was achieved not by standing on the shoulders of giants, but by being hewn from the crooked timber of humanity. And surely that alone makes this achievement all the more impressive and one of which we can all be proud.

Acknowledgements

When Rosalind Franklin was first told about James Watson's and Francis Crick's proposed double-helical structure of DNA (in which her own work had played a key role), she is said to have remarked that 'We all stand on each other's shoulders'. In writing this book, I have stood on more than a few pairs of shoulders and I take this opportunity to thank them.

First, I thank Mr. Lawrence Glynn and the Leeds Diabetes UK support group, who invited me to give a talk from which the germ of the idea for this book grew, and Dr. H. J. Bodansky at St. James' University Hospital, Leeds, for stressing how insulin would now be a part of daily life, which prompted me to delve deeper into its story. That digging would not have gone far without the help of Jennifer Toews, Natalya Rattan, and Danielle Van Wagner, who kindly dug up scans for me from Banting Archives, Fisher Library University of Toronto, and Katherine Marshall, who found correspondence between Charles Best and Sir Henry Dale in the Royal Society Archive. To Professor Viktor Jörgens and Herr Thomas Breitfeld, Geheimes Staatsarchiv, Preussischer Kulturbesitz, Berlin for helping shine a light onto the forgotten story of Georg Zuelzer, and Professor Jeffrey Friedman of the Rockefeller University for bringing to light another forgotten figure in the story of insulin, Israel Kleiner. I am also grateful to Professor Constantin Ionescu-Tirgoviste for providing information about Nicolai Paulescu and to Professors Pierre Lefebvre and Lorenzo Piemonte of the International Diabetes Federation for access to the archives of *Diabetes Voice* and discussion of the controversy over him.

On the theme of darker aspects of the insulin story, I thank Dr. Sharon Nightingale, Leeds Hospitals Trust, for first drawing my attention to the use of insulin in psychiatric treatment and Dr. Mike Green for casting some light onto the grisly subject of the use of insulin as a murder weapon.

To all the library staff who have patiently fielded my pestering and incessant requests for arcane documents, particularly Ms. Rebecca Turpin and all the staff at Special Collections, Brotherton Library, University of Leeds, and Mrs. June Longley and Mrs. Christine Forkins of Rawdon Library, Leeds.

My thanks go to Professor G. E. Blair who not only guided me through my fumbling efforts to unravel the molecular mysteries of gene regulation by adenoviruses during my PhD years, but also, and perhaps more importantly, first brought to my attention the work of Archer Martin and Richard Synge. Thanks also to Dr. Christine Holdstock, Dr. Maryon Dougill, and Mrs. Noreen Barrett, who are all former employees of WIRA and happily shared their recollections. Thanks also goes to Mr. George Nicol for inviting me to speak to the Forum

2000 group where I first met the late Dr. Ian Holme, who told me about the intellectual punch up between Archer Martin and William Astbury over the structure of wool proteins, and to Clare and Daniel Gordon for pointing me in the direction of old photographs of the WIRA site. And of course I thank Dr. Charlotte Synge and Mr. Matthew Synge for kindly sharing recollections of their father.

Martin and Synge would have remained nothing more than names had it not been for the help of Mr. David Allen, Archivist of the Royal Society of Chemistry, Dr. Louise Clark of the Reading Room at Cambridge (who kindly provided me with a copy of Synge's PhD thesis), and Dr. Adam Green and his team at the archives of Trinity College, Cambridge (who tirelessly dug out boxes of Synge's papers and his PhD thesis for me to root through). I am grateful to the Leeds Philosophical and Literary Society who, thanks to their generous grants, made both these research trips and the inclusion of certain figures possible. Thanks to Jen Zwiernick for her advice on how to find out whether Synge's political views had brought him to the attention of the security services, and to Professor Gareth Williams, whose clinical writings on diabetes were invaluable, as was his book *A Monstrous Commotion*, about the intriguing question of Synge's claim to have seen the Loch Ness Monster. Alas, despite the best efforts of Zoe Thomas, who searched the BBC Archive to see whether any recordings still existed of the 1958 documentary in which Synge gave his account of having seen the fabled beast, we can only wonder.

Unlike Synge, Archer Martin left much less of a paper trail behind him, so my thanks go to Judith Wright, senior archivist, at the Boots Archive for helping rectify this. I am also grateful to all the staff of the Bodleian Library at the University Oxford who handled my many requests to view their vast collection of papers relating to Dorothy Hodgkin. A special thanks also to Georgina Ferry, author of *Dorothy Hodgkin: A Life*, for kindly sharing insights and material from her book, and Professor Patricia Fara, University of Cambridge, for her illuminating paper about what we can learn from the portraits of Hodgkin.

Fred Sanger similarly has a stack of archival documents, and I thank Amelia Walker, Emma Jones, Kate Symonds, and Will Greenacre at the Wellcome Collection for helping me find a needle in an archival haystack, as well as Professor George Brownlee, author of *Fred Sanger: Double Nobel Laureate*, for kindly sharing his insights and material with me. Thanks also go to Dr. Lee R. Hiltzik of the Rockefeller Archive Center for help with photos of Oswald Avery, and to Matthew Cobb, Professor of Zoology at the University of Manchester for discussions around Avery, the genetic code, and information theory.

For help with the Genentech story, my thanks go to Michael Maire Lange of the Bancroft Archives at the University of Berkeley and Professor Sally Smith-Hughes for her invaluable help in gaining permission for me to cite from these archives. Also, thanks go to Heather Gloe of Genentech Corporate Relations for kindly providing a photograph of the wonderful statue commemorating Herb Boyer and Bob Swanson's historic first meeting, and Professor Arthur Riggs for kindly providing photos of the team involved in the cloning of human insulin.

It is with immense gratitude that I send a big thanks to my editor Sonke Adlung and all the team at OUP, particularly Michele Marietta for her tireless work and tenacity with copy editing the manuscript, as well as Kate Pool at the Society of Authors for her patience with my (and sometimes bizarre) questions regarding permissions and her invaluable advice in this matter. To anyone I have not mentioned, I can only offer a sincere and hearty apology—but rest assured this does not mean that I have forgotten you.

Finally, it would be remiss of me to write a book on the importance of insulin and not thank everyone involved with ensuring that I have a regular supply, so a very special thanks to Dr. Joanna Walker and the staff of Ireland Wood Surgery, together with Peter and all his hardworking team at the Ireland Wood Pharmacy, along with Dr. Steve Gilbey and all the diabetes team, past and present, at Wharfedale Hospital, Otley, West Yorkshire.

I would also like to acknowledge a tremendous debt of gratitude to the late Professor Michael Bliss - upon whose shoulders this book rests.

Introduction—Taming the Tiger

As the colour on the test strip began to change, so too did the expression on my doctor's face. It was at that moment that I knew that I was in big trouble. For the past few weeks I had been experiencing symptoms of general low-level irritability and lethargy, all of which I had simply dismissed as being nothing more than inevitable middle age. But when these symptoms began to be accompanied by a sudden rapid weight loss, raging thirst, and a ravenous craving for sugar, faint alarm bells began to ring.

At the same time, distant memories of undergraduate lectures in biochemistry began to stir. I had always hoped that groggy hours spent in lectures on glycolysis and metabolism would one day somehow come in handy. But I could never have imagined the particular set of circumstances in which they would finally prove their worth.

Spurred into action by half remembered passages from textbooks, I visited my GP who, horrified at my test results, made an urgent telephone call to St. James's Hospital.

'You need to go there', he told me, as he hastily scribbled some notes.

'You've made an appointment for me later in the week?' I enquired.

'No', he replied firmly. 'You leave this surgery now and you go straight to the hospital'.

A series of blood tests at the hospital confirmed my hunch. There was so much sugar in my blood it was if it had turned to treacle. Poisonous compounds called ketones were raging through my system which, if untreated, would acidify my blood and eventually put me potentially into a permanent coma. My body was in a state of metabolic meltdown—brought on by the onset of diabetes.

I had remembered enough to know that there were two forms of this condition. My first thought was that, having hit middle age, this must be type 2—the one which, were I to eat slightly less, drink slightly less, and run a bit more, might have some chance of being brought under control. But instead, I emerged from the hospital that day armed with a blood glucose testing kit and several syringes laden with insulin. I had, inexplicably, developed type 1 diabetes. It was little consolation to know that this rarer form of the condition usually presents in childhood or adolescence but had, in my case, had the decency to delay its unwelcome appearance until my middle years. And from now on, insulin would be my constant companion—the biochemical crutch on which I would need to lean.

There is a common misperception that diabetes is less serious than other chronic conditions—that life for a diabetic patient is just a matter of having to go easy on the cakes. However, in his book *Diabetes: The Biography*, diabetologist Professor Robert Tattersall offers a very different perspective:

> ... a patient of mine who had had it for many years compared it to living with a tiger, since, as he said: 'If you look after it, and never turn your back on it, you can live with a tiger. If you neglect it, it will pounce on you and rip you to shreds'.[1]

This is no mere exaggeration for poetic effect. Patients with both forms of the condition face the possibility of blindness, neuropathy, amputations, kidney failure, and increased predisposition to cardiovascular disease and stroke. While there is still no cure for diabetes, it *can* be managed. For type 1 patients, management means multiple daily injections of the hormone insulin—usually with each meal—and the importance of this regime cannot be understated. One of the first consultants that I saw following my own diagnosis explained the gravity of the situation with stark and honest advice: 'Take your insulin and you should be OK. Fail to take it, and you're dead'.

He wasn't just being melodramatic. Prior to the discovery of insulin, a diagnosis of type 1 diabetes was a death sentence. Without insulin, the tissues of the body cannot obtain glucose that they need as a vital supply of energy from the blood. Deprived of their regular fuel source, they resort to burning fat and protein, which in turn leads to emaciation and the production of toxic compounds called ketones, which acidify the blood, resulting in slow suffocation, coma, and death.

Thanks to the discovery of insulin, no diabetic patient today needs to suffer this grisly fate. I am personally very grateful for its discovery, but there are lessons for us all in what technology can—and, perhaps more importantly, cannot—do for us. Although insulin can save lives, it does not guarantee that those lives will be easy. Reprieved of early certain death, the diabetic patient faces a lifetime of daily injections, blood sugar tests (often in the most inconvenient, or even embarrassing, of situations) and an ever-present nagging worry about the many serious complications that can arise.

The story of insulin is a drama in three acts. Act One takes place in Toronto in the early 1920s, where Fred Banting, a struggling but ambitious young doctor, discovered that extracts of pancreatic tissue—described by one of his colleagues as being simply 'thick brown muck'—could save the lives of diabetic patients. But when Banting was awarded the Nobel Prize for this discovery, he was far from happy—and he quickly learned that fame came with a heavy price.

Act Two takes place in the 1940s in West Yorkshire and the setting is the unlikely field of wool chemistry where, in a converted stable billowing with chloroform fumes, the scientists Archer Martin and Richard Synge developed a method that would eventually reveal insulin not to be 'thick brown muck', but a protein with a precise chemical structure. Although developed for the analysis of wool, this method (known as partition chromatography) also offered the first hint at how DNA carries the genetic message—a discovery that was crucial to the events of the final act in the tale. Act Three takes place under the sunnier skies of the US

West Coast in the 1970s. It was there and then that scientists raced to use the newly discovered tools of genetic engineering to be the first to synthesize human insulin, despite facing vocal protest and opposition from a public and media frightened by this new technology.

As a result of this race, several of the scientists involved became overnight multimillionaires, and patients with type 1 diabetes can now manage their condition with human insulin instead of material recovered from cows or pigs. But as impressive as this feat was, was it genuinely a medical breakthrough? Was it perhaps just a piece of scientific showmanship? Could it really be considered a 'miracle cure' for diabetes, as so many press reports of the time hailed it? For that matter, is talk of miracle cures and medical breakthroughs helpful, or does it run the risk of inflating unrealistic expectations about what we can expect science and technology to do for us?

Admittedly, questions of this sort are not always foremost in my mind as I insert a hypodermic needle into myself several times each day. After all, why should I— or anyone—care about who discovered insulin, or how it was discovered, as long as it saves lives? One of the main characters in this book may offer an answer here.

Richard Synge was a biochemist who worked on a problem that seemed far removed from both the world of biomedical breakthroughs and the deep questions of biology. His day job was analysing the chemical structure of wool for the purpose of improving textile fibres. Yet Synge was not averse to philosophical reflection about science, and what it was there to do. He once wrote that 'to contemplate the grander and unexpected advances of natural knowledge does good to the souls of all engaged in scientific work, and of many outside of it'.[2]

I think that Synge was correct. Insulin may well be vital for my body, but its story—and many more like it—are equally vital for my soul. And as we face the post-pandemic world, these stories may have something to tell us all, whether or not we have diabetes.

1

The Pissing Evil

The earliest, and also the most pronounced, symptom[1] of diabetes had first been recorded on a 3000-year-old Egyptian scroll. In the winter of 1872–1873, Georg Ebers, a German Egyptologist, was in Thebes undertaking an examination of some ancient tombs, when he was approached by a local who claimed to have found something that might interest him. Opening a tin box, the Egyptian produced a papyrus found in about 1858 that had been stuck between the legs of a mummy in the Necropolis at Thebes.[2] The scroll was perfectly preserved, and Ebers immediately recognized this as a priceless archaeological discovery, as did the current owner. Perhaps aware that a representative of the British Museum was currently in Luxor seeking to acquire the papyrus, its owner realized that he was in a sellers' market, and priced the artefact accordingly. Ebers must have been delighted, therefore, when a wealthy German merchant intervened to advance him the money for its purchase.

The 'Ebers papyrus' currently resides in the Library of the University of Leipzig. Dating from around 1550 BCE, it is one of the oldest medical documents in the world and describes the preparation of medicines for a number of different pathological conditions, one of which is a treatment 'to drive away the too much emptying of urine'.[3]

In the second century CE, the physician Aretaeus of Cappadocia called this condition 'diabetes', derived from the ancient Greek word for 'to siphon' or 'to flow', and he offered a vivid description of its most pronounced symptom:[4,5]

> It consists of a liquefaction of the flesh and bones into urine ... The kidneys and bladder, the usual passageways of fluid, do not cease emitting urine and the outpouring is profuse and without limit. It is just as though aqueducts were opened wide. The development of this disease is gradual (chronic) but short will be the life of the man in whom the disease is fully developed. Emaciation proceeds quickly and death occurs rapidly. Moreover, life for the patient is tedious and full of pain. The desire for drink grows ever stronger, but no matter what quantity (of fluids) he drinks, satisfaction never occurs, and he passes more urine than he drinks. He cannot be stopped either from drinking or urinating.[6]

However, the most candid, and colourful, description of diabetes and its symptoms comes from the English physician Thomas Willis (1625–1675) who, in one section of his posthumously published work *Pharmaceutice Rationalis*, described diabetes simply as 'the Pissing Evil'.[7]

In the course of his work, Willis visited the markets held in the small towns around Oxford and offered to diagnose ailments by analysing samples of urine brought to him.[8] Known as 'casting waters', this was hardly the height of medical sophistication. However, for diabetes it proved to be an effective method of diagnosis, thanks to one particular property of the diabetic patient's urine known to physicians since antiquity. The Hindu physicians Charaka (third century BCE) and Sushruta (sixth century BCE) described a condition characterized by urine that tasted sweet. Diagnosis occurred either by tasting the patient's urine or observing that ants gathered around it; from the latter, it became known by the Indian name of 'Madhumeha' meaning 'urine of honey'.[9,10,11,12]

Liverpool physician, Matthew Dobson (1735–1784) first showed why the urine produced in such great quantities by diabetic patients tasted so sweet. In 1776, Dobson published a paper describing a patient with diabetes who, on admission to the hospital four years earlier, had been 'emaciated, weak and dejected, his thirst was unquenchable and his skin dry, hard and harsh to the touch like rough parchment'.[13] The most pronounced symptom, however, was that the patient 'passed twenty-eight pints of urine every 24 hours', and which Dobson noted 'had a sweetish smell and was very sweet to the taste'.[14] Curious, Dobson warmed the urine with a gentle heat and observed that:

> There remained after the evaporation, a white cake ... This cake was granulated and broke easily between the fingers, it smelled sweet like brown sugar, neither could it, by the taste, be distinguished from sugar except that the sweetness left a slight sense of coolness on the palate.[15]

Within only a few years of Dobson having described this 'white cake', the clinician Francis Home (1719–1813) showed that the urine of a diabetic patient was so sweet that it could actually be fermented into what he described as 'a tolerable small beer'.[16] This characteristic and distinct sweetness of the urine led the Edinburgh doctor John Rollo to add the suffix 'mellitus', derived from the Greek and Latin words for 'honey', to the name diabetes.[17] Rollo also devised the diet with which most patients would be treated during the nineteenth century. Reasoning that the high levels of sugar in the blood and urine of diabetic patients were formed in the stomach after eating vegetables, he concluded that the solution was to feed these patients with 'Game or old meats which have been long kept; and as far as the stomach may bear, fat and rancid old meats, as pork'.[18] It is probably no surprise that many of Rollo's patients did not adopt this diet with much enthusiasm, with Rollo recording that one particular patient was 'strongly remonstrated with, and he was told the consequence of repeated deviations'.[19]

The consequences of which Rollo warned his patients were far more serious than simply a frequent need to urinate. Rollo had himself observed the symptoms in his patients of the most common and serious complications of diabetes, including nerve damage caused by high levels of blood sugar. By the end of the nineteenth century, diabetes was also associated with retina and kidney damage; these complications continue for diabetic patients today. According to figures released by the charity Diabetes UK, treating the long-term complications of diabetes that arise from prolonged elevated levels of blood sugar cost 10% of the NHS annual budget, with diabetes being cited as the biggest cause in the UK of both blindness and lower limb amputations due to neuropathy.[20] Accordingly, the General Assembly of the United Nations describes diabetes as 'a chronic, debilitating and costly disease associated with severe complications, which poses severe risks for families, states and the entire world'.[21]

Before the discovery of insulin, however, these serious long-term complications of diabetes were less of an issue, simply for the grim reason that patients did not live long enough to experience them. Although the cause of the illness remained a mystery, by the early nineteenth century important discoveries were being made about its nature. In 1815, the French chemist Michel Chevreul (1786–1889) discovered that the cause of the sweetness in diabetic urine was grape sugar (glucose); although this discovery marked the beginning of understanding diabetes as a chemical disorder, it also presented a paradox.[22] How could the level of glucose have reached such high levels in these patients? The accepted orthodoxy at the time was that only plants were capable of synthesizing glucose and that glucose could only be present in the blood of animals as a result of having eaten plant products.

Yet, when the eminent French physiologist Claude Bernard (1813–1878) began to tackle this question, he was surprised to find that glucose was present in the blood of animals that had been starved. Moreover, he also observed that blood leaving the liver of a dog had higher levels of sugar than the blood entering it.[23] This suggested that the dog's liver was somehow making glucose. This hypothesis flew in the face of accepted thinking.

The reality, as Bernard discovered, was not so straightforward. On examining the livers of his dogs, Bernard found that, even though they had been fed only on meat, they contained a starchy substance. He hypothesized that this substance might act as a store from which sugar could be released into the blood and so he called it 'glycogen', meaning 'sugar forming'.[24] He then went on to explore how the nervous system might control this process. In a famous experiment, he found that pricking the region of the fourth ventricle in a dog's brain flooded the blood with glucose from the liver's stores of glycogen, making the animal diabetic.

Bernard's work had important implications for identifying the cause of diabetes. Its origins appeared to lie within the nervous system and, as Professor of Physiology J. J. R. Macleod at the University of Toronto observed, this might explain why certain professions appeared to be more prone to diabetes:

Diabetes is common in locomotive engineers and in the captains of ocean liners—
that is, men who in the performance of their daily duties are frequently put under
a severe nerve strain. It is apparently increasing in men engaged in occupations
that demand mental concentration and strain, such as in professional and business
work.[25]

But while he entertained the possibility that diabetes was ultimately a disorder of
the nervous system, Macleod was also open to alternatives. This willingness to
consider that the true cause of diabetes might lie elsewhere eventually led to him
becoming one of the biggest names in the history of diabetes research.

In 1889 Oskar Minkowski and Joseph von Mering at the University of Strasbourg conducted research that suggested diabetes might not be a disorder of
the nervous system. While researching how the pancreas was involved in fat
metabolism, Minkowski and Mering took the radical step of removing the pancreas from a healthy dog.

On the day after the surgery was performed, Mering was urgently called out
of town. On his own for over a week, Minkowski worked on, and observed that,
despite having been house-trained, the depancreatized dog was now leaving pud-
dles of urine all over the laboratory floor. Thinking that that this was because
the poor animal had not been allowed enough outdoor exercise, Minkowski first
thought was to reprimand his servant for not having let the dog out frequently
enough. In response, the poor servant protested, 'I do, but the animal is queer; as
soon as it comes back it passes water again even if it has just done so outside.'[26]

Curious about the dog's behaviour and the servant's explanation, Minkowski
measured the concentration of sugar in the dog's urine, and found it elevated well
above normal levels.[27] The depancreatized dog was exhibiting a classic symptom
of diabetes mellitus; thus, Minkowski suggested that the pancreas played a central role in the disease and queried how this might occur. One possibility was that
the pancreas was essential for the removal of toxins that would otherwise cause
diabetes and that, in a diabetic patient, this function was impaired. Another possibility, however, was that, under normal conditions, the pancreas secreted some
anti-diabetic substance that was missing in the diseased state.

Minkowski was not the first physician to suggest a connection between diabetes and the pancreas, nor did his work mean that a pancreatic origin for the
condition became universally accepted overnight.[28] Minkowski tried to show that
diabetes was not a disorder of the nervous system, but rather was due to some failure of the pancreas. His experiments attempted to demonstrate that transplants
of pancreatic tissue might prevent diabetes in a dog.[29] Similar experiments by
Édouard Hédon (1863–1933) showed that a small graft of pancreatic tissue could
have an anti-diabetic effect in a dog. However, it was Bernard's great rival, French
physiologist Charles-Edouard Brown-Séquard (1817–1893), who suggested how
it might do this.[30] From his earlier work on the adrenal gland, Brown-Séquard
had speculated that some glands might secrete substances not externally via a
duct, but internally directly into the bloodstream.[31] In 1905, the English physician

Ernest Starling (1866–1927) gave these 'internal secretions' the new name of 'hormones', derived from the Greek term meaning to 'stir up'.[32] Starling described their action as being like chemical messengers between different parts of the body, and Brown-Séquard had already foreseen that they might have tremendous therapeutic potential:

> ... if we could safely introduce the principle of the internal secretion of a gland taken from a living animal into the blood of men suffering from the lack of that secretion, important therapeutic effects would thereby be obtained.[33]

With regard to this possible therapeutic application, Brown-Séquard had one particular gland in mind. In 1889, Brown-Séquard was experimenting with bodily secretions of a rather different kind. He claimed that 'the feebleness of old men is in part due to the diminution of the functions of the testicles' and that 'well organised men, especially from 20–35 years of age, who remain absolutely free from sexual intercourse or any other causes of expenditure of seminal fluid, are in a state of excitement, giving them a great though abnormal physical and mental activity'.[34,35] From these observations, Brown-Séquard concluded that the testes produced what he termed a 'dynamogenic' substance, which, if extracted and injected into men, might rejuvenate them.

So great was his confidence in this idea that he happily used himself as a test subject. In a paper presented in 1889 to the Societé de Biologie of Paris, he said that his general vigour, which had once been considerable, was now significantly diminished:

> ... after only two hours of experimental work at the laboratory, although I sat down, I was left exhausted ... I was for many years so tired that I had to go to bed almost as soon as I had taken a hasty meal.[36]

At the age of 72, this loss of energy might not come as a surprise. However, thanks to a somewhat unusual medical intervention, Brown-Séquard divulged that:

> Experimental work at the laboratory tires me little now. I can to the great astonishment of my assistant, remain standing for hours together without feeling the need of sitting down .... I can also, without difficulty, and even without thinking about it, go up and down stairs almost running, a thing which I always did before the age of sixty.[37]

Over a fortnight, Brown-Séquard had injected himself daily with extracts of ground testicles from guinea pigs or dogs. As further examples of the rejuvenating powers of these extracts, he went on to describe how he could now move weights that were six or seven kilos heavier than before and 'all that had become difficult or impossible for him owing to advancing age became once more easy'.[38] For reasons best known to himself, he also chose to share the fact that the average length of his jet of urine had now increased by 25% and that he could now once more defecate without the need of laxatives.[39,40]

The *British Medical Journal* described Brown-Séquard's paper as 'communi-cations of a most extraordinary nature', likening them to' the wild imaginings of medieval philosophers in search of the elixir vitae'.[41] But while his report may have left some readers incredulous, it left others in a state of moral outrage:

> ... the idea of injecting the seminal fluid of dogs and rabbits into human beings [is] disgusting ... and when the treatment also involves the practice of masturbation ... it is time for the medical profession in England to repudiate it.[42]

Readers of the *British Medical Journal* might have been seething with disgust at Brown-Séquard's experiments, but its editors were soon singing his praises. Another scientist, M. Variot, appeared to have confirmed Brown-Séquard's extraordinary claims for testicular extracts and in 1891, George Murray had shown that a 46-year-old female patient with myxoedema, a disorder arising from underactivity of the thyroid gland, could be successfully treated by the injection of extracts made from the thyroid gland of a sheep.[43,44] In response to Murray's impressive results, the *British Medical Journal* declared:

> It is now some years since Brown-Séquard announced the wonderful effects which followed the subcutaneous injection of testicular extracts as exemplified in his own person; and though many jeered at him as the discoverer of the secret of perpetual youth, the notion has steadfastly gained ground that there is, after all, something in it. Since also, the success that has followed the injection of thyroid extract in myxoedema, we can hardly wonder that this belief has increased.[45]

Brown-Séquard championed the idea that 'all glands with an external secretion have at the same time, like the testicles, an internal secretion'. Of these, he said that there was little doubt that '... the pancreas, like the testicles, ovaries, the kid-neys, has an internal secretion, which is even more important than its external one'.[46,47,48]

He also recognized that a pancreatic extract that contained this internal secretion might have immense therapeutic potential:

> When the pancreas has been suppressed (ablation, ligature of its veins, etc.) there is diabetes. Would there be no diabetes, if after suppression of the functions of the gland there were daily injections of an extract of this gland taken from a healthy animal?[49,50]

He further argued that attempts to use pancreatic extracts to treat diabetes might be more successful if they were augmented with his recommended testicular liq-uid. However, it was far more likely that their failure was because the pancreas was still not fully understood.[51]

It was only in 1848 that Claude Bernard showed the pancreas secreted enzymes essential for digestion via a channel, called the Wirsungian duct, that led directly into the gut.[52] Twenty-one years later, Paul Langerhans, a German medical stu-dent, observed that the pancreas had another vital function—one that was central to understanding diabetes. Langerhans noted that pancreatic tissue was composed of two very different types of cell.[53] The first of these, known as the acinar cells, comprised most of the pancreatic tissue and secreted digestive enzymes. But Langerhans also observed clusters of a second type of cell that were interspersed, like small islands, within this main mass of tissue. This new type of pancreatic tissue was later named the 'Islets of Langerhans' in his honour.[54]

Study of this new tissue was proving to be a fruitful subject. When Russian PhD student Leonid Sobolev observed that the Islets lacked ducts and were instead infiltrated with an intimate network of blood capillaries, he speculated that they might well be the sites of production for the hypothetical anti-diabetic internal secretion.[55,56] Sobolev also knew that, while total pancreatectomy of a dog resulted in diabetes, tying the Wirsungian duct shut (to atrophy the digestive enzyme-producing cells) did not. He considered that the Islet cells remained intact after ligation of the duct, and therefore that it was these cells that produced the enigmatic—and elusive—anti-diabetic substance. His conclusions were supported by his observation that in patients with diabetes, the number of Islet cells were either completely absent or significantly reduced. At the same time, Eugene Opie at Johns Hopkins University in Baltimore also observed the connection between damage to the Islet cells and the onset of diabetes.[57]

By the beginning of the twentieth century, while still hypothetical, the substance now had a name. Belgian scientist Jean de Meyer christened it 'insuline', from the Latin for 'island'. The hunt was now on to demonstrate its existence.[58] The medical rewards of isolating insulin and then applying it in therapy were boundless, for the current methods of treating diabetes did little more than delay a slow and painful death.

Glucose is a Janus-faced molecule. Although it is essential for cells as a metabolic fuel, it can also be toxic and damaging. When carbohydrates—long chains of glucose molecules joined together that are found in foods like pasta, rice, potatoes, and bread—are eaten, the bonds between the glucose molecules are broken down by digestive enzymes in the gut. The glucose then passes from the gut into the bloodstream for transport to the muscles and brain (which use glucose as an essential metabolic fuel) and to the liver for storage. In normal, healthy individuals, the presence of insulin allows the glucose to pass back out of the blood and into the tissues where it is required.

In diabetic patients, this system breaks down, which traps glucose in the blood, and prevents its transfer into the tissues where it is needed. There are two forms of diabetes: type 1 and type 2. In both cases, insulin is at the heart of the problem, leading to elevated blood sugar levels and the same long-term complications.[59] In the more common form, now known as type 2 (formerly non-insulin dependent diabetes mellitus), the pancreas produces insulin at normal, or even elevated levels, but the body's ability to respond to it has become reduced. Why this happens is still not fully understood and there are thought to be a host of possible mechanisms leading to its onset. What is clear, however, is that obesity, lack of exercise, and poor diet can all be contributory factors to this form of diabetes, all of which might explain why it is currently on the rise.[60]

Type 2 can be managed with medication like metformin, which acts to increase the body's sensitivity to insulin. But for patients with the less-common type 1 form of the disease, this is not an option, for the simple reason that the pancreas either doesn't produce enough insulin in the first place, or has stopped making it at all. Why this happens is still not clear, but current thinking is that it is caused by damage to the Islet cells by an auto-immune response. For these patients, the only option to manage their condition is to administer daily injections of insulin.

In both cases, the end result is similar. Sugar is no longer able to pass out of the blood and reach the tissues where it is required; it remains instead in the blood, where its concentration steadily increases. It is this rise in blood sugar levels that explains the frequent need to urinate: the kidneys go into overdrive in an attempt to flush the excess sugar out of the blood. Being unable to use glucose as a fuel, the body then turns to alternative sources of metabolic energy. Stores of fat, and then protein, are broken down, resulting in rapid weight loss and the incomplete metabolism of fats. The latter leads in turn to the production of toxic compounds called ketones that cause a sweet, fruity smell and a steady rise in the acidity of the blood. As the blood becomes more acidic, the lungs struggle to function. Lung failure leads to slow, internal suffocation, resulting in drowsiness, coma, and death.

Before the discover of insulin, doctors were powerless to treat diabetes. Some administered opium in the hope that it might relieve the condition, but all it did was foster a dependence.[61] The physician Frederick William Pavy (1829–1911) described diabetes as 'one of the most inscrutable of diseases' and his colleagues pitied him for having devoted his life to its study. Sir William Gull (1816–1890), his fellow clinician at Guy's Hospital, asked, 'What sin has Pavy committed, or his fathers before him, that he should be condemned to spend his life seeking for the cure of an incurable disease?'[62,63]

Pavy did not dispute that diabetes was incurable. But he also knew that the death of a patient, while sadly inevitable, could be delayed. John Rollo's harsh dietary regime of eating only meat and no vegetables suggested that the main factor in managing diabetes was to limit the amount of sugar in the body. But thanks to the development of chemical methods that enabled the measurement of urinary sugar, Pavy was able to take this approach even further. Using these methods, he was able to quantify the effect of different diets on his patients and show that a diet consisting of little or no carbohydrate was successful in clearing the urine of sugar.[64] A French doctor, Apollinaire Bouchardat (1806–1886), made a similar observation and noted that, during the Siege of Paris in the Franco–Prussian War of 1870, conditions of starvation and rationing caused the glucose to disappear from the urine of some of his diabetic patients.[65] Furthermore, he noted that their tolerance for carbohydrates appeared to be improved by physical exercise, leading him to warn his diabetic patients: 'You shall earn your bread by the sweat of your brow.'[66]

The most prominent and vociferous advocate of the diet-based approach was the US doctor Frederick Madison Allen (1876–1964). As a leading diabetologist, Allen was concerned that some patients were being treated not with a restricted diet, but rather by being overfed. This idea had its origins thanks to French clinician Pierre Piorry (1794–1879) who, in 1850, had reasoned that since the urine of all diabetic patients had such high levels of sugar, then the complications of the disease must be caused by loss of sugar from the body. From this, Piorry concluded that the best way to remedy this situation was therefore to feed the patient a diet high in sugar.[67,68] Unsurprisingly, this approach did not have a happy ending for many of Piorry's patients, and the idea that diabetic patients needed to be overfed persisted into the early twentieth century.[69,70]

Although one obituary writer credited Allen with making 'the greatest advance in the treatment of diabetes prior to the discovery of insulin', his patients may have felt differently.[71] For although his course of treatment certainly delayed complications and death from diabetes, it involved starving the patients until their urine was completely clear of glucose, at which point small quantities of food could slowly be reintroduced. Although it left them constantly hungry and badly emaciated Allen's treatment did appear to prolong the life of his patients—at least until they died from starvation.[72] Another problem with Allen's method was simply that it seemed blissfully ignorant of human nature. Once discharged from hospital, many patients (perhaps understandably) succumbed to temptation and broke their diet regime by surreptitiously eating contraband food. For others, the temptation was too great even before they were out of the hospital doors. In one particularly extreme—and tragic—case, a twelve-year old boy already blind due to diabetes and still in hospital had become so desperate for some nutritional variety that he had resorted to pilfering bird seed from a canary cage.[73] When his 'crime' was discovered, the staff cut his food supply even further, and he eventually died of starvation.

Allen appeared to have little sympathy for such behaviour, dismissing it as evidence of a lack of character. Described as a 'stern, cold, tireless scientist, utterly convinced of the validity of his approach', he dismissed those who could not adhere to his harsh regime as being of 'the habitually unfaithful type'.[74] In drawing a connection between failure of treatment and lack of moral character in diabetic patients, he was far from alone. In 1921, in a paper evaluating the Allen method as a means for managing diabetes, New York doctor John R. Williams highlighted the importance of strength of character:

> All kinds of patients have been treated, the courageous and the cowardly, the enterprising and the slothful, the attentive and the interested, and the indifferent. These types are mentioned because the character of the individual has much to do with the outcome. To successfully contend with diabetes a patient must not only be wisely advised, but he must also possess courage and a willingness to learn and assist, to create new dietary habits and eliminate old ones.[75]

Having established the importance of character and mental discipline in managing diabetes by dietary methods, Williams went on to conclude that 'many cases unquestionably die because of lack of courage' and that 'there is little question but that many failures ascribed to the treatment are due to lack of faithfulness on the part of the patient'.[76]

This kind of moralizing about diabetic patients was common. One correspondent to the *Journal of the American Medical Association* wrote that:

The great majority of patients who suffer with this disease do so because, through ignorance, they have grossly abused their systems with the quantity, quality and kind of food used [...] Every physician who has treated many diabetic patients knows that, on the subject of eating, the diabetic is mentally unbalanced.[77]

For this correspondent, there was no difference between diabetes and 'chronic morphinism or alcoholism'. Furthermore, they stressed 'the futility of drugs in the treatment of this disease; futile, because harmful habits cannot be cured by drugs'.[78]

Not all diabetes specialists were prone to such finger wagging. The charm and warmth with which the Boston clinician Dr. Elliott Joslin (1869–1962) (Figure 1) treated his patients made for a striking contrast with the stern figures of Allen and Williams. Joslin placed great emphasis on the importance of educating patients

Figure 1 *Boston diabetes clinician Elliott P. Joslin (1869–1962).*

Credit: Insulin Collection, F.G. Banting (Frederick Grant, Sir) Papers, Thomas Fisher Rare Book Library, University of Toronto; MS. COLL. 76 (Banting), Box 63, Folder 3A. Reproduced with Courtesy of the Thomas Fisher Rare Book Library, University of Toronto. Online at: https://insulin.library.utoronto.ca/ islandora/object/insulin%3AP10132.

about their condition and, perhaps most importantly, imbued them with 'a sense of hope'.[79,80]

Despite being an advocate of the starvation-diet approach to treating diabetes, Joslin was also aware that this could, in no possible sense of the word, be considered to be a cure. In 1921, Joslin was visited by eminent German diabetes specialist Carl H. von Noorden (1858–1944), who had pioneered his own particular treatment for diabetes: this method involved feeding his patients oatmeal gruel. Joslin showed his guest some of his latest patients who were undergoing starvation treatment. One of these patients was a seventeen-year-old girl who now weighed only 24.5 kg and was now little more than a skeleton. On seeing her, Joslin recalled that von Noorden turned away in disgust.[81,82] Many of Joslin's patients were children or teenagers and he questioned the value of imposing a starvation diet on a life that would be unnecessarily cruel and short: 'Why not let the poor child eat and be happy while life lasts?'[83]

Figure 2 *Fred Banting (1891–1941) taken on 27th December 1922.*

Credit: Insulin Collection, F.G. Banting (Frederick Grant, Sir) Papers, Thomas Fisher Rare Book Library, University of Toronto; MS. COLL. 76 (Banting), Box 63, Folder 3A. Reproduced with Courtesy of the Thomas Fisher Rare Book Library, University of Toronto. Online at: https://insulin.library.utoronto.ca/islandora/object/insulin%3AP10042.

Yet even as he spoke these words, Joslin knew that a medical revolution was underway that would make starvation diets a thing of the past, along with leeches and bloodletting. In 1913, Allen had poured scorn on the notion that injections of pancreatic extracts might be used to treat diabetes, describing them as 'both useless and harmful'.[84] Yet within a decade, Allen retracted his statement, and was soon singing the praises of Fred Banting (Figure 2) and his colleagues in Toronto, whose extracts of pancreatic tissue were saving the lives of diabetic patients. This discovery would save countless lives and earn Banting the 1923 Nobel Prize in Physiology or Medicine. Yet, when Banting first learned that he had been awarded the most coveted accolade in science, his response was hardly happy. While the award certainly brought Banting fame and recognition, it would ultimately prove to be a blessing, and curse, for him.

2

Thick Brown Muck

When Fred Banting first heard the news from a friend over the phone that he had been awarded the Nobel Prize, he was furious. Telling his friend to 'Go to Hell', he slammed down the receiver and grabbed his copy of the morning newspaper to check that this was not all some kind of joke at his expense. But sure enough, it was all there in black and white. What enraged Banting about the headlines was not so much that he had been awarded the Nobel Prize, but also so had his boss, John J. R. Macleod, Professor of Physiology at the University of Toronto (Figure 3). For Banting, the award also going to Macleod was nothing short of a travesty. As far as Banting was concerned, Macleod had no right whatsoever to have any claim on the prize. Outraged, he jumped into his car and headed straight for the University of Toronto, intent on correcting this injustice.[1] Banting later recalled that his intention had been 'to tell Macleod what I thought of him', but one onlooker feared for Macleod's safety, observing that Banting 'was furious … He could have torn the whole building down … Oh, he was helling and damning'.[2]

Given that Banting had a notoriously short temper and on occasion even seemed to relish the prospect of a fight, Macleod might have good reason to fear for his safety. But before Banting could reach Macleod, he was stopped by his colleague J. G. Fitzgerald, who tried to calm him down. Raging at Fitzgerald, Banting shouted that he was going to turn the prize down and that Macleod had no right whatsoever to share it with him:

> I defied Fitzgerald to name one idea in the whole research from beginning to end that had originated in Macleod's brain—or to name one experiment that he had done with his own hands.[3]

But Banting's own sense of personal grievance played a small role in how his behaviour could be perceived. As the first Canadian to be awarded the Nobel Prize, Banting had brought a tremendous sense of honour to his nation. Was he really now about to throw all this away? Knowing Banting to be a patriot, Colonel Albert Gooderham, a member of the University of Toronto Board of Governors, appealed to his sense of national pride: how would Banting's fellow Canadians feel if he carried out this threat? Perhaps more importantly, how might they feel if

Figure 3 *Professor John James Rickard Macleod (1876–1935). [ca. 1928].*

Credit: Insulin Collection, University of Toronto Archives. B1995-0034. Reproduced with Courtesy of the Thomas Fisher Rare Book Library, University of Toronto. Online at https://insulin.library.utoronto.ca/islandora/object/insulin%3AP10134.

scientists appeared to be elevating their own personal grievances over life-saving research and the service of their nation?

Gooderham's insights into Banting's psychology worked. But while the nation's pride could be saved, the situation between Banting and Macleod was beyond repair. They had never enjoyed a particularly harmonious relationship; long before the furore over the award of the Nobel Prize had erupted, Banting had already harboured a mistrust of Macleod. Yet but for Macleod, Banting may have remained a struggling GP in provincial Ontario and never taken those first steps on the road to the Nobel Prize.

Banting had grown up on a farm in Alliston in rural Ontario about forty miles north of Toronto. He began his academic career in 1910 by enrolling on a course in General Arts at Victoria College at the University of Toronto. Having failed an examination in French at the end of the academic year, he was forced to retake the whole academic year, which prompted him to abandon the arts and study medicine instead. After dropping out of Victoria College he spent the spring and summer working on his father's farm, conscious that he needed to establish a secure career path for himself. On his application to the Medical School of the

University of Toronto, he falsified his date of birth to make himself appear a year younger; he was accepted onto the course in September 1912.

However, when the British Empire entered the First World War on August 4, 1914, Banting's strong sense of patriotism as a Commonwealth citizen stirred him into action. On August 5, he tried to enlist in the Army but was rejected due to poor eyesight, and a second application later that year also failed. Undeterred, Banting persisted and, having successfully enlisted in 1915 as a private in the Canadian Army Medical Corps, he found himself posted to a training camp at Niagara Falls.[4]

Advised to specialize in surgery, Banting soon found himself on a learning curve with a steep and unforgiving gradient. Once, when a patient required surgery for an abscess on his throat and no qualified staff were available, Banting performed the operation, although he was still only in the fourth year of his medical course.

Despite Banting's lack of experience, the operation was a success. In summer 1916, he went to Toronto to hone his skills under the guidance of the distinguished surgeon Dr. Clarence Starr. His career was at last taking shape and upon completing his final exams in October he promptly reported for military duty the day after graduating.

In March 1917, Banting set sail from Halifax, Nova Scotia bound for the Granville Canadian Special Hospital at Ramsgate, Kent, where he began helping Starr to pioneer a new technique for the suture of nerves. When the wind blew in the right direction, it carried sounds of the heavy guns on the front lines in France, but for most of the time the war still felt distant. However, as he remarked in a letter to an old flame, he would occasionally receive a stark reminder of the brutality of the conflict:

> Air raids are always interesting of course, but do little harm. Such things make one forget the war but those patients with arms & legs off always recall it very quickly.[5]

Despite working in a busy surgical ward, Banting still had plenty of time on his hands and complained that 'there was more wasted time in the army than almost any occupation or profession in the world'.[6] Putting these spare hours to good use, he began studying every night to qualify for membership to the Royal College of Physicians and the Royal College of Surgeons. But despite his long hours of hard study, when he travelled to London three months later to sit his oral exam in Obstetrics and Gynaecology, he was bitterly disappointed at the outcome:

> It was the first time I had ever failed an examination and my pride was injured. So I resorted to a bold experiment ... to prove that the system of examinations in England was wrong ... I burned in my fireplace every note and synopsis and sold or gave away every book on the subject. I decided that I would not study one more minute on this subject.[7]

Resolving to adopt a stoic approach to this disappointment, Banting decided to concentrate all his efforts on qualifying to become a Fellow of the Royal College of Surgeons (FRCS). Although Banting hated having to memorize the names of various bones, muscles, and nerves—all of which he dismissed as a waste of time, he did recognize that the prestige of this qualification would greatly enhance his professional status. But there was one honour that he craved even more than membership of a distinguished professional body. For as he told his mother in a letter home, '... I have been recommended for the Military Cross. Just at present I'd sooner have it than the F.R.C.S'.[8]

The chance to distinguish himself came soon enough. Having now been promoted to the rank of Captain, he was posted to the No. 13 Canadian Field Ambulance in the Amiens-Arras sector, where he worked as a relief officer at a dressing station.[9] Here Banting's task was to receive casualties from first-aid posts near the front line and, having cleaned, closed, and dressed their wounds, then send them on to base hospitals. It was a task fraught with danger but for Banting it had far more appeal than spending long hours committing anatomical features to memory for a professional exam:

> ... I would sooner be the medical officer of a front line [sic] battalion than any other place in the army. One feels that ones training and ones work was of real value to ones fellow man. It is on only rare occasions that a doctor can honestly feel that his skill has saved or prolonged a human life. A Battalion Medical Officer has this experience sometimes frequently.[10]

But of course, being the medical officer of a front-line battalion meant also that danger was never far away. In September 1918, Banting and a fellow officer were scouting out possible sites for new aid posts in preparation for the Battle of Cambrai when they came under fire:

> ... the Germans put over a shell. We both flattened ourselves automatically on the side of the road. I had flopped in the ditch where the remains of a dead mule having been liquefied by sun, rain and time had run down from the carcass above. I got up plastered with dead mule and smelling like a glue factory. I never quite forgave him for the way he laughed but his laugh was cut short by another shell and we both did a hundred yard [sic] dash to an old cellar. Such experiences draw men together ... .[11]

On another occasion, Banting found himself unable to clear his station of casualties in the face of an imminent German counterattack. Unwilling to abandon his patients, Banting remained at his post and quickly found himself face to face with an enemy soldier who had appeared in the doorway of his battlefield hospital. It was at this moment that one of Banting's patients made a timely expression of gratitude for his loyalty in not deserting them. A shot rang out and the enemy soldier dropped to the floor, killed by a sergeant whose foot Banting had just amputated.

Despite the rigours of the Western Front, Banting kept up his habit of writing letters home to his mother, which he had done since leaving home in 1911.[12]

But for all his reassurances, his letter of September 29, 1918 is likely to have caused his mother some concern. On the previous day, Banting had been working hard all day clearing the casualties of the 44th Battalion when he heard the news that the medical officer of the 46th Battalion had been hit. The 46th had taken over the assault and so, rounding up a team of stretcher bearers, Banting had set out towards the front line intending to get help to as many of the wounded as possible. But while trying to reach the Battalion's advance dressing station Banting was hit in the arm by flying shrapnel.

Despite being wounded, Banting insisted on staying at the front line, but his injury was too serious. While on the hospital train across France he struggled to write to his mother using a near-illegible scrawl, a result of writing with his left hand—he even attempted to draw an accurate picture of the piece of shrapnel:

> My Dearest Mother, This letter will b [sic] short left hand. I was slightly wounded yesterday in the right forearm. Had operation last night and shrapnel .·. removed from between bones. No fracture but ulna bone damaged. I feel petty [sic] good. Only tired. I have just had a big hot lovely dinner. Everyone is as kind as can be. Now please don't worry. I am the luckest [sic] boy in France.[13]

Unsure of the train's final destination, Banting ended the letter by saying 'I don't know where I am going'.[14] He would spend the rest of the war recovering in Manchester, UK, and Scotland. Although Banting eventually returned to Canada in 1919 as a war hero, having been decorated with the Military Cross for courage under fire, he learned quickly that the ribbons on his chest would be no guarantee of an easy life once out of active service.[15]

With Banting having now been discharged from the Army, his fiancée Edith Roach had hopes that they might now finally marry. Edith had been waiting patiently for this moment since their engagement in 1916, but now Banting had reservations. Edith was working as a teacher in a high school and earning a good salary. Banting meanwhile was painfully aware that he was now in his late twenties, with little money, and without a secure medical post. Being of a rather socially conservative outlook, Banting found the prospect of having to rely on financial support from his wife while he completed his training as a surgeon rather humiliating. In a frank admission to himself, he recognized that his relationship with Edith was over—'we had been separated so long that she had other interests'.[16]

Presumably as a distraction from these mounting concerns, Banting threw himself into his work. To finish his training in surgery, he returned to the Hospital for Sick Children in Toronto where he worked as a registrar and senior house surgeon. He began to harbour hopes that, following in the footsteps of his mentor Starr, he too might pioneer new surgical techniques.

However, in 1920, he learned that no new position in surgery was available for him to continue at the hospital. It was to be the start of a downward spiral of disappointments. Banting left the Hospital for Sick Children in June 1920 and borrowed some money to buy a house in London, Ontario, where he hoped to set up his own private medical practice. Then he sat and waited for the first patients to walk through his door. The days passed and still no one came. Eventually, after four weeks of waiting and becoming increasingly frustrated, Banting treated his first patient. According to his biographer Michael Bliss, Banting's accounts book for that particular day, July 29th, show that he earned a $2 fee for 'baby-feeding'.[17] However, writing in 1940, Banting described his first patient as an 'honest soldier' who simply wanted a prescription for alcohol so that he could entertain some friends who were visiting. Taking pity on the man, Banting willingly wrote the prescription but reflected that he thought himself to be 'rather highly trained for the Bar Keeping business'.[18]

Things were not working out as he had hoped. He was bored, lonely, and deeply in debt. The few patients who did come to his practice were so poor that he could not bring himself to send them a bill.[19] To earn some desperately needed extra cash he took a part-time job as a demonstrator teaching classes in anatomy at Western University and although this brought in $8–10 a week, he later recalled still having to cook his meals over a bunsen burner in his dispensary and forego trips to the cinema.[20]

His one relief from these ongoing worries was a hobby that also had the added benefit of being relatively cheap. Walking past a shop one day, Banting spotted a painting of some men and a boat in the window and mused to himself whether he might be able to paint a similar picture—and, better still, sell it. By his own admission he knew 'absolutely nothing about painting', but nevertheless, he entered the shop and emerged shortly afterwards armed with a selection of paints and brushes with which he began to make copies of pictures from magazines and books.[21] He said that painting gave him his 'happiest hours' and they might well have remained so had he not harboured ambitions that they could become anything more than a therapeutic and absorbing hobby.[22]

Whether in the field of surgery or painting, Banting craved to make his mark on the world. Confident that his pictures might be good enough to become a potential source of income, Banting took a few of them to an art dealer but was in for a rude awakening:

> He passed some scathing remarks and laughed at my best efforts. I felt disappointed and offended. When a person is down people set on him.[23]

His ability as an artist was not the only area in which Banting's estimation of his own expertise was starkly at odds with reality. Having bought what he described as 'a worthless old forth [sic] or fifth hand car for about five times what it was worth', the vehicle broke down after a mere 250 miles.[24] Banting attempted to fix it, but it would go no further: 'The man from whom I bought it from was a dishonest rogue and drove a huge car', he raged, 'I would have liked to punch his fat, prosperous face—but it would not have been worth it'.[25]

Banting was increasingly feeling that the whole world was set against him for, in addition to scornful art dealers who did not appreciate his talent, and dishonest car traders who ripped him off, London was home to what he described as 'a group of very undesirable type of doctors. They were not well trained but they were prosperous with many patients'.[26] Banting was convinced that these doctors had resolved to "'Starve him out of London"—and they did'.[27] Writing in 1940, he described how 'All of these experiences and worries and idleness made me dislike the practise [sic] of medicine.... . I was anxious to get away from it all'.[28]

On the evening of Sunday October 31st, Banting thought that he had seen the first glimpse of how he might finally do this. Having taken on some part-time teaching of university classes simply in order to make ends meet, he had spent the day reading textbooks as preparation for a class that he was due to teach the following day on the physiology and function of the pancreas. With his lecture prepared, went to bed with a copy of a leading surgical journal intending to read a few articles before falling sleep. It proved to be a fortuitous choice of bedtime reading, for the very first article was to change the course of his career—and his life.

For a lay person, the title of the paper by Moses Barron, 'The Relation of the Islets of Langerhans to Diabetes with Special Reference to Cases of Pancreatic Lithiasis', might seem to be the perfect antidote to insomnia. But for Banting it caused a sudden revelation: he believed that within its dull text lay a holy grail for physiologists—the key to finding the anti-diabetic internal secretion made by the pancreas.

Banting was convinced that the reason that no one had yet been able to isolate this anti-diabetic secretion was because it was being destroyed during the process of extraction by digestive enzymes such as trypsin that were also made in the pancreas. What was therefore required was some means of preventing the action of these enzymes, and Banting was convinced that this paper pointed the way.

The paper showed that when the duct connecting the pancreas to the gut is blocked by insoluble calcium deposits, the acinar cells that produce the digestive enzymes rapidly degenerate while the Islet cells—which were believed to produce the hypothetical anti-diabetic compound—were left intact. This immediately suggested a way by which the Islet cells could be separated from the enzyme-producing tissue.[29] As Banting went to bed after having closed the journal, he reflected on the article and 'thought about my miseries and how I would like to get out of debt and away from worry'.[30] Sleep did not come easily that night—but for the first time in months, this was due to excitement, and not worry. Sitting long into the early hours of the morning, Banting hit on the idea of artificially blocking the duct—perhaps by ligating or tying it shut. Then, once the enzyme-producing cells had died and withered away, the Islet cells could be isolated to prepare an extract that was free of destructive enzymes.[31]

Banting mentioned his idea to Professor F. R. Miller for whom he taught classes at Western University in London, Ontario. Miller, a neurophysiologist, knew little about carbohydrate metabolism or diabetes and suggested instead that Banting approach John J. R. Macleod, Professor of Physiology at the University of Toronto. Macleod was currently working on various aspects of respiration but was also a recognized authority on the metabolism of carbohydrates and had written a book on diabetes and its pathology.[32]

On November 8th, Banting stepped into Macleod's office eager to present his idea. But according to Banting's recollection, Macleod's response was not the one for which he had been hoping. After seeming initially receptive to what Banting had to say, Macleod's attention drifted to reading the letters on his desk. He then explained that many other researchers had worked for years in far better equipped laboratories and still failed to find any evidence of an anti-diabetic secretion made by the pancreas—what made Banting think he could succeed where so many others had failed?[33]

In fairness to Macleod, his caution was understandable. Banting was not the first enthusiastic young researcher to have approached Macleod confident of success in hunting down the enigmatic anti-diabetic hormone. In 1912, Macleod had been contacted by Ernest Lyman Scott, a researcher at the University of Chicago, whose doctoral thesis had concluded with the bold claim that he had prepared an extract from pancreatic tissue that could significantly reduce blood sugar levels.[34] But by the time that Scott's results were published in the *American Journal of Physiology*, the key sentence which described his important conclusion had been omitted and replaced with a far more modest claim.[35] Scott's wife later maintained that what she called this 'damning sentence' had been inserted by Scott's supervisor, the Professor of Physiology Anton J. Carlson. According to Scott's wife, Carlson had never considered her husband to be anything more than 'a fourth-generation Ohio farm boy' and had sought to put him in his place by diluting his claims.[36]

Banting felt that he was receiving a very similar treatment from Macleod. 'A hot iron gives off steam when cold water is thrown on it', he later wrote when remembering this encounter. He then added: 'It was the first time that I had ever seen the famous professor and I was not overpowered with either the man or his knowledge of research'.[37]

However, Macleod's own account of this very first meeting between himself and Banting offers a rather different perspective. Rather than pouring scorn on Banting's ideas, Macleod recalled that he had praised them and insisted that he had always been as supportive as possible:

> Dr. Banting deserves complete credit; if he had not contributed this idea and undertaken to test it experimentally, the discovery of Insulin would probably not as yet have been made. On the other hand, with the knowledge which he possessed of the methods for attacking such a problem he could certainly not have made such rapid progress without careful guidance and assistance. In every field of research innumerable suggestions are offered as to the new pathways that should be explored, and it takes judgement and long years of experience to know which of these suggestions should be encouraged and assisted by affording the means to put them to the test. I believe I am perfectly safe in saying that there are very few investigators in the field of diabetes, at the time Dr. Banting undertook it, who would have thought the experiment with duct-tied pancreas worth a serious trial.[38]

Macleod recognized that, although Banting had some experience of surgery, the skill required to perform the necessary techniques was currently well beyond his level of capability. He went on to warn Banting that there would be no point in starting this work unless Banting was willing to devote all his time to it for at least several months. If Banting could agree to this, then Macleod said he would be willing to 'place every facility at his disposal and show him how the investigation should be planned and conducted'.[39]

Although willing to give Banting a chance, Macleod already had his own clear ideas about how the research should be planned and conducted, and more importantly what its aims should be. Banting had high hopes that his research would result in the isolation of a life-saving new substance, but Macleod had far more modest aims in mind. Banting recalled how, sitting back in his chair with his eyes closed, Macleod began to talk. And as Macleod talked, Banting recalled that he continually repeated the single phrase 'negative results would be of great physiological value'.[40]

This did not give Banting confidence and left him distinctly unimpressed. Macleod seemed to think that Banting's proposed research would indeed be useful—but only in so far as it would act as a negative control to show that it was actually possible to prevent the destructive effects of digestive enzymes on pancreatic extracts. His interview with Macleod seemed to be a repeat of his humiliation at the hands of the London art dealer. After all, Banting had passed up the opportunity of accepting appointments in surgery to take up Macleod's offer—and for what? Banting wanted to shoot for the stars, not act as an overqualified technician grinding out 'negative results of great Physiological [sic] importance'.[41]

Determined to prove Macleod wrong, Banting returned to his clinical practice in London, Ontario and considered the choices that lay before him. Should he take a gamble and give up what had become a relatively stable and secure income from his clinical practice to embark upon a precarious career in a field of research that had no guarantee of success? Ultimately, his final decision appears to have been dictated by affairs of the heart, not the head. At some point in late 1920 or early 1921, Edith broke off their engagement. Around the same time, Banting was exhibiting a restlessness that contrasted sharply with the focus and determination that he had shown when pondering the mysteries of the pancreatic secretion. For a while he toyed with the idea of joining the medical service of the Indian Army and also applied for the post of medical officer on an expedition to drill for oil in the Northwest Territories, only to be informed that although he had been a good candidate for the post the expedition had decided it no longer needed a medical officer.

Then, an invitation arrived from Macleod to come and work in his lab, and Banting accepted.[42] On May 14, 1921 he collected a box of cigars as a parting gift from his students and stepped aboard the train from London to Toronto to begin working in Macleod's laboratory. According to his biographer Michael Bliss, he might well have had a happier life had he remained in London.[43]

The starting point for Banting's research project was the hypothesis that all previous attempts to isolate the anti-diabetic secretion had failed due to the action of digestive enzymes produced by the pancreas. The main task therefore was to prevent the action of these enzymes. Inspired by what he had read in Barron's paper about how natural blockages of the pancreatic ducts cause the cells producing these enzymes to die off, Banting was convinced that this same effect could be achieved by artificially tying shut the pancreatic duct of a dog. When, after a period of several weeks, all the enzyme-producing cells had withered away, the only remaining cells would be the Islet cells, which made the anti-diabetic substance. Preparations made from this remaining tissue should then be free from digestive enzymes. If these extracts did indeed contain the anti-diabetic factor, injection of them into a dog that had been rendered diabetic by pancreatectomy should cause a significant drop in the concentration of sugar in the animal's blood (Figure 4).

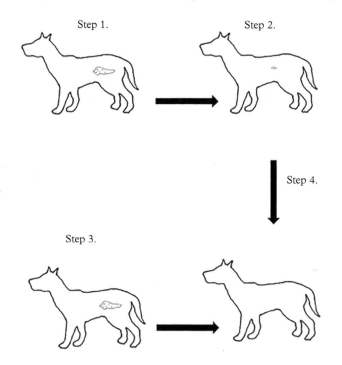

Figure 4 *Banting's Proposed Plan of Research. Step 1) Artificially block the duct of the pancreas in a dog. Step 2) This causes the bulk of the pancreatic tissue which produces digestive enzymes to degenerate, but leaves the insulin-producing Islet cells intact. Step 3) Sacrifice a dog using pancreatectomy to induce diabetes. Step 4) Inject extract made from degenerated tissue of first dog and inject this into the diabetic dog. Diagram by K. T. Hall.*

On paper at least, it looked like a straightforward plan of research. But as was all too apparent to Macleod, Banting 'had only a superficial text-book knowledge of the work that had been done on the effects of pancreatic extracts in diabetes'.[44] Not only did he lack the necessary surgical skills to perform a pancreatectomy, but also he was unfamiliar with the techniques for carrying out a measurement of blood sugar levels—measurements crucial to demonstrate that pancreatic extracts had an anti-diabetic effect.

To address this serious problem, Macleod himself taught Banting how to perform a pancreatectomy and introduced him to two final year students, Clark Noble and Charles Best (Figure 5), who were to help with the measurements of blood sugar levels. As the initial plan was for Banting to spend two months working in Macleod's lab, Clark and Noble would each spend a month working with him. Whoever went first would work for a month with Banting and then enjoy an unbroken summer holiday, and the decision was made by flipping a coin.[45] When it landed, Best was the winner. And thanks to the toss of that coin he was to win much more than a few uninterrupted weeks in the summer sun.

Figure 5 *Charles Best (1899–1978).*

Credit: Insulin Collection, C.H. Best (Charles Herbert) Papers, Thomas Fisher Rare Book Library, University of Toronto MS. COLL. 241 (Best) Box 47, Folder 3.*Reproduced with Courtesy of the Thomas Fisher Rare Book Library, University of Toronto. Online at https://insulin.library.utoronto.ca/ islandora/object/insulin%3AP10070.*

Like Crick and Watson, the duo of Banting and Best are now recognized as such a legendary partnership in the history of science that the eminent British physiologist Sir Henry Dale later described their work together as a 'historic collaboration' (Figure 6).[46] However, when they first began working together in May 1921, conditions were less than ideal. First, Banting and Best had to face the far more mundane task of getting down on their hands and knees to scrub the floor.[47] A description given by Banting of their early working conditions hardly conjures up the image of a state-of-the-art medical research facility:

The place where we were operating was not fit to be called an operating room. Aseptic work had not been done in it for some years. The floor could not be scrubbed properly, or the water would go through on the laboratories below. The walls could not be washed for they were papered and then yellow washed. There were dirty windows above the unsterilisable wooden operating table. The operating linen consisted of towels with holes in them.[48]

Figure 6 *Photograph of Frederick Banting (right) and Charles Best (left) with one of their dogs on the roof of the Medical Building (August 1921).*

Credit: Insulin Collection, Best (Charles Herbert) Papers, Thomas Fisher Rare Book Library, University of Toronto; MS. COLL. 241 (Best) Box 109, Folder 4.*Reproduced with Courtesy of the Thomas Fisher Rare Book Library, University of Toronto. Online at https://insulin.library.utoronto.ca/islandora/object/insulin%3AP10077.*

While Sir Henry Dale also praised their collaboration as being one of 'intimate understanding', it is clear that, in those early days at least, it was not always an amicable one.[49] While Best was away for a couple of weeks at a training camp in Niagara, Banting decided to repeat some of the measurements that Best had made.[50] He was not impressed to find that the glassware Best had been using was filthy and concluded that this might well explain the serious discrepancies he found in Best's data.[51] When Best called in at the lab, having returned from visiting his girlfriend, Banting made his feelings on the matter clear:

I was waiting for him, and on sight gave him a severe talking to. He thought that he was both God's and Macleod's appointed, but when I was finished with him he was not sure. I told him that if he was going to work with me that he would have to show some interest, that his work was totally unsatisfactory that he lacked accuracy and was too sloppy, and I ended up by telling him that before doing another thing he must throw down the sink every solution that he had been using, wash every bit of glass-ware—make up new solutions that were truly 'normal'. When I finished setting him out he gazed fiercely at me for some moments in silence. He looked very defiant. His fists opened and closed. He was very vexed. I thought that he was going to fight and I measured the height of his jaw. He delayed and I feared that he was not going to fight. Suddenly he turned on his heel, went upstairs and I heard rough usage of glassware for some time. He worked all night (for the first time in his life but it was the beginning of many). In the morning he left and I inspected. Everything was spick and span. We understood each other much better after this encounter.[52]

Despite these early teething troubles, by the middle of June Macleod was sufficiently confident that Banting and Best could be left to carry out their investigations without his guidance and, leaving them with a full set of references to the relevant research literature, he departed for a holiday in Europe.[53] The grim entries in Banting's laboratory notebook during the weeks that followed, however, suggest that Macleod's confidence in their combined competence may have been a little premature. When Banting first attempted to perform a pancreatectomy without Macleod's assistance he found that it was fraught with difficulties, as was the surgical procedure of ligating the pancreatic duct. As the two researchers struggled to work in the heat of the Toronto summer, they found their dogs were dying at an alarming rate and, as Best explained in a letter to Macleod, this was not due to experimentally induced diabetes:

We have delayed writing you because until recently we have not been able to secure any significant results. Infection has been our great trouble. We have found it next to impossible to keep a wound clean during the very hot weather. Conditions in the animal room also are not very good, as you know. We have gone through seven dogs as controls ... . We have had very heavy casualties among the 'duct tied' dogs during the hot weather and have now only two survivors in good health.[54]

By early July, fourteen out of the nineteen dogs that Banting and Best had used were dead from infections, overdoses of anaesthetic, and complications during surgery.[55] As Best later recalled, this forced them to take some rather desperate measures so that their work might continue:

Suggestions have been made by poorly informed authors that dogs were appropriated from the street with very little ceremony. This is not true, but there were occasions when we made a tour through various parts of the city and bargained with owners of animals. They were paid for by funds which we took from our own pockets.[56]

Not that their pockets were particularly laden with cash in the first place. Neither of them received a salary for their work and Banting had to either lodge with his cousin or stay at a boarding house that he had used during his student years. To add to their frustrations, not only were they running out of dogs, but also Banting's method of making extracts that were free from digestive enzymes by sealing the pancreatic duct shut was not working out as well as he had hoped. Having tied the pancreatic ducts shut, he and Best then had to wait several weeks for the organ to degenerate sufficiently to allow the preparation of an extract that was free from enzymes. The heat and humidity of the hot summer made healing of the surgical wounds difficult, but there were even bigger problems.[57] After waiting patiently for several weeks, Banting found that five out of seven animals showed no signs at all of any degeneration of the pancreas having taken place.[58] One of the test animals, Dog 410, appeared to possess an uncanny insight into the practical competence— or more accurately, the lack thereof—of his masters. A lab notebook containing entries by both Banting and Best records that, when an attempt was made to take a blood sample from the dog on July 27th, the animal, 'jumped from table [sic] and tried to escape'.[59] Given what was to follow, it was an understandable response.

For having established that repeated injections of the pancreatic extracts could lower the blood sugar levels in a diabetic dog, Banting and Best now wondered whether they could prevent the sudden rise in blood sugar that would occur following the ingestion of a solution of glucose. But when they attempted to feed the dog a solution of sugar, they mistakenly inserted the feeding tube into its lung instead of its stomach, and in so doing, nearly drowned the poor animal before it made a complete recovery fifteen minutes later.[60]

Accounts like this make for uncomfortable reading and Banting was sensitive to charges of cruelty or callousness in the treatment of his animals, fondly recalling one of his particular dogs, 'wrapped in my old lab coat, dim light, I pat her head, the end of her tale [sic] wags. I bend over and take her in my arms. She is comfortable. I clear her eyes with my handkerchief, she looks at me and wags her tail'.[61]

And in response to accusations of cruelty from anti-vivisectionists he offered the following defence:

> There have been times when I have been severely criticised by the anti-vivisectionists. Despite this criticism I have always been in sympathy with much of their work and I find myself very tolerably disposed towards those of them who have sufficient intelligence to prevent them from becoming too extreme ... . I have always felt that a dog should be as far as possible treated like a human with regard to operation, aftercare, nursing, drawing of blood, etc. and that nothing should be done to cause pain unless necessary and that an anaesthetic should be used as it would in the case of a private patient ... no laboratory man should be employed to care for animals unless he has an actual fondness for them.[62]

Despite the litany of disasters with their lab animals, Banting and Best had cause for optimism by early August. At 1 p.m. on August 4th, yet another dog (408) was injected with some pancreatic extract. Two hours later the concentration of sugar in the blood had fallen by almost fifty per cent before starting to climb again. At 9 p.m. that same evening, some more extract was injected, resulting in a similar pattern—the blood sugar fell before starting to climb again. At midnight, an entry written in Best's hand recorded that the dog was 'in good condition'.[63]

By early afternoon the following day, the situation was less promising. The blood sugar levels had continued to rise and injections of extract made from both liver and spleen, administered as a control, showed no effect with the dog left in 'poor condition' and so lethargic that it would drink but not eat.[64] Fearing that the dog would not live for much longer, Banting and Best injected it with a fresh batch of pancreatic extract that they described in their lab notes as 'Isletin' and, suddenly, things began to look up.

Over the course of the afternoon and into the evening, the dog's blood sugar levels fell and the animal became brighter and more responsive.[65] On the evening of August 6th, Banting worked all through the night, giving more injections of Isletin and recording his observations on the hour (Figure 7). By 8 the next morning he was delighted to see that the dog could now rise to its feet and stand up, despite having appeared to be lifeless and on the verge of death only three hours earlier.[66] Moreover, its blood sugar levels had fallen by nearly two thirds.

Sadly, this recovery was short lived. By noon, the dog was dead. In their lab notes, Best recorded that this was due to 'widespread infection and large quantities of pus in lower abdominal and pelvic regions ... abdominal wall infected .... Cause of death—infection'.[67] But this did not deter Banting from writing to Macleod to tell him about what he believed to be a major discovery:

> I have so much to tell you and ask you about that I scarcely know where to begin. I think you will be pleased when you see how the problem is unrolling from one end and rolling up at the other. At present I can honestly state my opinion that (1) the extract invariably causes a decrease in the percentage of blood sugar in diabetic dogs (2) that it is active at least for four days if keep [sic] cold (3) it is destroyed by boiling (4) that extracts of spleen and liver at least, purposed under similar conditions have no such activity ... the number of problems that are presenting themselves is becoming greater and greater.[68]

Banting could barely contain his excitement. Confident that the effect of the pancreatic extracts was real, his mind was racing with questions. In his letter he outlined sixteen key questions that he now wanted to answer, including how to determine the chemical nature of the substance. How did it work? And, most importantly of all, how might it be applied clinically?

When Macleod's reply finally arrived, he was as ever, the voice of caution, urging Banting to learn first to walk before breaking into a sprint:

> The results of your experiment are certainly very encouraging but in order that there may be no possibility of mistake I would suggest that you continue along the

Figure 7 *Photograph of laboratory 221. This was the laboratory in which Banting and Best carried out some of their research in 1921–22. It is not clear whether this was the original laboratory in which Banting and Best first began their work and made these blood sugar measurements (having had first to clean and scrub the floors), or the one that was given to them when Macleod returned from Europe. Taken in April 1929, the photograph is simply commemorated as 'the laboratory where insulin was discovered.'*

Credit: Insulin Collection, University of Toronto Archives, Thomas Fisher Rare Book Library, University of Toronto. Original glass negative in U of T Archives: A65-0004/204 MS. COLL. 76 (Banting), Box 63, Folder 3B.*Reproduced with Courtesy of the Thomas Fisher Rare Book Library, University of Toronto. Online at https://insulin.library.utoronto.ca/islandora/object/insulin%3AP10043.*

> same lines without at the present taking up any of the problems which you sug-
> gest in your letter. You know that if you can prove <u>to the satisfaction of everyone</u>
> [emphasis in original] that such extracts really have the power to reduce blood
> sugar in pancreatic diabetes, you will have achieved a great deal. Kleiner & others
> who have published somewhat similar results have not convinced others because
> their proofs were not adequate. It is very easy often in science to satisfy ones
> own self about some point but its very hard to build up a stronghold of proof
> which others cannot pull down. Now, for example, supposing I wanted to be
> one of those critics I would say that your results on dog 408 were not absolutely
> convincing … .[69]

Playing devil's advocate, Macleod went on to point out some of the criticisms that Banting's claims might face. Firstly, he pointed out that Banting had given each injection of extract at similar times on each day. How could Banting be absolutely sure that the changes in blood sugar levels were indeed due to the action of some factor in the tissue extract and not simply due to physiological variations arising from diurnal rhythms? And why had the injection of the same volume of extract on two consecutive days caused significantly different changes in blood sugar levels? Most importantly of all, a crucial control experiment had been omitted. How could Banting be sure that the drop in the concentration of blood sugar that he observed when extract was injected was actually due to the action of the extract and not simply a dilution effect resulting from the large volume of liquid injected?

Unsurprisingly, Banting was disappointed, and Charles Best later described Macleod's response as showing an 'amazing lack of enthusiasm'.[70] Some accounts of the story give the impression that Macleod poured scorn on Banting and Best's efforts, but on the basis of the available evidence, this seems unfair, for after having drawn Banting's attention to the weaknesses of his work and the criticisms it would be sure to face, Macleod then went on to offer constructive advice and praised Banting's decision to include control experiments using boiled pancreatic extract and preparations made from spleen and liver as 'admirable'.[71]

In all likelihood, Macleod was seeking not to dismiss Banting and Best's efforts, but simply to take seriously his responsibility as their supervisor. Far from intending to undermine Banting and sap his morale, Macleod's comments seem to have been more in keeping with the spirit of the military motto to 'train hard—fight easy'. Macleod knew that the stakes were high with this work. The therapeutic potential of an anti-diabetic hormone was huge: any claims to have discovered it would therefore be subjected to rigorous scrutiny by the medical community. As a stark reminder of this, Macleod warned Banting to remember the fate of Israel Kleiner, a researcher who had made similar claims only a few years earlier(Figure 8).

In fairness to Kleiner, his failure to become immortalized in textbooks as the discoverer of insulin had not been solely due to weak and unconvincing scientific evidence, as Macleod believed. But his story nevertheless served as a good cautionary tale. Working at the Rockefeller Institute in 1913, Kleiner and his colleague Samuel Meltzer had found that when glucose was injected into a diabetic dog, the level of sugar in the animal's blood showed a sharp rise that was not seen in a healthy animal. But when the diabetic animal was injected with a glucose solution that also contained a preparation of pancreatic extract, this rise in blood sugar was prevented.[72] The implication was clear—there was some agent present within the pancreatic extract that acted to lower blood sugar. Within only two years, Kleiner and Meltzer's work had landed them on the front page of *The New York Times*, with the headline 'Find Diabetes Cause: Now Seek a Remedy'.[73]

Despite the hyperbole of its headline, the newspaper article went on to note that Kleiner and Meltzer had been modest and sober in drawing conclusions about what their work meant. They had, it told readers, refrained 'from asserting any value for their discovery beyond what is shown by the results of the experiments' but the article demanded to know more—'in what direction will the discovery lead? Will it have a practical value?'[74]

Kleiner's answer remained reserved and cautious. Rather than offering the possibility of a cure, he felt that the main value of his work was simply to refute alternative theories about the origin of diabetes:

> We are not justified in saying that the idea gained in these experiments may be utilised in a practical manner. We hope to understand much more of the origin of diabetes and the relation of the pancreas to the disease and to clear up certain

Figure 8 *Israel S. Kleiner (1885–1966).*

Credit: Insulin Collection, Best (Charles Herbert) Papers, Thomas Fisher Rare Book Library, University of Toronto MS. COLL. 241 (Best) Box 109, Folder 34.*Reproduced with Courtesy of the Thomas Fisher Rare Book Library, University of Toronto.*

> preliminary questions from further work ... . The experiments would indicate the incorrectness of the theory that diabetes is of a nervous origin. Further work may show some hint of a means for reaching the source of defective pancreatic action, which might open the way to a more effective method of treatment for diabetes.[75]

In June 1914, Kleiner travelled to Europe to learn more about some new experimental methods that would allow him to develop his work further. But the timing of the trip could not have been more unfortunate. As the news broke of the assassination in Sarajevo of Archduke Franz Ferdinand, Kleiner found the Paris hotel in which he was staying besieged by hordes of people chanting for war.[76]

Once safely back in the USA and far from the chaos that was engulfing Europe, Kleiner continued his work on the anti-diabetic factor. But although the war in Europe may have seemed remote, its impact would soon be felt. In

1915, Samuel Meltzer, the director of Kleiner's laboratory and his collaborator in the research into the pancreatic factor, had given a lecture in which he had voiced his opposition to US involvement in the war. Now Meltzer was becoming ever more active in establishing the Medical Brotherhood for the Furtherance of International Morality, an anti-war organization. In the following year, Kleiner published a much larger and more promising study, but with Meltzer becoming more involved in anti-war activity, Kleiner found himself working increasingly alone.

When the USA finally entered the conflict in 1917, things got worse for Kleiner. With the rapid mobilization of conscripts for military service, there was growing concern about the threat and spread of infectious diseases in crowded training camps. In response to these fears, the Director of the Rockefeller Institute, Simon Flexner, redirected all research to focus on infectious diseases. Although Kleiner was able to continue his work on diabetes, he found himself becoming ever more isolated as the scientific efforts of the Rockefeller became redirected towards supplying the Army with anti-sera for the treatment of meningitis, pneumonia, and dysentery.

Kleiner struggled on alone, but in 1917 he had to break off from research completely. His undergraduate mentor, Lafayette Mendel, with whom he had first begun to conduct research into the metabolism of glucose, had been called into military service and so Kleiner temporarily left the Rockefeller to fill his post at Yale.

With an apparent gift for understatement, Kleiner described the disruptions brought by the war as having simply 'interfered with' his quest to find the anti-diabetic hormone. When the war came to an end, he resumed his research, and in 1919 published a paper that has since been called his 'masterpiece'.[77] By now the caution that he had expressed in the pages of *The New York Times* only a few years earlier was gone, and he was much more confident about the medical potential of his work:

> The fact that these pancreas emulsions lower blood sugar in experimental diabetes without causing marked toxic effects indicates a possible therapeutic application to human beings.[78]

But by the time the ink was dry on this paper, Kleiner's quest to find the anti-diabetic hormone was over. His collaborator Samuel Meltzer with whom he had blazed the trail in the search for the pancreatic hormone was planning his retirement and Simon Flexner, the Institute's Director, saw this as an ideal opportunity to wind up Meltzer's research programme and let his staff go. Kleiner may well have hit the front pages of *The New York Times* only a few years earlier, but as a letter written from Flexner to Meltzer in 1918 makes quite clear, this would not spare him from the impending cull of staff:

His [Kleiner's] work is not essential to you or the Institute's war program … I believe it would be well for Kleiner to go into teaching and this might be the time to make the change … He is not a man the Institute wishes to attach itself to permanently. He should be encouraged, I think, to look elsewhere for a more permanent position.[79]

In his reply, Meltzer begged that Kleiner be allowed to stay for at least a couple more years, pointing out that he was having to support his immediate family along with his widowed mother-in-law. Meltzer lamented that Kleiner was a victim of his own decency, his unwillingness to blow his own trumpet, and his hesitation to fight to keep his position. He warned Flexner that 'Kleiner has had to pay the penalty for being a gentleman and of a modest retired disposition … his leaving will be a definite loss to us'.[80]

Shakespeare may well have observed that 'in peace there's nothing so becomes a man as modest stillness and humility', but as far as Meltzer was concerned, these virtues were working against Kleiner.[81] In the end, Meltzer's pleas were in vain; not even a headline in *The New York Times* and the monumental therapeutic implications of his work were enough to save Kleiner, and he was informed that there would no longer be a position for him at the Rockefeller after June 1919.

Nearly four decades later, as Kleiner celebrated his seventieth birthday, his former colleague, the distinguished clinical chemist Donald Van Slyke (1883–1971) tried to reassure him that, in blazing the trail to find the anti-diabetic hormone, 'The honor of clearly showing the way remains yours'.[82] But it seems that Van Slyke's kind words brought little consolation. A few years later, when Kleiner gave a lecture to the American Association of Clinical Chemists, his reluctance to elaborate on why his early work on insulin had come to such an abrupt end suggests that the memories of that time were still too painful. All he would say was 'Why we did not continue and attempt to isolate the antidiabetic factor is a long story and has no place in the present discussion'.[83]

If Banting was going to do better than Kleiner, he first had one urgent—and huge—problem to overcome. His supply of extract made from duct-ligated animals was running out. To add to the frustration, these extracts had been starting to show real promise. Thanks to regular injections with this material, a yellow collie known as Dog 92 was now skipping around the lab—healthy and happy despite having had her pancreas surgically removed. In stark contrast, the control animal which had received no extract had died. The results were exactly what would be expected if the pancreatic extracts contained an active anti-diabetic agent.

To have defeat snatched from the jaws of victory due to a shortage of material must have been infuriating enough, but even more so was that the surgical procedure of tying the pancreatic duct of a dog shut simply took too long. Banting had pinned all his hopes on this procedure, believing it to be essential in order to kill off the tissue that produced digestive enzymes. They needed to find an alternative to this entire convoluted and time-consuming process.

In the end, the solution was obvious—why not make tissue extracts using the whole pancreas? This process was quicker and easier than Banting's original method, but it came with a price: what Banting gained in speed he would lose in originality over predecessors such as Kleiner.

If Banting's hypothesis was correct and the anti-diabetic factor was being destroyed by pancreatic enzymes, then extracts made using whole pancreas should be weaker than those prepared from tissue that had degenerated due to the duct-ligation process. But when Banting and Best injected the first extracts made from whole pancreas into Dog 92, this was not what they found—a chart in their original lab notes shows that an injection on August 17th of extract made from whole pancreas caused a similar, if not slightly greater, drop in blood sugar levels than one given on August 16th that had been made from degenerated material (Figure 9a).

Banting and Best later included this same chart (Figure 9b) in their first published work in a scientific journal, and while they acknowledged that 'extracts of whole pancreas do have a reducing effect on blood sugar', they insisted that it was 'obvious from the chart that the whole gland extract is much weaker than that from the degenerated gland'.[84]

Had Banting's hypothesis that extracts made from whole pancreas contained destructive enzymes been correct, then the material injected on August 17th would have been expected to have much less effect than that of the degenerated material. But in addition to suggesting that Banting and Best were prone to confirmation-bias, the chart is interesting for another reason.

In the legend that accompanies the original chart found in the lab notebook, the extract injected on August 16th is described as being an extract of degenerated pancreas made from Dog 390, with the addition of acid and the enzyme trypsin. As trypsin has proteolytic activity, this was presumably yet another test of Banting's hypothesis that the anti-diabetic factor was being degraded by digestive enzymes. If Banting was correct, then the presence of trypsin should *reduce* the efficacy of this particular batch of extract. But this is not what happened. The batch with trypsin resulted a similar reduction in blood sugar *as one containing degenerated gland extract and no enzyme* that had been given on August 15th!

This flew in the face of Banting's working hypothesis, and might go some way towards explaining why, when this graph was finally published in the *Journal of Laboratory and Clinical Medicine*, there is no mention of the addition of trypsin on August 16th. Instead, the extract is simply described as 'Degenerated pancreas + 0.1% HCl'.

The result was a direct contradiction of Banting's cherished hypothesis on which he had built his entire programme of research, and it is little wonder that he chose to ignore it. Perhaps it was too painful to accept that all those laborious weeks of tying the ducts in dogs and sitting around waiting for the degeneration of pancreatic tissue had been unnecessary.

Refusing to accept what their data was telling them, Banting and Best remained fixated on overcoming the problem of digestive pancreatic enzymes instead of the

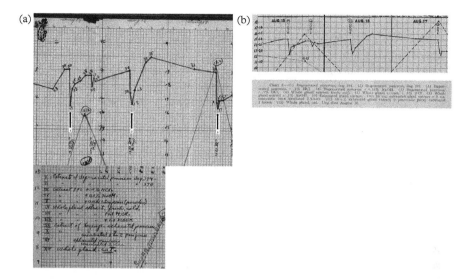

Figure 9 *a) Section from a chart found in Banting and Best's original lab note books showing the effects of various preparations of pancreatic extracts on the blood sugar levels of a diabetic dog (Dog 92). Banting's hypothesis was that the antidiabetic factor was being destroyed by digestive enzymes secreted by the pancreas, and that this could be overcome by using pancreatic material in which the tissues that produced these enzymes had been allowed to degenerate. If this hypothesis was correct, then the extracts administered on 16th August (V) (Arrow 2) which contained the proteolytic enzyme trypsin, and the one given on 17th August (VI) (Arrow 3) which had been made from whole gland extract would have been expected to be less effective at reducing blood sugar levels than extract administered on 15th August (III) (Arrow 1) that contained no trypsin and had been made from degenerated tissue only. But in complete contradiction to Banting's hypothesis, extracts V and VI were just as effective at lowering blood sugar as extract made from degenerated material.*

b) The same chart as it was presented in Banting and Best's first scientific publication. ('The Internal Secretion of the Pancreas', F.G. Banting & C.H. Best, The Journal of Laboratory and Clinical Medicine (1922), 7; 251–266; p.257, Chart 2.) The legend which accompanies the chart in this paper now states that the 10cc of extract (5) injected on 16th August was simply 'Degenerated pancreas + .1% HCl'. Crucially, there is no longer any mention of the addition of any trypsin to this extract. Moreover, although Banting and Best conclude that 'It is obvious from the chart that the whole gland extract is much weaker than that from the degenerated gland.' the data tell a rather different story. The chart clearly shows that a preparation (6) of 'Whole gland extract, fresh, cold' on 17th August appears to cause a similar, if not slightly more pronounced drop in blood sugar levels, than preparation (5) made from degenerated pancreas, making their conclusion far from as 'obvious' as they claimed.

Credit: Insulin Collection, F.G. Banting (Frederick Grant, Sir) Papers, Thomas Fisher Rare Book Library, University of Toronto; MS. COLL. 76 (Banting) Mapcase. Reproduced with Courtesy of the Thomas Fisher Rare Book Library, University of Toronto. Online at https://insulin.library.utoronto.ca/islandora/object/insulin%3AM10002.

Credit: Insulin Collection, F.G. Banting (Frederick Grant, Sir) Papers, Thomas Fisher Rare Book Library, University of Toronto; MS. COLL. 76 (Banting), Box 62, Folder 13 Offprint. Reprinted from the Journal of Laboratory and Clinical Medicine, vol. 7, no. 5 (Feb. 1922). Reproduced with Courtesy of Elsevier and the Thomas Fisher Rare Book Library, University of Toronto. Online at https://insulin.library.utoronto.ca/islandora/object/insulin%3AP10080.

messy business of duct-ligation. Furthermore, they believed that, thanks to the hormone secretin, they might just have found a much quicker solution than the messy business of duct-ligation.

For secretin was known to stimulate the production of these enzymes by the pancreas. Banting reasoned that, by injecting an animal with this hormone, it ought to be possible to send the pancreas into overdrive and so exhaust the production of these enzymes, thereby allowing an enzyme-free extract to be made.

At first things looked promising. By the time that Banting and Best had obtained their first extracts using this new method, Dog 92 had fallen ill with diabetes. But after the first administration of the new extracts, it was soon running around the lab wagging its tail—a moment which Banting later remembered as one of the greatest days of his life. Unfortunately, his jubilation did not last long. At the end of August, despite receiving more secretin-prepared extract, Dog 92 died. Banting was left distraught—not because yet another experiment had failed but rather because he had become emotionally attached to the dog:

> I shall never forget that dog as long as I shall live. I have seen patients die and I have never shed a tear. But when that dog died I wanted to be alone for the tears would fall despite anything I could do. I was ashamed. I hid my face from Best, but now I am not ashamed. Diabetes, the world over owes much to the true brave faithful loving little dog who played her heroic role as part of her day's work. She did it so trustingly and cheerfully that I would have given everything I had to restore her to normal ... For the moment suffering humanity sank into oblivion in the face of the necessity of prolonging the life of this dog. I had no responsibility for the diabetes of humanity but I was responsible for the condition of this dog. No human has ever been more true, no friend has been more understanding. No death has touched me more deeply than the death of this dog.[85]

At the end of the summer, when Macleod returned from his trip to Europe, Banting was eager to tell him all about the exciting results that he and Best had obtained. But he also had a number of requests that he felt were essential if the work was to continue. Firstly, they needed more dogs, as well as a technician who could look after them; secondly, they needed their own laboratory, as they had been having to carry out some of their work in a passageway which was now filling up with students returning for the start of term. Finally, as he was heavily in debt, Banting asked for a salary.

Macleod, however, took a different view of the situation. He was concerned to find that Banting had already used more than the ten dogs originally agreed upon for the experiments. He also felt that to grant Banting's requests would have a detrimental impact upon other work in his laboratory. But when Macleod tried to explain these points, Banting's temper quickly flared:

> He [Macleod] said that others had obtained equally good results and nothing had come of it. I then stood up and told him that I could not agree with him in this statement and that I was determined to continue the work and that if the

University of Toronto did not consider the work of sufficient merit to support further investigative work that I would go elsewhere. Whereupon he became angry and said "As far as you are concerned I am the University of Toronto."[86]

Taking his leave of Macleod, Banting turned to Best and articulated one very clear resolution to his young colleague—'I'll show that little son of a bitch that he is not the University of Toronto'.[87]

Banting took the rest of the day off, simply in order to calm down. But before doing so, he met with Professor Velvyn Henderson and described his bad-tempered exchange with Macleod. Banting later said that he would have left Toronto that very same day to join a lab in the USA had it not been for Henderson, and that it was only thanks to his power of persuasion and calming influence that insulin was eventually discovered in Canada and not the United States.[88] After spending the remainder of the day in the cinema, Banting returned to work the next day where, on meeting Macleod in a corridor he found him to be 'cordial and even artificially friendly'.[89] Macleod showed Banting a room which, although small, could be converted into a laboratory and also explained that he had found a technician who could work part-time. But best of all was the news that, if Banting was willing to do some teaching, he could be paid a salary, thanks to funds made available by Professor Henderson.

However, Macleod's concessions to Banting's requests did little to ease the growing tension between them, and the situation soon took a turn for the worse. By mid-November Macleod felt that the work that Banting and Best had done over the summer was ready to be presented to the Physiological Journal Club, an informal gathering where researchers within the University could discuss preliminary work. Unfortunately, it did not go quite as well as he had hoped. The plan was that Macleod would introduce the subject area and its history, while Best would present the charts showing the effect of the pancreatic extracts on blood sugar levels. Meanwhile, Banting would discuss their conclusions and how it was an improvement upon previous research in this area.[90] With his thirtieth birthday looming on the horizon, Banting was beginning to feel an acute need to make his mark professionally. But if he had hoped that this meeting would provide him with an opportunity to do so, he was disappointed. As Banting later recalled with some bitterness, in the days immediately following the meeting, it was not his name that was on everyone's lips:

> Professor Macleod in his remarks gave everything that I was going to say and used the pronoun "we" throughout. The following day students were talking about the remarkable work of Professor Macleod.[91]

Banting did not express these concerns to Macleod at the time, but they festered away quietly inside him. From a 1940 assessment he gave regarding Macleod's character, time did little to heal these wounds:

Macleod on the other hand was never to be trusted. He was the most selfish man I have ever known. He sought at every possible opportunity to advance himself. If you told Macleod anything in the morning it was in print or in a lecture in his name by evening. He was grasping, selfish, deceptive, self-seeking, and empty of truth—yet he was clever as a speaker and a writer … He loved acclaim and applause. He had a selfish overpowering ambition. He was unscrupulous and would steal an idea or credit for work from any possible source. Like all bullies Macleod was a coward and a skulking weakling if things did not go his way.[92]

When Macleod later became aware of Banting's suspicions, he dismissed them as being completely 'unwarranted'.[93] Keen to refute any charge that he was trying to steal their glory, he pointed out that he had declined to have his name included on Banting and Best's first scientific paper as he 'did not wish to fly under borrowed colours'.[94]

He also gave Banting and Best the credit for having had the insight to abandon the convoluted and time-consuming process preparing degenerated tissue by duct-ligation in favour of using extracts prepared from whole pancreas. By now, it was evident that, in addition to being laborious and time consuming, the method of using degenerated tissue prepared by duct-ligation was nowhere near as effective as Banting had hoped. When Banting and Best published their first paper on this work, they reported that, although these extracts had been able to reduce blood and urinary sugar, it was nevertheless 'very obvious that the results of our experimental work, as reported in this paper do not at present justify the therapeutic administration of degenerated gland extracts to cases of diabetes mellitus in the clinic'.[95]

Preparing extracts from whole pancreas was a better option, but still not good enough. They needed a method that was faster and—crucially—cheaper. Wrestling with the problem late one night Banting hit upon what he thought might well be the solution. According to the scientific literature, foetal pancreatic tissue contained far more Islet cells than in the mature animal, but it also had another crucial advantage. Digestive enzymes were not produced in this tissue during the first four months of foetal development. Extracts made during this time should therefore be free of the enzymes that Banting believed were degrading the antidiabetic agent.

Obtaining canine foetuses for the work would be far too much trouble, but thanks to the meat industry, there was another solution. Visiting an abbatoir in northwest Toronto, Banting and Best found a ready supply of bovine foetuses that would otherwise have been discarded as waste. Making extracts from this tissue proved to be far more straightforward than the previous convoluted procedure on dogs.

At the same time as Banting hit on the idea of using foetal tissue, he and Best had another epiphany. This was all thanks to alcohol—not so much its influence upon their spirits, but rather its chemical properties as a solvent. Until now, Banting and Best had prepared their extracts in saline solution, from which the water had to be boiled off in order to concentrate the active agent. But this process suffered from one huge flaw: the process of boiling reduced the activity of the final extract. Alcohol, by contrast, could be simply evaporated off without causing any loss of activity in the extract.

This was a major step forward in improving the process, but it was not one for which Banting and Best could really take the credit. Nine years earlier, as part of his doctoral research, the 'Ohio farmboy' Ernest Lyman Scott had already prepared pancreatic extracts using this method.[96] Like Banting, Scott had also believed that the biggest challenge to overcome in the extraction process was to avoid the destruction of the anti-diabetic agent by digestive enzymes. But he was convinced that these enzymes could be inactivated by using alcohol in the extraction process. Citing Scott's work, Banting and Best believed that alcohol might have another key advantage. By using alcohol in the extraction they believed that it might be possible to prepare enzyme-free extracts from adult bovine tissue, which had the advantage that it could be obtained in far greater quantities and would be much easier to work with than foetal tissue.[97]

As Banting and Best breathed a sigh of relief at this discovery, they probably tried not to dwell on the thought of all the time spent previously working with the duct-tied dogs. Hindsight may be 20/20, but the view is not always pleasant. Using the method of alcohol extraction, Banting and Best now prepared material from foetal, and then from adult, bovine pancreatic tissue. When daily injections of these extracts were shown to prolong the life of a diabetic dog for seventy days, things seemed at last to be looking up.[98] But alcohol would prove to be not just a blessing, but a curse. The substitution of alcohol for saline into their extraction protocol might sound like a mere technical detail. Instead, it became a major aspect in the ensuing controversy that would later rage around who really deserved the credit for the discovery of insulin.

The first ominous rumblings of this controversy began to stir towards the end of 1921. Rumours about the work in Toronto were beginning to circulate within the community of diabetes specialists. When Dr. Elliott Joslin wrote to Macleod asking whether 'there is a grain of hopefulness in these experiments which I can give to patients', Macleod gave a careful reply, saying 'I would hesitate to attempt the application of these results in the treatment of human Diabetes until we are absolutely certain of them'.[99,100]

Despite his caution, however, Macleod confided to Joslin that 'I may say privately that I believe we have something of real value'.[101] It would prove to be a historic understatement, and by the end of 1921, he felt sufficiently confident in Banting and Best's work for them to give a formal presentation of it to a scientific audience. A few days after the Christmas holiday Macleod travelled with Banting and Best to a conference of the American Physiological Society (APS) at Yale University in New Haven. Here, the great and the good of North American diabetes research had gathered and it was crucial for Banting and Best to make a good impression. But Banting and Best were newcomers to this world—two young researchers whose names were as-yet unknown to the luminaries of the North American medical establishment. For this reason, Banting suggested that Macleod, who was a member of the APS, participate in their presentation to lend them some credibility and kudos as they stood before an audience of medical heavyweights. Macleod agreed, but insisted that Banting himself should give the actual presentation. With hindsight, he may well have come to regret this decision and Banting very quickly found himself wishing that he had never invited Macleod to participate.

As Banting stood up to give his presentation, and gazed out at the faces seated before him, the prestige of his audience took its toll on him. 'When I was called upon to present our work', he later wrote, 'I became almost paralyzed. I could not remember nor could I think. I had never spoken to an audience of this kind before—I was overawed. I did not present it well'.[102] As Macleod watched Banting dry up, he feared rejection of all they had worked for by those gathered at the presentation. He recalled that Banting was 'very nervous, and it was evident that he had not succeeded in convincing all of his audience that the results obtained proved the presence of an internal secretion of the pancreas'. Banting's nerves were turning what should have been a glorious moment—a medical milestone–into a disaster. Determined to snatch victory from the jaws of defeat, Macleod took decisive action. When the time came for Banting to take questions from the audience, Macleod calmly stepped in and took charge of the remainder of the presentation, taking care to explain the significance of their work, while Banting looked on.[103]

Macleod thought he was acting with the best of intentions and in the interests of the greater good. As with his criticisms of Banting and Best's first experiments, Macleod was acting as he believed a responsible supervisor should. But Banting saw things very differently. For him, this was nothing less than an aggressive coup to take over his work. It confirmed all his festering fears and suspicions that Macleod was out steal his glory and, to add insult to injury, he had been brazen enough to do it in front of the elite of North American medicine. In a report written to the Board of Governors about the incident, Banting expressed his sense of gross injustice:

> ... although I gave the paper he [Macleod] discussed it using the term 'we' throughout the discussion. I was the only one who gave a paper to the Physiological Section who was not asked to respond to his paper. Professor Macleod was chairman at this meeting. It must be remembered that at this time Professor Macleod had not done one single experiment, nor had he contributed one idea of value except estimation of haemoglobin before and after extract.[104]

Macleod protested that Banting had misinterpreted his actions. Faced with the accusation that he had deliberately tried to undermine Banting and steal his glory, Macleod offered an olive branch saying, 'If this was so, it was entirely unintentional on my part, my object being to persuade the audience of the real value of the investigations'.[105]

But Banting would not be appeased. He was fuming with rage and when recalling the incident many years later, it was evident that time had done little to abate his anger towards Macleod:

> I did not sleep a wink on the train that night—I did not even go to my berth but sat up in the smoker condemning MacLeod as an imposter and myself as a nincompoop. I decided that I must first learn to write clearly, precisely, legally, explicitly, and then be able to talk convincingly, freely and unhesitatingly. I knew MacLeod for what he was, a talker and a writer. Apart from his pen and his tongue he would not even be a lab man for he had no original ideas, he had no skill with his hands in an experiment. He only knew what he read or was told and then he could rewrite or retell it as though he were a scientist and a discoverer. It was foolish to spend weeks and months working day and night at experiments and then have them told beautifully by someone else who had the art as though they were his ideas and works.[106]

The Yale meeting had not, however, been a complete disaster. On returning to his hotel room, Macleod received a telephone call from Dr. George Clowes (1877–1958), research director at Eli Lilly and Company, a manufacturer of pharmaceuticals based in Indianapolis. Clowes had been so keen to attend the meeting that he had left Indiana early on Christmas Day in order to travel there—much to the disappointment of his young family. He had been sitting in the audience during the talk by Banting, Best, and Macleod and, although he described Banting as having 'presented his material somewhat haltingly', Clowes also recognized that it was of major importance. 'Anyone who was at all cognizant with the subject', he later wrote, 'must have realized that a great discovery had been made'.[107]

But Clowes also knew that many hurdles remained before Banting and Best's extracts could use as a viable clinical therapy for diabetes. One of the most formidable challenges was how to produce pancreatic extracts in sufficient quantity. To prepare an effective extract at the lab bench in sufficient quantity for a couple of scientific publications was one thing: to produce enough material to meet the ongoing clinical need of patients was quite another. Production on this scale was well beyond the capability of a university department—only a private pharmaceutical company had the necessary facilities, expertise, and capital. But when Clowes called to offer Lilly's assistance in scaling up production of the extracts, Macleod was, as ever, cautious in his response. Replying that although he would certainly bear Clowes' offer in mind, Macleod went on to explain he felt that the work was currently not yet at a sufficient stage for commercial development.[108] After all, it had not even been tested in human patients. This was all about to change dramatically, but the motives for doing so lay less in an altruistic wish to advance science and improve the lot of humankind, and more in the murkier crevices of the human condition.

Banting was still simmering with humiliation and rage after the debacle at the Yale meeting. With overtures from newly interested parties such as Clowes and Lilly, he feared that insulin was slipping from his grasp. He desperately needed to do something bold that would stamp his authority on this work and in so doing ensure that his name—not Macleod's—would forever be associated with the discovery of insulin. And he knew exactly what to do to achieve this.

The extract had to be shown to work not just in diabetic dogs, but in human patients. But when Banting approached Professor Duncan Graham, Eaton Professor of Medicine to ask about the possibility of organizing some clinical trials of the extract in human patients, he was met with short shrift. As far as Graham was concerned, Banting had no qualifications whatsoever to carry out such a trial and Graham's first priority was to protect his patients from such cavalier experimentation. 'What right have you to treat diabetics? How many of them have you ever treated?' he snapped at Banting.[109]

The answer to this question was simple. On December 20, 1921, Banting had made a phone call to Joe Gilchrist, an old friend from medical school who had been diabetic since 1917. Later that same day Banting recorded in his notes that 'we gave him [Gilchrist] extract that we knew to be potent'.[110] But the next day, to their utter disappointment, the extract had 'no beneficial result' whatsoever.[111] Had Banting and Best known anything about the chemical nature of insulin, this result would not have come as a surprise. The extract had been administered to Gilchrist orally and, because insulin is a protein, had therefore been quickly degraded by digestive enzymes in the stomach. But fortune smiled on Banting when, only a few weeks later, his work was thrown a lifeline.

When Leonard Thompson's (Figure 10) father brought his dying son into Toronto General Hospital, he could have had little idea that he was about to make medical history. Leonard had first been diagnosed with diabetes in 1919 and had been admitted to a diabetic clinic for charity cases where he was placed on the Allen diet. As a result of this treatment, by the time his desperate father brought him into Duncan Graham's office, he was described as being 'poorly nourished, pale, weight 65 pounds, hair falling out, odour of acetone on his breath … appeared dull, talked rather slowly, quite willing to lie about all day'.[112] Recalling this moment years later, Dr. Graham's secretary said that the only time she had ever seen another human being look so severely emaciated was in pictures of famine victims or the survivors of concentration camps. One senior medical student spoke for everyone present when he said simply 'All of us knew that he was doomed'.[113]

On the afternoon of January 11, 1922, Ed Jeffrey, a young doctor, injected Leonard with 15cc of pancreatic extract that Jeffrey's boss Dr. Walter Campbell likened to 'thick brown muck'.[114] Banting probably paid no heed to such descriptions, as long as the extract worked. But its description in Leonard Thompson's medical notes would have set Banting's blood boiling: although the extract had been prepared by Charles Best, it was recorded in Thompson's medical notes as 'MacLeod's Serum'.[115,116]

The effects of this first injection were disappointing. For although Leonard's blood sugar levels fell by about twenty-five per cent, and there was a slight reduction in the excretion of sugar in his urine, he continued to produce ketones—a sure sign that there had been no real anti-diabetic effect. But most serious of all was

LEONARD THOMPSON
First patient to receive insulin in
Toronto.

Figure 10 *Leonard Thompson (1908–1935).*

Credit: Insulin Collection, F.G. Banting (Frederick Grant, Sir) Papers, Thomas Fisher Rare Book Library, University of Toronto; MS. COLL. 76 (Banting), Box 12, Folder 1.*Reproduced with Courtesy of the Thomas Fisher Rare Book Library, University of Toronto. Online at https://insulin.library.utoronto.ca/ islandora/object/insulin%3AP10046.*

that the extract had triggered a toxic reaction resulting in the eruption of abscesses at the injection site. Reporting on this work in the *Canadian Medical Association Journal*, Banting and Best were forced to draw the dismal and blunt conclusion that 'no clinical benefit was evidenced' by the injection of their extract.[117]

Two weeks later, on January 23rd, Leonard Thompson was given two more injections of pancreatic extract. This time, the result was starkly different. Leonard 'became brighter, more active, looked better and said he felt stronger'.[118] His blood sugar levels were markedly reduced, as was the excretion of sugar in the urine, while measurements of the respiratory quotient showed that his body had begun to metabolize carbohydrate efficiently. But perhaps the most important result of all was that there were no toxic side-effects—no formation of abscesses. Something had changed in those two weeks to make this new extract much more effective and free of toxic side-effects so that it saved the boy from early death: this particular batch of extract had not been made by Banting and Best, but by someone else.

Although James Bertram Collip (Figure 11) was only a year younger than Banting, he had already been appointed Professor and placed in charge of a new Department of Biochemistry at the University of Edmonton, Alberta, where he had amassed an impressive record of publications through his research into the blood chemistry of vertebrates and invertebrates, as well as the action of various substances on the cardio-inhibitory and respiratory centres of the brain stem.[119] In order to gain some experience working at other institutes, Collip had applied in 1920 for a travelling grant from the Rockefeller Foundation, thanks to which he had come to Toronto in late 1921 to spend half a year working with Macleod.

Figure 11 *Photograph of James Bertram Collip (1892–1965) ca. 1920.*

Credit: Insulin Collection, Collip (James Bertram) Papers, Thomas Fisher Rare Book Library, University of Toronto MS. COLL. 269 (Collip), Item 1. This photograph is reproduced from the original held by Barbara Collip-Wyatt. *Reproduced with Courtesy of the Thomas Fisher Rare Book Library, University of Toronto. Online at https://insulin.library.utoronto.ca/islandora/object/insulin%3AP10005.*

Collip's expertise in biochemistry proved to be invaluable. The whole premise of Banting and Best's work had been that the failure of all previous attempts to isolate the internal secretion of the pancreas was because the active agent had been degraded by digestive enzymes. But as far back as 1875, German researcher Rudolf Heidenhain (1834–1897) had already shown that this was a red herring. Heidenhain had found that, although the pancreas *did* produce the proteolytic enzyme trypsin, it was only as an *inactive precursor* called trypsinogen, which did not become converted into active trypsin until it entered the intestine. This meant that freshly isolated intact pancreatic tissue should actually contain *no protein degrading activity*.[120] The real reason for the inconsistency and unreliability of previous attempts to isolate the anti-diabetic hormone was not that insulin was being chewed up by digestive pancreatic enzymes, but rather that the pancreatic extracts used had simply been too full of residual impurities such as proteins, lipids, and bits of cellular debris.

The use of alcohol in the extraction process—which Banting and Best had adopted—had helped to address this problem to a degree, but Collip had realized how the purity of the extract could be improved even further. As Collip increased the concentration of alcohol in the extract, he found that the impurities began to precipitate out of solution and could be removed simply by centrifugation. Furthermore, when the concentration of alcohol was increased to a certain level, the active agent itself precipitated out of solution. It could then be redissolved to give a final product of much higher purity and potency than the extracts made by Best.[121] In a letter written to Dr. Henry Marshall Tory, President of the University of Alberta, only a couple of days after the Leonard Thompson had been injected for the second time, Collip described his excitement at this discovery:

> You can imagine my delight when around midnight one day last week, I discovered a way to get the active principle free from all the muck ... On the advice of Professor Macleod and others, I am keeping the process an absolute secret. We have decided not to patent it but to offer it to the University ... When it has been proven ... the method will be published in full for the world at large.[122]

This achievement alone was enough to establish Collip as a key member of the Toronto group, but the value of his contribution did not stop there. While Banting and Best had been struggling with extracts that had no effect, Collip had shown that his extracts could not only reduce blood sugar levels, but also inhibit the production of toxic ketones in diabetic animals. They also stimulated the formation of glycogen, which acts as a store of glucose in the liver. All these findings strengthened the evidence that the pancreatic extracts contained an agent that had a powerful anti-diabetic effect.

But while Collip's expertise as a biochemist was making the extracts less toxic, his mere presence was having the opposite effect among members of Macleod's team. Collip was once described by the President of Harvard University as 'a bold explorer among the tangled complexities of the internal secretions' but nothing could have been so tangled or complex as the ever-more acrimonious politics that now blighted the Toronto team.[123]

Banting claimed that, at the meeting of the American Physiology Society in New Haven at the end of 1921, he, Best, and Collip had all shaken hands in a 'gentleman's agreement'. According to this agreement, Collip was to be responsible for purifying the extract while Banting and Best would continue investigating its physiological effects. Most importantly, all three of them had agreed that all results and discoveries would be shared openly.[124] But by the time that Collip's extract was injected into Leonard Thompson, this informal arrangement was coming under severe strain.

Adding to the pressure was that their work had by now caught the eye of the media. After being approached by a journalist from *The Toronto Star* newspaper, Macleod had given an interview in which he stressed that any talk that their work might offer a cure for diabetes should be treated with caution. But this message appears to have fallen on deaf ears, for on January 14, 1922, only a couple of days after Leonard Thompson had first been injected, the interview was published under the headline 'Work on Diabetes Shows Progress Against Disease: Toronto Medical Men Hoping That Cure is Close at Hand'.[125]

Macleod was not pleased. He feared that such exaggerated claims might arouse 'false hope in the thousands of people who are suffering with this dread disease', saying:

> We've really no hope to offer anyone at all as yet ... We don't know anything yet that would warrant a hope for a cure. But we are working intensively at this thing with a hope that some day we may be able to help on a little bit.[126]

But it was too late. The story was out and the months that followed produced a flood of sensationalist headlines. *The Toronto Star Weekly* hailed the discovery as being 'epoch making ... unlike the wild rumours that come from Paris' and, alongside photographs of Banting, Best, Macleod, and Collip, asked 'Have they robbed diabetes of its terror?'[127,128]

Like Macleod, Banting was also not pleased to find himself in the media spotlight. With headlines such as 'Toronto Doctors On Track of Diabetes Cure' in the Toronto Star, he and Best shared Macleod's concerns 'about the harm that would be caused by exaggeration through arousing false hopes in patients'.[129,130] Speaking to the Biology Club at the University of Toronto, Banting stressed 'that this can never be a cure for diabetes but that all results so far favour the idea that its use may someday relieve those suffering from this complaint'.[131]

But what irritated him more than the sensationalist headlines was that, in the original newspaper story, Macleod had repeatedly used the pronoun 'we' when referring to the work. It may have only been a small word, but for Banting, it was one that Macleod had no right to use. When finally challenged on this issue, Macleod repeated that he had no plans to steal Banting's rightful glory. This assurance did little to calm Banting's fears, and by now Macleod was no longer the only person whom he had come to distrust.

Collip had, for some time, been meeting Macleod for lunch and bringing him up to date on the progress of his work. These lunchtime meetings fuelled Banting's fears that Macleod might not be the only person intent on stealing a glory that he felt was rightfully his. When Collip apparently declined to share the technical details of his purification process, Banting recalled how his simmering suspicions finally boiled over:

The worst blow fell one evening toward the end of January. Collip had become less + less communicative and finally after about a week's absence he came into our little room about 5.30 one evening. He stepped inside the door and said 'Well fellows I've got it'. I turned and said, 'Fine, congratulations. How did you do it?' Collip replied, 'I have decided not to tell you'. His face was white as a sheet. He made as if to go. I grabbed him in one hand by the overcoat where it met in front and almost lifting him I sat him down hard on the chair. I do not remember all that was said but I remember telling him that it was a good job he was so much smaller—otherwise I would 'knock hell out of him'. He told us that he had talked it over with Macleod and that Macleod agreed with him that he should not tell how by what means he had purified the extract.[132]

Best recalled the same incident and feared for Collip's safety:

Banting was thoroughly angry and Collip was fortunate not to be seriously hurt. I was disturbed for fear that Banting would do something which we would both tremendously regret later and I can remember restraining Banting with all the force at my command.[133]

The gentleman's agreement was unravelling at an alarming rate. But why had Collip acted in this manner?

Best claimed that Collip had announced his intention to file a patent on the purification process which, if true, would have been a stark violation of the agreement reached at Yale:[134] it is little wonder that Banting was furious. Banting's biographer Michael Bliss, meanwhile, has suggested that Collip and Macleod now distrusted Banting after he had failed to consult with them about his plan to test the extract on Leonard Thompson. As far as they were concerned, it was Banting who had betrayed the gentleman's agreement by rushing headlong into testing the extract on a human patient without their knowledge—and so they chose to respond in kind. They may have feared that, by divulging the technical details of the purification process to Banting, he would then claim the credit for himself. Whatever the truth, insulin—or more accurately the glory that it promised—seemed to expose human flaws and turn people against each other.

To prevent the situation from deteriorating any further, a formal agreement was drawn up between Banting, Best, Collip, Macleod, and the Connaught Anti-Toxin Laboratories of the University of Toronto, who were to begin scaling up the purification process.[135] At around this time, a number of clinicians began to test the extract on patients. Macleod carried out further physiological investigations of its effects while Collip continued to work on the purification. But as the whole insulin project began to gather momentum, Banting feared that he was being left behind. His role, together with Best, was now simply to provide depancreatised dogs, a task which left Banting feeling as if he had now been relegated to being a technician. Insulin was slipping from his grasp, and this took its toll on him:

It was an extremely trying time for me. Best was still intimate with Macleod and the others about the laboratory. I was out of the picture entirely. Macleod had

> taken over the whole physiological investigations, Collip had taken over the bio-chemistry. Professor Graham and Dr. Campbell had taken over the whole clinical aspect of the investigation. None of them wanted anything to do with me ... the whole affair was too much for my nervous system. The only means by which I could get to sleep was to take alcohol. I did not always have enough money to buy it. On two occasions I actually stole a half a litre of pure alcohol from the laboratory, diluted it with water and drank it in order to sleep. I do not think that there was one night during the month of March 1922 that I went to bed sober.[136]

Towards the end of March, Best called around one evening and found Banting in a rare state of sufficient sobriety to be brought up to date on the latest developments in the lab: things were not going well. Macleod was frustrated because he was not getting enough extract, Collip was struggling to repeat his earlier success, and as a result further clinical trials were being delayed. On a more positive note, $5,000 dollars had been offered to set up a new laboratory if Banting was willing to come and work in it. However, Banting had sunk so deeply into despair that not even this offer could entice him back. He wanted nothing more than simply to leave and 'look for a place where there were decent people to live with'.[137]

When Best asked what would become of him if Banting were to leave, the reply was blunt and bitter: 'Your friend Macleod will look out for you'. Best's response was equally brusque: 'If you get out, I get out'.

If Best hoped his brusqueness would shock Banting back into action, he was correct. Banting later described Best's reply as being 'possibly the only thing that could have changed my attitude'. Draining his glass, he swore an oath abstaining from alcohol until the anti-diabetic agent 'circulates in diabetic veins'. Then, with new resolve he declared: 'Shake on it, Charlie. We start tomorrow morning at 9.00 o'clock where we left off'.[138]

Returning to work with renewed vigour, Banting spent the next month preparing a new paper with Best, Collip, and Macleod, and the clinicians involved in the early trials. The paper summarized all the previous work of the Toronto group and, when Macleod first presented it on May 3, 1922 at a meeting of the Association of American Physicians in Washington D.C. he finished on a characteristic note of caution:

> While these observations demonstrate conclusively that the pancreatic extracts, which we employed, contain some substances of great potency in controlling car-bohydrate and fat metabolism ... in patients suffering from diabetes mellitus, we cannot as yet state their exact value in clinical practice.[139]

But his audience had already seen beyond Macleod's caution and glimpsed the far-reaching implications that this work would have for the treatment of diabetes. On finishing his presentation, he received a standing vote of thanks—something which, according to Dr. Elliott Joslin, had not occurred for twenty years.[140] Joslin went on to describe the work as an 'epoch-making discovery', while Dr. R. T. Woodyatt of Chicago hailed it as having marked:[141]

… the beginning of a new phase in the study and treatment of diabetes. It would be difficult to overestimate the ultimate significance of such a step. Heretofore we have managed diabetes by providing dietetic crutches. Out of this work there will develop in time a specific treatment.[142]

It was during this address that the 'thick brown muck' that had saved Leonard Thompson's life received its new name. Banting and Best had originally suggested the name 'Isletin' but in his address Macleod proposed that it be known as 'insulin'—derived from the Latin term for 'island'.[143] Although this name had already been suggested by the Belgian clinician and physiologist Jean de Meyer in 1909, Macleod's use of the name in his paper at this meeting is usually heralded as being the moment that the discovery of insulin was officially announced to the scientific world.[144] Curiously, however, neither Banting nor Best were actually present at this historic moment. Despite now being reasonably well paid, both claimed that they could not afford the train fare—a claim which has since been called into question.[145] Maybe Banting was trying to snub Macleod. Perhaps he simply loathed him so much that he could not bear to share a podium with him. Alternatively, he might have been quietly hoping that the success of his work would speak for itself in establishing him, and not Macleod, as the discoverer of insulin.

If so, his confidence appeared to be well founded. In early July, Banting received a letter from Mrs. Antoinette Hughes, the wife of Charles Hughes, U.S. Secretary of State had first asking whether it might be possible to bring their fifteen-year-old daughter Elizabeth to Toronto in the hope of receiving some insulin.[146] By mid-August, *The London Times* reported that Elizabeth's treatment was showing 'encouraging results' and was in no doubt about who deserved the credit for this discovery:

The fifteen-year-old daughter of Mr. Hughes, the United States Secretary of State, is in hospital in Toronto undergoing treatment for diabetes by Dr. F. G. Banting, who is the originator of the insulin treatment.[147]

But if Banting thought that this meant his problems were over, he was to be gravely mistaken. In fact, they were about to get a whole lot worse.

3

The Vision of Ezekiel

Macleod's announcement of the discovery of insulin was met with a standing ovation from his audience in Washington D.C. But even as the applause echoed round
the hall, he knew that there were problems brewing behind the scenes. Following
Leonard Thompson's recovery, six more patients had been successfully treated
with insulin at Toronto General Hospital. However, now Macleod found himself
having to justify the decision to rush these results into print. Writing to Professor W. B. Cannon at Harvard, Macleod explained why he felt his hand had been
forced:

> We have had to publish in the *Journal of the Canadian Medical Association* a report
> of several clinical cases because the newspapers got wind of what we are doing and
> through some agents of their own had enough information of a haphazard type
> from which they could at any time piece [sic] together a garbled account of the
> work. They kept constantly prodding us for further information until at last we
> were compelled to publish an account in our most substantial medical journal.[1]

Pursuit by journalists was annoying, but in the next sentence, Macleod admitted
that '… we have run into serious difficulties in producing active extracts in bulk'.[2]

The problem was greater than Macleod had let on. The successful isolation
of insulin at the lab bench was a success, but to be of any clinical use, insulin
production was required on an industrial scale. Support and funding from the
Connaught Anti-Toxin laboratory— established by the University of Toronto for
the manufacture of vaccines—provided facilities for James Collip to scale up the
production of insulin. But despite the ample provision of equipment and funds,
Collip now found, to his dismay, that he could not repeat his earlier success. He
was unable even to produce insulin at the lab bench.

The consequences were tragic. Back in February, one young female patient
had shown significant improvement after receiving injections of Collip's extract.
But when the supply ran out and Collip, despite his best efforts, was unable to
make any more, the girl slipped into acidosis, followed by coma. In desperation,
the doctors began to treat her with huge doses of a partially purified extract. At
first, it seemed to work. The coma reversed and the girl temporarily regained

consciousness, but died shortly afterwards. It was a tragedy that Best, in particular, would not forget.

By the time that Macleod gave his address in Washington D.C., there was at last some hope. A number of technical modifications to the process of preparing extracts meant that insulin production had started again, and the Toronto group breathed a collective sigh of relief.[3]

However, respite was short. Scaling up the production of insulin using these new modifications was proving to be expensive—and dangerous. The new method used acetone as a solvent to prepare the pancreatic extracts and which then had to be removed by evaporation. This was achieved by placing the extracts in a specially constructed wind tunnel where the heat produced by electrical coils in its roof evaporated the acetone. The solvent vapour was then sucked down the tunnel by an old exhaust fan procured from the building's heating system—on one occasion its rattle caused a bottle of highly explosive picric acid to fall from the shelf of a nearby laboratory. Had it ignited the highly flammable acetone fumes now billowing through the corridors, the resulting death and destruction would have negated the few millilitres of insulin produced.[4] Additionally, the growing attention from the media caused the demand for insulin to rapidly outstrip the supply. Despite their best efforts, the Toronto group simply could not scale up production to produce insulin in sufficient quantities to meet demand.

But help was soon to follow. George Clowes, who, following Banting's disastrous presentation at Yale, had contacted Macleod to offer the assistance of Eli Lilly and Company was not a man who took 'No' for an answer and had maintained his efforts to persuade Macleod that collaboration was to everyone's advantage. Macleod remained hesitant. In reply to a letter from Clowes, Macleod repeated his intention to 'carry the thing as far as we can ourselves here and then publish all we know', but reassured Clowes that 'if the time comes for collaboration your firm will receive first consideration'.[5]

A month after writing the letter, Macleod had thrown caution to the wind. After hearing Macleod's address in Washington D.C., Clowes insisted to Macleod 'the necessity of starting work on the problem of large-scale production of this product with as little delay as possible'.[6] Macleod agreed. Near the end of May a telegram from Clowes and the other officers of the company pledged 'whole-hearted co-operation with a view of making the product available for the medical profession at the earliest possible date and in the largest possible amount'.[7] The deal was done.

It was a noble—and necessary—ambition. In order to take the University of Toronto breakthrough discovery forward, big investment was needed to scale up production, standardize the product, and manage its distribution. By the end of June, an agreement between Eli Lilly and the University of Toronto Board of Governors was signed that granted Eli Lilly an exclusive licence to manufacture insulin in the United States for a period of one year.[8]

Both Clowes and Macleod were worried that, without such an agreement in place, 'an unscrupulous concern might foist upon the public an extract the potency of which is uncertain and untested' and their fears were not without good reason.[9] By now, it was becoming clear that insulin could save lives, but could also bring them swiftly and violently to an end.

Accounts differ as to who first made this crucial discovery. Clark Noble may have lost out to Best over the toss of a coin to blaze the trail with Banting, but in a 1971 memoir, Noble described his role in events that would shape the treatment of diabetes in a profound way. He was studying the physiological effects of pancreatic extracts on test rabbits and recalled that many of these animals were found to be either dead in their cages the next morning, or suffering convulsions:

> That evening and night I worked continually doing as many B[lood] S[ugars] as I could, and noting their convulsions or stuporous state. I will never forget first thing in the morning, Prof. Macleod coming into my room, and, standing with pipe in hand remarking after I had notified him all that had transpired 'ah Noble most amazing – now let's give them glucose' – something which I had failed to think of.[10]

Macleod's suggestion of giving the animals glucose was in response to Noble's findings that their blood sugar levels had been dramatically reduced. Yet within only a few minutes of being injected with a sugar solution, they showed a remarkable recovery.

Noble described witnessing this recovery as a 'thrilling event', but whether he and Macleod were the first to observe it is debatable.[11] It is quite possible that the idea of giving the rabbits a sugar solution to aid their recovery had occurred to Macleod as a result of similar observations made by Collip. Having seen that pancreatic extracts could lower the blood sugar levels of a diabetic animal, Collip had wondered what their effect might be on a healthy one. When he tested his theory on rabbits, the effects were both dramatic and disturbing:[12]

> Some time after the injection of insulin the rabbits often show characteristic symptoms. A preliminary period of hyperexcitability gives place to a comatose condition in which the animal lies on its side, breathing rapidly (often periodically) with sluggish conjunctival reflex and widely dilated pupils (rectal temperature normal). On the slightest stimulation, as shaking of the floor, violent clonic convulsions supervene in which the animal either throws itself over and over or lies on its side with head markedly retracted and the limbs moving rapidly as in running. These convulsive seizures usually last 1–2 minutes, and they often come on without any apparent stimulation, when the interval between them is about 15 minutes. They frequently terminate in death from respiratory failure.[13]

Collip first thought that this might be a toxic reaction to the insulin, but a sample of the rabbit's blood tested almost nil for glucose.[14] He concluded therefore that his extracts had been so potent at lowering blood sugar levels that the animal had been plunged into a state of hypoglycaemia.[15] When Collip then dissolved some sugar and injected it into rabbits that were comatose or convulsive due to hypoglycaemia, he found that they made a sudden recovery.

Regardless of who first made this discovery, it shaped the thinking of those who managed type 1 diabetes. Patients, meanwhile, would be eternally thankful that most hypoglycaemic episodes can be treated not with a hypodermic injection of glucose, but simply a handful of sweets. Today, all diabetic patients know that they must carry with them at all times such a supply of rapid sugar which can be used in the event of a 'hypo'. They are also taught to recognize the first warning signs of such an episode.[16] Writing in 1925, the diabetes specialist Dr. Otto Leyton (1874–1938) listed these as including '(a) apprehension; (b) coarse tremor; (c) sweating; (d) intense muscular weakness; (e) difficulty in articulation; (f) convulsions; (g) coma'.[17] In addition, Leyton noted that 'Some patients during a hypoglycaemic attack behave as if they were intoxicated with alcohol or demented'—an observation drawn from his own professional experience:[18]

> One of my most sober patients had some friends to a meal and began by pressing them to help themselves more liberally to pepper; then, most excitedly in a loud voice, he hurled insulting epithets at his wife, who recognizing the condition besought him to take some sugar. He replied that of course she wanted him to take sugar, a thing the doctor had forbidden him, so that she might get rid of him and marry someone else. The patient was forced to take sugar; the symptoms passed within half an hour and he had no recollection of what had occurred.[19]

For other diabetes clinicians, the experience of a hypo was far closer to home. Professor Charles Fletcher (1911–1995) was diagnosed with Type 1 diabetes at the age of 30, and described his own experiences of hypoglycaemia as 'curious':

> In one attack, many years ago, I awoke at night with a feeling that I had an extremely important message for mankind. I wrote it down by torchlight, took sugar, and went to sleep. Next morning I found my message was 'Let's all commit suicide as a protest and jump into a mass grave'. Recovery from coma on injection of intravenous glucose is unpleasant. I feel that I am being dragged back from a heavenly existence to a horrid and sordid world, and struggle against it. Recently hypoglycaemia has given me a feeling that Armageddon is near. Once I clutched my wife saying, 'The world is coming to an end, and I want to hold on to you'. 'Alright', she said, 'but drink some Lucozade first', and the world was saved.[20]

The discovery that hypoglycaemia could be induced by insulin soon led to it being used in clinical fields other than diabetes. Between 1928 and 1933, the Austrian doctor Manfred Sakel (1900–1957) developed what became known as 'insulin coma therapy' while working at a private clinic in Berlin. Many of Sakel's patients were morphine addicts and he was intrigued to observe that both their withdrawal symptoms and their dependency on morphine disappeared after they had been plunged into a hypoglycaemic coma by the injection of insulin.[21] When Sakel returned to Vienna he began using what he called 'insulin coma therapy' (ICT) for the treatment of psychosis and schizophrenia.[22] This involved administering insulin to psychiatric patients at such high doses that their blood glucose levels would plummet and plunge them into unconsciousness. It was a procedure that was often distressing not only for patients, but also their clinicians—one of whom recalled, '…with all these people—tossing, moaning, twitching, shouting, grasping—I felt as though I were in the midst of Hell as drawn by Gustave Doré for Dante's Divine Comedy'.[23]

Sakel's patients would remain in a coma between a few minutes to a few hours before being roused back to consciousness by nasal or intravenous administration of a sugar solution. Psychiatrists claimed that, on emerging from the insulin-induced coma, their patients' symptoms temporarily disappeared, leaving them in what was termed a 'lucid period'. The aim of the treatment was then to extend this 'lucid period' by administering insulin shocks five or six times a week until either the patient was deemed to have recovered or was considered incurable.

When ICT came to the attention of the international medical community, it was quickly adopted and soon became an established treatment for schizophrenia, with an article in Collier's magazine hailing it as a wonder-therapy:

> Insulin shock, the new violent method of dealing with certain forms of insanity, is as dramatic as medieval magic. And it really works. Dr. Manfred Sakel, its inventor, has brought hope to hundreds of thousands of persons otherwise condemned to a life of constant nightmare.[24]

Despite this optimism, however, the treatment was not without risks. ICT could leave patients with lasting brain damage and, due to the anabolic action of insulin, they also tended to become severely obese as a result of the treatment.[25] In Sylvia Plath's *The Bell Jar*, her protagonist Esther Greenwood spends some time in a psychiatric hospital and describes to her fellow patient Valerie how the nurses inject her with a needle three times a day before giving her a sweet drink. Valerie explains that this is insulin treatment, which she herself had undergone and as a result was left looking 'just as if I was going to have a baby'.[26]

Plath had obviously done her homework on the effects of insulin. Certain patients did seem to be resistant to ICT and Banting himself published some work in which he attempted to identify a substance in the blood that counteracted insulin's effects.[27] But despite these problems, ICT remained in use as a standard therapy for schizophrenia until the 1950s. One suggested explanation for its wide adoption has been that the clinical use of insulin enabled psychiatrists to align their field more closely with mainstream medicine.[28] With insulin as a tool, psychiatrists could now actively intervene in a patient's condition in the same way that specialists in other areas of medicine did, e.g. as a cardiologist operating on a heart.

But in 1953, young doctor Harold Bourne challenged the use of ICT. In a paper called 'The Insulin Myth', Bourne reviewed the difficulties in diagnosing schizophrenia and assessing the effectiveness of ICT. According to Bourne, one major problem was that the selection of patients for ICT was biased, since only those patients who showed signs of making a good recovery were selected for treatment. Another factor was that ICT was administered in a designated 'Insulin Unit' where patients received a much higher level of attention and care from doctors and nurses. Bourne suggested that the recovery of patients might have less to do with insulin and more to do with these higher levels of care.[29] As a result, Bourne concluded that 'the evidence for the value of insulin treatment is unconvincing ... there is no proof of any specific therapeutic effect, and the long-term prognosis is in no way influenced'.[30]

At around the same time as Bourne's critique, a host of new psychopharmacological drugs became available, e.g. phenothiazines, that soon replaced ICT as a treatment for schizophrenia. Within a few years, ICT was engulfed by an 'aura of shame and silence' with several psychiatrists feeling embarrassed to admit using a procedure that was now perceived as inhumane.[31]

Alongside insulin shock therapy, Collip's discovery of insulin shock had an even darker legacy. In 1957, a jury found Kenneth Barlow of Bradford, West Yorkshire to be guilty of murdering his wife Elizabeth, and he was sentenced to life imprisonment. Although Barlow had protested that his wife had died after accidentally drowning in the bath, post-mortem studies using new antibody detection techniques revealed that she had been deliberately injected with a lethal dose of insulin.[32]

The Barlow case was the first documented instance of insulin being used in a case of murder, and although the authorities at the University of Toronto might not have anticipated their discovery being put to criminal use, they had certainly recognized its possible dangers. Anticipating this problem, the University of Toronto Board of Governors formed a body known as the Insulin Committee, composed of a small number of clinicians with the responsibility for overseeing the production of insulin and regulating its distribution. Under the terms of the agreement signed between the University of Toronto and Lilly, any requests for insulin from clinicians would have to be approved by the Insulin Committee. As one of the key figures on this committee, Macleod corresponded on a regular basis with Clowes to make decisions about which clinicians should receive a supply of the new drug.[33] With insulin still only available in limited supply, tough choices had to be made about who should receive it. The initial list of candidates was restricted to a small number of eminent diabetes specialists at large hospitals, but eligibility to be included on this list was not always based on scientific standing alone. On one occasion, Clowes informed Macleod that two members of the Insulin Committee, Elliott Joslin and Frederick M. Allen, had both recommended that Dr. A. I. Ringer of the Montefiore Hospital in New York City, not be supplied with insulin. This request was based on the grounds that 'while Ringer had done brilliant work from the scientific end, his clinical treatment of diabetics was very unsatisfactory'.[34]

Macleod, however, pointed out to Clowes that a hidden agenda might well be lurking behind the objection:

> With regard to supplying Ringer I wish to urge that this be done. Of course both Allen and Joslyn [sic] will be inclined to consider Ringer out of their class as Clinical Investigators. He has criticised Allen's work on several occasions and you know perfectly well that Allen cannot stand criticisms.[35]

Alongside the complex politics of academic and clinical life, the Insulin Committee faced more practical headaches, such as how to ensure that the batches of insulin being made at Connaught labs were of the same potency as those being made by Lilly—how should the drug be standardized?[36] How could it be prevented from deteriorating? How could more protein be eliminated?[37] How could it be shipped abroad?[38]

Looming over all these challenges, however, was the even more formidable problem of human nature. For despite the weekly quota of insulin being set at 100 units, Clowes wrote to Macleod telling him that one diabetes specialist in New York had requested at least 40 units of insulin daily to be used solely for the treatment of one particular patient. Clowes complained that 'It is appalling to realise the greed of certain individuals who, for the sake of their own indulgence would sacrifice the lives of others whose need is infinitely greater than their own'.[39]

Clowes' insight into human nature had also led him to make another suggestion. Shortly after Macleod had agreed to the deal with Lilly, Clowes had begun consulting lawyers about applying for a patent on insulin and on the process by which it was made.[40] A month later, Clowes had written to Macleod stressing the urgency and importance of taking this action:

> I cannot too greatly impress on you the necessity of giving this patent question very serious consideration ... knowing that there are plenty of unscrupulous people in the world I am very anxious that no stone should be left unturned to safeguard your interests as well as ours and to prevent undesirable people getting any sort of hold on insulin[41]

Clowes knew that Macleod was uneasy about taking out a patent, and offered him some words of reassurance:

> To tell you the honest truth, I don't think much of it myself. It was drawn up by one of our patent attorneys who apparently was anxious to get everything possible into it and whose zeal was superior to his discretion. It is really a purely technical procedure and in my opinion has little or no value ...[42]

But as far as the University of Toronto was concerned, applying for a patent on insulin was far more than a mere 'technical procedure'—it was a step that might be necessary to save lives. In a public statement made a year later, after the patent had been approved, the University of Toronto defended its actions, warning that the unregulated production of insulin might pose a 'real source of danger' to patients. Without the protection of a patent, 'the market would by now be flooded with preparations of unknown potency and durability, and serious accidents would inevitably have resulted because of overdosage'.[43] The statement went on to offer reassurance that the patent was 'not to be used for the purpose of restricting the preparation of this or similar extracts elsewhere, or by other persons'.[44] Rather, it was a defensive measure, taken in order to ensure that insulin remained readily available by preventing other parties from filing similar patents, which might then allow them to monopolize production of insulin and exercise control over its price.[45] To add to these official reassurances, Banting made clear in a newspaper interview that he had never gained a single dollar from the patent—he, Collip, and Best had sold their patents to the University of Toronto for $1, despite never receiving their payment.[46]

Notwithstanding all these reassurances, there may have been another, slightly less-altruistic motive in filing a patent. Although it might well prevent quack doctors in back rooms from cooking up concoctions of dubious therapeutic quality, it would also be a rock-solid guarantee that members of the Toronto group were recognized as the discoverers of insulin. At the moment, this recognition could not be taken for granted, for while Banting was still losing sleep over fears that Macleod was out to steal his thunder, another rival had now stepped into the ring.

Word had reached Banting that John Murlin at the University of Rochester was claiming to have prepared pancreatic extracts with a proven anti-diabetic effect. This wasn't actually the first time that Banting had encountered Murlin. Back in the summer of 1921, when Macleod had told Banting of Israel Kleiner's failure, he had offered John Murlin as another cautionary tale: 'Do not on any account add alkali to the extract ... Murlin tried that and came to grief because alkali injections per se depress blood sugar'.[47]

Between 1913–1916 at Cornell University Medical College, Murlin and his colleague Benjamin Kramer had found that the injection of pancreatic and duodenal extracts could reduce the excretion of glucose in animals from which the pancreas had been removed.[48] Further work revealed that this anti-diabetic effect could also be achieved simply by injecting an alkaline solution of sodium carbonate.[49] Murlin and Kramer could not believe their luck, and later recalled that 'we danced around the laboratory with hilarious joy believing we had found the real international [sic] secretion'.[50] They had, it seemed, discovered the elusive secret of how the pancreas controls blood sugar levels. While the rest of the medical world was fixated on finding a theoretical pancreatic hormone, Murlin and Kramer had shown that elevated blood sugar levels could be returned to normal simply by adjusting the alkalinity of the blood. Murlin speculated that this occurred due to an increase in the permeability of the kidneys to glucose caused by the change in blood pH, and that the 'administration of alkali to diabetics may be of importance in more ways than one'.[51,52]

Having worked long into the night to make this discovery, they 'went home about 2 a.m., happy as the proverbial larks'.[53] But just as had happened with Israel Kleiner, the outbreak of the First World War derailed Murlin's work on diabetes and while on active service with the US Army, he was unable to do any further research in this area.[54]

In October 1921, apparently unaware of the work being carried out in Toronto, Murlin returned to his work on diabetes. But in May 1922, when Macleod announced the discovery of insulin in Washington D.C., Murlin realized that he had serious competition and was spurred into action. Going back through the results that he obtained before his work had been disrupted by the war, he now published a much more detailed account of them, pointing out that 'they lend support to the evidence furnished by Banting and Best'.[55] He also added a frank admission about his earlier conclusions:

> … we were misled by the belief that the alkali (Na_2CO_3) had been wholly responsible for the effect … This singular coincidence diverted our attention for several years to the alkali, although we were convinced in 1916 that the pancreatic extract played a part in the various signs of improvement reported.[56]

Somewhat understandably, Murlin had no wish to be remembered as having missed out on the discovery of insulin due to having been misled up a blind alley. The ideal of science may well hold that being wrong is as valuable as being right, but the reality is rather different. Grant funding, publication in prestigious journals, and departmental promotions do not come from failure. Thus, Murlin seized the opportunity to remind the medical world of what he felt had been his own vital contribution to the discovery:

> As matters stand, the credit for final and complete demonstration of the hormone (or enzyme) belongs to Banting and Best. It may be permissible however to remark that the experiments reported by Murlin and Kramer in 1916 cannot be explained except upon the hypothesis that the active principle was present in the extracts employed by them. It is not to detract in the slightest degree from the brilliance of Banting and Best's achievement, but to confirm that we are reporting in more detail than has hitherto been done the observations made nearly ten years ago.[57]

Murlin hoped that this statement would speak for itself. While he did not deny Banting's priority in demonstrating that pancreatic extracts were effective in human patients, Murlin was confident that the credit for first showing this effect in animals belonged to him.[58]

With reports in *The London Times* hailing Banting as 'the originator of the insulin treatment', he might well have hoped that his status as its discoverer was beyond question.[59] But when he saw the *Toronto Star* headline that 'Dr. John Murlin of Rochester Says His Treatment Was Discovered Before Banting', Banting felt his support crumble.[60] The article went on to explain that Banting and

Murlin had both arrived at their discoveries independently of each other's work and described this as being 'one of these odd coincidences which are frequently cropping out in the world of research'.[61] This was little consolation to Banting. Nor was a letter from Allen in October enough to reassure Banting that 'Dr. Murlin is publishing some things about his extract work but he concedes full priority to you'.[62]

When Murlin learned that Banting was upset, he put pen to paper in an attempt to set matters straight:

> It has come to my attention that you have been displeased with a nauseating newspaper article ... I want you to know that this article displeased me much more than it can possibly have affected you ... To my surprise and horror when the article appeared it was captioned by a very sensationalist headline which was wholly misleading ... In order to assure you that I have no disposition to withhold full credit for your contribution in the field I quote a brief statement from an article soon to appear.[63]

Murlin went on to reassure Banting that, in an address he had given to the Rochester Medical Association on the previous evening, 'I gave you full credit for your great work on diabetes'.[64] Along with his letter, Murlin included a newspaper clipping that contained a verbatim account of the address he had given. But while the words of Murlin's letter suggested that he conceded priority to Banting, the headline 'Discovery of Extract That Has Power to Restore Capacity Lost in Diabetes is Made Public By Dr. John R. Murlin' (Figure 12) and the article told a very different story:

> It is a mistake to suppose the pancreatic extracts had not been prepared before or to suppose that they had not been used with some success on human cases. It is no discredit to Banting and Best, rather the contrary, to say that more than twenty scientific experimenters had made extracts which diminished the excretion of sugar in both depancreatized dogs and in human cases of severe diabetes before they had done so ... we claim the credit for having proved first that pancreatic extract can restore the lost power to oxidise sugar to animals rendered diabetic by removal of pancreas. This we have confirmed many times over the past summer and this we maintain is the crux of the whole problem.[65]

Banting was not alone in his fears about Murlin's motive. When word reached George Clowes that Murlin was acting as an advisor to Chicago-based pharmaceutical company Wilson and Company, he wrote to Macleod warning him:

> It is so obvious that the points in the process that Murlin borrows from the Toronto group are the essence of the whole thing ... the reference to previous work is simply used as an excuse for his acting as an advisor to Wilson and Company in an attempt on their part to manufacture Iletin by a process which is to all intents and purposes that which was developed by the Toronto group.[66]

Figure 12 *Newspaper headline - 'Discovery of extract that has power to restore capacity lost in diabetes is made public by Dr. John R. Murlin', Democrat and chronicle (Rochester, N.Y.); 11th November 1922; Includes photographs of Dr. Harry B. clough, Dr. John R. Murlin, Dr. C. Clyde Sutter, and Dr. C. B. F. Gibbs. This is the newspaper clipping that was sent to Banting by Murlin.*

In order to avoid what he called any 'unpleasant antagonism', Murlin was quite open with Banting and Macleod that he was acting as an advisor to a pharmaceutical company about the possibility of scaling up production of pancreatic extracts for clinical use.[67] When he asked Banting and Macleod for their consent to allow him to produce insulin under a different name and by a slightly different process, that it was 'For the benefit of humanity ... that more than one method of preparation should be tried' seemed unassailable.[68]

Murlin said that he had no intention of patenting his work, but instead simply wanted the freedom to work independently with another company. He reminded Macleod that, ultimately, 'we all have the same aim, namely the advancement of scientific medicine'.[69] However, he also issued a gentle challenge:

I scarcely need to remind you that I made a partially successful extract of pancreas as long ago as 1913 ...Naturally, we have profited a great deal by the publications from your laboratory and the method we are now using appropriates a suggestion or two from the method as published by Collip. However, we had a powerfully effective extract before your methods were known to us ...[70]

A letter from Murlin to the Insulin Committee gave even greater cause for concern. Despite his reassurances to Macleod, Murlin now admitted that he was indeed considering filing a patent on his own work and assigning the rights to Wilson and Company. To Macleod and his colleagues, it was a clear and brazen challenge, 'so worded as to indicate that Dr. Murlin hopes to secure priority over the Toronto group'.[71]

The gloves had come off, and in a letter to Charles Riches, the patent attorney whom the University of Toronto had appointed to take charge of their application, Murlin bluntly questioned the basis of their claim to have priority by pointing out 'that the essential step in the production of insulin, namely of alcoholic extraction was not original with Banting and Best nor Collip but was originated by Ernest L. Scott in 1911'.[72]

George Clowes urged Macleod to take the threat from Murlin seriously:

> I learned from rather influential sources in Rochester that Merlin [sic] has quite convinced the Rochester people that he was the original discoverer of this product that he now proposes to put on the market ... I think it would be advisable for you or some other member of the Toronto group to carefully review the literature, particularly the Merlin [sic] papers, from this standpoint.[73]

But Clowes' assessment of the threat posed by Murlin was not entirely accurate. At least one of Murlin's colleagues at Rochester, John Williams, remained unconvinced of his claims and made this abundantly clear to Macleod and Clowes in a letter marked 'CONFIDENTIAL':

> I do not think either of you fully appreciate the significance of this man's [Murlin] position. There can be no question but that he is playing a desperate game and that he stands to rise or fall on the recognition that he receives. I have heard several express the opinion that if he is discredited by scientific men, he will lose his position here. On the other hand, if his work is accepted it means a domination in science in the local University and a position of great influence. The stake therefore is big and the man is desperate. The most important recognition that he can obtain obviously is that of the Toronto group.[74]

At the same time that Murlin was considering taking out his own patent, the application by the University of Toronto was mired in a number of technical complications, not the least of which was that Banting was not actually named on it. Banting had expressed great reluctance about being included on the patent, and so the initial application had been restricted to cover only the process of making pancreatic extracts, with Collip and Best named as inventors.[75] Patent attorney Charles Riches recognized that this omission would make the patent significantly weaker in the face of the challenge from Murlin and urged that the Insulin Committee amend it to include Banting's name and thereby widen its breadth.[76]

With the lawyers now warning that 'this should be taken care of quickly ... before others file applications which may involve interference complications [sic]', Banting finally agreed to be included on a revised application—but he was not without great misgivings.[77] For although Macleod was confident in his belief that the patent 'shall be used for no other purpose than to safeguard the manufacture and distribution of Insulin and to guard against undesirable commercial exploitation', Banting felt that he had now compromised his own professional integrity:[78]

> I signed the Insulin patents for Canada, and other countries with some reluctancy, though I feel myself safeguarded by my assigning to the University all rights and claims arising therefore, and also by the fact that I did so at the request of the Insulin Committee ... The act of taking out a patent for what we hope will prove to be a remedial agent of considerable value has awakened a great deal of criticism, more especially since I am a graduate in Medicine and bound by its ethical code, as contained in the Hypocratic [sic] Oath taken by me on obtaining the M.R.C.S. This criticism I feel I will have to meet for a long time and may never live down.[79]

To counter the perceived threat from Murlin, Banting now called in a favour. Writing to Charles Hughes, the U.S. Secretary of State, whose daughter had come to Toronto for treatment that summer, Banting asked whether he might be able to speed up the patent application process for insulin.[80] Within a matter of only a few days, Banting, who by now was helping to run clinical trials in a diabetic clinic, was told by Hughes that he had been in touch with the Patent Commissioner and that the application would be expedited.[81]

A successful patent application patent on insulin ought to see off any challenge from Murlin, but as far as Macleod was concerned, offensive manoeuvres were required: the forthcoming convention of federated American Societies for Experimental Biology was the perfect battleground upon which the contest over priority would be settled. 'I can well see that Murlin proposes to fight his main battle for priority at this meeting', he wrote to Clowes, assuring him, 'and I will be prepared for a good fight'.[82]

Reporting on the symposium, *The Mail and Empire* newspaper gave Banting the credit for what it described as being 'undoubtedly one of the most important discoveries of the present century'.[83] The article went on to add that 'Additional interest was added by reason of the presence of J. R. Murlin, of the University of Rochester, who claims to have made discoveries in the same direction'.[84] As Murlin rose to give his own research paper, everyone present was eager to hear his view, but any hopes for drama were spoiled. Murlin was magnanimous in defeat:

> I wish to pay my tribute of admiration to the work that has been done here in Toronto. If you have ever been hot on the trail of something and then were thrown off the trail by accident and returned to it to find someone else two or three jumps ahead, you will understand with what feelings I am speaking. Like the Eels that Molly skinned, we have been torn between pain and admiration.[85]

The headline of the article left its readers in no doubt that the matter was settled: 'Biologists Acclaim Victory Over Diabetes: Prof. Murlin Humorously Admits Dr. Banting Forestalled Him'.[86] Another newspaper reported that, in a scene that it described as being 'unparalleled in the history of the federation', Banting received a standing ovation that lasted three minutes.[87]

While Banting was receiving a standing ovation in Toronto, on the other side of the Atlantic at a meeting of the Societé de Biologie in Paris, distinguished French endocrinologist Eugene Gley (1857–1930) was requesting that a sealed envelope that he had deposited in 1905 be opened. In it was a note written by Gley entitled 'Sur la sécrétion interne du pancréas et son utilisation thérapeutique', in which he described experiments that he had performed in 1900–1901 showing that the injection of extracts made from degenerated pancreatic tissue caused a reduction of urinary sugar in diabetic dogs. Why he never published this work at the time or even followed it up remains unclear.[88]

Despite his theatrical gesture of asking for the sealed envelope to be opened in public, Gley made no claim to have priority in the discovery of insulin. John Murlin had no such intentions, and although he may have left his audience laughing at the meeting in Toronto, behind the scenes he was much less willing to accept defeat with such good humour. The following year, John Williams, his colleague at Rochester, told Macleod:

> Murlin in his recent address before a large number of laymen at the Rochester Chemical Society meeting following his usual tack, stated that he was working on the problem at the same time as were you (this is of course untrue) and that you obtained priority because you rushed into print.[89]

Williams had actually tried some of Murlin's extracts in clinical trials, and his verdict was damning:

> He had not made an extract which in any way served to correct diabetes. On the contrary, the product which he was making was a highly toxic, extremely irritating preparation which caused abscesses and much irritation as to necessitate its discontinuance. It had not the slightest value.[90]

This victory meant an awful lot to Banting. For despite the adulation in newspaper headlines, he still felt that he had to battle constantly for recognition. Although he had seen off Murlin's threat, there was still the problem of Macleod. While Macleod himself gave Banting full credit for the discovery of insulin, others in the medical world took a very different view.

In August, a letter from the distinguished British physiologist Sir William Bayliss appeared in *The London Times*. Bayliss had done pioneering work during the First World War on the use of intravenous saline to treat shock, and claimed that the contribution of Macleod to the discovery of insulin had not been 'sufficiently recognized' and went on to say that this landmark 'was the result of the painstaking and lengthy investigations of Professor Macleod, which have extended over many years, and it is to him that the chief credit should be given'.[91] Banting, meanwhile, was described as having been merely 'one of the collaborators of Professor Macleod'.[92]

When Bayliss's claims were quoted shortly afterwards in *The Toronto Star*, Macleod found a journalist knocking on his door to ask for some clarification on the situation. Quickly, Macleod wrote a very diplomatic letter to him in order to clarify the situation:

> After discussing the matter thoroughly with him, I saw that Banting, although unfamiliar with the literature and quite inexperienced in the methods of investigation involved in such a research, was willing to undertake it under my guidance and to give all of his time for several months to the work. I therefore planned with him the investigation and I offered him the assistance of one of my best research students, Mr. C. H. Best, who had been working with me on blood sugar problems. I also arranged for his having every facility of my department for carrying out the work and as it proceeded I gave all the guidance and judgment I could to render the results definite and conclusive. At his request I allowed my name to appear with his and Best's on the preliminary communication before the American Physiological Society and at the meeting I found it necessary to 'jump in' and defend the several criticisms that were made of the work ... I have directed and actively participated in all of these physiological investigations, and we all hope that when they are published, as will be the case in the near future, that they will be considered of sufficient merit to spread the laurels among all those who have participated in the work ... I may say that I greatly appreciate the friendly spirit which prompted you to write this letter to *The Times* and I will leave it to your judgment as to whether you ought to clear up any misunderstanding as to the question of priority that may have arisen from it.[93]

Fearing that 'a disagreeable controversy was threatened in our local evening newspapers', Macleod gave an interview to the *Toronto Daily Star* in which he gave Banting full credit for the discovery of insulin under the headline 'Gives Dr. Banting Credit for "Insulin": Prof. J. J. R. MacLeod Replies to *Times* Letter of Sir William Bayliss. How it was Done. Head of Department Declines Credit for New Diabetes Treatment'.[94,95] When Bayliss learned of this, he wrote to Macleod explaining that very little information about the work done in Toronto had been available to him when he had written his original letter. But although he asked Macleod to pass on his sincere apologies to Banting for this misunderstanding, the situation was well beyond repair.[96] While Macleod gave Banting full credit for the discovery of insulin in the newspaper columns, in his private correspondence he was far less complimentary. Writing to Archibald Byron Macallum, Professor of Biochemistry at McGill University, Montreal, Macleod fumed:

> As the result of a most unfortunate letter of Bayliss in 'The Times' ... Banting has been going about stirring up unbelievable trouble here. To try to stop this I gave

a statement to the Press giving full credit for his work on the effect of extracts of duct-ligated pancreas. Now he claims that I should in the same way give him full credit for all the work that has been done subsequent to this experiment. This I will of course not do since he has participated very little in this work, and not at all in the past six months ... [97]

With demand for insulin now rapidly outstripping supply, Macleod had turned his attention to what he believed could be a vital new source. In 1903, zoologist John Rennie (1865–1928) at the University of Aberdeen had discovered that, in certain species of bony fish, the insulin-producing Islet cells were not only separate from the rest of the pancreas but also relatively large, making them particularly easy to isolate. Four years later, Rennie and his colleague Thomas Fraser (1872–1951) treated four diabetic patients with extracts made using material obtained from the local fish markets.[98]

But making the extracts was only half the battle. For when they finally published their work, Rennie and Fraser admitted to having experienced 'some difficulty' in persuading their patients to continue with the treatment.[99] However, the biggest disappointment was that the results were simply inconsistent, leading Rennie and Fraser to a blunt and sober appraisal of their work that 'the experiments as they stand cannot be regarded as conclusive'.[100] As a result, Rennie is remembered not for having discovered insulin, but rather for the identification of a mite that causes disease among bees on the Isle of Wight.[101]

Macleod was familiar with Rennie's work, and he understood why it had failed. Firstly, Rennie and Fraser had boiled their extracts, which would have destroyed the activity of the insulin, and secondly, in all but one case, the extract had been given orally. In the case of the single patient in whom the extract was administered by hypodermic injection, the treatment had to be stopped due to side effects.[102] Armed with these insights, Macleod believed that fish might still be an abundant source of insulin.

Monkfish showed particular promise for two reasons. Firstly, it had particularly large Islet cells, and secondly it was usually discarded from a catch as diners had little appetite for it due to its ugliness. Arrangements made by Macleod with the fishing industry in Halifax, Nova Scotia to procure a steady supply of monkfish tissue led to hopes among trawlermen that insulin might well provide a lucrative new source of income for them.[103]

For a while things looked hopeful. With the quantity of material recovered from UK abattoirs proving to be insufficient, the Medical Research Council (MRC) began to consider the possibility of using fish as an alternative source.[104] Meanwhile, at Toronto General Hospital patients were treated successfully using preparations made from Macleod's monkfish extracts.[105] However, despite the triumphant headline of a local Halifax newspaper, 'From Ugly Fish to Conquer Death', the attempt failed.[106]

Scientifically, the preparation of insulin from fish had been a success. But alongside logistical complications, Macleod was unable to find any companies with sufficient interest in the commercial production of insulin from fish.[107] In a letter written at the end of 1923 to Sir Henry Dale of the MRC, Macleod had sent instructions on the preparation of insulin from fish, as well as some projected costs. Presumably he hoped that this might still become a viable method of preparation in the UK, but in his letter he conceded that 'Although fish material may be useful in Canada it does not seem to me likely that it would replace abattoir material which is here so easily collected'.[108]

There was another problem. Archibald Macallum was keen that Macleod should publish his results on insulin prepared from fish in a prestigious scientific journal, but as Macleod pointed out, this would most likely incur Banting's fury:

> In view of all this I believe that it would only serve to fan the fires still more—and they are unbearably hot at present—if I were to publish my recent researches in the Transactions of the Royal Society, dearly though I should love to do so. I find that Banting has succeeded so well in sowing the seeds of distrust in me that it will be necessary for me not to take any step that could possibly be misinterpreted. If I sent this to the Royal Society he would immediately say 'I told you so, Macleod all along was endeavouring to minimize the importance of my work by its publication in ordinary journals whilst he placed his in the most conspicuous one he could think of' and if I should be elected to the Society after this article appeared he would claim that I sailed in under false colours.[109]

The situation was also taking its toll on Banting. 'I have been feeling very weary of late', he wrote that October, 'but if I can only last for six months longer I will look forward to having a little holiday'.[110]

Alongside his own fears of being unrecognized, Banting was concerned that the contribution of Best had been overlooked, and to put this right he made a statement to *The Globe* in which, referring to Best's contribution, he stated 'I would never have been able to do anything had it not been for him'.[111]

With the imminent arrival of a delegation from the UK MRC, who were interested to learn more about insulin and how it might be used in therapy, the furore erupting in the pages of the local newspapers was hardly welcome publicity for the University of Toronto. Desperate to make a good impression on the MRC representatives and seriously concerned that the entire venture might be derailed by the 'unsatisfactory relationships between Prof. Macleod, Dr. Banting and Mr. Best regarding the contribution which each has made to this discovery', Colonel Albert Gooderham, Chair of the Insulin Committee intervened.[112] Pleading the case 'to get everything working in harmony', Gooderham requested that Banting, Best, and Macleod each submit a written statement that summarized their contribution to the work.[113] Collip, who had by now returned to Alberta, was not asked, but Gooderham instead requested a summary of his work from Banting. It was a curious request, considering that Banting had once almost physically assaulted Collip. Furthermore, it suggests that perhaps Gooderham was unaware of just how toxic the situation had become. Macleod, however, sent Collip a copy of his own statement as reassurance that his vital contribution would not be overlooked. In his letter, he confided to Collip just how badly the stress of the situation was taking its toll on him:

> You may have seen in the newspapers that I gave a statement to the 'Star' giving Banting and Best full credit for their researches with duct-ligated pancreatic extracts. I did this to correct an error made in this connection by Bayliss in a letter to 'The Times'. It appears that Banting is not satisfied with this but considers that I should give him full credit for the discovery of insulin as it is now used in the treatment of Diabetes. This I have declined to do and I have prepared a written statement of all the steps which have led up to the completed discovery and at his request I have submitted this to Colonel Gooderham who has volunteered to straighten things out. I will ask the Colonel's approval to send a copy of this statement to you ... I have been so bewildered with this fresh outbreak of Banting's that I have not been able to compose my mind sufficiently to correspond more fully with you over your work.[114]

Collip was pleased to hear that Macleod would make a personal statement on his behalf, testifying to the crucial role he had played. He also drafted a list to Macleod of what he considered to be ten key achievements in the work. These included the demonstration that insulin could cause hypoglycaemia in healthy rabbits and, of course, the removal of protein impurities. In his letter, he also made his feelings about the conduct of Banting and Best quite clear:

> There are some people at Toronto who felt I had no business to do physiological work ... The result was that when I made a definite discovery, my confreres instead of being pleased were quite frankly provoked that I had had the good fortune to conceive the experiment and to carry it out. My own feelings now in the matter are that the research and its aftermath has been a disgusting business.[115]

Although well intended, Gooderham's plan was little more than a sticking plaster slapped on a severed artery. Nevertheless, it worked well enough to leave Dr. Henry Dale, one of the visiting MRC representatives, suitably impressed with the work being done in Toronto. Dale was quickly persuaded that insulin was no mere 'quack cure' and, following his visit, it was agreed that, under the direction of the Insulin Committee, the MRC should take control of directing the production and licensing of what *The London Times* described as 'Canada's gift' to Great Britain.[116]

Dale became a powerful champion of Banting and Best's work and soon they were all too grateful for his support in fighting their corner. A letter had appeared in the *British Medical Journal* (*BMJ*) that cast doubt on the quality of Banting's and Best's work. Dr. F. Roberts of the University of Cambridge had scrutinized Banting's and Best's first paper and pointed out a number of major errors in their early experiments. The most important of these was that the premise upon which their entire research project had been built was flawed. This was that the anti-diabetic hormone was vulnerable to destruction by digestive pancreatic enzymes.[117] As Roberts went on to point out, even though the pancreas did indeed produce a digestive enzyme, this was synthesized as an inactive precursor and could therefore not degrade the insulin produced by the Islet cells. Banting had, it seemed, built an entire programme of research around a non-existent problem.

Roberts also argued that, far from being a novel approach, Banting's method of making extracts from degenerated tissue after duct ligation was simply unnecessary. The real problem, according to Roberts, was not to prevent the action of digestive enzymes, but rather to remove traces of impurities that might cause toxic reactions.

But Roberts' final point was a *coup de grace*. Through his meticulous study of their early papers, Roberts had spotted the graph that Banting and Best claimed to show that extracts made from whole pancreas were weaker than those made from degenerated material, but which actually showed the *opposite* to be true. As Roberts pointed out, the very data presented by Banting and Best contradicted their own hypothesis. Contrary to most of the medical community who lauded Banting and Best, Roberts reached a very different—and rather scathing—conclusion:

> The conclusion we come to, therefore, is this: the production of insulin originated in a wrongly conceived, wrongly conducted, and wrongly interpreted series of experiments. Through gross misreading of these experiments interest in the pancreatic carbohydrate function has been revived, with the result that apparently beneficial results have been obtained in certain cases of human diabetes. Whether insulin will fulfil its promise time alone will show; but I venture to believe that whatever success the remedy will have will be found to be due to the fact that the hormone has been obtained free from anaphylaxis-producing and other toxic substances.[118]

As far as Roberts was concerned, Banting's and Best's only achievement had been to show that digestive enzymes produced by the pancreas did not inhibit the action of insulin. And although he did not say it explicitly, the inference from Roberts' criticisms was that it was Collip, and not Banting, who had made the truly novel contribution of successfully purifying insulin so it could be given without producing a toxic reaction.

After his trip to Toronto, Dale wished Banting and Best every success, and he rallied to defend them against what he condemned as 'a destructive and somewhat censorious criticism' from Roberts:

> Nobody can deny that a discovery of first-rate importance has been made, and, if it proves to have resulted from a stumble into the right road, where it crossed the course laid down by a faulty conception, surely the case is not unique in the history of science. The world could afford to exchange a whole library of criticism for one such productive blunder, and it is a poor thing to attempt belittlement of a great achievement by scornful exposure of errors in its inception.[119]

Although he had not stated it explicitly, Roberts had raised an important and awkward question: when had insulin truly been discovered? What sequence of events constituted its discovery? Was it in the summer of 1921, when Banting and Best had shown that the injection of pancreatic extracts could lower blood and urinary sugar levels in a diabetic dog? Or was it in January 1922, when Collip's purified extract had been shown to reduce blood sugar in a human patient without causing a toxic reaction?

This was more than academic pedantry. For alongside the question of when, exactly, had insulin been discovered, Roberts had raised a second, perhaps more awkward question: who deserved the credit for this achievement? As far as the University of Toronto was concerned, however, the matter of priority was settled. On January 1, 1923, Banting, Best, and Collip, who were all now named as inventors on the patent, assigned their rights to the governors of the University of Toronto.[120,121]

Macleod reassured the medical community that this had been done in order to protect patients from what he called the 'uncontrolled commercial exploitation of insulin, with the consequence of its improper use in general practice'.[122]

Joslin, who was receiving insulin from Lilly for use in his Boston clinic, said that clinicians should be thankful for the University of Toronto's wisdom in taking this decision:

> It is a mercy that at present insulin becomes inert when given by mouth, and that its use is restricted to a syringe. The medical profession cannot be too grateful to the Insulin Committee of Toronto for limiting the supply and the distribution for a few months. Consider for a moment what would happen if morphin was the drug discovered and then was at once sold over the counter. The useless pancreatic preparations of the past required no supervision; insulin does, because it is so potent.[123]

Joslin and the other handful of diabetes specialists who were receiving the Lilly insulin now found themselves in a predicament. As news of insulin spread, they were increasingly overwhelmed with desperate requests from diabetic patients and their families. Letters arrived for Banting, addressed to 'To the Dr. who cures diabetis', to 'Dr. Fred', and to 'Monsieur le Professeur, University of Medicine, who has found a way of curing diabetis'.[124] By July there were even reports that diabetic patients were camping at the doors of the laboratory in the hope of getting their hands on some of the precious material.[125]

Banting's new-found fame also attracted therapeutic application of insulin some unwanted attention. Joe Gilchrist, who was helping Banting run a diabetic clinic for military veterans at the Christie Street Hospital, gave a lecture on the therapeutic application of insulin, which had come to the attention of the Anti-Vivisection Society of Toronto. In the lecture, Gilchrist had cited the case of one patient who, as a result of receiving too much insulin, had tried to climb the walls. Gilchrist was quick to add that the patient had soon recovered with the administration of sugar, but this did little to pacify the Anti-Vivisection Society. Believing that war veterans were being coerced into being test subjects in Banting's clinic, they referred the matter to the Parkdale branch of Great War Veterans Association (GWVA).

Outraged, the GWVA passed a motion opposing the work being done at the Christie Street Hospital:

> To us it seems only in accordance with the prevalent grateful treatment of men whose frames have been appallingly racked on the battlefield that they should be administered extracts that throw them into convulsions or cause them to climb the walls of the experimental [sic] torture chambers ...[126]

While the GWVA stressed that it did not wish to block insulin research, the Canadian Anti-Vivisection Society waded in with even stronger words:

> The diabetes 'remedy' has never been proved and is said to have caused great harm in many cases ... We object to money used in this way both on moral and scientific grounds. Moreover, we have statements from the highest medical and scientific sources, that the use of serums, and animal organ extracts, is having a deleterious effect on the public health and that deaths from diseases so treated have in no case shown a decrease and in many diseases have shown an increase.[127]

Banting issued an emphatic denial of the claims and Sir Robert Falconer, the President of the University of Toronto, came quickly to his defence. The *Toronto Daily Star* described Sir Robert as a man who 'is seldom angry' but when speaking of the Anti-Vivisection Society of Canada, he was willing to make an exception:

> These people don't realise what the word humanity means. They have a half squinted view of humanity. That's exactly what it is. Is there anything could make a man's anger rise faster than that sort of thing? Such a discovery saves thousands of people untold anguish ...[128]

Even more testimony came from Banting's own patients. When the *Toronto Daily Star* interviewed six of the nine soldiers undergoing the treatment, all were said to be 'unanimous in their censure' of the charges against Banting and 'scoffed at the idea of compulsory treatments'.[129] Asked whether he would be willing to receive treatment with insulin again, Lieutenant R. S. Jackson had no reservations: 'I consider myself that it's given me a chance for life ... You bet I would'.[130] When asked the same question, his fellow patients all gave the same reply, leaving the newspaper to conclude that:

> The only person who cannot understand patients running risks in treatments is one who has never seen or does not understand the terrible ravages of the disease called diabetes which will make a man willing to do or endure anything to obtain even a measure of relief.[131]

For those patients lucky enough to be treated, the results were stunning. At his Boston clinic, Joslin could not praise Banting highly enough:

All that I imagined and more than I hoped could be accomplished Dr. F. G. Banting has done ... Nothing which I have done for patients in the last twenty-five years has accomplished in months what insulin has done for them in weeks.[132]

Nor were patients with type 1 diabetes the only ones to benefit from the discovery of insulin. In 1934, Dr. George Minot (1885–1950) of the Harvard Medical School and his colleagues George Hoyt Whipple and William P. Murphy were awarded the Nobel Prize in Physiology or Medicine for their discovery of a treatment for pernicious anaemia. Caused by lack of Vitamin B_{12}, this condition had been fatal until the demonstration by Minot in 1926 that it could be treated simply by including raw liver in the patient's diet. When Minot received the Nobel prize for this discovery, there was one person for whom he reserved very special thanks—'I would have been dead 14 years ago but for Banting'.[133]

Minot had first been diagnosed with diabetes in late 1921 and was treated by his Harvard colleague Joslin. As insulin was not yet available, Joslin had no choice but to put Minot on a starvation diet, fully aware that at best this would only delay his death. But although Minot's diagnosis was described as a 'staggering blow', its timing was fortunate.[134] As a result of Joslin's close association with the Toronto team and their discoveries, Minot became one of the first patients to receive insulin in the trials conducted by Joslin in 1923. Taking three injections of insulin a day, Minot was able to manage his condition sufficiently well to go on and win a Nobel prize that would save numerous other lives.

Describing the effects of insulin on patients like Minot, Joslin said 'it still remains a wonder that this limpid liquid injected under the skin twice a day can metamorphose a frail baby, child, adult, or old man or woman to their nearly normal counterparts'.[135] As a man with a deep religious faith, the power of insulin had a profound effect on Joslin:

By Christmas of 1922 I had witnessed so many other near resurrections that I realised I was seeing enacted before my very eyes Ezekiel's vision of the valley of dry bones—Ezekiel XXXVII, 2-10: '... and behold, there were very many in the open valley; and, lo, they were very dry. And he said unto me, Son of Man, can these bones live? And ... lo, the sinews and the flesh came up upon them, and the skin covered them above: but there was no breath in them.... Thus saith the Lord God: Come from the four winds, O breath, and breathe upon these slain, that they may live ... and the breath came into them, and they lived, and stood up upon their feet, an exceeding great army'.[136]

But at the same time as he was invoking biblical images of resurrection, Joslin was sounding a warning note about the danger of raising false hopes: no one should be under the misapprehension that insulin was a cure for diabetes.[137] In response to sensationalist headlines from *The London Times*, which hailed insulin as 'A Cure for Diabetes', Joslin offered the sobering example of one of his own patients whose symptoms of depression he maintained could be 'attributed to false hopes aroused by newspapers'.[138,139]

Joslin stressed that insulin was only effective if used in conjunction with a tightly controlled dietary regime and urged that both doctors and patients be educated in its use. He warned that 'Insulin is a benefit, but for the ignorant, who likewise may be rich or poor and young or old, it can be dangerous, and a gift of free insulin to such is like casting away pearls'.[140]

The clinicians involved in these early trials had to navigate a minefield of questions. What should be the aim of treatment with insulin? Was it to completely abolish the excretion of sugar in the urine or was some degree of glycosuria permissible? Over what time period should insulin be administered, and how long before the ingestion of food? How did dietary intake affect the action of insulin? Was the efficiency of insulin affected by the age of the patient and how long they had been diabetic?

One of the most urgent questions was how to standardize the potency of insulin and, having done so, to agree on what dose should be administered. Again, Collip's work was vital here. His work on insulin-induced hypoglycaemia in rabbits allowed the definition of a standard unit of insulin activity—a crucial development because it allowed clinicians to quantify the doses administered to patients. This not only allowed meaningful comparisons of results from different experiments and trials, but also enabled clinicians to fine tune the dosage for a given patient—and hopefully avoid the danger of an accidental overdose.

For Clowes, there was an extra headache. Under the agreement signed with the University of Toronto, Lilly had agreed to make all their insulin freely available to the handful of clinics approved by the Insulin Committee for the period of one year. But this meant that the company was bearing the cost of production with no return. Therefore, Clowes was delighted when George Walden, one of his chemists, made a crucial discovery.

Walden had found that, simply by adjusting the pH of the pancreatic extract to a value called the isoelectric point, at which the insulin molecule has no overall electric charge, the insulin precipitated out of solution. It could then be easily separated and redissolved at high purity. Using Walden's method, insulin could now be prepared on a large scale without any need for expensive distillation equipment—or more importantly, risk of explosions due to home-made wind tunnels and volatile vapours. Most importantly—for Eli Lilly and Company—Walden's innovation made the whole process a lot cheaper.

At the same time as Lilly was making these technical improvements, work was going ahead to get insulin production up and running on the other side of the Atlantic. *The London Times* hailed the decision by the University of Toronto to grant a license to the MRC for the production of insulin.[141] However, the decision also raised a host of challenging scientific, political, and social questions.

The *BMJ* lauded insulin as 'a scientific discovery of immense importance', but it also warned that the credit for this discovery would count for little if the hopes of the public were raised so high only to be dashed by insufficient and unreliable supply or, even worse, if it proved to be fatal.[142] Similarly, when the MRC first accepted the offer of a license from the University of Toronto, *The*

Lancet praised their decision to make clear their policy on the production and regulation of insulin as being 'humane and wise ... because the raising of false hopes will be prevented'.[143] But the *BMJ* also urged caution 'against a too hasty assumption that the practical therapeutic problems have been completely solved' and warned that:[144]

> No definite expectation is yet warranted that a case of diabetes can be cured by insulin, so as to be placed beyond the need for its further use. The indications point rather to the probability that regular administration, continued in most cases throughout life, will be needed to maintain improvement.[145]

This note of caution was wise, for while the new drug brought hope to patients, it raised a multitude of questions for clinicians. The first was that the efficiency of the product could not be guaranteed. Different batches of material showed a high degree of variability in their efficacy, which could have devastating consequences if a patient received too little—or too much—insulin as a result. There was therefore an urgent need to standardize the product so that reliable, accurate doses could be used in clinical treatment. Acknowledging these challenges, the MRC warned that 'to leave the production of the remedial extract to unassisted and uncontrolled commercial enterprise would be to imperil the credit and success of the treatment, arid entail for many sufferers not only disappointment but serious danger'.[146]

To address these crucial issues, Harold Ward Dudley and his fellow researchers at the National Institute for Medical Research (NIMR) laboratories in Hampstead began working to find ways of simplifying and improving the production process, as well as a means of accurately measuring and standardizing the potency of the drug.[147] The speed and urgency with which the MRC coordinated this work and the way in which the public were regularly updated with its progress was commendable. In the end, thanks to 'rapid and satisfactory progress' made to improve the purification process at the NIMR, the MRC announced early in 1923 that it was now involved in negotiations with several British pharmaceutical companies to start large-scale commercial production of insulin.[148,149,150]

Successful mass production allowed the MRC to start treating patients at five London hospitals, as well as the Sheffield University Department of Pharmacology and Physiology, the Sheffield Royal Infirmary, and the Edinburgh Royal Infirmary.[151,152,153] When this work was first reported in the *BMJ*, the doctors involved described their results as being 'immediately satisfactory' but they were also under no illusions that this should be considered a cure, and went on to stress that the daily life of a diabetic patient would still be a difficult one:

> No substitute has yet been found for the hypodermic method of giving insulin. The injections have to repeated at least once each day. They can control all the ordinary phenomena of diabetes, but the effect of each injection is only transient. As yet there is no satisfactory evidence—time has hardly allowed of that—to show that diabetes in its most serious forms can be cured by insulin. It may be that the great majority of severe cases cannot hope for more than prolongation of life.[154]

With triumphant headlines such as 'Success of New Insulin Cure', it was quite apparent that this call for caution had been lost on the newspapers.[155] In the face of such sensationalist claims clinicians were keen to emphasize that insulin should not be considered as a cure—and successful though it was, this success did not come cheap.

In a letter to the *BMJ*, Dr. Otto Leyton warned that some of the promises made for insulin 'may be mirage' thanks to its cost proving 'prohibitive if made in London unless wealthy diabetics are prepared to pay for the needy'.[156] One reason for high production costs was the difficulty in obtaining two crucial raw materials—alcohol and pancreatic tissue. The supply of alcohol in the large quantities needed was limited by prohibitive restrictions while an embargo on the import of Canadian cattle restricted the amount of pancreatic tissue available.[157] Leyton suggested moving production from the UK to Durban, where 'alcohol is ninepence a gallon, whilst the pancreas of an ox can be bought for less than a shilling'.[158,159]

In April 1923, *The BMJ* made the triumphant announcement that the first commercially prepared insulin had become available in Britain.[160] One hundred units (equivalent to ten doses) of insulin prepared by the company British Drug Houses (BDH) cost 25 shillings, which, when converted to decimal currency and adjusted for inflation would today cost roughly £76.[161] The cost of the drug, and who should pay for it, quickly became a contentious issue.

A letter in *The Times* claimed that insulin was being sold for a third of this price in the US and that the price was being kept deliberately high by the pharmaceutical companies in order to recoup their costs.[162] The letter concluded with a stark warning that 'The really poor diabetics may die whilst the price remains prohibitive'. In response, the Ministry of Health argued that insulin was more expensive in the UK because it was not yet possible to produce it in such large quantities as in the USA, where the resulting economy of scale brought down the cost.[163] This may have offered some reassurance, but the grim conclusion of the original letter raised an important point—how were poorer patients supposed to pay for this new drug? J. A. Nixon of the Bristol Royal Infirmary made a bold and radical proposal:

> One drawback to the use of insulin is the expense it entails. The drug itself is costly and the repeated blood sugar estimations add to the bill. At present only the millionaire or the insurance patient can afford insulin treatment. There is no reason why this should continue. As already said, the Ontario Provincial Board of Health issues insulin free to those who cannot afford to pay for it ... Free blood-sugar tests and free insulin for diabetics might be seriously considered by the Ministry of Health.[164]

Otto Leyton agreed and stressed that:

> Personally, I do not treat patients in hospital with insulin unless I am satisfied that they will be able to obtain a supply after leaving. It is too cruel to let an individual learn that there is some substance which will allow him to live in comparative

comfort, but because he has not money to buy it he must perish miserably. Some fund should be started to supply poor people with insulin...

[...]

[but] the State should not supply it. The administrators of the fund might stipulate that those whom they assist to pay for insulin must have no progeny: the State could not make this their stipulation, and without it the diabetic population would increase, because the tendency to diabetes mellitus is undoubtedly hereditary. In the past the multiplication has been checked automatically, because a man suffering severely from diabetes is impotent, whilst a woman, even if she conceive, rarely bears a living child.[165]

In the days before the National Health Service, the cost of medical treatment for manual workers in the UK or those in nonmanual labour who earned a salary of less than £250 a year was covered by a scheme of national health insurance known as the Panel system. But not everyone thought this was a fair system by any means and most of those who suffered under it were the middle classes. One letter to *The Times* complained that, although a patient whose treatment was covered under the National Health Insurance Act might receive insulin at no cost, their employer who had no such cover might have to pay £180 a year for their own supply or else risk death.[166] Women, children, and the unemployed were also not covered by the system and would therefore have had to pay for insulin from a private doctor.[167,168] Whether any of them adopted the following recommendation from one particular clinician remains unknown:

The personal hygiene of a diabetic patient is of the first importance. Sources of worry should be avoided, and he should lead an even, quiet life, if possible in an equable climate. Flannel or silk should be worn next to the skin, and the greatest care should be taken to promote its action ... An occasional Turkish bath is useful.[169]

Others complained that the issue of price was but 'one feature of the shameful manner in which this great relief to human suffering has been dealt with in England'.[170] That a single body—the MRC—controlled and regulated insulin production was felt to be equally shameful. The argument that unregulated production of insulin would put patients' lives in danger left them unconvinced. Alfred Clark, Professor of Pharmacology at the University of London, warned that 'the trade in secret or quack remedies is an unmixed evil, but it is important to remember how large and powerful is the quack medicine trade'.[171]

Opponents argued that, instead of protecting diabetic patients from quack home-brewed insulin, the control of its production by the MRC was choking the supply and thus actually putting patients' lives in danger. Why not allow clinicians the freedom to make their own insulin free from the constraints of patents and licensing? Those calling for unregulated production of insulin pointed to continental Europe as a paragon of bold buccaneering pharmaceutical libertarianism while Britain remained choked by the dead hand of bureaucracy. One such champion of freedom was Mr. Vivian Gabriel who, in a letter to *The Times*, called for Britain to emulate the example of Dr. Leon Blum (1878–1930) who, in his laboratory at the University of Strasbourg, was happily producing insulin free from interfering regulation:

> ... it seems little short of scandalous that a great human benefit of this kind should have been restricted by a patent or in any way held up. Did Jenner, Lister or Pasteur patent their free gifts to mankind? Did they advance the bureaucratic argument of our Medical Research Council that mistakes might have been made in the application of the remedy, or that it was necessary to standardize it and to have a sufficient quantity for everyone before allowing issue to the public. Dr. Blum made his insulin with his own hands and began with one patient. It is obvious that any medical man in England could have done the same.[172]

In response to this charge, BDH's managing director C. A. Hill gently pointed out that, while Dr. Blum's solitary efforts were noble, they were unlikely to meet the growing demand for insulin and that BDH was so successful in increasing production of insulin that they were ready to begin exporting it to a number of different countries in Europe and the Commonwealth.[173]

Hill was also keen to stress that, thanks to this success, the price of insulin was now falling—a point also made by Minister of Health Neville Chamberlain when he was challenged in the House of Commons by Viscount Ednam, the Unionist MP for Hornsey.[174] The price of insulin was a hot political issue. When Ednam asked Chamberlain whether he was aware that diabetic treatment cost no less than 10 shillings a day, Chamberlain replied that he was confident that the Panel system was sufficient. Pressing his attack, Ednam then asked if he was willing to provide insulin at a significantly reduced cost to those who were not covered by the National Health Insurance Act. He further demanded whether the time had not come to finally impose controls on its price in order to make it affordable for all. Chamberlain's opinion was that the price had fallen far enough but expressed hope that it might fall further. Ednam retorted that the current price was 'still far above the figure which an ordinary person can afford'.[175]

Some warned that making such a precious resource too easily available could be counterproductive. When commercially produced insulin was first available in April 1923, the MRC warned that it was not available to those whose condition could be controlled by a restricted diet, with the justification that there should be 'no luxury use of insulin till supplies are abundant'.[176] K. S. Hetzel of University College Hospital offered a cautionary example of what might happen if insulin was too easily available. Hetzel described a 43-year-old patient who, despite being diabetic, had:

> ... lost nearly six months' work at £3 10s a week; he had spent £25 on a holiday, which did him no good, and much money on extra food which was harming him, while the insulin supplied, almost uselessly because of this prodigal consumption

of food, had cost the National Health Insurance funds over £50. The sole fault was in the neglect to control his diet.[177]

In drawing lessons from this tale, Hetzel warned that 'all possible economy must be exercised in the use of insulin' if its benefits were to be realized and that this could only be done if insulin treatment went hand in hand with strict dietary control.

An editorial piece in *The Lancet* echoed Hetzel's concerns with the stern pronouncement that 'Doles of all kinds are objectionable. Those who have are already taxed sufficiently for those who have not ... Things which are to be had for nothing are generally used wastefully'.[178] However, the same article then concluded:

> The entire hospital system is a plain monument to the ideal that the sick have a call on the service of the sound, that the needy are entitled to the help of those who have what they require, and that those who are sick as well as needy are a charge upon everyone. So long as mankind clings to its immemorial axiom that life on any terms is better than no life at all, it will be unthinkable that a potent and valuable remedy should, as a piece of actual practice, be withheld from anyone merely because he has not got enough money to pay for it. There were inhumanity and disappointment and pain enough through the period when everyone with diabetic children and friends knew all about it and very few could get it. Now that it is available in ample amounts, there ought to be no more difficulties.[179]

With insulin still in short supply but demand nevertheless growing, the *BMJ* reported that the new drug 'shall be distributed so as to reach those patients whose need it greatest, and so as to be placed in the hands by which it can be administered to the greatest effect and with the least waste'.[180] But whose hands were best for this—those of a doctor, or those of a patient? An editorial in *The Times* responded: 'Unfortunately the administration of the substance requires a measure of technical skill, and consequently cannot safely be entrusted to other than a doctor's hands'.[181]

But there was disagreement within the medical community itself about who should be entrusted with this responsibility. Some clinicians felt that the new drug was so potentially dangerous that it should only be administered by specialists in diabetes. Others were confident that insulin could be placed safely in the hands of general practitioners. At an address to the Royal Society of Medicine, Professor Hugh Maclean reassured his audience that 'insulin may be quite safely used by the general practitioner, and I have no hesitation whatever in saying that it is indeed the duty of the general practitioner to use this remedy'.[182] But he also went on to remind them that 'Insulin, like fire, is a good servant but a bad master', and that for this reason, GPs must approach its use gradually and with strict precautions.[183]

One very good reason for such a gradual and cautious approach was that insulin raised as many questions as it answered. For example, what constituted a successful treatment—the restoration of sugar in the blood to normal physiological levels, or the absence of glucose and ketones from the urine? Should a patient being treated with insulin still be kept on a diet of strictly limited carbohydrates, or could this regime be relaxed? How many times in the course of a day should a patient be injected with insulin—and how often should this be done before a meal? What dose should be administered in a single injection—and did this vary from one patient to another? Did injecting oneself for an evening meal run the risk of blood sugar levels plunging the patient into a hypoglycaemic coma in their sleep? Each question seemed only to spawn a host of others. Faced with these huge challenges, one despairing clinician lamented that diabetes might prove to be fatal not only for patients, but also for doctors involved in their treatment, as 'it bade fair to bring many practitioners to a premature grave, so multitudinous, bewildering, and worrying were the problems involved'.[184]

Banting added his voice to the debate in an address to the British Medical Association, saying that he saw 'no reason why diabetic patients should not be treated as well by the general practitioner as by the specialist'.[185] But he went further. He added that, in Toronto, patients had even been trained to administer injections themselves. While this is common practice today, at the time it was highly controversial. An editorial in *The Times* accepted that 'it was perhaps inevitable that some people should wish to treat themselves and should ask for instruction in the methods now employed', but it then went on to warn that 'Nothing is less to be desired than the indiscriminate use of this potent drug by ignorant people'.[186,187]

A letter published in the same edition described one patient's experience of using insulin and argued the case for why they should be allowed to administer it themselves:

> I was so sceptical after my first injection—they jam in a sharp tube an inch or two under the skin—that I even felt a little virtuous. My reward was a slice of good honest bread at luncheon instead of those abominable biscuits of the taste and consistency of sawdust. And scones at tea. It gave one a homely feeling to get back to the amenities of life. So far so good. But was it worth it? The diabetic patient is generally disciplined into being something of a fatalist. Very few people one meets know much about diabetes. I should explain that the disease doesn't hurt; its general effect is a series of privations. One has to give up eating most of the things one likes; then games have to be abandoned, progressively, first tennis, then golf; one's walks are curtailed; one doesn't feel in the least sociable ... The disease amounts to a painless wilting of the body and mind; a premature old age.[188]

The patient went to praise insulin as being 'a force of magical activity', thanks to which they were now enjoying a new lease of life. However, the patient was also aware that this new life was far from easy. Alongside the discomfort of multiple daily injections of insulin described as a 'ding dong battle' to control blood sugar levels, were the challenges of preparing and maintaining syringes with which to deliver it. The older type of syringe made of metal with a leather plunger had by now been replaced by glass syringes that had to be sterilized by boiling before each use, and often broke. Other problems included the plunger becoming jammed by residues that had accumulated from the methylated spirits in which the syringes were stored to keep them sterile. On top of this, the needles of the syringes became blunt over time and there were calls by physicians that every patient should be taught how to sharpen them.[189]

Patients also needed to be taught how to carry out measurements of their blood sugar levels in order to check that their insulin was working effectively. While these estimations of blood sugar concentration were essential, they also added significantly to the cost of treatment and were prohibitive to those who could not afford it. The letter concluded with a powerful case for giving patients the responsibility to manage their own condition:

> This cannot go on indefinitely; too many thousand lives are at stake. The release of insulin to the patient, for giving his own injection, will have to come … For the man in an advanced stage of diabetes the penalty of being without insulin is death. Schoolboys are, or used to be, punished in the bulk for a single undiscovered offender; but one cannot apply the rule to adults and condemn a thousand sufferers because one or two of them through the abuse of insulin might die; while, without insulin, die all must.[190]

In the United States, Joslin had faith not only in 'the vision of Ezekiel' but also in the competence of his patients to use it. His confidence that 'intelligent patients can be taught the use of diet and insulin in a week' was born out by the case of Jack Eastwood, a patient who had first developed type 1 diabetes in 1925 when he was thirteen years old.[191] After being diagnosed, Eastwood was kept on a strict diet with only limited intake of carbohydrates and two daily injections of a fixed dose of insulin given in the morning and evening for the duration of the following year.[192] The following year, having won a scholarship to St. Paul's School, Eastwood found himself making daily trips by train to London where the opportunity to eat lunch in a restaurant was an irresistible temptation. The problem was that his rigid regime of insulin injections made this difficult, so being a resourceful young man, Eastwood resolved to do something about it:

> I felt the need to learn more for myself both about diabetes in general and about my own case in particular. I therefore began regularly doing my own insulin injections and urine tests and reading as much as I could on the subject, and it was not long before I was able with considerable accuracy to guess the weight of a normal helping of most of the commonest foodstuffs and to work out the carbohydrate, protein and fat content of any meal that I ate, realising that it was essential for me to do this if I was ever going to have a meal in a restaurant or, indeed, anywhere other than at home.[193]

While most other teenage boys of that time might be filling notebooks with cricket or football scores, Eastwood began keeping daily charts on which he recorded the details of each injection, the food eaten at every meal, the results of urine tests, and other notes and comments. His discipline and attention to detail were to pay enormous dividends. Writing in 1986, at the age of seventy-four, Eastwood calculated that during the course of his life he had performed over 50,000 injections of insulin. But he described these as little more than 'a minor inconvenience' and even recalled how he had covertly injected himself on a London bus without attracting the attention of any fellow passengers.[194]

His approach had allowed him to eat 'almost anything I wanted' and the freedom to play 'an active part in the life of all the schools in which I worked and all the communities in which I lived'. But most importantly of all, he had not suffered from any of the usual complications that plague the life of diabetic patients.

With his piles of notebooks recording dietary intake and insulin injections, Jack Eastwood was blazing a trail. Today, programmes like DAFNE (Dose-Adjusted-For-Normal-Eating) play a central role in the modern approach to managing diabetes by educating patients about insulin dosage and the carbohydrate content of meals. But the importance of educating patients had been recognized at a very early stage. From the very first use of insulin in his Boston clinic, Joslin had been a passionate advocate of educating patients in how to manage their condition:

> If insulin is to be of permanent help in diabetes, it must be usable by diabetics in their own dwellings ... In the second month of treatment Thomas D. died in a coma, apparently as a result of having omitted insulin in his home. He felt so much better, had so much confidence in the good that he had received from insulin that he probably believed the omission of it for a few days would be of little consequence. We have not forgotten the lesson which his case teaches. If insulin is to be used in the home, the patients must not only be educated in dietetic treatment but they must be educated in treatment with insulin. This has led to happy results and to results not at first appreciated. The increased attention given to education and to insulin has called forth greater cooperation on the part of the patients, and it is quite evident that the increased cost of treatment due to insulin will be largely offset by decreased cost of medical attendance and the shortening of hospital stay.[195]

The campaign for the education of patients also had a vocal champion on the other side of the Atlantic. Dr. R. D. Lawrence (1892–1968) had very good reasons to be so passionate about this issue. As a young doctor at King's College Hospital, London, he was chiselling at a mastoid during an operation when a chip of bone struck his cornea, resulting in sepsis.[196] This was bad enough but there was worse to come for during his resulting treatment, he was found to be suffering from type 1 diabetes. At this point, Macleod had not yet announced the discovery of insulin and the only treatment for diabetes was to delay inevitable death by using a severely restricted diet. Resigned to the inevitable outcome of his condition, Lawrence took himself off into exile. As a lover of Italian culture, he headed to Florence to see out what time remained to him.

It was while trying to find solace in the cradle of the Italian Renaissance that Lawrence experienced his own kind of rebirth: it was in Florence that he received a telegram telling him about the discovery of insulin. In response, Lawrence drove across Europe and, on his arrival back in Britain, became one of the first patients to be treated with insulin—by which time his diabetic neuropathy had become so pronounced that he could no longer 'get the matches out of the box' to light a celebratory cigarette.[197]

Having been granted a new lease of life, Lawrence resolved to devote the rest of his career not only to the treatment of diabetes, but also to the education of fellow patients. He was adamant that continuous injections of insulin, although necessary, were not sufficient for the successful management of diabetes. Patients themselves needed to be educated about their condition and how to control it— particularly with respect to monitoring their intake of carbohydrates and adjusting their insulin dose in response. Lawrence became an evangelist for this message and in 1929 published *The Diabetic A.B.C.*, which used a colour-coded scheme to give simple and clear instructions about the dietary content of meals.

Lawrence also had another grand vision of how to improve the lives of diabetics, and he shared it with an ally from outside the medical community. Author H. G. Wells is more usually associated with visions of time machines or marauding Martian tripods, but having developed diabetes in 1933, the condition had become a subject close to his heart.[198] The impact of this blow prompted Wells to write to *The Times* on behalf of the many thousands of patients who, like himself:

> ... would either be dying slowly or uncomfortably or be already dead if it were not for the work of a small group of experimentalists and practitioners who have brought this particular maladjustment under control, and none of us can feel anything but the liveliest gratitude for that work.[199]

Wells went on to describe the poor state of the diabetic department at King's College. With growing numbers of diabetic patients arriving for treatment, conditions there were becoming cramped. Wells described the equipment and benches in the congested laboratory as looking 'already heroically worn'. He appealed to wealthy diabetic patients to donate to the department to raise £800 for its renovation.

Wells was also surprised that there was no charitable body to promote the interests of diabetic patients and offer them support. Together with Lawrence, he sought to remedy this situation. At a festival dinner held in November to raise funds for King's College Hospital, Wells dined with other prospective members of the new organization on a specially prepared menu containing no sugar.[200] By spring of the following year, a new diabetes clinic was opened at King's College Hospital thanks to Well's earlier appeal.

Furthermore, only a few months before the clinic opened, Wells had written again to *The Times* announcing his intention to 'form a Diabetic Association, open ultimately to all diabetics, rich or poor, for mutual aid and assistance, and to promote the study, the diffusion of knowledge and the proper treatment of diabetes in this country'.[201,202]

Wells had high hopes that the Diabetic Association might do more than just help patients with diabetes. He hoped that it might become an engine of social progress by providing 'a pattern for other organizations for bringing together and utilizing the common interest of people with other diatheses'.[203] He threw himself wholeheartedly into the venture and, in the words of one of his biographers, was not 'a man simply to bestow his name on a letterhead, [but] he drummed up support, chaired meetings and wrote letters to the press'.[204] In his role as its first president, he campaigned on the pages of *The Times* for financial support to enable the new association to provide holidays for diabetic children.[205]

In his original appeal for funding to support the diabetic clinic at King's College Hospital, Wells had said that donations were an opportunity by which patients could express their gratitude 'in a direct and effective fashion' to the researchers in Toronto who had done so much for them.[206] By the time that Wells made this appeal, Banting was no stranger to such expressions of gratitude. For the past decade he had enjoyed a flood of tributes and accolades.

In March 1923, Allen sang the praises of Banting to the rafters:

> Insulin is performing miracles in diabetes ... Banting holds indisputable priority in the discovery of insulin which will rank among the leading achievements of modern medicine. I think the failure to recognise his contribution in some fitting way would be held up as a reproach to Canada.[207]

Allen's concern that Banting's achievement might be overlooked by his home-land proved to be misfounded. In the spring of 1923, the Ontario Government passed an act of legislature allowing the University of Toronto to establish a permanent Professorial Chair in Medicine for Banting with an accompanying stipend of $10,000 per annum to cover his salary and expenses.[208] Nearly two years after his ill-tempered exchange in which he had begged Macleod for more resources and money, Banting had finally come good on his promise to show 'that son of a bitch that he is not the University of Toronto'.

In the space of three years, Banting had gone from being a struggling provincial doctor to being the star of the medical establishment. And although the award of a Professorial Chair was welcome, some of his admirers felt that he deserved far more. Writing in support of Banting, Dr. G. W. Ross took his case to Mackenzie King, the Canadian Prime Minister:

> If I might venture to express an opinion, and I have every reason to believe that it is one generally held here, I feel that the Ontario Government has done a good deal for Dr. Banting, but that the magnitude of his discovery warrants a larger measure of consideration from the Canadian people than this represents.[209]

Shortly afterwards, Banting received a letter from Prime Minister King informing him that the Canadian Parliament had passed a resolution to award him an annuity of $7500 a year:

> You will be pleased, I am sure, at the general expression of approval not only in Parliament but by the Press, of our own and other Countries, of the action of the Government in seeking in this manner to give to you some expression of the nation's gratitude for the distinguished services you have already rendered to

medical science and humanity by the discovery of the insulin of diabetes and the public spirited way in which you have placed the application of your discovery at the disposal of the public.

I cannot sincerely reaffirm the hope which I ventured to express at the time of introducing the resolution in the House of Commons which the provision which our Country has thus made in recognition of your services may prove such as will enable you through further research to add to the benefits which your discoveries in the field of medicine have already brought to your fellow-men.[210]

Banting was suddenly a national hero. In June, he was invited to open the Canadian National Exhibition (CNE), an honour reserved for 'a distinguished Canadian or British citizen.[211,212] The organizers of the CNE felt that Banting easily fulfilled these criteria:

You, perhaps more than any other individual, certainly of our own country, fulfil the desired requirements, for the nature of your discovery is both Scientific and International in its appeal, and indeed is of such an outstanding character as to merit the approbation of your scientific conferees throughout the world and the eternal gratitude of humanity at large.[213]

In a speech given at a lunch of the CNE Directors, Banting took the opportunity to address an issue that had been bothering him. Some early newspaper reports had described Best as being merely 'the assistant to Dr. Banting in this work' and Banting was determined to now put this right. Giving full credit to Best, he stressed that, 'Work on insulin is not the product of any one man. I am glad I have been able to take part in it, but the beneficent results that have flowed from the use of insulin could not have from the work of any one man'.[214,215]

But despite this public display of magnanimity, the gnawing fears about Macleod refused to go away. Having achieved not just national, but also international, renown, Banting travelled to England that summer to attend a number of scientific meetings, one of which was a gathering of the International Physiological Society where Macleod also happened to be present. After listening to a lecture given by Macleod, Banting was left seething: 'There could be no doubt in the minds of the listeners that Macleod was the discoverer of the physiological principles of insulin'.[216,217]

The next destination on Banting's UK tour was a meeting of the British Medical Association in Portsmouth, where any fears of being eclipsed by Macleod ought really to have been dispelled. Addressing the delegates, BMA President Sir Thomas Horder reminded them that insulin should be considered as a 'walking stick to a lame pancreas' rather than a cure for diabetes but lauded the 'great value of Dr. Banting's discovery'. Another delegate prophesied that, in years to come, this particular meeting would go down in the annals of medical history as those who had been present would forever recall with pride that 'I heard Dr. Banting's address on insulin'.[218]

One person who would have been less than impressed with Banting's lecture on insulin was Collip. For when the lecture appeared in print in the *BMJ*, the section describing purification of the extract using alcohol had misspelled Collip as 'Cobb'. It was an oversight that foreshadowed worse to come.

A mere glance at the newspapers should have been enough to reassure Banting that his place in medical history as the discoverer of insulin was now secure. In fact, in the pages of the press, a whole mythology was starting to grow around the story of how Banting had come to make this discovery. A 1923 article in *The Times* described Banting as being so dedicated to his research into diabetes that:

> ... he determined to sacrifice everything to it. Practice, house, furniture, every-thing was disposed of and he went to the laboratory in Toronto where his discovery was made. He was then in his early twenties. There were vast obsta-cles in the way and many difficult techniques had to be mastered. Yet the young doctor surmounted all troubles and carried his idea to a successful conclusion.[219]

Banting had risen to the starry heights of celebrity status. *The London Times* reported every honour and accolade he received, even keeping its readers informed of such details as his travel arrangements to sail from Liverpool or that 'the discoverer of insulin' had got married.[220,221] Had *Hello* magazine existed then, Banting might well have graced its pages. As a result of this new-found celebrity status, after leaving the meeting in Portsmouth, Banting returned to London where he received a sudden and unexpected invitation for which he was ill-prepared.

Banting had been summoned to Buckingham Palace for an audience with King George V, but his immediate concern was to find something suitable to wear. Dashing to a shop at closing time on the day before he was due to meet the King, he managed to buy a hat and gloves, and (to his relief) a suit he had ordered arrived just after 9 o'clock on the following morning, leaving him with only an hour to spare before his royal engagement.

This last-minute frenzy could have done little for his nerves, but on his arrival at the Palace he took some consolation from the fact that the shoes worn by the young man who met him and introduced himself as second secretary to the King, 'were not as clean as my own'.[222]

Banting was led upstairs and taken down a wide corridor through a set of heavy double doors where he was presented to the King. The two men were then left alone, which, as Banting later recalled, did not set him at ease:

> ... I did not like to be alone with him for if anything happened to Him I would be blamed. I had exactly the same sensation as if I had in my hand a fragile vessel of inestimable value. His Majesty was most gracious and in a moment I was com-pletely at ease and we were talking about doctors, hospitals & research work. I was amazed at the amount [of] knowledge the man had. And I was amazed at the fact that he knew the income of some of the most eminent surgeons and the cost of

maintaining some of the hospitals. He gave one the feeling that he was genuinely interested.[223]

Later, as Banting was leaving the Palace, he was approached by a journalist—a profession for which he had little respect:

> If there is one thing in modern civilisation that disturbs a research man', said Banting, 'it is the newspapers. If there is anything that I fear, if there is anything that I loathe, if there is anything that I despise as unfair, untrustworthy, as undependable and as unscrupulous, it is the modern newspapers.[224]

Banting's actions spoke even more about his contempt for the Press than his words. Failing to recognize Banting—perhaps mistaking him for a member of the Palace staff—the journalist asked, 'Is Dr. Banting coming soon?' In reply, Banting looked back over his shoulder and said simply, 'Pretty soon'. Then, walking past a line of cameras and journalists armed with notebooks, Banting stepped quickly into a taxi. Only then did the throng realize that they had missed their scoop. Savouring his moment of victory, Banting turned and waved his hand at them through the rear window of the taxi as it pulled away.

After an audience with King George V, Banting apparently no longer had any insecurities about Macleod. A few months earlier, he had even written to Joslin requesting that Macleod's name be included on some key publications:

> I wish you could mention all the names and especially that of Prof. Macleod. I feel so different about so many things now. I was selfish and I did not get the proper perspective. I have recently learned that Prof. Macleod is being criticised and I am very sorry and would do anything to save him from it because he is such a wonderful man and has done so much for insulin and diabetics.[225]

But one morning in October 1923, all such feelings for Macleod went right out of the window:

> As I went in the door I picked up my newspaper, tucked it under my arm and hearing the telephone in my office I hurried to answer. The excited voice of a friend was saying something about 'congratulations' 'been trying to get you' 'where have you been?' 'Have you seen the newspapers?' I said 'Please calm yourself and tell me what you are talking about'. 'You damned fool, didn't you know you and Macleod got the Nobel Prize?" I said "Go to hell" and hung up the receiver. I opened the paper and there it was. Macleod! Macleod! Macleod![226]

The announcement in that morning's newspaper, 'Nobel Prize to Banting' should have described the greatest moment in Banting's scientific career (Figure 13). But the next sentence ruined any jubilation he might have felt: 'Dr. J. J. R. Macleod also'.[227]

Banting wasn't the only one who was fuming and raging at the decision of the Nobel committee—there were others who considered the award to be a grave

Figure 13 *'Toronto Doctors Honored: Nobel Prize to Banting. Dr. J.J.R. Macleod also.' The headline from October 1923 that sent Banting's blood pressure through the roof.*

Credit: Insulin Collection, F.G. Banting (Frederick Grant, Sir) Papers, Thomas Fisher Rare Book Library, University of Toronto, MS. COLL 76 (Banting) Scrapbook 1, Box 2, Page 81; Reproduced with Courtesy of the Thomas Fisher Rare Book Library, University of Toronto. Online at:https://insulin.library.utoronto.ca/ islandora/object/insulin%3AC10111.

mistake. However, their grievance wasn't just that the award had been given to Macleod, but that it had also been given to Banting. They were determined to set the record straight.

4

A Greek Tragedy

Raising a glass at a banquet held in honour of Banting and Macleod, University of Toronto President Sir Robert Falconer foresaw a golden future. Thanks to the ongoing work of Banting and Macleod, Falconer anticipated a scenario in which the university was destined for greatness:

> Unselfish idealism will crown a school of medicine as a home of glorious accomplishment. It may have to wait for its reward, but in due time it will come in the gratitude of the poor to whom its glad tidings of recovery are preached; of the captives released from the bondage of the tyrant disease; of the blind on recovery of their sight; of those bruised in the dungeons of despair who have been set at liberty. Such idealism has been magnificently displayed in the discovery and progress of Insulin. May that idealism continue in co-operative effort to keep the University of Toronto as the chief hearth for its subsequent development .... I propose we drink a toast.[1]

The banquet seating plan told a different story—Banting and Macleod were seated thirteen seats apart.[2] While Falconer sang their praises to the rafters, not everyone joined the adulation. Thomas Wingate Todd, Macleod's former colleague at the School of Medicine at Western Reserve University, Cleveland, was appalled at Banting's Nobel Prize. Writing to Macleod, he expressed his disgust at the decision of the Nobel Committee:

> On Friday morning there appeared a note in the Plain Dealer stating that the Nobel Prize had been awarded to Banting; this threw us into acute depression: there was no mention of your name in the wretched announcement ... To give Banting the prize at all shows a deplorable lack of scientific judgement on the part of the Committee but worse than that it shows also a lack of scientific common sense. It encourages the all-too-common lay belief in inspiration and indeed sets the stamp of approval at what the layman considers the high court of appeal. Apart altogether form the indignity which it offers yourself it is a very severe and calamitous setback for medical science.[3]

In stark contrast to Falconer's rosy vision for the University of Toronto School of Medicine, the Nobel Prize was dividing the medical establishment into two bitterly opposed camps. While members of one faction, e.g. Thomas Wingate Todd, supported Macleod, others, e.g. clinician Henry Rawle Geyelin, were vocal in their support for Banting:

> I want to offer you my most affectionate congratulations on receiving the Nobel prize for medicine. I am disgusted, however, with your corecipient. First, I do not understand why he should receive it and in the second place I do not understand how he had the nerve to accept it. I read an editorial in the Philadelphia Public Ledger of November 3rd, which spoke of you and Dr. McLeod [sic] as the 'joint discoverers' of insulin. This made my blood boil.[4]

The Nobel Prize seemed to be a poisoned chalice—one that had stirred passions far beyond the North American medical community. In Berlin, German clinician Georg Zuelzer (Figure 14) was dismayed by the news that Banting and Macleod had received the award. Nor would he have been pleased to hear Banting's address at the opening of the Canadian National Exhibition, in which he assured his audience that the discovery of insulin had been a team effort. For Zuelzer, the discovery of insulin was thanks only to one person:

Figure 14 *Georg Zuelzer (1870–1949).*

Credit: Photographer: Suse Byk-Published by:'Zeitbilder' 15/1930. Vintage property of ullstein bild © Getty Images. Reproduced under license from Getty Images.

> Be it known that I, Georg Zuelzer, doctor of medicine, citizen of the German Empire, residing at Berlin, Kingdom of Prussia, German Empire, have invented a new and useful Improvement in Pancreas Preparations Suitable for the Treatment of Diabetes, of which the following is a specification.[5]

Zuelzer made this bold statement in the opening paragraph of a patent that he had filed with the US Patent Office in 1908. In it, he described the preparation of an extract of pancreas that was 'particularly suitable for the treatment of diabetes and is administered advantageously by injecting the same into the venous system'.[6] Significantly, a key part of the process that he described was the use of alcohol to remove protein impurities from this extract. When the patent was finally published in 1912, Zuelzer felt confident that it would establish him as the discoverer of the anti-diabetic hormone. Therefore, his anger when he learned of the Nobel Prize is hardly surprising.[7]

Born in Berlin in 1870, Zuelzer was the son of a Professor of Medicine and chose to follow in his father's footsteps. He first became interested in diabetes whilst working with Ferdinand Blum (1865–1959), a distinguished endocrinologist who had discovered that an injection of adrenaline resulted in the excretion of high levels of urinary sugar.[8] This discovery of what was called 'adrenal diabetes' led Zuelzer to speculate that the origin of diabetes might be due to an imbalance between adrenaline and the anti-diabetic hormone. His hypothesis was incorrect, but it nevertheless steered him in the direction of what he later called 'the undiscovered country of insulin'.[9] Following the 1889 discovery by Minkowski and Mering that removal of a dog's pancreas resulted in the animal becoming diabetic, Zuelzer dedicated his efforts to isolating the hypothetical anti-diabetic hormone. However, the existence of this substance remained a controversial subject. Distinguished German physiologist Eduard Friedrich Wilhelm Pflüger (1858–1931) believed diabetes to be a disorder of the nervous system, and was so sceptical about the existence of this hypothetical chemical messenger that he issued a challenge to those who thought otherwise. Throwing down the gauntlet, Pflüger, who has been described as 'notoriously argumentative', demanded evidence that injections of pancreatic extract into a diabetic dog could consistently alleviate its symptoms.[10]

Keen to make a name for himself by proving Pflüger wrong, Zuelzer gladly took up his challenge.[11] The stakes were high, for although Pflüger had instigated Zuelzer's research, he also had a reputation for pouring scathing criticism on those he deemed to be opponents: when Pflüger spoke, people listened. It has since been suggested that this might explain why Zuelzer's results were not more widely discussed amongst the medical establishment of Berlin, but in time poor Zuelzer would have much more to worry about than whether or not he had Pflüger's blessing.[12]

Zuelzer began his first experiments in 1903 and showed that injections of pancreatic extract into rabbits could reduce the excretion of glucose into the urine caused by the administration of adrenaline. But he did not—and indeed could not—measure the concentration of sugar in the animals' blood as easily as

Banting and Best were later able to do. At the time that Zuelzer was carrying out his research, the methods for determining the concentration of blood sugar required at least 25 ml of blood—a volume so large that it made such measurements impractical. Things had changed by the time that Banting and Best began their experiments only thirteen years later. By then, new methods existed that allowed the measurement of sugar levels in volumes of blood as small as 0.2 ml.[13] What now seems a mere technical detail was to have a major negative effect on Zuelzer's research.

Despite this technical limitation, Zuelzer had grasped a crucial insight that eluded Banting and Best until Leonard Thompson had erupted in abscesses. Zuelzer recognized that pancreatic extracts were contaminated with protein impurities that might cause toxic effects.[14] If he was to achieve his goal of using these extracts as a therapeutic agent, then these impurities must somehow be first removed.[15]

Zuelzer's research was difficult. To secure a reliable and abundant supply of raw material from which to make his extracts, he had first to scour the local abattoirs of Berlin before sunrise, much to the bemusement of abattoir staff, who regarded the pancreatic tissue Zuelzer sought as offal.[16] With his powers of persuasion and the occasional goodwill of the abattoir workers, Zuelzer obtained a regular supply of material, but its quality was often questionable. As the animals in the abattoir had often not been fed for a long time, their levels of insulin were low. Even when Zuelzer did manage to produce sufficient extract, its activity was weak.[17]

And in addition to procuring material, there was the pressing need to secure financial support. Zuelzer tried to persuade a number of pharmaceutical companies to invest their resources and capital in his work. Eventually, he was successful in securing investment from Schering, formerly Chemische Fabrik Auf Actien. Sadly, very few details about Zuelzer's methods have survived. But from the details given in his patent, it is clear that he had recognized the importance of using alcohol to remove impurities from the pancreatic extract—well before this idea had occurred to the Toronto team.[18] When he injected extracts prepared by this method into diabetic dogs, Zuelzer found that the concentration of sugar and ketones in the animals' urine were significantly reduced. It was a promising result, but as these experiments had involved only two test animals, Zuelzer was clear that his hypothesis was still far from proved. Then came an unmissable opportunity.

Thanks to his contacts at various Berlin clinics, Zuelzer was able to test his extract on six diabetic patients ranging in age from 6 to 65 years old. The first, a 50-year-old man, had been admitted to hospital for the amputation of a gangrenous toe due to diabetes, and whose condition had now become so severe that he was comatose. Today such treatment would require rigorous scrutiny by an ethics committee, but at the time, and given the severity of the patient's condition, perhaps Zuelzer and his colleagues judged that he had little to lose.[19]

In several of these patients, the injection of pancreatic extract resulted in a drop in the concentration of sugar and ketones in the urine and Zuelzer even claimed that it could bring a patient out of diabetic coma which led him to call the substance, 'Acomatol'. This was a promising sign, but despite Zuelzer's use of alcohol to remove impurities, the extracts still produced side-effects of fever, shivering, and vomiting. Yet, Zuelzer was so confident of the potential of his extracts to be used as an anti-diabetic therapy that he went ahead and published his work.

In June 1908, Zuelzer presented this work at a scientific meeting in Berlin, and only two years later, it appeared in an early textbook on endocrinology published by Professor Arthur Biedl, one of the founders of the discipline.[20] Zuelzer's work seemed set to gain further recognition when some of his extracts were tested in the Department of Internal Medicine in Breslau, where Oskar Minkowski had just been appointed a Professorial Chair. After testing Zuelzer's extracts on three dogs and three diabetic patients, Minkowski's collaborator, Joseph Forschbach agreed that Zuelzer's work seemed promising:

> My experiments in dogs and humans confirm the important fact that Zuelzer has, for the first time, successfully produced from the pancreas an extract that, by intravenous application, over short or long periods of time resulted in a reduction of the excretion of sugar in cases where the diet has remained unchanged.[21]

Zuelzer must have been delighted that his work was being repeated—and apparently supported—by such a luminary in the field as Minkowski. But unfortunately, Forschbach's paper came to the sober conclusion that:

> For any practical application, the aforementioned side-effects naturally pose an insurmountable barrier. Out of a fear that an even worse outcome might result following injection of the pancreatic extract, I refrained from carrying out any further experiments with patients.[22]

There were further obstacles. Schering had decided that the costs of producing Zuelzer's extract were simply too high, and they withdrew their support for any further work.[23] Zuelzer regarded this decision as being short-sighted and, adamant that his work still showed promise, he continued alone. Reluctant to resort once more to haggling with abattoir staff for discarded tissue, he instead now followed the example of Rennie and Fraser in Aberdeen and placed his hopes in fish as an alternative source of pancreatic material.[24]

But while Rennie and Fraser had dismissed their results as being inconclusive, they had at least developed a preparation from fish that could be administered to patients. This was more than could be said for Zuelzer, who got no further than a rejected grant application for 500 marks from an endowment bequeathed by the Countess Bose to the Medical Faculty of the University of Berlin.[25]

With dashed hopes, Zuelzer returned to pestering (and sometimes bribing) the staff of local abattoirs in order to procure a reliable supply of material with which to continue his experiments.[26] In 1911, his stoicism and determination finally paid off. The director of the company Hoffman-La Roche recognized that Zuelzer's work might be of great clinical importance and offered not only to build him a small laboratory at Berlin's Hasenheide hospital, but also to provide Zuelzer with an extra pair of hands for assistance.[27]

Dr. Camille Reuter (1886–1974) was a chemist working at Roche's research laboratory in Grenznach, Germany and was assigned to work with Zuelzer to provide invaluable expertise at the lab bench. This proved to be a shrewd decision on the part of the director of Hoffman-La Roche who, by February 1914, was so impressed by Zuelzer and Reuter's method of making pancreatic extracts that he ordered the company's main laboratory in Grenznach to begin large-scale production of Zuelzer's 'Acomatol', or as it later became known 'the German Insulin'.[28]

By August 1914, Reuter had managed to process 114 kg of pancreatic tissue, much of which Zuelzer had obtained from horses that had been transported from East Prussia to the abattoirs of Berlin.[29] But to his frustration, extracts prepared on such a large scale were still not free from side effects.[30] Curiously however, these side effects were strikingly different from those of Zuelzer's first experiments in 1908. For instead of causing the fever, shivering and vomiting that Zuelzer had observed in his earlier work, they caused severe convulsions in test animals.[31] Zuelzer had never observed this effect with his previous preparations, and little could he have realized at the time just how significant it was.

Zuelzer and Reuter were baffled by these new side effects. Zuelzer suspected that, perhaps in response to suggestions from his supervisors at the production facility in Grenznach, Reuter had somehow altered their established method for preparation of the extracts. Reuter, meanwhile, speculated that the side effects might be due to contamination by residual copper present in some of the vessels used during the preparation.[32]

Despite these problems, Roche filed a patent on Zuelzer and Reuter's method and when Reuter improved the purification process and found that the extract caused a significant drop in the blood sugar levels of a dog, things seemed to be looking up.[33] However, when he presented this work to the board of Roche, they felt that the effect of the extract was too short-lived and that patients could surely never be persuaded to inject themselves several times a day. Unconvinced, they advised Reuter to pursue work on an oral anti-diabetic agent.[34] It was a decision they would regret.

Then disaster struck – and not just for Zuelzer and Reuter, but for the entire world. For just as the board of Roche were pouring cold water on Zuelzer's and Reuter's work, Europe entered the First World War. Research came to an abrupt halt as hospitals were given over to military use, and Zuelzer served as a doctor at Kobryn on the Eastern Front, where he found himself preoccupied not with diabetes but with the typhus that was rampant among the soldiers. When hostilities finally ceased, Zuelzer tried again to breathe some life into Acomatol. But in the economic chaos that engulfed Germany in the aftermath of the First World War, no pharmaceutical company was willing to take such a gamble.[35]

The lack of support was disheartening, but the most crushing disappointment came when Zuelzer learned that the Nobel Prize had been awarded to Banting and Macleod. To add insult to injury, Minkowski published a review of the work on insulin, in which he praised the Canadian team, saying:

> The credit of obtaining from the pancreas the first practically applicable product on a large scale, which when applied correctly to diabetic patients is harmless and free from side-effects must without doubt go to the Canadian researchers in Macleod's school of Banting, Best and Collip. If Zuelzer maintains that he was the first to apply alcohol in the extraction method, as these authors did, in order to separate the agent that was active in carbohydrate metabolism from destructive digestive enzymes in the pancreatic tissue then the objection can be raised that the use of alcohol, which can now even be abandoned, was not the most significant result of the Canadians, but that the use of acid in the extraction was of far greater significance and that most decisive of all was that the discoverers of Insulin did not allow themselves to be discouraged by toxic side-effects but instead correctly recognised that these were merely the result of too high a dosage of the preparation … It is without doubt thanks to them that today we have in our hands this precious material for the treatment of Diabetes and to be hoped that from their discovery further benefits for its practical application and our theoretical understanding of it will be found. It is only fitting therefore that we give them the credit for their part in this honour without bearing grudges.[36]

But for Zuelzer, this was no mere grudge, as he made quite clear in a riposte to Minkowski published in the same journal:

> Herr Minkowski has finally touched on the historical question and therefore compelled me to express my view. It is without doubt that the Americans were the first to have mass-produced insulin and that they were also undisputedly the first to establish a model of a research clinic. It is however a misunderstanding if Herr Minkowski supposes that my pancreas-hormone of 1914 still had toxic side-effects and was of no practical use. In the protocol of the established methods that are still used today without alteration to produce my Acomatol, Trypsin is removed using acidified salt-solution.
>
> […]
>
> But more importantly—from a historical point of view—is the question of whether the American researchers knew of my work–did they know that in 1908 I had already, through my Acomatol, made two severe cases of coma acid-free in 24 hours? That my Acomatol was already mentioned in the work of Noordschen in 1917 on the sugar sickness, suggests that the Americans must have known about it. It is little wonder then that, considering they had a staff of 24 workers and almost limitless resources, they were able by using my work as a foundation, to reach their end-goal faster than I, whose 12-year work with one chemist was interrupted by the war and its aftermath, could. I believe that I have the right to be described as the founder of the therapeutic application of the pancreas hormone, because it was my theory of an antagonistic relationship between Adrenalin and the Pancreatic hormone which I first proposed nearly 20 years ago and is since to be found in all textbooks, that first steered me in a direct course to the undiscovered territory of 'Insulin'.[37]

As if this was not galling enough, there came one final agonising twist of fate. On hearing the news of the Nobel Prize, Zuelzer wrote an article in which he summarized his own work, hoping that it would clarify who deserved the credit for the discovery:

> Without intending to deny the painstaking efforts of Dr. Branting [sic] or to lessen his merits for the medical science, I venture to say that the <u>actual discovery of the preparation has been made by me a decade ago under the name of 'Acomatol'</u> ... It is a great pity indeed that Germany on account of her complete impoverishment is not able any more to use this epoch-making preparation for the benefit of sufferers. I trust however that the above stated history of my studies and the preparation of Acomatol granted me the priority right of the discovery.[38]

Despite being confident in his claim to have priority for this discovery, however, Zuelzer went on to admit that his own method of purification was still far from perfect. Or so, at least, it had seemed to him at the time:

> Testing the preparation in my laboratory I found that the animals inoculated with it were attacked by convulsive fits. Having experienced this phenomenon never before, I reflected that with the extraction process of the curative some spasmodic poison must have been extracted.[39]

By now, the truth had finally dawned on Zuelzer about just how close he had been to discovering insulin. On hearing of Collip's discovery that insulin could plunge rabbits into hypoglycaemic shock, giving rise to symptoms of convulsive fits, Zuelzer realized to his dismay that this was exactly what he and Reuter had observed with their own preparations:

> The results of American research has since made clear to us that the cramps were only the sign of a highly effective preparation, that caused a pronounced drop in blood sugar, and in fact when prepared today using the former method is extremely effective in patients and free from any harmful side effects.[40]

And as Zuelzer explained to Macleod, far from being side effects due to toxic impurities, these symptoms had been the result of a highly pure and potent preparation of insulin.

> Today American scientists have proved that with a sudden sharp reduction of the saccharine content by means of Insulin convulsive fits are experienced which can however quickly be neutralised by the application of sugar. I had therefore at that time received a preparation of special effectiveness which I had left untouched on account of the alarming phenomena noticeable during the testing. A reproduction of the drug was unfortunately out of the question on account of the war.[41]

The penny had finally dropped. Far from being riddled with toxins, the extracts made by Zuelzer and Reuter had been so pure that they had actually been plunging test animals into insulin shock.

Zuelzer paid a heavy price for what he described as this 'outspoken inventor's misfortune'.[42] Had he only been able to perform frequent measurements of blood sugar in his test animals, he might have appreciated what was going on. As the scale of his error began to sink in, he realized that he desperately needed to make his story more widely known—and quickly. In November 1923, he wrote to *Medizinische Klinik*, another German medical journal, arguing that he had first discovered insulin in 1907, with the demonstration that pancreatic extracts could counteract the action of adrenaline on the excretion of sugar. As far as he was concerned, the Toronto researchers had merely delivered technical improvements to a discovery and a method of purification for which he rightfully deserved the credit.[43]

In his article, Zuelzer stressed that he had been working under difficult conditions. The supply of material and financial support had all been unreliable, and then, of course, there had been the chaos of the war. He also emphasized the fact that he now realized that the convulsions observed in his test animals had been due, not to side effects arising from impurities, but rather to insulin of increased potency and purity.[44] In a letter to the Nobel Committee, his pleas were met with a resounding silence.[45]

Reuter also tried to argue their case. After failing to persuade the board of Roche to follow up the promising results that he and Zuelzer had obtained in August 1914, Reuter had filed the records of this work in the Roche archives later that year. In 1924, prompted by the news from Canada, Reuter finally published this work in a journal and presented it at a scientific meeting in Luxembourg. There, he revealed that the extracts he had prepared in 1914 had not only been effective in dogs, but had also caused a significant reduction in the blood sugar levels of a diabetic patient.[46] Only a year later, Hoffman-La Roche were producing insulin under the name 'Iloglandol'.[47] The company's initial scepticism towards Zuelzer and Reuter's research and the failure to publish this work in 1914 has been described as 'an inexcusable mistake' that cost Europe a Nobel Prize and lost Hoffman-La Roche the opportunity to establish themselves as a world leader in the insulin market.[48]

Still, Zuelzer had at least had the foresight to file a patent on his process, which should surely have been sufficient to guarantee his priority for the discovery of insulin. However, Zuelzer proved to be wrong on this, too. Having been first filed in May 1908, Zuelzer's patent on Acomatol had been subjected to a number of amendments and revisions before finally being granted in the USA in 1912. Concerned that their own patent application on insulin might infringe that of Zuelzer, the University of Toronto Insulin Committee had assigned attorney Charles Riches to investigate the matter. Riches was reminded that 'you should try to consider yourself in the position of Counsel for Zuelzer, in order that the opinion you give to the Committee may be as unfavourable to the Committee's interests as could conceivably be held'.[49] This he duly did, and his resulting verdict was damning.

He concluded that the number of amendments and revisions that had been made to Zuelzer's original 1908 application were 'unquestionably new matter and should have been excluded from the application'.[50,51] He reassured the

University of Toronto that they need not worry about infringement and in so doing, hammered the final nail into the coffin for Zuelzer's hopes of securing priority:

> There are numerous decisions of the Courts holding patents of this kind to be invalid. Under the circumstances I do not believe that Claim 2 of Zuelzer's patent could be sustained and in the second place, it if was sustained I am of the opinion that it is not infringed.[52]

Despite Zuelzer's hopes being crushed, his work did gain publicity among the scientific community, although perhaps not in the way he would have wished. In an interview given to the *Toronto Daily Star* in January 1923, Macleod said that Zuelzer's work 'must be called an abandoned treatment ... although it was a touch and go thing'.[53] Furthermore, the article not only dismissed Zuelzer's preparation as having 'caused overwhelming toxic or poisonous effects', but also misspelled his name, referring to him as 'Zeugler the German'.[54]

A couple of years later, Zuelzer's work did come to the attention of the Nobel Institute—but only as a historical footnote. For when Macleod gave his Nobel lecture in Stockholm on May 26, 1925, he reminded his audience that his path to the discovery of insulin had already been trodden by others:

> In 1907 Zuelzer published results which must be considered, in the light of what we now know, as really demonstrating the presence of the antidiabetic hormone in alcoholic extracts of pancreas. But unfortunately, even although several diabetic patients were benefitted by administration of the extracts, the investigations were not sufficiently completed to convince others, and, apparently, Zuelzer himself was discouraged in continuing them because of toxic reactions.[55]

When word reached Zuelzer about what Macleod had said, he wasted no time in putting Macleod straight on the matter:

> From the amiable informations of Professor Benedict I have seen, that you were so kind as to appreciate in your Nobel Price [sic] Lecture my investigations regarding the Insulin. There is, however, a misunderstanding, as I did not give up my experiments owing to my being discouraged. I should very much like this error to disappear from the literature [sic]. I went on with my insulin-experiments unswervingly up to the outbreak of the War, with the result that my preparation 'Acomatol' was just ready for use in July 1914, and only showed the one drawback that it did not keep. The War put an end to all scientific matters and when it was over I could not find on account of the depreciation of our currency, a chemical establishment willing to take up the manufacture, when I told the people that it would still take a quarter of half a year until I could definitely solve the problem of keeping.
> [...]
> When the Insulin now suddenly appeared from America, I had in 8 days time prepared my Acomatol according to my own method and obtained with it exactly the same results as you did with the Insulin.[56]

Zuelzer was deeply grateful to receive Macleod's assurance that he would 'take whatever opportunity may offer to correct this statement in the future'.[57,58] Others were also starting to change their minds about Zuelzer and what he had achieved. Closer to home, Minkowski now made a dramatic U-turn:

> I reproach myself for not having tried at the time to investigate the causes of these side effects in view of their undoubted effects on the excretion of sugar, and to have been satisfied simply to conclude therefore that the preparation was unsuitable for the treatment of patients.[59]

But although Minkowski's words seem to suggest some sympathy with Zuelzer, there was yet another twist of the knife. By this time, insulin production was already underway in Europe, the trail having been blazed by the Spanish clinician Rosendo Carrasco-Formiguera who, having studied at Harvard in 1921–1922 and been present at Banting's disastrous presentation at New Haven, had returned to Barcelona to treat a four-year-old patient with some homemade insulin.[60] In Germany meanwhile, Minkowski was appointed by the University of Toronto to take charge of insulin production. As the pioneer of insulin research in Germany, Zuelzer might well have been expected to be at given a prominent role in this work, but neither the correspondence between Macleod and Minkowski, nor documents held by the Insulin Committee regarding this matter, contain any mention of Zuelzer being involved.[61] And when Macleod suggested to Minkowski the names of a number of German clinicians who might form a committee to oversee the production and distribution of insulin there, Zuelzer was not among them.[62] Given this catalogue of misfortunes, it is little wonder that Zuelzer's story has been compared with a Greek tragedy.[63] For a brief moment he had held an active, potent preparation of insulin in his hands, only to have it snatched away by circumstances that had largely been beyond his control.

Over the sound of cheers for Banting and Best, some in the medical world were trying to make Zuelzer's voice heard. Writing in the *British Medical Journal*, Dr. P. J. Cammidge pointed out that, in comparison with Banting's and Best's work, 'The most satisfactory results appear to have been obtained by Zuelzer and his associates, who employed an expressed extract of the gland treated with alcohol'. Even Banting himself made the stark admission to his audience at the Cameron Lecture in 1928 that the results he and Best had obtained 'were not as encouraging as those obtained by Zuelzer in 1908'.[64]

Zuelzer might have been forgiven for spending the rest of his days consumed with bitterness at having been overlooked. However, rather than fester with resentment, he drew on his time as a military doctor during the First World War and pursued his interest in whether quinine—which had been used to successfully treat malaria—might also be used to treat typhus, which had been rife among the soldiers. He also wondered if it might be used to treat scarlet fever—an area in which he had developed more than a professional interest, having contracted this condition while serving on the Eastern Front.[65]

Although he was denied the Nobel Prize for his work on diabetes, Zuelzer's scientific work on scarlet fever did not go unrecognized. In 1932, he was appointed Chair for Special Pathology and Therapy of Infectious Diseases by the Prussian Minister for Science and Art. Yet only a year later the position was suddenly withdrawn. No explanation was offered at the time but with hindsight the reason is tragically clear: Zuelzer was Jewish and had become a victim of the Nazi regime. Zuelzer himself quickly sensed the darkening political mood and in 1934 he emigrated to the USA. Tragically, several other members of his family were not so fortunate, including his cousin, biologist Dr. Margarete Zuelzer, who was murdered in the Holocaust.[66]

Having started a new life in New York, Zuelzer worked as the Assistant Director of Research at the Israel Zion Hospital before setting up in private practice.[67] When he died in 1949, an obituary in *The New York Times* made no mention of Zuelzer's work on insulin and described him as having been a 'heart specialist', while one of his daughters claimed that he had discovered new hormonal treatments for cardiovascular disorders and weight loss.[68,69] It is safe to assume that, given the choice, Zuelzer would have eschewed both these accolades in favour of winning the Nobel Prize for discovering insulin. Perhaps he would have found some consolation in knowing that today his epitaph in Troy, Michigan reads: 'Dr. Georg Ludwig Zuelzer, the first physician to bring diabetic patients out of coma with his extracted pancreas preparation'.[70]

5

The Wasp's Nest

Zuelzer was not alone in feeling that his efforts in discovering insulin had been overshadowed and ignored. There were others, closer to home, who felt aggrieved at the news that Banting and Macleod had received the Nobel Prize. In a letter to the *Journal of the American Medical Association*, Ernest Lyman Scott gently reminded the world that, although he may once have been dismissed as a 'fourth-generation Ohio farm boy', he had also discovered insulin while a doctoral student at the University of Chicago:

> ... it seems that priority of isolation and in the development of the fundamental principles involved in extraction clearly belong to the work reported from the laboratories of the University of Chicago ... I wish again to emphasize that it is no part of my intention to withdraw any part of the credit from Dr. Banting and his co-workers for the work they have done in extending our knowledge of the physiologic and therapeutic properties of insulin ... It is only because it has come to my attention that others are attempting to claim priority for the principles involved in making the extract that I call attention to the papers cited in this note.[1]

Scott claimed that he had been reluctant to write this letter and had done so only when pressed by his family. But although his tone was diplomatic, he made it quite clear that 'the discovery of the curative power of "insulin" has been open from January, 1912 to any one who cared to repeat and extend my work'.[2] Like Zuelzer, Scott felt that Banting's work was less a novel discovery, and more 'the first repetition of my own work'.[3] Scott finally found national recognition when, after his retirement in 1942, he turned his attention to horticulture and became the first President of the National Chrysanthemum Society of America.[4]

It is doubtful that John Murlin, whose media attention and threat to take out his own patent on insulin had, for a time, caused some concern in Toronto, could have found consolation in such an honour. Even after thirty-three years since the award of the Nobel Prize, Murlin still expressed anger. In 1956, he published an even-more detailed account of his early research carried out in 1913–1916. This time, however, he pressed his case for having discovered insulin with slightly more force, emphasizing that the extracts he had made had been free from toxic side effects:

> Experiments performed by us long ago, and now presented for the first time in these pages, demonstrated that the pancreatic hormone lowers sugar concentration in the urine of depancreatized dogs … We feel that these facts merit consideration in the history of insulin.[5]

Like Scott, Murlin's tone was diplomatic, but his point was clear. His son took a similar approach and, after Murlin's death, continued to argue that his father's research 'could very well have been a major cornerstone for the investigation completed by Banting and Best, yet little is ever mentioned about, or credit given to, this pioneer work'.[6]

After all this time, why did Murlin and his son feel the urgent need for a 'further clarification of the history of the discovery of insulin'?[7] Perhaps they felt that others were attempting to rewrite the history of the discovery of insulin in their own interest. According to Michael Bliss, the main culprit behind these efforts was yet another person who felt overlooked in their role in the history of insulin.[8]

When Charles Best first heard the news that Banting was to be awarded an annuity from the Canadian Government, he was not pleased. He complained in a letter to Banting: 'It was rather disconcerting to me after the way my side of the story has been supported, especially by you, to have the Government acknowledge you as the discoverer, with no reference whatever to myself. However, this is an old story'.[9]

But while appearing willing to let this pass as simply 'an old story', Best's actions over the years indicated otherwise. In a letter to his cousin Freddie Hipwell, Banting confessed that he felt guilty at the injustice done to Best: 'I have just had a letter from Charles Best in which he is very disconcerted that the Dominion Government did not enumerate him. I wish they had instead of me. I scarcely know how to answer him. It worries me'.[10]

Guilt wasn't Banting's only concern. Writing to Dr. Henry Geyelin, a New York doctor who was conducting trials with insulin, Banting expressed concerns that the affair might damage the reputation the University of Toronto and the public perception of science:

> While I feel that the whole thing has been a great injustice to Best, and whereas I cannot understand Professor Macleod in this matter, I would beg of you not to publish this letter because the University of Toronto and Science in general would be discredited for their rangling [sic] … any additional controversy would do only harm, since nothing can actually be done about the award.[11]

Yet, Banting saw one clear way in which he could make amends to Best and absolve his conscience. When the award of the Nobel prize was announced, he sent a telegram to Best, who was working in Boston, saying: 'Nobel trustees have conferred prize on Macleod and me. You are with me in my share always'.[12] And he meant this quite literally.

The announcement by Banting that he would give Best half of his $20,000 prize money made the front pages of the newspapers. Banting took the opportunity to praise Best for his crucial part in getting the first attempt at larger scale production of insulin up and running at Connaught Labs.[13,14,15] Banting's public gesture of support was greatly appreciated by Best's father, who wrote to him, saying bitterly 'The whole world knows that the award was misplaced'.[16] But if Banting hoped that his generosity would satisfy Best, he was to be disappointed.

As the controversies around insulin became ever more complex and bitter, Sir Henry Dale of the UK Medical Research Council (MRC) asked Macleod, 'I suppose there are moments when you also wish that you had never heard of insulin?'[17] For Banting, this was certainly becoming the case. After spending years craving recognition and glory, Banting was learning the truth of being in the limelight, and about the media, which were not always complimentary about him:

> The Nobel Prize for medicine has just been awarded to Dr. Banting for his discovery of insulin, a pancreatic extract derived through the brutal torture of dogs, many of which Dr. Banting himself picked up in the streets of Toronto secretly in the dead of night. He was aided in this noble practice by a chum with whom he has divided his prize money in recognition of his services in connection with insulin. The account of this 'secret black and midnight' luring of dogs into the torture chamber, given by Dr. Banting himself before a London audience of doctors and reporters, met with great applause and no one seemed to think there was anything vile and unmanly and despicable in enticing a wandering pet or a poor homeless mongrel by soft words and the implied promise of food into a hell from which there should be no return.[18]

The relentless media attention, Best's simmering resentment, his own gnawing pernicious fears about Macleod, and the general growing toxicity around who had really discovered insulin were all taking their toll on Banting. In increasing despair, he wrote to Best:

> All I want in the world at present is to get down to some work quietly and uninterruptedly in a lab. Any person can have any damned thing they like if I can only be left alone. I have some new remote ideas in a new field and am going to give up practice and everything pertaining to Insulin, and am sick of it all.[19]

But while Banting sought peace and quiet, the world desired a medical hero. In September 1930, Banting was in the headlines again when the University of Toronto opened the Banting Research Institute, a building that, according to *The Mail and Empire*, was 'inspired by the genius of the man, whose name it bears'.[20] His fame seemed to have gathered an unstoppable momentum, and four years later he was honoured in King George V's birthday list with a knighthood.[21,22] Even a couple of his paintings, once the object of scorn and derision, were now on public display in San Francisco.[23] But despite now being 'Sir Fred', Banting seemed to be deeply uncomfortable with his new-found fame. During the opening ceremony of the Banting Institute, one newspaper described him as having 'Shifted restlessly in his chair and smiled uncomfortably though thick-lensed glasses like a small boy caught stealing jam'.[24]

Media attention was also becoming a problem for Charles Best, albeit for very different reasons to those suffered by Banting. For while Banting was being hailed in the headlines as a genius and medical hero, Best's role appeared less clear. On some occasions he was described as having been merely Banting's 'chief assistant', while on others—such as when he took over the Chair of Physiology vacated by Macleod—he was elevated to being 'the co-discoverer of insulin'.[25,26] Yet only a year later, in a newspaper article reporting on the opening of the Banting Institute, Best was not mentioned until the tenth paragraph, which informed readers that the discovery of insulin by Banting had been made 'with the aid of Dr. C. H. Best'.[27]

Best's contribution hadn't gone entirely unrecognized. When Macleod retired as Professor of Physiology and Best was appointed as his replacement, Banting told the Press:

> I do not know of a worthier or more capable man. The only possible criticism that might be brought up against his appointment is the fact that his usefulness is so great in the world of research [sic]. He has so many other duties which has fulfilled that the question has been raised whether the new post will leave him free enough to carry on in research, and whether it is best for him to devote his energies to research or to take over the administration of an important department. However, personally, I believe he is capable of doing both ... I do not think the University of Toronto has ever produced a more brilliant graduate in physiology.[28]

Despite having been promoted to this prestigious position and Banting's kind words in the media, a quiet envy continued to fester within Best. Although Banting offered public praise for Best in public, in private he felt Best's attitude to be a little irksome. When several positions became vacant following the death of a senior member of staff in 1940, Best hoped to be appointed—yet, he was denied all of them. As an olive branch in the wake of this disappointment, the President of the University of Toronto suggested that perhaps a new Institute of Physiology could be established with Best appointed as its head. Best jumped at the chance and immediately asked whether some of the funding for this venture could be provided from a research foundation that had been set up in honour of Banting. As this would mean diverting vital funds from other researchers within the department it was a controversial suggestion and one guaranteed to cause upset. Perhaps Best thought that the suggestion was justified by his status as the co-discoverer of insulin, but on hearing of the proposal, Banting had a far more blunt assessment—'Best is naïve in his abject selfishness'.[29]

Best resented having to walk in Banting's shadow. After working so closely with Banting during the intense summer of 1921, it was frustrating to feel so ignored. But as the result of a tragic turn of events, Best's situation was about to change dramatically.

For the world was once again at war, and after having lobbied strenuously for some time to travel to Britain in order to discuss the war effort, Best suddenly announced that he would be unable to make the trip. His decision was most likely due to work commitments and the poor health of his father, but it was met with disdain by Banting:

> I must say privately that I am inwardly disappointed in Charlie Best. He has no guts. He has the opportunity for which he has long been bellyaching to go to England & he has passed it up. I think it has worried him and he blames his family and the conditions at home but essentially it is a matter of guts. He has not the required number.[30]

By now, Banting had moved on from insulin. When he was contacted by an editor to write an article on insulin and diabetes, he replied:

> I simply cannot do it. I am not interested in the subject ... I have nothing to say that has not already been said a thousand times by people who have the gift of writing or talking. My knowledge of insulin is out of date. I have not done an experiment on it in the past 14 years. My whole attention is upon other problems in medical research.[31]

One of the problems that now occupied Banting's attention was the conquest of yet another serious disease. Hoping to repeat his success with insulin, he had turned his attention to cancer. While visiting England in 1925 he had become interested in work on Rous sarcoma, a type of tumour known as found in chickens and named after Peyton Rous (1879–1970), who had first identified it in 1911.[32] Rous had discovered that the tumour could be transmitted between chickens via the injection of cell free extracts prepared from a diseased bird. This had led him to specu-late that it might be caused by a type of virus. The very existence of viruses was still being debated at the time, but Banting was more interested in whether Rous sarcoma held the secret to finding a cure for cancer. If some means of blocking transmission of the tumour could be found—perhaps through a vaccine—might this offer a means by which cancer could be managed and controlled as effectively as insulin had done for diabetes?[33]

However, after five years transplanting over a thousand Plymouth Rock hens with tumour-inducing filtrate, piles of lab notebooks showed nothing but nega-tive results.[34] And if this work did not appeal to Banting's lab assistants, there were always alternative lines of research in Banting's lab, including silicosis, the collection of soiled nappies to explore a hypothesis that infant faeces might also hold some clues to cancer, or discovering the biochemical secrets behind the growth-inducing properties of royal jelly.[35]

Despite his odd research subject, Banting did possess insight. Unable to sleep one night, Banting reflected on how his position as Director of the Department of Medical Research was an unwelcome distraction from research at the lab bench that he had once enjoyed:

> The lab has changed. I will have to get a new laboratory. When I go to it I find that it is not a lab but an office. There are a pile of letters to answer, phone numbers to call up, people waiting to have an interview, routine work that must be done.

Some person wants me to give him some money, someone wants a signature, someone wants to know what to do about a friend of a great aunt's cousin who has cancer, or who has gone insane. Someone has a cure for diarrhoea, cancer or anterior polyio [sic] myelitis. Some anti-vivisectionist damns. Some of the staff are sick or want a raise in salary or want a holiday. Some newspaperman wants an exclusive story ... [36]

To add to his woes, he was also discovering that being a scientific celebrity came at a cost:

Some visitor from China, the USA, England has arrived and 'cannot visit Canada without seeing the distinguished discoverer of Insulin!' ... And such is life! Would anyone care to live it as I do? A letter from my home town asks for a scholarship for the school—a letter from a church asks for a donation—a [sic] artist wants to sell his pictures—a dear friend of the family who has known me 'all my life' wants a loan—an editor of a magazine wants an article—a movie man wants a few feet of film of me in the lab, a broadcaster wants the outstanding incidence of my life to put into a skit—the National Research Council wants me to approve the minutes of the last meeting. What a life! At times I feel like a wrung out dishrag— exhausted with giving and nothing left to draw from. At the end of each day I am tired and would like to sleep but cannot unless soothed to rest with alcohol.[37]

With the outbreak of the Second World War, Banting's fame took him into a new field of research. As one of Canada's most eminent scientists, Banting became involved with research into aviation medicine for the war effort and visited Great Britain to see how Canada could best deliver assistance. While in London he had delivered a call to arms, declaring that 'the Empire is fighting Hitlerism, and Hitlerism is my enemy too'.[38] And he had no qualms about the extreme lengths to which the Allies might have to go if they were to defeat the Nazis:

Whether we like it or not similar and more diabolical inventions must be made by our scientists. Traditions and sportsmanship must be put on the shelf while we are dealing with Mr. Hitler and his crew. After the war virtues may be restored. There is no one who believes more firmly than I do that Science was meant for the benefit of mankind, but our enemy has used science for the invention of the most destructive weapons of warfare. Therefore we must do likewise. Hitler will win this war if we do not. No amount of bravery, no amount of human flesh will win against scientific weapons of destruction. Self-preservation demands that we use the same weapons.[39]

The 'diabolical inventions' to which Banting referred were most likely chemical weapons—a field with which Banting already had some experience. Having received the formula of an antidote to mustard gas from a German refugee, Banting used himself as a subject to test its effectiveness. He confirmed that the antidote was effective after burning a small area of his own skin with the gas. The antidote was sent to the MRC in Great Britain. In 1940, Banting spent three weeks in Britain touring laboratories, including those at Porton Down, where he became concerned that, although the British were conducting research into gas warfare, they did not appear to be taking the threat of biological warfare using microbes very seriously. He wrote a memorandum on the subject and, on his return to Canada, began conducting experiments into how such agents might be dispersed. In his diary he recorded a meeting with the Minister for National Defence in which he argued the case for a starting a programme to develop biological weapons. While he had wept over the death of his lab dogs, Banting seemed to have no scruples about using weapons of mass destruction:

> I placed the matter squarely & fairly before him—as I saw it. We were beyond the purely experimental stage—it was a matter of production on the pilot plant plan. Not for protection but for obtaining the means by which we could retaliate 100-fold if the Germans used bacterial warfare ... Our job lies clearly before us. We have to kill 3 or 4 million young huns—without mercy—without feeling. The job of self-preservation is uttermost. It has to be done by whatever means seems best under the circumstances.[40]

When Best announced that he was postponing his trip to Britain, Banting took his place. Shortly before his departure, he is said to have given a very frank opinion of Best that was starkly at odds with the adulation he had given his colleague in the Press:

> This mission is risky. If I don't come back and they give my Chair to that son-of-a-bitch Best, I'll never rest in my grave.[41]

Banting's words were to prove tragically prescient. On the evening of February 20, 1941, he boarded a Hudson bomber on an airfield at Gander, Newfoundland as a lone civilian passenger alongside pilot Joseph Mackey, radio operator Bill Snailham, and Flying Officer William Bird. Shortly after take-off, as the plane was about fifty miles north-east of Gander and out over the Atlantic Ocean, Mackey noticed a problem with the cooling system. Shutting off his starboard engine, Mackey turned the aircraft around in the hope that he could make a safe landing back to Gander. When the cooling system on the port engine then also failed, he realized that this would be impossible. Jettisoning as much excess fuel as he could, Mackey prepared for a crash landing. For some time afterwards, a rumour circulated that the aircraft had been sabotaged by Nazi agents with the intention of assassinating Banting, who was now a high-profile target thanks to his scientific prestige and close involvement with the war effort.[42] While this might have made a gripping Hollywood script, the truth was more prosaic: the new cooling systems on the Hudson, were unreliable and had become prone to rupture in extreme cold weather.[43]

Despite having been ordered by Mackey to bail out, Banting and the two other crew members were still on board when the Hudson T-9449 hit the ground on the shore of a large pond about twelve miles south-east of Musgrave Harbour in Newfoundland. Having braced before the crash, Mackey regained consciousness a few moments after impact to find both Snailham and Bird dead. Banting was lying unconscious on the floor of the main cabin. Having failed to brace, he had been thrown forward, smashing the left side of his head and breaking his arm, but he was still alive.

Mackey moved Banting to a bunk in the cabin and made bandages for him using parachute silk.[44] As Banting drifted in and out of consciousness, he asked Mackey to take a dictation. What followed was a delirious stream of technical details that Mackey said were 'on medical problems beyond my comprehension'.[45] Nevertheless, Mackey sat with Banting through the night, sometimes just going through the motions of writing his words down 'in order to quiet him'.[46]

As the next day dawned, it was clear to Mackey that Banting would not survive without swift medical help. At noon, with Banting having fallen unconscious once more, Mackey broke the map-board in half to fashion a pair of crude snowshoes and set out into the surrounding wilderness. When he eventually returned at dusk, he spotted a body lying in the snow about fifteen feet from the wreckage of the plane. It was Banting. At some point that afternoon, he had regained consciousness, struggled up from his bunk and stumbled out into the snow where he had collapsed and died. In the failing daylight, Mackey reflected: 'Here lay dead a great man, a man I did not know, a man of importance in the world. He had been the only other living thing, speaking to me in urgent riddles. Now he was still'.[47]

At his funeral, the casket carrying Banting's body was accompanied on a gun carriage through the centre of Toronto by a two-hundred-strong military escort. As he was lowered into his grave, dressed in the uniform of a major in the Canadian army with his war medals on his chest, three volleys were fired and trumpeters played 'The Last Post' and 'Reveille'.[48,49,50] The once-struggling doctor from provincial Ontario had truly become a national hero (Figure 15).

With Banting dead, only two members of the original Toronto research team remained alive. Macleod had died in 1935, having left Toronto seven years earlier to return to his native Aberdeen, where he became Regius Professor. It is said that he never talked about his days in Toronto and that, when asked by a friend who had accompanied him to the railway station on the day of his departure why he was shuffling his feet, Macleod is said to have replied 'I'm wiping away the dirt of this city'.[51]

In an obituary to Macleod, James Collip wrote of him with admiration:

> It was typical of him that he would not allow an enthusiastic colleague or assistant to embark on an investigation without pointing out all the technical or theoretical difficulties which would probably be encountered, and yet would supply encouragement and all the practical assistance and his disposal.[52]

Collip had good reason to sing the praises of Macleod. When Banting had announced that he would share half of his prize money with Best, news reporters had quickly come knocking on Macleod's door wanting to know what his own

Figure 15 *Photo spread from The Toronto Daily Star covering Banting's funeral – 5th March 1941.*

Credit: Insulin Collection, F.G. Banting (Frederick Grant, Sir) Papers, Thomas Fisher Rare Book Library, University of Toronto, MS. COLL 76 (Banting) Scrapbook 2, Box 2, Page 116. Reproduced with Courtesy of the Thomas Fisher Rare Book Library, University of Toronto.

intentions were regarding his half of the prize money. Macleod had quickly dismissed them saying, 'My decision will in no way be influenced by the action of others ... I am a Scotchman, and I never make up my mind in a hurry'.[53]

But his actions told a rather different story. Within only a matter of days, the newspapers reported that Macleod had come to a decision.[54] He would share half of his prize money with Collip, who thanked him shortly afterwards for sending 'The fattest check I have ever received or dare ever hope to receive in the future'.[55,56]

It was a magnanimous gesture, but one that may not have been motivated entirely by altruism and a sense of injustice. In a letter written earlier that year to Collip about various aspects of ongoing research, Macleod asked whether he had seen an article by W. M. Bayliss in *Nature* regarding the discovery of insulin. 'I think it puts things pretty straight and you will see that you are credited for your share in the work', Macleod reassured him, before adding bitterly, 'Banting is meanwhile greatly in the limelight here and seems to bask in it'.[57]

The publicity surrounding Macleod's announcement to share his prize money was without doubt a powerful way of making the world aware of the key role that Collip had played. But reading between the lines of a letter to a colleague, it was also a gift of an opportunity to rob Banting of some of the limelight that so irked Macleod:

> By dividing my share of it with Collip I think I have succeeded in getting people here to realise that his contribution to the work as a whole was not incommensurate with that of Banting. It is of course sad that it should require such drastic methods to persuade people of this fact but it could not be helped, it was the only thing to do under the circumstances.[58]

Macleod refused to be outshone by the man who was rapidly becoming his nemesis. Perhaps, also, Macleod was trying to absolve himself of his own perceived betrayal of Collip.

For in early 1922, when Collip had left Toronto and returned to Alberta, he had hoped to continue his work on the development of insulin. But the University of Alberta were unable to find the necessary funds for him to do so. Macleod took up his cause, writing to University of Toronto and the Carnegie Corporation President Sir Robert Falconer to argue that some of the funding given to the University of Toronto for the development of insulin be assigned to Collip on the grounds that 'he had a full share in the isolation of insulin'.[59,60]

Collip hoped that the work he had done in Toronto on insulin might make him eligible for election to the Royal Society of Canada, but was surprised to find that when the election was held, his name was absent from the ballot paper for nominations.[61] In a letter to Macleod, Collip reassured him that he was 'not losing any sleep over this', but had he known what was going on behind the scenes, he might not have been so sanguine.[62]

The problem was that elections of membership to the Royal Society had become yet one more front in the ongoing battle over who really deserved the credit for the discovery of insulin. Collip was keen to know how he might help Macleod with this situation. But the advice offered to Macleod by Archibald B. MacCallum, a professor at McGill University, Montreal, on this issue was not particularly flattering to Collip. MacCallum recommended, 'Perhaps the best way for him [Collip] now is to lie low, like Br'er Rabbit, for a time. He is not a good tactician and he might blunder very seriously were he to interfere now'.[63]

Collip had become a pawn in the career politics of both the Canadian medical establishment and the Canadian Parliament. MacCallum confided to Macleod he had heard that members of the Canadian Parliament representing the Western Provinces were proposing to vote for an annuity for Collip in the same way as had been done for Banting. Collip was a hero to the Western provinces, and hailed on the pages of the *Edmonton Journal* as 'the co-discoverer of insulin'. But Macallum insinuated that the move was as much about using Collip as a means by which the Western Canadian Provinces could assert themselves against their Eastern neighbours as it was about honouring his work.[64,65]

MacCallum also warned Macleod that, although Collip apparently had support within the Canadian Parliament, he had enemies within the medical establishment.

He warned Macleod about a group of clinicians in Toronto who 'will shrink from nothing to advance their own interests by crying down everybody else opposed to them. I am very sorry to recognise that Dr. Banting has too many affiliations with that set'.[66] MacCallum explained to Macleod that, although he had considered putting Collip's name forward for election, he had decided against it for fear of angering certain individuals within this group, whom he described as being 'out to down Collip every time'.[67]

One member of this cabal was John Fitzgerald, whose timely intervention may well have spared Macleod a black eye when Banting first heard news of the Nobel Prize. When Fitzgerald nominated Banting for election to the Royal Society, his proposal was rejected on the grounds that Banting should be given more time to complete independent research. Fitzgerald would only agree to this, but only on one condition—that Collip's name also be withdrawn as a candidate.[68] On hearing this demand, Macleod, despite having completed the nomination papers for Collip, found himself suddenly reluctant to submit them:

> It seems to me that it would be highly undesirable, with only vacancies to fill, that these two names should appear as candidates. I believe that Collip's election is overdue and that his work distinctly deserves this recognition, but at the same time I believe that it would only serve to stir matters up again if it should come a vote between Banting and Collip. It would therefore be desirable, in my judgement if Dr. Collip will agree to it, that his name be withdrawn for the present year.[69]

This was Collip's second attempt at being elected to the Royal Society and Macleod warned him that, if it were to fail, it would be damaging to his career. He tried diplomatically to explain to Collip that it would be better to withdraw his nomination that to suffer the indignity of defeat against Banting:

> To be turned down a second time would be serious ... We cannot shut our eyes to the fact that through his notoriety Banting would receive the majority of the unenlightened votes. I am explaining these matters to you so that you may understand why it is that I cannot do anything at present to make your election a certainty.[70]

But despite Macleod's attempt, the reality of the situation was that—on this occasion at least—Collip's career hopes had been dashed by the machinations of Banting's allies in combination with Macleod's instincts for political survival by keeping a truce between rival factions.

Whether or not Macleod's decision to share his prize money with Collip was at least in part done to assuage his own guilt, Collip was nevertheless grateful. At the same time, he did not want to be perceived as a charity case or a mere pawn in an ongoing struggle between Banting and Macleod. He wrote to a friend: 'As to the part I played in the development of insulin, I can tell you that the method for the isolation of insulin in sufficiently pure state for human use was accomplished entirely by myself'.[71]

When Collip's letter came to the attention of the press, Banting's initial response to his confident claim was blunt: 'I have nothing to say'.[72] However, within a few years, Banting had softened his stance towards Collip. On hearing that Collip had been in Toronto recently, Banting wrote to him saying 'I regretted to hear that you hesitated to come to see me when you were in Toronto. As far as I am concerned the past is buried, and I hope that you will not let our past differences interfere with our future'.[73]

Collip's reluctance to pay Banting a visit while in Toronto was hardly surprising, given that their working relationship had almost ended in physical assault. But in his reply, it was clear that he appreciated Banting's offer of an olive branch: 'Personally I regret very much the unfortunate misunderstandings of the hectic winter '22 ... I would be only too glad to put them in the background of oblivion. With a hearty handshake in spirit and the best of luck to you'.[74]

When Banting died, Collip even wrote an obituary to him saying, somewhat euphemistically, that although their early relationship had for a time been 'strained by certain misunderstandings', he had 'lost a colleague for whom I had a most profound respect and admiration, and also a very dear friend'.[75,76] Speaking with modesty about his own role in the early work on insulin, he said that it had been 'only that which any well-trained biochemist could be expected to contribute, and was indeed very trivial by comparison with Banting's contribution'.[77]

These were not just empty words spoken for the sake of politeness. On the day before he boarded his fateful flight from Newfoundland to Britain, Banting stopped off in Montreal and paid a visit to Collip in his laboratory. When Collip asked whether there was anything he still needed for the trip, Banting replied that he had no warm gloves. Returning home, Collip dug out a pair of sheepskin gloves and, after his daughter Barbara had sewn up their finger ends, he called round with them to the hotel where Banting was staying.[78] The rest of the evening was spent enjoying a 'long and intimate talk' before saying their farewells.[79] Time had indeed healed old wounds.

But for Charles Best, however, the passing of the years had exactly the opposite effect. Far from healing old wounds, they festered. When Banting died, Best said that his 'name will live on for ever in the hearts of successive generations of diabetics and in the minds of young investigators who will be stimulated by his brilliant and fearless career'.[80] And by the time those words were written, Best had already taken steps to ensure that it would be his name - and not that of Collip - etched into the minds of successive generations of young researchers.

After returning to Alberta in 1922, Collip had written to Macleod thanking him for the hospitality he had shown during his six months in Toronto. He also added that, despite initial tensions, he was confident that he had parted on good terms with Best:

I want to thank you again for the many kindnesses you showed me while in Toronto. I feel that I have had a wonderful year and one which I will always remember. I had a heart to heart talk with Best before leaving and I think we parted good friends.[81]

This would prove to be a gross misjudgement. Collip was described as being modest 'to the point of shyness … [and] was reserved in large groups and shunned publicity'.[82] Public speaking made him so uncomfortable that on one occasion, having just finished an address to the Toronto Academy of Medicine, he was so desperate to return to his seat that he leapt away from the lectern, not realizing that he was still attached to it by his microphone lead.[83] With such a quiet disposition, it was little wonder that Macleod assumed the task of fighting for Collip's recognition. But with Macleod now dead, Collip no longer had a champion.

Macleod had also been similarly supportive of Best. After leaving Toronto for Aberdeen, Macleod had written to Falconer saying that he could think of no one more suitable than Best to succeed him as Professor of Physiology at the University of Toronto.[84] Congratulating his protégé on taking over his former role, however, Macleod offered Best an apology—'I am afraid I have left you a heritage not unmixed with what, to some, may appear to be encumbrances'.[85] But for Best, the greatest encumbrance was Collip, and he was determined to address this.

In 1935, Best told Macleod that he and his colleague Professor Norman Taylor of the Department of Physiology at the University of Toronto were co-authoring a textbook for medical students called *The Physiological Basis of Medical Practice* and which Best hoped 'will probably serve a useful purpose in so far as the teaching in our own department goes'.[86] But by the time the book was published in 1937, he may well have hoped it would serve another purpose entirely.

Thomas Kuhn (1922–1996), the physicist turned philosopher of science, once warned about placing too much faith in the historical accounts of discovery found in science textbooks. Kuhn himself may not be a household name, but his idea of a 'paradigm shift' is now a common term - particularly in the corporate culture of big business. Introduced in his book *The Structure of Scientific Revolutions*, Kuhn used the phrase to describe a monumental upheaval in scientific knowledge, but his criticisms about science textbooks are less familiar. Kuhn's problem was not with the science that they contained, but their presentation of its history. He suggested the problem was that the aim of science textbooks was to be 'persuasive and pedagogic' about a scientific idea, rather than to reveal the historical processes by which it had emerged. According to Kuhn, the historical accounts found in science textbooks are as superficial in their insights into the tradition and culture of science as a glossy tourist brochure is about those of a foreign country.[87]

An early edition of Best and Taylor's textbook bears out Kuhn's point perfectly. In the preface to the first edition, the authors said that their aim was to 'promote continuity of physiological teaching throughout the pre-clinical and clinical years of the undergraduate course'. But alongside pedagogical aims, their account of the discovery of insulin was clearly also intended to persuade students about who really deserved the credit for this achievement:

> While other workers, among whom Hedon, Zuelzer, and Scott may be mentioned, obtained very suggestive results, which in some cases were probably due to the presence of insulin, Banting and Best working in Macleod's laboratory (1922) were the first to obtain a preparation containing the antidiabetic hormone in a form which consistently alleviated all signs of diabetes in completely depancreatized dogs.[88]

Best recalled in later years that Taylor had written about seventy per cent of the book, but it is difficult to believe that Best himself would not have written and approved the section on insulin.[89]

Furthermore, the passage on the discovery of insulin, although small, was significant—not so much for what it said, but for what it did not. It contained no mention that it was Collip's purified extract that had saved Leonard Thompson's life.

Also of significance was a diagram that accompanied this passage (Figure 16a). This showed the effect of two injections of pancreatic extracts given on August 20 and 21, 1921, on the blood sugar of a diabetic dog. It had been taken from the very same chart that Banting and Best had presented in the *Journal of Laboratory and Clinical Medicine* in 1922, and from which, as explained in Chapter 2, had already undergone a significant change from the original chart found in their lab notes.

Now, it appeared to have undergone yet another significant change. According to the legend that accompanies this figure in the textbook, the extract administered on August 20 was prepared from 'degenerated pancreas'. However, the legends that accompanied both the original chart in their lab notebook and the copy of this that appeared in their first paper state that this material had been prepared from 'exhausted pancreas' (Figure 16b). The difference may well be significant. 'Exhausted pancreas' referred to material that had been prepared by first treating animals with the hormone secretin in an attempt to exhaust the production of digestive enzymes that were supposedly destroying the insulin. But by the time that the textbook was published, it had already become clear that pancreatic extracts did not include active proteolytic enzymes and using secretin to overcome their effects was just a red herring.

This might seem like mere semantics, but it may be significant. It was clear that there had never really been any need for Banting and Best to mess about with degenerated pancreatic material, and that extracts made from whole pancreas were just as effective. The suggestion to make material from degenerated pancreas had given Banting's original plan its novelty, and so this was presumably included for authenticity. But to have included the secretin experiments would have been an admission that he and Banting had, for a time been misled, and might have tarnished what would otherwise have been a polished, aesthetically pleasing account of discovery. A simple explanation in which he and Banting had been on the right lines from the very start would be much more persuasive and convincing.

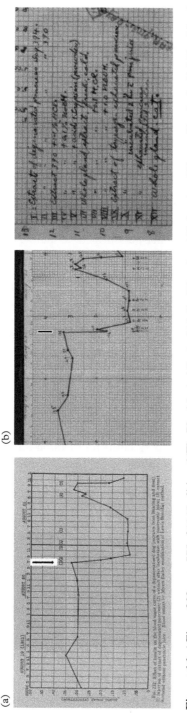

(a)

(b)

Figure 16 *a) Figure 232 on p. 575 of 1945 edition of 'The Physiological Basis of Medical Practice: A University of Toronto Text in Applied Physiology' by Charles Best and Norman B. Taylor. According to this chart, an injection of extract prepared from what is described as 'degenerated pancreas' on 20th August (arrow) caused a significant drop in the blood sugar level of a diabetic dog. b) But the original chart from Banting and Best's lab notebooks describes this same injection given on 20th August as having been performed using extract IX 'Extract of trypsinogen exhausted pancreas'.*

The textbook actually proved to be so phenomenally successful and popular that it is still used today and is known to generations of medical students simply as 'Best and Taylor'. Sadly, if Best was hoping that textbooks would guarantee his scientific immortality, he would have been disappointed to find that, in the 2012 edition of one popular undergraduate biochemistry textbook, the discovery of insulin was credited not to Frederick Banting and Charles Best, but rather to *George* Best (1946–2005)—a name far more likely to be associated with the phenomenally talented but equally troubled footballing legend who played for Manchester United, than with a medical pioneer. Kuhn, it seems, was once again correct about the pitfalls of trusting the historical accounts found in science textbooks.[90] But there were other fronts on which to win a place in scientific history.

Although the passage describing the discovery of insulin in Best and Taylor's textbook did mention Zuelzer, Collip's name was conspicuous by its absence. Best, in his obituary to Banting, had acknowledged that Collip 'had made several important contributions within a short period of time' but significantly he had not elaborated on the details of exactly what these contributions were—nor how crucial they had been.[91] Anyone reading this account would think that the discovery of insulin had been due to the sole efforts of Banting and Best alone.

By 1946, Best's stock among the medical community was on the rise, and when he addressed a meeting of the New York Diabetes Association, the session chair Dr. Edward Tolstoi described him as a speaker whose reputation was so great that he did 'not require much of an introduction'.[92] In his address, Best revealed for the first time that when he was younger, he had lost a much loved aunt to diabetes.[93] With this revelation, Best's achievements now took on a new dramatic aura, for not only had he made a life-saving medical discovery but had done so as a result of having embarked on a personal crusade.

But while Best's aunt had made a sudden appearance in the story, another pivotal figure was missing entirely. Collip received no mention—and members of the audience might well be forgiven for thinking that the idea of using alcohol, which had proven to be such a crucial step in preparing and purifying the extracts, had all been thanks to Best:

> The use of alcohol in the extraction of whole pancreas has an interesting history. It was, of course, obvious but perhaps not so apparent then as it would be now. When these experiments were initiated I thought that I had hit upon a very good idea. My friend and classmate, Henry Borsook, now of the California Institute of Technology, helped me determine the concentration of alcohol necessary to precipitate pancreatic proteolytic enzymes. I can still visualize Banting when he saw the results of the alcoholic extracts of whole pancreas. He said: 'Do you know, Charley, I believe that Professor Macleod said something about using alcohol as a precipitating agent for proteolytic enzymes'. Various incorrect accounts have been given about this phase of our work by people who certainly had no first-hand knowledge of the situation.[94]

At an address to the American Diabetes Association that same year, Best did indeed acknowledge 'an appropriate time to pay my tribute to the late Professor

J. J. R. Macleod, and to Professor J. B. Collip for their contributions to the development of insulin'.[95] But he did not elaborate on exactly just how important those contributions had been, and again gave the impression that the discovery of insulin had been made by himself and Banting well before Collip had first arrived in Toronto in September 1921.[96] Crucially, he also neglected to mention that Leonard Thompson's life was saved only after he had been injected with Collip's material—not that prepared by himself and Banting:

> The first diabetic patient to receive the material, January 12th, 1922, was Leonard Thompson, a diabetic boy in the medical ward of the Toronto General Hospital. The actual clinical result was not striking, but has, of course considerable historical significance. It was my privilege to make this extract. In fact, I depancreatized the likely looking steer and carried through the fractionation and testing up to the stage when the material was given to the human subjects.[97]

According to Best, the pivotal moment in the discovery of insulin had been on 11th Jan 1922 when Leonard Thompson was injected with material prepared by himself and Banting, and over the next few years Best continued to defend his version of events.

When he gave the prestigious Banting Lecture six years later, Best may have sensed some murmuring dissent in certain quarters of the medical community about this account. At the lecture, he remarked that 'there have been a few interesting statements about the early developments of insulin which I completely fail to recognize'.[98] Collip might well have said the same thing about Best's own account of events. During this address, Best once again failed to mention that the side effects suffered by Leonard Thompson when he had first been injected with Banting and Best's preparation, had been absent when the boy was injected two weeks later with material prepared by Collip.[99]

Aware that Collip's contribution was a thorn in his side, Best reinforced his own position by stressing that, as far as he was concerned, the discovery of insulin had been made by himself and Banting well before Collip's arrival on the scene:

> We worked completely alone during the four summer months of 1921 without any verbal or written advice from any senior investigator and until well after the salient facts, consistent dramatic lowering of blood and urinary sugars and complete recovery of our moribund depancreatized dogs, had been repeatedly demonstrated.[100]

But when Banting's biographer, the Canadian historian Michael Bliss made what was the first—and exhaustive study—of Banting and Best's original laboratory notebooks from that summer, he found that their results had not been quite so clear cut as was being portrayed by Best:

> Banting and Best's first 75 injections of extract of supposedly degenerated or 'exhausted' pancreas, using nine dogs, produced 42 favourable results, 22 unfavourable ones, and 11 inconclusive observations. This is an impressive statistical picture in its own right, impressive enough to justify the work on the one

hand and explain the researchers' overly enthusiastic claims on the other. In the face of so many good results, the tendency was to forget or ignore the bad ones.[101]

But why let this spoil a good story? Two young researchers valiantly struggling alone against all the odds to make a medical breakthrough was an image that gripped the imagination.

It also brought Best a steady stream of accolades and honours. By 1953, he had already received a CBE and been elected a Fellow of the Royal Society, but now he finally received the one honour he had always craved. In an inaugural address at the opening of the new Charles Best Institute, Dale of the MRC, with whom Best had done his post-graduate research, praised the achievements of Banting and Best to the rooftops:

> The collaboration was to be one of intimate understanding, with no question between the two participants of any but an equal sharing of its success. Matters having been thus arranged, Professor MacLeod, still quite naturally sceptical of any successful outcome to the enterprise, left Toronto to spend the summer in Europe; so that it was in an otherwise deserted Department that the two young and inexperienced but determined enthusiasts, working at tremendous pressure through the hot summer months of 1921, taking turns to sleep on occasion, solved the main problem without further aid from, or communication with, anybody. As a result they had clear evidence of the existence of insulin, and of the possibility of obtaining it in a separate solution, and of eliciting its effects by artificial injection, by the time MacLeod returned from Europe.[102]

An attentive listener to Sir Henry's address might well have wondered why he felt the need to stress that Banting and Best had 'solved the main problem without further aid from, or communication with, anybody'. But for one listener who was paying particular attention to this point, the answer was painfully clear. Collip sat in the audience listening to Dale's version of events, and he was not pleased—a response which seemed to come as something of a surprise to Dale, who afterwards told Best:

> I just caught sight of Collip after the ceremony of opening the door of your new Institute, and I remember thinking at the time that his greeting of me seemed a little bleak and reluctant. It did pass through my mind to wonder whether he was offended that nothing had been said of his particular contribution to making insulin early available for clinical trial. From Peters I gather that severe umbrage has, in fact, been taken on that account ... he was only concerned that I appeared to have put my head, unwittingly, into a wasp's nest, though it was one which I hoped had long ago been dug out, destroyed and happily forgotten ... I would go a long way out of my normal course to prevent any kind of resentment.[103]

In order to prevent such resentment by Collip and maybe also to soothe his own guilt, Dale proposed that, when this speech was finally to appear in print, the following paragraph should be added:

I wish that I had time, indeed, for more than a passing tribute to the importance of the part played at an early stage in the discovery and elaboration of a method for separating insulin from the complex extract in a form suitable for clinical trial and use, by J. B. Collip, now a world-famous contributor to a much wider field of endocrinology.[104]

Keen to avoid further offence and get his facts straight as he prepared to give the Banting lecture in 1954, Dale wrote to Best asking for clarification about exactly who had done what in Toronto:

I feel I ought, very reluctantly, to ask you to give me the straight appraisement of the whole situation ... From you, I would really like to know exactly what the position was, not because it will in any way seriously alter my view concerning the primary and central importance of what you achieved. If, however, I have tended at all to under-estimate the importance of what Collip did, I want to be put right and I know that I can trust you to give me a clear and unprejudiced statement of the position. I hate to bother you again about the matter, but I just want to feel if anybody tries to discuss such matters with me, that I know fully and exactly what happened.[105]

Best replied that he had 'the highest admiration for Collip's ability and would give him a great deal of credit for the improvement in the methods of purifying insulin'.[106] He even went so far as to acknowledge that the extracts prepared by himself and which had been used for the first injection of Leonard Thompson had soon been 'superceeded [sic] by Collip's improved extract'.[107] His closing remarks, however, were much less complimentary:

I have to confess that even after all these years the revival of the memory that Prof. Macleod and later Collip instead of being grateful for the privilege of helping to develop a great advance, used their superior experience and skill, with considerable success, in the attempt to appropriate some of the credit for a discovery which was not truly theirs, still makes me warm with resentment. I must state, also, that I have only to think of the understanding and the fairness of scientific colleagues in many countries who have read our reports carefully, to replace resentment with a much better feeling.[108]

At around this time, Best was presented with a golden opportunity to bring his story to a much wider audience—and in so doing, secure his place in history. Having walked in the shadow of its southern neighbour for so long, the discovery of insulin was a momentous achievement for Canada and a source of great national pride. To celebrate this triumph, the National Film Board (NFB) of Canada wanted to make a feature film, and the Metropolitan Life Insurance Company proposed to sponsor a filmstrip telling the story. Both companies approached Dr. W. R. Feasby, a colleague and great admirer of Best, as a consultant for their respective projects and the drafts of scripts were sent to both Feasby and Best for comments and review.

The prospect of a film gave Best the perfect chance to seize the high ground. One of his concerns was that he should not be portrayed as Banting's assistant but rather as having volunteered to be his 'partner'.[109] In a letter to NFB script writer Leslie Macfarlane, Best proposed the inclusion of a scene in which Macleod is cast as the antagonist in the narrative by warning them at the very outset of their work 'that the experiment will probably fail as many other investigators with the same idea have failed'.[110] Best then suggested the addition of a scene in which he himself explains to Banting that all previous attempts to isolate insulin had failed due to flawed chemical procedures for the measurement of blood glucose levels.[111] The aim was clearly to show that their working relationship had been one of equals and to refute any notion that, as a final year student, he had merely been Banting's junior.

Despite Banting's own admission that he had, on occasion, sunk into alcohol-fuelled gloom during their work together, Best objected to a scene which portrayed this and instead dismissed such insinuations as having been 'no doubt spread about by Collip'.[112] But the film also gave him the chance to tackle a far more crucial problem posed by alcohol. When MacFarlane proposed the inclusion of a scene in which Macleod was to give final instructions and technical advice to Banting and Best before departing for his summer holiday in Europe, Best was keen that one particular issue should not be mentioned: 'I think, Mr. MacFarlane that I would not mention the alcohol matter there. It is rather hazy in my memory and, if it is not followed up later, there is very little point of bringing it in here'.[113] The inclusion of this scene might well give the impression that the use of alcohol in the purification had been Macleod's idea; surely this would rob Best of a glory that only few pages later he claimed to be rightfully his:

> The idea of using alcohol had actually been in the literature for many, many years but as you realise no success had resulted from that idea, that is, using alcohol to extract the hypothetical internal secretion, and I was actually the first to do it.[114]

Had Zuelzer and Scott read this claim, they would have been livid. Not only did Best claim that the idea to use alcohol had been his, but he also seized upon this as yet another opportunity to diminish the contribution of Collip. He stressed to Macfarlane that 'the complete proof of the presence of insulin' had been achieved by himself Banting when they had been carrying out experiments on dogs together during the summer of 1921—well before Collip's arrival.[115]

When Best sent Macfarlane a list of what he considered to be the main events portrayed in the film, he was clear that the climax of the film should be 'Our joy at hearing the encouraging result—from the Toronto General Hospital, Leonard Thompson, Jan 11 1922'.[116] Presumably, the depiction of Leonard Thompson breaking out in abscesses as a result of this first injection, while more authentic, would have detracted somewhat from the sense of triumph that Best wished the film to portray.

Best's proposed finale also conveniently ignored the fact that it had been the second injection of Leonard Thompson on January 23—with material prepared by Collip—that had been the real moment of therapeutic success. When Collip first learned about the plans to make a film, he was baffled. Writing to Feasby, he complained that 'It is difficult for me to understand why there should be so much pressure brought for publicizing in a dramatic way the development of insulin some 35 years ago'.[117] Collip felt that to make a film about the discovery of insulin was way too premature—especially when there remained so much acrimony over who really deserved the credit.

When Collip's concerns came to the attention of the NFB, their suspicions were aroused and, concerned that they might not be getting the whole story, they assigned one of their commissioners, Dr. A. W. Trueman to investigate the matter further. Having travelled to Toronto, Trueman interviewed Best and Feasby in the hope of obtaining some much-needed clarification on the matter. His questions were blunt and to the point. 'Who gave the first insulin? And, who made it? When was it given? What did Collip contribute? When did he actually work at this?'[118]

In reply, Best was forthright in his account:

> ... he asked me these key questions, and I answered them with the truth as I know it; that Banting and I had worked alone from May 'till December, and Collip had come in with this end, and he had nothing to do with the making the [sic] first extract given to a human being, and that he had made a contribution which was later used in some of the processes for making insulin, but completely eliminated from it—other processes, and it was not an essential step that everybody had to use at all. It wasn't. I can think of half a dozen other people who immediately used some other procedure for the purification of material which Banting and I had made.[119]

Faced with the possibility that the whole project might be shelved due to this complication, Feasby had a simple solution:

> ... the alternative, which would avoid all this difficulty, is what Dr. Best has always wanted us to do, and I think he has been quite cross with us, sometimes, because we didn't, was to stop the film at the end of the discovery ... you could stop the story on January 11th 1922. Right there and then, and then you don't have to bring anybody else into it except Macleod, Banting and Best. You don't have to mention Collip ... it leaves the enemy the opening.[120]

When Trueman asked Best specifically about Collip and his contribution:

> Well, I think he is timid—scared of his own shadow and afraid of what is going to come out, and I have always felt that a lot of his reactions were governed by—well, during the insulin time he was obviously a bit sly and a bit of an opportunist, and he took up with Macleod as soon as there was any division in the ranks[121]

While Best expressed what he described as 'admiration' for Collip's work in other areas of endocrinology, he felt that his contribution to the discovery of insulin had been minimal.[122] He wrote that 'The problem of purification was not completed by Collip ... it took many, many more investigators and years of work before the insulin was perfectly pure'.[123]

Feasby, meanwhile, took a different line of attack. In his interview with Trueman, he was dismissive of the fact that the injection of Leonard Thompson with Best's extract had resulted in abscess formation. It was, he said, just 'an unlucky accident that it didn't prove that the extract was impure at all, merely proved that he [Leonard Thompson] was highly susceptible to infection, and maybe the syringe and the needle were dirty'.[124]

With a sense of controversy steadily growing around this story, the Metropolitan Life Insurance Company pulled the plug on their sponsorship for the proposed filmstrip. Meanwhile, despite having already invested $15,000 in the project, the NFB scaled down their plans in favour of something far less grand. Called *The Quest*, this short film became something of an embarrassment in Canadian medical circles, with the head of Canada's National Research Council, Dr. E. W. Steacie, concluding 'it was a pity that Dr. Best, a man of undeniably great gifts, had devoted so much time to building up his own part in the insulin discovery far beyond its actual importance'.[125]

While Best and Feasby were focusing their efforts on diminishing Collip's role, others were rallying to his side. In 1954, the clinician Joseph H. Pratt reviewed the early attempts to isolate insulin, including those of Georg Zuelzer. Like most others at the time, Pratt was unaware that Zuelzer had actually obtained a preparation of such high purity that it had resulted in hypoglycaemic shock. Not knowing this, he therefore concluded that the success of the Toronto group had been due to 'the precipitation of insulin by 95 per cent or absolute alcohol, a procedure which left the toxic products in solution. This was the insulin which gave the Toronto investigators their deservedly great reputation ...'.[126]

As far as Pratt was concerned, the crucial step in the discovery of insulin had been the innovation of using alcohol to prepare a non-toxic extract and he was in no doubt as to who should be given the credit for this innovation:

> What the group of chemists and physiologists in Berlin working with Zuelzer had failed to accomplish in four years, the young Canadian chemist did in a few weeks. Success came so quickly to Collip that neither he nor his co-workers Macleod, Banting and Best realized the great importance of his discovery that the precipitation of insulin by absolute alcohol freed it largely from contaminating products and made possible the development of a preparation that could be successfully employed in the treatment of diabetes ... Credit for the discovery of a preparation of insulin that could be used in treatment belongs to the Toronto investigators Banting, Best, Collip and Macleod working as a team ... It should be emphasized, however, that it was Collip who was the first to prepare an extract of the pancreas

that contained the internal secretion sufficiently freed from toxic products to be employed with success in the treatment of diabetes.[127]

Perhaps it was in response to Pratt that Best now changed his strategy. Instead of omitting Collip from the story, he began to give him public credit for his work in isolating insulin. But there was a twist, for with these new accolades Best came not to praise Collip but to bury him. In an address given in 1956, he praised Collip for having 'made rapid strides in the fractionation and concentration of the material, and contributed brilliantly in other ways to the development of the insulin researches'.[128] But this was false praise, built as a scaffold upon which to hang Collip's reputation. For highlighting the importance of Collip's contribution, was merely laying the groundwork to point the finger of blame when Collip had encountered what Best described as 'serious difficulties in the preparation of active material'.[129] Best told his audience that 'Several of the patients, including a young girl, in whom my wife and I were particularly interested died from lack of insulin having been dramatically improved by the first injections which they received'.[130] Reading between the lines, it was clear who Best blamed for this tragedy. When Best was awarded the Dale medal in London three years later, he was more explicit in his accusation: 'We had passed through a stage when Collip had lost the secret of making larger lots of insulin and some of the patients, who had been successfully treated, died from lack of the hormone'.[131]

The frustration that Banting, Best, and Macleod must have felt when Collip was unable to reproduce his earlier success in purifying insulin is understandable. But, so too, is Collip's failure. The purification of biochemical substances requires delicate chemical procedures, reliable equipment, and perhaps, most important of all, knowledge of their chemical nature. Hundreds of thousands of dollars had already been invested in attempts to purify the hormone thyroxine in the hope that it might be applied therapeutically, and these too were proving unsatisfactory. What hope was there that a lone chemist working with limited resources might be able to consistently and reliably isolate a substance whose chemical composition remained completely unknown?[132]

Collip had very much been fumbling in the dark, which makes his success in purifying insulin all the more impressive. But according to Best, this success was eclipsed by Collip's apparent ineptitude. In turn, this allowed Best to portray himself as the hero who saved the day by stepping into the breach when Collip eventually left Toronto to resume his post at the University of Alberta:

The failure of the supply of insulin created a major crisis. Fred Banting insisted that I should give up my study of physiological problems and take up the large-scale preparation of insulin. Dr. Collip had returned to his professorial duties and no one else was available for this work. The struggle to recover the secret of making insulin in sufficient quantities for clinical use was for me the most difficult and trying part of the whole insulin investigations. There was no time to approach the problem systematically and the only thing that seemed worth-while was to wage a night and day struggle in the hope that we might hit upon success.[133]

In response to Pratt's challenge, Feasby—who would go on to write a biography of Best—came to his defence:

> ... in February 1922 when Collip left the group Best had to resume responsibility for the production of insulin. Some of the patients originally treated died in the few weeks before insulin became available again. Best's modifications consisted essentially in the use of acid acetone as an extractive and of a fine wind tunnel for rapid concentration of extracts. This made processing at a lower temperature possible and provided all the insulin used in Toronto for experimental and clinical work in the late winter, spring, and summer of 1922.[134]

But unfortunately, Feasby had got his dates wrong. Collip had not left Toronto in February 1922, but had instead remained until May. This had given him enough time to make several important contributions to the recovery of the production of insulin, such as the use of acetone as a solvent instead of alcohol. But such inconvenient details were drowned out by the drama of Best's story and his growing prestige on the medical circuit. When the American Diabetes Association proposed to grant Collip the Banting Medal, Best used his growing influence to urge them not to do so until Pratt's article had been published. He then suggested that, if Collip did get a medal, he should not be invited to give the customary lecture.[135]

By the time that he gave the prestigious Oslerian Oration in London in 1957, Best appeared to have finally solved the problem of Collip. The solution was simply to backdate the moment at which insulin had been discovered to a point in time before Collip had even arrived in the lab: 'The seven months of harmonious, intensive work which Fred Banting and I carried on together in 1921 was the period which we both considered to be that of the Discovery of Insulin'.[136]

With this masterstroke, Best had hit upon a cast-iron strategy. Or at least, so it seemed, until 1969 when a letter appeared in the *British Medical Journal* which confronted Best with some rather inconvenient and very awkward facts.

6

Be Careful What You Wish For

The letter was from Dr. Ian Murray, a Scottish diabetologist who, while research-ing a book on insulin had made a discovery that intrigued him—and which ought to have been of great concern to Best. In his letter to the *British Medical Journal*, Murray pointed out that the discovery of insulin, and its subsequent isolation, were actually two very separate and distinct historical events. And although Murray readily accepted that Banting, Best, and Collip had isolated and purified insulin, he believed that the credit for its actual discovery was due to a Romanian scientist by the name of Nicolai Paulescu (Figure 17).[1]

Murray had found some of Paulescu's research papers, which described how in 1921, at the same time as Banting and Best were struggling in the summer heat against a mounting death toll of laboratory dogs and inconsistent results, Paulescu had published the results of very similar experiments. Even more, he had apparently carried them out with far greater success.[2] Intrigued by this discovery, Murray wrote to Professor Ion Pavel in Bucharest, who had been one of Paulescu's students and was one of the few Eastern European specialists in diabetes known in the West:

> Some time ago I read with the greatest interest a paper published in 1921 by Prof. Paulescu of Bucharest. It seems to me that he had isolated insulin before Banting had even started his investigations, and that if some pharmaceutical firm had taken advantage of his discovery—as Eli Lilly did after Banting and Best's experiments—the work of the Canadian workers would have been superfluous. Nevertheless virtually no recognition is now accorded to Paulescu. I would like to do what I could to rectify this oversight if the opinion which I have expressed is justified ... I hope you share my view that Paulescu's work deserves wider recognition ...[3]

Murray's timing could hardly have been better. Not only did Pavel share his view that Paulescu deserved wider recognition, but also he had already decid-ed to address this situation. With 1971 marking the fiftieth anniversary of the discovery of insulin, the International Diabetes Federation (IDF) was making preparations to celebrate this medical landmark. This presented Pavel with a golden opportunity to restore Paulescu to what he felt was his rightful place.

1869 -1931

Figure 17 *Nicolai Paulescu (1869–1931).*

Born in 1869 in Bucharest, Paulescu had already shown an impressive breadth of intellectual accomplishments by the time that he had finished high school. By the time he left for Paris to study medicine in 1888, he could speak French, read Latin and Greek, play the piano and organ, and was a fine artist. In Paris he took up a prestigious post working for the distinguished physician Etienne Lancereaux (1829–1910) at the Hotel Dieu Hospital, where he first became interested in isolating the enigmatic anti-diabetic substance made by the pancreas.

Lancereaux was eager that Paulescu should remain with him in Paris, but Paulescu's loyalties lay elsewhere. Turning down Lancereaux's offer, as well as posts in Switzerland and the USA, Paulescu left Paris in 1900 to return to

Romania. Paulescu was an ardent Romanian nationalist who held passionate anti-Communist views, and even after his death in 1931 this made him a figure of hate for the regime that took power in Romania after the Second World War. The new Communist government despised Paulescu as a dangerous reactionary and as a result, few original sources about his work survive today. In honour of his mentor, Dr. V. Trifu, one of Paulescu's former students, had carefully preserved all his notes and correspondence. However, when he learned that a search of his home by the state security services was imminent, Trifu took dramatic action. Knowing that he was already under surveillance himself for his own political sympathies, Trifu burned all Paulescu's documents.[4]

Although the original lab notes were destroyed, Paulescu had already published detailed reports of his work in several scientific journals before he died. After returning from Paris to Romania, he had conducted experiments into the role of the pancreas in diabetes. His work suffered a severe setback due to the occupation of Bucharest by German troops in 1916. During this time, the Medical Faculty was closed and Paulescu spent his time at home writing a three-volume textbook of physiology in which he described his work up to this point. After the First World War he resumed his work and, in the autumn of 1920, carried out a new set of experiments, the results of which he presented at four meetings of the Biological Society of Bucharest in April 21–June 23 1921. These results showed that not only could pancreatic extracts significantly reduce the levels of glucose, ketones, and urea in the blood and urine of animals rendered diabetic by pancreatectomy, but also, crucially, that they could consistently reduce the levels of blood sugar in a normal animal.[5,6,7,8]

These four papers were published in July 1921, and at the end of August—eight months before Banting and Best's first publication—Paulescu followed up with a more extensive paper, including crucial controls to show that this effect was specific to pancreatic extracts and was not observed with preparations made from other organs.[9] Confident that he had discovered the anti-diabetic substance, Paulescu filed a patent in April 1922 for the production of what he called 'Pancreine'. He then turned his attention to the challenges of scaling up the production of his preparations and improving their purity by using alcohol to remove protein contaminants.[10]

Then came the bombshell. Paulescu was stunned to learn from press reports that Banting and Macleod had been awarded the Nobel Prize for the discovery of insulin. As the dust settled, he wrote a furious letter to the President of the Nobel Committee:

> Dear President,
> I learned from the newspapers that the Nobel Prize Commission for Medicine has awarded Messrs Banting and Macleod of Toronto (Canada) a scientific reward for the discovery of the anti-diabetic effects of the pancreatic extract called by them insulin. I seek the opportunity to protest against the fact that this distinction was

accorded to some people who did not deserve it ... they did nothing other than repeat what I had earlier stated about the reduction of glycaemia and glycosuria, of blood and urinary urea, of ketonemia and ketonuria following the i.v. injection of the pancreatic extract into a diabetic animal. Your committee was therefore led to make a mistake in honouring some people who felt it proper to exploit and appropriate another's work.[11]

Paulescu couldn't even console himself by thinking he had been overlooked simply because he was an outsider, whose name was unknown to the Nobel committee and wider scientific circles. Ernest Lyman Scott, whose doctoral thesis ten years earlier had described the isolation of anti-diabetic extracts, was well acquainted with Paulescu's work. When Scott had first read Paulescu's papers in the *Archives Internationales de Physiologie*, he was so impressed that he wrote to congratulate him.[12] Meanwhile, John Murlin, who had originally been 'misled' into thinking that alkaline carbonate had an anti-diabetic effect, hailed Paulescu's 'favourable' results as having provided 'the immediate stimulus for the resumption of this work ...They were distinctly encouraging'.[13]

Nor had Banting and Best been completely unaware of Paulescu's work. In February 1923, well before hearing the news of the Nobel Prize, Paulescu had written to Banting saying:

Dear esteemed Professor Banting,
I understand from the Paris medical journals that you have published, in collaboration with a number of other experimenters, research on insulin ... Permit me to send you on this subject, one of my own publications, with which you are probably already familiar.[14]

Banting never responded to Paulescu's letter but was already aware of the scientific papers that Paulescu had sent him. Although these were published in French, Best had made an attempt to translate them and in his accompanying notes remarked that Paulescu 'Proves that the % blood sugar of a normal animal is lowered by injection of extract'.[15]

When Banting and Best published their first paper in 1922, they cited Paulescu's experiments. However, in attempting the translation, Best's grasp of French had let both himself—and Paulescu—down badly. While translating a key passage in which Paulescu discussed his results, Best misread Paulescu's original phrase 'non plus' ('no more') for 'non bon' ('no good').[16] As a result, Banting and Best wrote in their own paper that 'He (Paulescu) states that injections in peripheral veins produces *no* effect and his experiments show that second injections do not produce such marked effects as the first'.[17]

According to Professor Pavel, the effect of this single mistranslated sentence was devastating to Paulescu. It had given the Nobel Committee the impression that Paulescu had done nothing particularly worthy of note and so left his achievements eclipsed by those of Banting and Best.[18]

In desperation, Paulescu now wrote to other scientists to enlist their support for his cause. Some of them had already acknowledged the importance of his achievements, but when he wrote to the Medical Academy of Paris, he was simply told that, since neither Paulescu nor the Canadians had published their work in the Academy's Bulletin, 'it is difficult to establish a claim of priority'.[19]

Three years before he died, Paulescu expressed his feelings in the final volume of the physiology textbook he had been writing:

> Thus, while some have robbed me, others have tried to block my efforts to redress this injustice. This is happening in the midst of honest Men of Science! I feel it is my duty to inform my students of such shameful events so that when they dedicate themselves to scientific work they may know what to expect from some unscrupulous colleagues. Before all these, I believed, I learnt that a scientist could work in complete security—since I was convinced that the published data would silence any doubt that it might be contested. Unfortunately today, I am forced to confess that I erred completely in this respect. But I find it impossible to support another vice—even more ignoble—which is the true theft of another's scientific work ... I insist and demand, therefore, that an impartial international tribunal be established to investigate scientific fraud.[20]

But his call for justice did not go unheeded. In the late 1960s, Professor Ion Pavel took up the cause of his former mentor and, following the contact from Ian Murray, wrote a letter to Charles Best enquiring about his mistranslation. But when it arrived, Pavel found Best's response to be disappointing:

> Dear Doctor Pavel,
> In answer to your letter of the 8th of October, I am very pleased to learn that you plan to celebrate the 50th anniversary of Professor Paulescu's publication of his paper on the secretion of the pancreas. I regret very much that there was an error in our translation of Professor Paulescu's article. I cannot recollect, after this length of time, exactly what happened. As it was almost fifty years ago I do not remember whether we had relied on our own poor French or whether we had a translation made. In any case I would like to state how sorry I am for this unfortunate error and I trust that your efforts to honour Professor Paulescu will be rewarded with great success.[21]

As far as Pavel was concerned, Best had just squandered a golden opportunity to make amends for what he considered to be a grave injustice, and by doing so, set the record straight. Pavel even went so far as to suggest that the original mistranslation might not simply have been a mere accident arising from Banting and Best's poor grasp of French. Instead, Pavel suggested that it might well have been 'a deliberate confusion to discredit Paulescu's findings', which had been 'made to draw away the attention of the Nobel Committee, from the discovery of Paulescu'.[22]

Undeterred in his quest to restore Paulescu to his rightful place, Pavel wrote to Professor Arne Tiselius, Director of the Nobel Institute. Although Tiselius appeared to be more sympathetic, he felt that little could be done to rectify the situation:

> In my opinion, Paulescu was equally worth the award. As far as I know Paules-
> cu was not formally proposed, but naturally the Nobel Committee could have
> waited another year ... Unfortunately there is no mechanism by which the Nobel
> committee would do anything now in this or similar cases. Personally, I can only
> express the hope that in an eventual celebration of the 50th anniversary of the
> discovery of insulin due regard is paid to the pioneer work of Paulescu.[23]

As the sixties came to an end, Pavel had fresh cause for optimism. In 1970 at the 7th Congress of the IDF, it was proposed that a committee be established to determine who should be credited with priority in the discovery of insulin. But Pavel's optimism was to be short-lived. Even before the committee's report was published, Pavel expressed his doubts whether this committee would be entirely non-partisan in its judgement. For of the five committee members, one had links with the Toronto researchers, and there were no representatives from Romania to argue the case for Paulescu. Pavel later came to regret bitterly that he had not argued the case more strongly for a Romanian representative; he had been so confident at the time that the committee would find in favour of Paulescu that he had not felt it necessary to do so.[24]

According to the President of the IDF, the intention of the report published by the committee had not been 'to detract in any way from the contributions of Bant-ing, Best and Macleod in Toronto in 1921–1922 but rather to pay tribute to others whose published observations formed part of the background in which the inves-tigations of the group in Toronto began fifty years ago'.[25] Yet, when the report was finally published, Pavel described himself as feeling 'deeply disillusioned' by its conclusion:[26]

> Undoubtedly Professor N. C. Paulescu should be given special credit for the
> success with which his experimental observations were crowned. But more than
> experimental physiology was needed if insulin were to become available in the
> form, and on the scale, in which it was quickly needed for therapeutic use. The
> resources required involved not only the large-scale production of material of
> a refinement that ensured no irritant reaction on subcutaneous injection into a
> human being, but also the biological standardization of the hormon [sic] and the
> extensive testing of the standardized product.[27]

The report acknowledged, that while Paulescu might well have isolated insulin, this alone was not enough to merit the award of a Nobel Prize to him. For in the eyes of the IDF Committee, Banting and Macleod had overcome two further crucial challenges for which they were justifiably entitled to be awarded the prize.

The first of these was that they had successfully applied insulin to the treatment of human patients. Contesting this point, Pavel took pains to point out that the successful trials in humans had been thanks only to Collip. He also suggested that

Paulescu's failure to test his extracts on human patients was because he was concerned about the possibility of toxic side effects arising from impurities. According to Pavel, Paulescu's caution was rooted in concern for the safety of patients, in contrast with Banting, who was simply a glory seeker who had abandoned any such concerns in the interests of self-promotion.[28]

But even had Paulescu managed to get as far as testing his pancreine in patients, this might still not have been enough to secure fame and recognition. Around the same time as Paulescu was testing the effect of pancreatic extracts on diabetic dogs, another researcher in Eastern Europe had gone even further. In April 1924, Anatoly Kuznetsov presented a paper at a scientific meeting in Leningrad in which he claimed that his mentor, the late Professor Nikolai Kravkov (1865–1924), had not only successfully isolated an anti-diabetic extract from the pancreas, but also had successfully tested it in patients.[29] Unlike the extracts prepared by Paulescu, Zuelzer, or Banting, Best, and Collip, Kravkov's preparation, that he named pancreotoxine, was made by a rather different—and much simpler—method. Rather than prepare pancreatic extracts using alcohol, Kravkov had simply isolated the pancreas from a dog and perfused it with Ringers–Locke's solution. Through studies on dogs, rabbits, and frogs, Kravkov found that this perfusate had a hypoglycaemic effect and went on to obtain it in a dried form that was prescribed to patients in St. Petersburg hospitals.

Kravkov's widow maintained that, at the start of 1924, her husband had announced a plan to publish a comprehensive volume describing the production of pancreotoxine and its clinical applications, but it was not to be.[30] Aged only 59, Kravkov died that same year. And although Kravkov was a member of both the Imperial Military Medical Academy and the Russian Academy of Sciences, and the labels on bottles of pancreotoxine read 'Obtained according to Professor Kravkov's method', his name remained unknown beyond the borders of his native Russia.[31] Engulfed by the turmoil of the 1917 revolution and the civil war that followed, scientists in the fledgling Soviet Union had little opportunity for contact with their international peers.[32]

Meanwhile, Banting, Best, Collip, and Macleod remained unscathed by the social and political turmoil of wars and revolutions. They had also enjoyed one crucial extra advantage over Kravkov and, crucially, Paulescu: the support of a major pharmaceutical company. For the IDF, this was the second key reason why Banting and Macleod were the legitimate recipients of the Nobel Prize.

Pavel protested that the industrial infrastructure of Romania was so far behind that of the USA that Paulescu could never have hoped to achieve large scale production of 'Pancreine' as had happened when Eli Lily began their partnership with the Toronto research team to produce insulin.[33] But perhaps the greatest disappointment in the report for Pavel was that it failed utterly to address the issue of Banting's mistranslation of Paulescu's work and its consequences:

> When he realized his error, Banting might have been expected to seek an opportunity to correct it, especially since he had received extracts of the publications

of Paulescu which the latter had sent by post to Toronto. On the contrary, in subsequent articles, Banting chose either to refer to him in the following manner "Paulescu also briefly records favourable reports" or not to cite his work. All this painful controversy could have been avoided if, at the beginning, Banting had fairly stated his own investigations confirmed the previous work of Paulescu. One cannot get away from the idea that Banting misrepresented or filched the work of Paulescu. It is painful for me to write this, and I restrain myself from making more statements which are so critical of the Canadian author.[34]

Despite the disappointment of the report, Pavel's efforts were not in vain. In 2001, a statue of Paulescu was unveiled in Bucharest at a ceremony attended by Ion Iliescu, then-President of Romania and Sir George Alberti, President of the IDF.[35] Two years later, plans for the 18th Congress of the IDF included the unveiling of a bust and a plaque in honour of Paulescu at Hotel-Dieu in Paris, where he had worked in collaboration with his mentor Etienne Lancereaux. The ceremony was due to be held on August 27, followed by a conference the next day in honour of Paulescu. But just as it seemed that Paulescu had at last earned the recognition that many felt he deserved, the IDF suddenly called off the entire event.[36]

For it turned out that Paulescu's role in the story of insulin was not the only forgotten part of his past. Just as the IDF celebrations were to start, an article appeared in the French newspaper *Le Monde* revealing that, as well being a scientist, Paulescu had also been actively involved in extremist politics and was an active proponent of anti-Semitic views.[37] At around the same time, both the French Minister of Health and the Romanian Embassy in Paris received a fax in which Dr. Shimon Samuels, Director for International Liaison at the Simon Wiesenthal Centre, directed a similar and very serious charge. According to Dr. Samuels, 'whatever his scientific pretensions, Nicolae Paulescu was a founding father of the Rumanian fascist National Christian Defense League—a precursor to the notorious Iron Guard'.[38]

Dr. Samuels argued that 'Paulescu's articles inciting to murder contributed to the climate of hate, resulting in the massacre of half a million Jews in Holocaust-period Rumania' and it has since been claimed that he used his background in science to justify this violence.[39]

The League of National Christian Defence (LNCD), to which Dr. Shimons referred, was a political movement established in 1923 and the largest anti-Semitic organization in Romania at that time. Its founder Alexandru C. Cuza was a university professor whose entire political philosophy has been described by an international report into the Holocaust in Romania as having been 'built around a single issue, resting on a set of anti-Semitic convictions that he pursued steadfastly throughout his career'.[40] The LCND was just one of at least six different political movements founded by Cuza between 1895 and 1923 that espoused ultra-nationalism and violence against the Jewish population.[41]

Another of these was the National Christian Union, which Cuza founded with Paulescu in 1922 and which adopted the swastika as its official symbol. A report into the Holocaust in Romania, written in 2004, described how Paulescu actively supported and promoted Cuza's anti-Semitic ideology:

Paulescu was also self-trained in philosophy, which he sharpened into an anti-Semitic weapon, and like Cuza, authored pseudo-scientific works that served as vehicles for racial and religious hatred. Paulescu served as co-publisher and wrote regular articles for *Apararea National,* Cuza's newspaper starting in 1922. He wrote articles and books that sought to merge theology, medicine, and science into 'philosophical physiology' (fizologia filozofica), which was in reality simply a route through which he could express an obsessive anti-Semitism that made his views very appealing to Cuza.[42]

Anti-Semitism was, tragically, far from uncommon among Romanian intellectuals at the time, and Paulescu's political writings, with their vile references to the Jewish population as being 'parasites' and suggestions of their extermination, were to leave a horrific legacy.

Writing in 1913, Paulescu posed the following question in relation to Jewish citizens living in Romania:

> We Romanians are faced with a capital question: What shall we do with these uninvited guests who suddenly installed themselves in this country, or rather, these evil parasites who are both thieves and assassins?[43]

Tragically, Paulescu's vicious racist diatribes soon found their expression on the streets. In 1927, citing Paulescu's writing as his inspiration, law student Corneliu Zelea Codreanu (1899–1938) founded The Legion of the Archangel Michael (known simply as 'The Legion') with the aim of defending Romania—with violence if necessary—against what he claimed was the moral, political, and economic decay brought about by Judaism.[44,45] Its members were drawn from a wide range of social groups, including military officers, university academics, teachers, priests, magistrates, shopkeepers, and students who abandoned their studies to march in parades singing Legionary songs.[46,47,48]

A year after Paulescu's death in 1931, The Legion was renamed 'The Iron Guard' and, over the course of the next decade, they operated as a terrorist organization carrying out acts of violence against both the Jewish population and the political establishment of Romania. After orchestrating a number of political assassinations, including the murder of two Prime Ministers during the 1930s, the Iron Guard eventually seized power in Romania for a short while in 1940. During this time, synagogues were attacked, Torah scrolls burned, Jewish cemeteries desecrated, and torture centres established in which Jewish citizens were forced to hand over possessions, money, and property.[49]

Confronted with this shocking and appalling revelation about Paulescu's past, the IDF was forced to consider how, if at all, he and his work ought to be remembered. Wrestling with this issue, Professors George Alberti and Pierre Lefebvre of the IDF made the following statement:

> Although they did not hesitate to immediately cancel the ceremony, it seems to me that the Romanian authorities are extremely reluctant to publicly recognise

and condemn Paulescu's past ... One might ask, if by its conspiracy of silence, Romania has tried to manipulate international public opinion by hiding its anti-Semitic past behind a public veil ... The IDF does not wish to mix science and politics. But more information is needed before we can internationally laud an individual who has undoubtedly made a major scientific contribution, but who might have espoused a morally unacceptable position later in life.[50]

At a board meeting of the IDF in Copenhagen in 2004, Israeli diabetologist Professor Zvi Laron was commissioned with the task of producing a report to assess whether, in the light of these new revelations about Paulescu's past, it would still be appropriate for the IDF to go ahead with a plan to hold a lecture in his honour in 2006. The following year their decision was announced. After much reflection, the IDF declared that although it 'recognizes the contribution of Dr Paulescu in the discovery of insulin, the Federation would not be associated officially with Nicolae Paulescu and there would be no Paulescu Lecture at World Diabetes Congresses should such a request be received'.[51]

Speaking at a Symposium in Delphi in 2005, Professor Laron, who had led the commission, declared that the 'time has come to remove Paulescu out of the cathedral.'[52] With Laron's pronouncement, it seemed that Paulescu's dying wishes for international recognition had at last been granted.

Had these ugly revelations about Paulescu's past come to light during Best's lifetime, they might well have helped to extricate him from a rather awkward situation, for Best had devoted considerable time and effort into pressing the case that the moment at which insulin had been discovered was when pancreatic extracts had been shown to be effective in a diabetic dog—as he and Banting had done in the summer of 1921. His motive for this had been to diminish the contribution of Collip, but in so doing, he had strengthened the argument that it was actually Paulescu who had first discovered insulin. As the historian John Waller, put it, Best had well and truly 'painted himself into a corner'.[53]

Seven years before his death in 1978, Best was asked to comment on the contribution of researchers such as Paulescu and Zuelzer to the discovery of insulin. From his response it seems that he had paid heed to the fate of Israel Kleiner and concluded that modesty in science was not always a virtue: 'None of them convinced the world of what they had ...This is the most important thing in any discovery. You've got the convince the scientific world. And we did'.[54]

In his assessment of Best's motives, however, historian Michael Bliss arrived at a much more damning conclusion about Best's efforts to 'convince the scientific world':

> I never met Charles Best. From studying the written documents, I have not been able to make up my mind about the rationality of his sustained attempts to rewrite the history of the discovery of insulin. At times Best's distortions of the historic record seem to amount to a deliberate, unethical exercise in falsification which verges on scientific fraud. The obsessive pattern and often transparent sloppiness of Best's distortions, however, reinforced by many comments garnered in

interviews with former associates, suggests that there was perhaps less controlled calculation to his historical reflections than a structured retrospective study of them implies. In the later years of his life Charles Best appears to have been deeply insecure about and obsessed with his role in history. He appears to have had a profound hunger for recognition, a serious ego-problem, many thought, which overwhelmed his good sense even in the eyes of such steadfast friends as Sir Henry Dale. The fumbling attempts ... to manipulate the historical record would have been pathetic and hardly worthy of comment had they not been so grossly unjust to Best's former associates and, for a time, so influential.[55]

Bliss may well have been right that Best had been driven by ego and insecurity, but there may have been other forces at work, too. On March 13, 1964, following a lecture tour in the USA which had left him exhausted, he was admitted to hospital.[56]

A day earlier, a number of harmful pharmaceutical samples had been found lined up next to his bed.[57] According to a biography by his son, Best had been suffering from a severe depression for some time and after receiving electro-shock therapy he began to improve, but he was also aware that, as both his mother and father had suffered from it, the condition ran in his family.[58] By this time, Best was a world-renowned scientist who had accumulated an impressive list of honours. He had been invited to give prestigious lectures around the world, had received honorary degrees from numerous foreign universities, medals from esteemed institutions, and sat on the boards of organizations. Best may have won the fame he had always craved, but it seemed that none of these accolades had been enough to keep his demons at bay.

Whatever the motivations behind his efforts to 'convince the scientific world', Best had recognized something that would come to be of increasing importance in science. For in the age of the mass media, it was no longer enough for scientists to just confine themselves to the lab bench—they also needed to be step outside their laboratories and communicate the importance of their work to the wider world.

Best's insight proved to have particular relevance to the story of insulin. Six months after Best's death in March 1978, a team of scientists from City of Hope Hospital and a small company called Genentech gathered before a pack of journalists and cameras to 'convince the world' that they had achieved something of which Best had only dreamed.

In 1956, Best had already speculated that it might one day be possible to synthesize human insulin.[59] At that time, diabetic patients were only able to treat their condition by injecting themselves with insulin obtained from either bovine or porcine pancreatic tissue recovered from the meat industry. But there were growing concerns about the future reliability of these sources and even talk of an impending shortage of insulin. Now, thanks to new developments in genetic engineering, the Genentech team had coaxed an ordinary bacterium into making human insulin—and in so doing, made Best's vision at last a reality.

But none of this could have happened without one crucial question first being answered—what exactly was insulin? The 'thick brown muck' prepared by Banting, Best, Collip, and Macleod was saving lives, but its precise chemical nature remained a mystery. The handful of clinicians and scientists working on insulin being likened to 'explorers who had found an unknown continent and were trying to map it all in a few months'.[60]

In time, however, a map would be drawn and the result was a fundamental transformation in our understanding of biology. The origins of this seismic intellectual shift began far from the glamour of Genentech's press conference or life-saving medical research in a converted former stable. The consequences would be profound. To understand them we must first take a short detour into outer space.

7

'In Praise of Wool'

In May 2014, scientists at the European Space Agency (ESA) held their breath and crossed their fingers as they watched the screens in front of them. After a ten-year journey through space, the Rosetta probe, on which they had all been working for many years, had finally reached its destination and was about to enter orbit around a distant comet called 67P/Churyumov–Gerasimenko.

As the Philae landing vehicle released by Rosetta touched down on the surface of the comet, the ESA scientists were euphoric. But two years later, data from Rosetta gave them new cause for excitement. Sensitive equipment on board the Rosetta probe had detected traces of an organic compound called glycine in the dust around the comet.[1] The chemical formula $C_2H_5NO_2$ of glycine is just one of the twenty different types of chemical known as amino acids that are the building blocks of proteins (Figure 18).

Proteins are far from just being an essential dietary requirement. They are nanomachines capable of carrying out a vast array of essential functions in living systems. The protein haemoglobin carries oxygen in the blood; the protein rhodopsin traps photons of light in the retina of the eye and converts them into electrical impulses that travel down nerve cells to the brain; the enzymes of the digestive system are proteins that break down complex molecules into simple ones that can be absorbed across the lining of the gut into the blood.

The first evidence that insulin might also be a protein came in January 1926, when American chemist John Abel (1857–1938) used crutches to stagger to his feet to give the sixth Pasteur Lecture at the Chicago Institute of Medicine. Only a couple of weeks earlier, Abel had been hit by a car, resulting in the fracture of his leg in two places. But he was determined not to let what he dismissed as this 'minor inconvenience' prevent him from making an announcement that the *British Medical Journal* hailed as a 'brilliant investigation, which promises to open up a new insight into carbohydrate metabolism'.[2,3]

Abel was 67 years old when he had first begun his work on insulin—an age by which most research scientists today have long since left the laboratory bench (albeit perhaps reluctantly) to spend their days in an office working their way through piles of grant applications or enduring interminable committee meetings. In 1893, he had been appointed the first Professor of Pharmacology at the

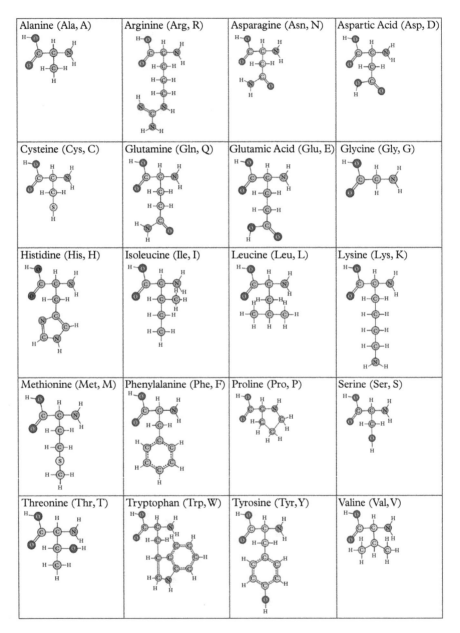

Figure 18 *Amino acids are the building blocks of proteins and have the general formula:*

Where R can be a number of different chemical groups as shown in the table. In the case of Glycine, which is the simplest of the amino acids, the -R group is a single atom of hydrogen (H). Each amino acid can be represented by a three letter or one letter code, which in the case of Glycine is 'Gly' or simply 'G'.

Diagram by K.T. Hall.

Johns Hopkins School of Medicine in Baltimore, where he developed an interest in hormones. After isolating a chemical derivative of adrenaline from the adrenal medulla, he narrowly missed out on obtaining the hormone in pure form and turned his attention to how the active agent of the posterior pituitary gland might be isolated.[4] By 1924, however, after several years of concentrated efforts on this problem with no results, Abel was discouraged.

Then came an unmissable opportunity: the California Institute of Technology (CalTech) had been awarded a $10,000 grant for research into insulin and, as one of the country's leading experts in the field of hormones, Abel was approached to see whether he would be willing to accept this role. His response, sent in a cable to CalTech's Professor Arthur Noyes said simply: 'Will attack insulin'.[5]

Abel purified several different fractions from crude pancreatic extracts, one of which was of particular interest to him. Known only as 'Fraction IV' it had two properties that distinguished it from the others: a high content of sulphur, and the ability to lower levels of blood sugar in test animals.[6] Comparing himself to Theseus in classical mythology, Abel described Fraction IV as being 'the thread that will lead out of the labyrinth'.[7] It would prove to an apt image, for it set Abel on the path to showing that, far from being just 'thick brown muck', insulin was a defined chemical entity.

Leaning on his crutches, Abel told the audience in Chicago that seeing the glistening crystals of insulin forming on the sides of a test tube had been one of the most beautiful sights in his life.[8] And no sooner had he obtained them than he swiftly sent samples to both Macleod and Banting, proudly informing them that they were the first researchers outside Baltimore to see crystals of pure insulin.[9,10]

When treated with Biuret or Millon's reagents that detect the presence of protein, the crystals gave a reaction that was 'unequivocal and positive' and the tiniest amount of them was sufficient to cause convulsive levels of hypoglycaemia in a test animal.[11]

This was all strong evidence that insulin was a protein, and in a letter to Macleod, Abel outlined his plan to blaze the trail even further:

> The compound promises to be one of great interest from a chemical point of view ... I hope to go on soon to the determination of the molecular composition and other points connected with this fascinating compound.[12]

Abel was not alone in attempting to determine the exact chemical nature of insulin. Others had tried before him, but the picture that was emerging was unclear. Researchers at Eli Lilly had isolated an extract from pancreatic tissue that gave a positive result when subjected to almost all the tests available for protein and was also destroyed by enzymes which specifically digest proteins.[13] But although these results might be interpreted as strong evidence that insulin was a protein, the Lilly researchers stopped short of drawing this conclusion, opting instead for caution:

> … the pancreatic substance containing insulin appears to be a complex mixture …
> from which it has been as yet impossible to isolate a simple substance or to detect a
> chemical reaction that is characteristic of the physiologically active constituent.[14]

Had others shown the same degree of caution, the situation might have been
less confusing. When John Murlin had been spurred by the news from Toronto into resuming his own research on pancreatic extracts, he foresaw that the next
milestone was to determine the 'nature of the active principle' in these extracts.[15]
Having lost out on the glory of discovering insulin to the Toronto team, Murlin
may have hoped that identifying its chemical nature would give him a second
chance to make history. Unfortunately, it was not to be.

Murlin had found that pancreatic extracts with lower protein content caused
a less toxic reaction when injected, and so concluded that all protein should be
removed from the preparation.[16] When he reported the successful isolation of
material that was both physiologically active and which, according to chemical
tests, showed no trace of protein he declared that 'the evidence accumulates therefore that insulin per se is not a true protein in the sense of being composed of
nothing but amino acids'.[17,18] To add to the confusion, Best and Macleod reported that, although insulin prepared from the pancreas of an ox or pig gave a positive
chemical test for protein, material obtained from the pancreas of skate did not.[19]

Was insulin a protein, or not? Chemist Hans Friedrich Jensen (1896–1959)
from the University of Göttingen joined Abel and together they did much to
pioneer the chemical study of insulin. As a result, Jensen had an answer to this
question—one that would make a politician proud:

> No observations have been made in the course of this work which could be inter-
> preted in favour of the assumption that the chemical nature of crystalline insulin
> is essentially different from that of a protein-like body. Bearing in mind, however,
> the extraordinary physiological properties of the substance which have no analo-
> gy, in so far as we know, among the great number of well defined animal proteins,
> we wished to avoid any preconceived attitude in this regard.[20]

There was good reason to be politically canny about the claim that insulin was a
protein. Through his work on the chemical nature of insulin, Abel had stepped
into an arena in which two rival groups of scientists were fighting an ongoing—and
sometimes quite acrimonious—dispute.

At least part of this animosity was, quite literally, hormonal. Hormones are
able to exert very specific physiological effects such as, in the case of insulin,
the lowering of blood glucose levels. Similarly, enzymes also exert very specific
physiological effects by acting upon particular chemical substrates, such as the
digestive enzyme amylase, which breaks down starch in the diet into glucose. But
what exactly the chemical nature of hormones and enzymes might be, and how it
enabled them to achieve these highly specific effects remained a mystery.

The most popular view was that held by the German scientist Richard Willstätter (1872–1942), who had been awarded the 1915 Nobel Prize in Chemistry for his work on plant pigments, and the Iron Cross Second Class for his development of protection against poison gas during the First World War.[21] Willstätter had done much work to show that enzymes are not tiny organisms, but chemical entities and, as a Nobel Laureate, his views on their chemical nature carried much weight—particularly his conviction that they could not possibly be made of protein.

Willstätter had good reasons for taking this stance. Ever since the end of the nineteenth century, it had been known that proteins were made of amino acids, like the glycine detected by the Rosetta probe. But while physical studies had shown proteins to be molecules of enormous size, amino acids were tiny. How, then, could these small amino acids form something as big as a protein? One popular idea was that proteins were formed by the loose aggregation of amino acids to form an amorphous blob, or 'colloid', with no specific molecular mass. But it was impossible to see how a substance such as a colloid could exert the kind of specific physiological effects carried out by hormones and enzymes, e.g. lowering blood sugar or catalysing chemical reactions in the body. Willstätter concluded that, at best, proteins could only be passive carriers whose role was to transport the actual agent—most likely a small molecule—that was actively causing the enzymatic or hormonal effect. When Willstätter spoke, people listened, and most of them agreed.

Abel, however, was one of the minority who did not. Reporting on his studies of crystalline insulin in 1927, he, showed his skill as a politician:

> We may conclude our remarks touching the possibility of the adsorption of an unknown hypothetical hormone on our crystalline compound by saying that all our efforts to establish anything of this nature have ended in failure. We should indeed have been highly delighted to isolate, or even to prove the existence of, an insulin of the potency that might justly be anticipated as inherent in a substance adsorbed on our crystals. Had we succeeded in doing this we would gladly, and with no repinings have transferred our allegiance ...[22]

Abel's case, however, was not helped by the fact that, for a time, he lost the ability to grow crystals of insulin. He blamed this on the quality of the samples that he was being sent from Eli Lilly, but a more likely explanation was found a few years later by Canadian chemist David Scott who, working at the Connaught Labs in Toronto, showed that zinc is essential for the crystallization of insulin.[23]

But Abel at least had an ally. James Batcheller Sumner's (1887–1955) willingness to take on seemingly lost causes might well have been forged through the adversity he had encountered early in his life. While out hunting grouse when he was only 17 years old, Sumner had been accidentally shot in the left arm by a companion.[24] As a result of the injury, Sumner's arm had to be amputated above the elbow and as he was left-handed, this left him with a serious challenge. But rather than crush his morale, the injury galvanized Sumner's determination to distinguish himself on the sports field. Having trained himself to use his right arm, Sumner pushed himself to excel in a diverse range of sports, including skating, skiing, swimming, canoeing, and billiards. And it wasn't only on the sports field that he excelled. Having graduated from Harvard with a degree in chemistry, he set his sights on a PhD in biochemistry, but was advised by one Professor that he should take up law instead, as 'a one-armed man could never make a success in chemistry'.[25] Sumner was to prove him to be spectacularly wrong.

In 1914, having completed his PhD, Sumner took up a position as Associate Professor of Biochemistry at Cornell University Medical College where, three years later, he first began working on the enzyme urease, which catalyses the conversion of urea, a metabolic waste product, into carbon dioxide and ammonia. After nine long years, just as John Abel was announcing that he had prepared crystalline insulin, Sumner published a short paper reporting that he had obtained crystals of urease that gave positive results when subjected to a number of chemical tests for the presence of protein. Moreover, these crystals retained their enzymic activity.[26]

This was strong evidence that the urease enzyme was made solely of protein—there was no elusive additional non-protein chemical agent, as Willstätter and his acolytes claimed. But Sumner had to fight to get this idea accepted. In 1927, Willstätter gave a couple of lectures at Cornell and had the opportunity to discuss with Sumner the exact methods that he had used in his own work. When Sumner explained that he had used a different experimental protocol to that used by Willstätter, the German professor threw up his arms, declaring 'What is the use of saying any more', before bowing and walking off.[27] The following year, one of Willstätter's supporters, Dr. Hans Pringsheim of the University of Berlin, visited Cornell, where it became very apparent that conversational small talk was not one of his strengths. For on being introduced to Sumner his first words were, 'I have read about your isolation of urease, which is not true'.[28] A few weeks later, after Sumner had delivered a lecture on his work, Pringsheim continued to challenge him, asking 'Why did you have the presumption to think that you could isolate an enzyme when so many of our great German chemists have failed?'[29]

Sumner's presumption served him well. In 1946, he was awarded the Nobel Prize in Chemistry for showing that an enzyme could indeed be crystallized and demonstrating that they were protein in nature. The award was jointly shared with John Northrop who, in 1930, had obtained crystals of the enzyme pepsin and also shown them to be made of protein, and with Wendell Meredith Stanley for his crystallization of proteins from the Tobacco Mosaic Virus. Their work was strong evidence that, far from being colloids, proteins were discrete molecules—and this, in turn, opened up the possibility that they could have specific physiological effects.

Like Sumner, Abel was convinced that his crystals were a homogenous substance made of protein and that there was no extra non-protein component responsible for the physiological effect of lowering blood sugar.[30] Yet, even as Sumner and Abel were presenting their respective work on the crystallization of urease and insulin, the idea that proteins were colloids was coming under attack from other quarters. In 1926, the Swedish scientist Theodor Svedberg received the Nobel Prize in Chemistry for his invention of the ultracentrifuge, which allowed solutions to be spun in a rotor at incredibly high speeds to generate a centripetal force several thousands of times stronger than the Earth's gravitational field. Under these conditions, a suspension of particles in solution will settle out, or 'sediment', at a rate that is proportional to their weight and size. This allows substances in a mixture to separate according to their weight. Using this method, Svedberg found, to his surprise, that the blood protein haemoglobin appeared not to be made up of colloidal lumps of varying size, but rather that it had a precise molecular mass—suggesting that it was a discrete molecular entity.

Further evidence in support of this view came from an unlikely source. In 1902, the German scientists Franz Hofmeister and Emil Fischer had both independently proposed that proteins were not ill-defined colloidal aggregates of amino acids, but rather that these amino-acids formed covalent bonds with each other to generate macromolecular chains of specific molecular weight (Figure 19).

X-ray studies of textile fibres made during the 1920s and early 1930s by German researchers Hermann Mark and Kurt Meyer and the physicist William Astbury (1898–1961) (Figure 20) of the Department of Textile Physics at the University of Leeds seemed to confirm this view.[31] Meyer and Mark's studies of fibroin, the main protein in silk, and Astbury's work on keratin, the major protein component in wool and hair, both showed these fibres to be long chains of amino acids bonded together.[32,33] Astbury memorably likened this configuration to that of a 'molecular centipede' and went on to show that the keratin proteins in wool could adopt a compacted or extended conformation, which explained in molecular terms what made wool fibres so attractive as a raw material to the textile industry—their elasticity.[34]

By the time Sumner was giving his Nobel speech in Stockholm, the importance of proteins had already become apparent. Struck by the fact that so many physiologically important substances such as digestive enzymes, haemoglobin, and insulin were all proteins, the scientist Max Bergmann (1886–1944) felt that this could not be mere coincidence. 'Wherever life phenomena occur', wrote Bergmann, 'proteins are involved in one way or another. Proteins are therefore regarded as being the chemical requisite of life'.[35]

From his chemical analyses of insulin, Jensen had concluded that the secret of the molecules' activity must 'originate in the manner in which the component

(a)

(b)

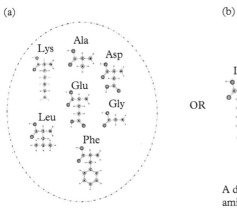

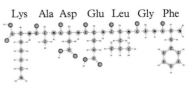

OR

A definite chemical structure formed by amino acids bonding together with each other to form a chain, like beads on a necklace?

An amorphous blob formed by the loose association of amino acids?

Figure 19 *(a) During the first half of the 20th century, proteins were widely believed to be a loose amorphous aggregate of amino acids. But Emil Fischer and Franz Hofmeister proposed an alternative (b) in which proteins were long chain molecules formed by the linking together of adjacent amino acids through the formation of a chemical bond (known as the peptide bond) between their amino (NH2) and carboxyl (COOH) groups:*

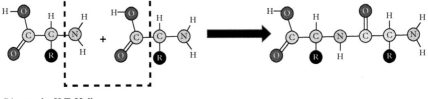

Diagram by K.T. Hall.

amino acids are linked'.[36] This was to be a crucial insight, and one which had also occurred to Bergmann:

> What is the basis for the fact that the protein pepsin possesses enzymic properties while other proteins, composed of the same amino acids, lack these enzymic characteristics? The same question may be posed in regard to the hormone character of insulin. All evidence points to the fact that the enzymic properties of pepsin, as well as the hormonic properties of insulin, are due to particular arrangements of amino acids in these proteins.[37]

Bergmann and Jensen had both recognized that the property of haemoglobin to carry oxygen, or of insulin to lower blood sugar levels, must reside in specific differences in the structure and chemical composition of these two proteins. If different amino acids really were joined together like beads on a string to make a protein chain, then Bergmann suggested that the task of the biochemist was 'to establish the molecular composition and architecture of the various proteins

Figure 20 *William Astbury (1898–1961).*

Credit: Reproduced with permission of Special Collections, Brotherton Library, University of Leeds.

with the same precision as the molecular composition and architecture of simpler molecules have been ascertained'.[38] The main question was if amino acids were strung together in chains like beads on a necklace, were they arranged in any particular order? If so, how was this determined?

In Leeds, physicist William Astbury was already convinced that he had glimpsed the first hint of an answer. Thanks to his work on textile fibres, Astbury was becoming internationally renowned as an expert in using X-ray methods to reveal the molecular structure of fibrous proteins, but he had not confined his studies merely to the keratin proteins in wool. From a combination of X-ray studies and chemical analysis of the protein gelatine, he proposed that every third residue along the polypeptide chain of gelatine might be occupied by glycine—the same amino acid detected by the Rosetta Probe.[39] This suggested a tantalizing possibility—that the arrangement of amino acids in a protein was dictated by mathematical laws—a possibility which, according to Astbury's friend and colleague Reginald Preston, was a kind of holy grail for him:

> ... he almost worshipped order in nature and was always exceedingly excited to think that he was working in the realm of living things in which the order was a bit obscure, but he was doing his little bit to discover it. He was convinced of the intense simplicity of this order and he had an almost mystical belief that it rested on a very simple structure of a protein which is universal.[40]

Astbury's lab at Leeds was once described as 'the X-ray Vatican' and the other high priests of protein research certainly shared his dogma that a hidden mathematical order existed at the level of their molecular structure.[41] Working at the Rockefeller Institute in New York City, Bergmann and his colleague Carl Niemann put this to the test by using acid to break the links between individual amino acids in proteins chains. Different amino acids were then selectively precipitated out of these mixtures as complex insoluble salts and their amounts measured. To return to the analogy of a protein as a necklace strung together from beads of different shapes, colours, and sizes, Bergmann and Niemann's approach was to snap the string between each individual bead with acid and then count the number of each different type of bead.

When Bergmann had first applied this approach to gelatine, he had made a startling discovery. Three amino acids—glycine, proline, and hydroxyproline—all occurred in the simple ratio 6:3:2, leading him to conclude that each different type of amino acid occurred with a regular frequency along the chain—every third amino acid in the chain was a glycine, every sixth a proline, and every ninth a hydroxyproline.[42] Applying the same approach to the blood protein fibrin, he and Niemann were excited once again to find 'impressive stoichiometrical relationships' between the proportions of different amino acids which suggested that each type of amino acid occurred throughout the length of the protein chain with 'its own particular periodicity'.[43] Further experiments on three more proteins—haemoglobin, egg albumin, and silk fibroin—gave similar results, leaving Bergmann and Niemann convinced that they had made a major discovery. They were confident that they had glimpsed the first tantalizing hint of 'a basic law'—a general principle or algorithm that governed the arrangement of amino acids in protein chains.[44] According to this law, the number of residues of a particular amino acid in a protein could be calculated from a simple formula based on multiples of 2 and 3 and each different type of amino acid occurred throughout the protein chain 'with a characteristic whole number frequency'.[45]

For a scientist like Astbury, who had become passionate about applying the methods of physics to biology, this was a dream come true. It was starting to look as if beneath biology's messiness, there lay the kind of mathematical order that physicists crave. Astbury was ecstatic:

> ... it seems clear that we are now on the fringe of something very fundamental indeed in protein theory, and the moral value alone of Bergmann and Niemann's discoveries will be immense.[46]

Astbury was famous not just for his work on biological fibres, but also for sometimes allowing his excitement to run ahead of his results. In a letter a few years later to biochemist Albert Chibnall, who was based at Imperial College and with whom he collaborated, Astbury wrote with excitement that 'I feel convinced now that 2's and 3's really do lie at the root of things'.[47] He also did not miss the opportunity to point out to Chibnall that, as far as he was concerned, it had been his work on gelatine that had been the inspiration for Bergmann and Niemann to begin their own studies into protein structure. Having once described himself as being the 'alpha and the omega' of protein fibre studies, it appears that modesty was not always his most pronounced trait.[48]

Not everyone, however, was so easily convinced. Richard Synge (Figure 21) was a young scientist who had just started his PhD at the University of Cambridge, and although the subject of his research seemed unlikely to set scientific pulses racing, it would, in time, not only demolish the ideas of Astbury, Bergmann, and Niemann, but also transform our understanding of biology. For although Synge was working on the chemistry of wool, the impact of his work would go much further and unravel not only the chemical mystery of insulin but also of DNA.

Figure 21 *Richard L. M. Synge (1914–1994).*

Credit: Elliott & Fry, 1952. © National Portrait Gallery, London. Reproduced under license by National Portrait Gallery, London.

Synge was the son of Laurence Millington Synge, a Liverpool stockbroker who had changed his surname by deed poll from 'Sing' to the older 'Synge' in 1920. The family could trace its ancestry back to the sixteenth century and consisted of an English and an Irish branch, the latter boasting the playwright J. M. Synge. The family name had originally been 'Millington', named after Millington Hall in Cheshire, where they had once lived, but according to family legend, after one of their ancestors had sung so beautifully in front of Henry VIII the monarch had demanded that they change their surname to 'Synge'.[49]

Before attending boarding school at age nine in preparation for Winchester College, Synge was educated at home by his mother. She once described him to a visitor as being 'unusual' and 'so absent minded (really thinking of other things) that he was not safe to be let walk the village street alone'. The visitor responded: 'clever and absent minded, you've got a genius on your hands'.[50]

Even at an early age, Synge demonstrated a phenomenal ability to retain facts and an attention to detail. On one occasion, while he was still quite young, he discovered some documents relating to local land sales that his father, who was High Sheriff of Cheshire, had left lying around. From a careful study of these papers, the young Synge showed that a local firm of solicitors had sold the same piece of land twice to the church. Another source of detail that held endless fascination for him were the routes and timetables of railways. On occasion, when taking the train to go home at the start of school holidays, he would send all his luggage by a different rail route in order to see whether he could arrive in Chester before his bags did.[51] By this method he hoped to test the accuracy of the timetables for the journey and later, as an adult, he wrote a detailed account of the journey from Peking to Moscow for *The Railway Magazine*, having returned from Australia via China and the trans-Siberian route.[52]

While at Winchester, Synge became friends with a boy called John Humphrey, with whom he shared a growing interest in organic chemistry. Their early experimental forays into this subject usually involved strategically placing the unstable compound nitrogen triiodide around spilled marmalade to terrorise wasps, or using it to annoy prefects, but as Humphrey's father had been one of the founders of the company ICI, this friendship gave Synge his first glimpse of how chemistry might be applied to more serious matters.[53] Being sworn to secrecy, Synge and Humphrey were shown drawings of the interior of the reaction towers at the company's plant in Winnington. In these towers, sodium carbonate was produced by the Solvay process, which used the counter-current flow of fluids—a principle which would later prove to be an important step in Synge's own work.

In 1932, Synge won an Exhibition to read Classics at Trinity College, Cambridge—an achievement made even more impressive by his later confession that, immediately before sitting the exam, he had gone out for lunch with his uncle and downed half a bottle of wine. Reflecting on the high spirits with which he had entered the exam hall, he remarked that 'I shall never know if that half-bottle raised my performance to the level of an exhibition or reduced it to an exhibition from a scholarship'.[54] The following year, however, he read a newspaper report that made him change his mind about where his academic interest really ly lay. The report contained an address given by the biochemist Sir Frederick Gowland Hopkins (1861–1947) to the British Association for the Advancement of Science and it left Synge 'impressed by the idea that living things must have wonderfully precise and complicated working parts on the molecular scale and that the biochemist had the best chance to see how these are put together and do their work'.[55] The most important of these were the proteins, and Synge was now convinced that his future career lay not in the study of Pliny and Plutarch, but of proteins. Thus, shortly after his arrival in Cambridge, he switched from Classics to reading Natural Sciences before starting a PhD.

Although the title of Synge's doctoral thesis ('Some New Methods in Amino-Acid Analysis: The Amino-Acid Composition of Wool') hardly sounded as it was going to set the scientific world on fire, it was nevertheless of great economic importance.[56] There was a growing fear at the time that wool might be super-seded as one of the main raw materials of the textile industry by new synthetic fibres. To address these concerns, the wool growers of Australia, South Africa, and New Zealand had formed an organization called the International Wool Sec-retariat (IWS).[57] Thanks to the guidance of Sir Charles Martin (1866–1955), one of its scientific advisors and former Director of the Lister School of Preven-tive Medicine, the IWS was encouraged to pursue fundamental scientific research into wool. Supported by a generous grant from the IWS, the aim of Synge's doc-toral research was to develop a new method by which to analyse the amino-acid composition of the keratin fibres in wool. But from the very opening pages of his thesis, Synge was well aware that the implications of this work might well go far beyond improving the raw materials of the textile industry.

In his introduction, Synge made it clear that his aim was to test the idea pro-posed by Bergmann, Niemann, and Astbury that the different amino acids in proteins were arranged according to some mathematical law. Synge knew that this would first require an entirely new method of separating amino acids from a mixture. The current methods of estimating the amounts of amino acids, such as crystallization, or those used by Bergmann and Niemann which relied on pre-cipitating amino acids as insoluble salts, were too inaccurate to allow any firm conclusions to be drawn—a point which Synge made quite bluntly at the opening of his thesis:

> Owing to the technical backwardness of amino-acid analysis this hypothesis has not been rigidly tested, and these authors may have claimed more than the accuracy of the experimental evidence justified.[58]

In the final year of his degree, Synge worked for a short time on glycoproteins—proteins modified through the attachment of sugar molecules—and an important part of this work was finding an effective way of physically separating differ-ent types of sugar molecule. Under the tuition of 'wee Davie' Bell, an expert in carbohydrate chemistry, Synge learned how individual solutes could be effi-ciently separated from a mixture on the basis of their different solubility in two immiscible solvents. This insight—which relied on the simple principle that 'oil and water don't mix'—was one that now proved invaluable for his PhD—and beyond.

Using water and chloroform as solvents, Synge applied this same technique to the separation of amino acids from a mixture derived from wool proteins and quickly realized that the separation was far more effective if the amino acids were chemically modified by attaching an acetyl group to them. However, carrying out this procedure by hand was laborious, tedious, and time consuming; thus, Sir Charles put Synge in contact with another young scientist called Archer Martin (Figure 22) (no relation) who he thought might be able to help.

Figure 22 *Archer John Porter Martin (1910–2002).*

Credit: Elliott & Fry, 1950. © National Portrait Gallery, London. Reproduced under license by National Portrait Gallery.

Synge recalled Martin as having a range of eclectic interests, which, besides chemistry, included the occult, extra-sensory perception, psychiatry, and the theory of numbers.[59] Later in his life Martin described his career as simply having been that of 'a hobbyist who gets paid to have fun'.[60] Of all these hobbies, one had held a lifelong interest ever since adolescence when, like generations of teenagers before and since, Martin would spend long hours alone locked away behind closed doors. But his parents need not have worried about what was keeping their son so preoccupied for in Martin's case, this particular pastime was both innocent and somewhat unusual:

> I was fascinated by fractional distillation as a method whilst still a schoolboy, and built in the cellar of my home, which was my combined workshop and laboratory,

distillation columns, packed with coke of graded size, some five feet in height. They were made from coffee tins (obtained from the kitchen) with the bottoms removed, soldered together! Experience with them served me in good stead ... [61]

Ever since he was a young child, Martin had been fascinated with understanding how things worked, for as he later recalled:

I remember from as early as I can, being fascinated to watch any workmen that came to the house, to repair anything. I would stay with them as long as they were there and absorb everything that they did with great detail. I am told, though I don't remember it myself, that at the age of 5, I was replacing tap washings having the maids unscrew the tap which was beyond my strength, but directing the operation and putting the washers on.[62]

But it was chemistry that held a particular fascination for him, and as a child he found a novel means of funding his early research:

I got my beginnings in chemistry from my youngest sister who was five years older than myself. During the war, about 1915 or 1916, my sister and I kept hens, which we sold to my mother. My father believed that it was very bad to give any significant amount of money to children, so we were always very short of pocket money, and we had to earn it in some way. And so we kept hens; we spent the proceeds from selling the eggs on buying chemicals and doing chemistry. She, at the time, was 11 and had just started chemistry in school so she had a textbook in chemistry, and, together, we set up a laboratory in a tiny potting shed which must have been about 7 feet by 3 feet.[63]

Despite being unable to read until the age of eight years old due to dyslexia, Martin eventually won a place at the University of Cambridge. Thanks to his interest in the separation of chemicals he had hoped to study chemical engineering but, as Cambridge did not offer this course, he opted instead to study Natural Sciences. It was here that he was persuaded by the biologist J. B. S. Haldane—who held an open house for undergraduates on Sundays—to turn his attentions to biochemistry. Unfortunately however, Martin faced additional challenges, which took a toll on his studies:

I did get a degree in part two biochemistry, but it was a 2-2 which is a poor degree. I like to think that this was because in fact I was very depressed in the third year I was at Cambridge. I'd been subjected to depressions throughout my life. They would come for a year or 18 months and go away for a few years. During that period I was pretty useless. I didn't receive treatment until 1968; and by taking the right monoaminoxidase inhibitors, I've been brought onto a pretty normal keel.[64]

According to Synge's own recollections, Martin's bouts of depression had, on one occasion, become so severe that he had even contemplated suicide. This episode had been triggered when a senior lab member had purchased an expensive piece of equipment without having first secured either the necessary authorization or the funding to do so. When the equipment was delivered, the member of staff hit on the idea of charging his research students 'bench fees' in order to make up for the hole in his own accounts left by the purchase of the machine. Synge claimed that, faced with the prospect of financial ruin that would result from paying these fees, Martin had actually considered taking his own life.[65]

Following his graduation in 1932, Martin worked for a brief period at the Physical Chemistry Laboratory before moving to the Dunn Nutritional Laboratory in Cambridge, where he worked on the purification of vitamin E. During his time here he witnessed something that was to have a profound effect on him. Dr. A. Winterstein, a visiting scientist from Heidelberg, gave a demonstration of how the different pigments in a mixture of carotene could be separated into distinct coloured bands when passed in solution through a chalk column.[66] This method had first been demonstrated in 1906 when the botanist Mikhail Tsvett showed how, when passed down a column of calcium carbonate in a solution of petroleum ether, pigments in a mixture of chlorophyll resolved themselves into a series of distinct coloured bands along the length of the column according to their solubility, leading Tsvett to christen this process as 'chromatography'.[67,68]

On seeing this demonstration, Martin was struck by how the principle underlying the distillation columns which had fascinated him as a boy and the chromatogram was the same—both processes used the relative movement of two phases to achieve the separation of a solute. Inspired, he began using a chain of separating funnels to move the upper and lower phases of two immiscible solvents in opposite directions in the hope of achieving successful separation of both vitamin E and carotene pigments according to their distribution between these two phases. He quickly realized, however, that this particular problem would benefit from the application of a maxim he called 'Martin's Principle of Scientific Research', which is summarized simply as 'Nothing is too much trouble, if someone else does it'.[69] For Martin, it was immaterial whether this 'someone else' was another lab member or a machine, so he constructed a machine that could carry out the process.

When complete—only after what he described as 'considerable effort'—Martin's machine consisted of 45 half-inch tubes each about five feet long that were stacked vertically in a rack and connected by a pair of narrow tubes that ran between the top of one tube and the bottom of the next one in the series.[70] Each tube contained a ball valve to prevent the flow of liquid back into the previous tube and in a rare moment of poetic allusion, Martin likened the noise of the machine when it was in operation, to sounding 'like the sea on shingle'.[71]

After completing this work, Martin moved on to apply his methods to the isolation of the dietary factor which could prevent pellagra (a disease now known to be caused by deficiency of vitamin B3) in pigs, a task which required him to diversify his skills way beyond chemistry:

I kept 30 pigs for three years, until I finished my degree in 1938. By keeping pigs I mean I looked after them wholly, made up their diet, cleaned out the pens, and even rigged up my car to pump out the sump when the sump became full of sewage from the pigs.[72]

Being a swineherd was probably not what Martin had envisaged when he had started his research and he was no doubt relieved—and grateful—when Sir Charles suggested that, instead of slopping out pigs, he team up with Richard Synge.

Sir Charles hoped that Martin's technical skill at building an apparatus for chemical separation might help Synge with his analysis of the amino acids in wool. To begin their work Martin built a new machine based on the first, but which was able to cope with using water and chloroform as solvents.[73] In 1938, however, Sir Charles recommended that Martin move from Cambridge to Leeds to work at the laboratories of the Wool Industries Research Association, or WIRA. Originally known as the British Research Association for the Woollen and Worsted Industries before being (mercifully) simplified to WIRA, this was one of several bodies formed towards the end of the First World War by the British Government with the aim of applying basic science to industrial problems in response to fears that Britain might be losing its competitive edge.[74,75]

Leeds was the ideal location for WIRA. Ever since the Cistercian monks of Kirkstall Abbey had raised sheep and sold their fleeces to foreign merchants, wool and textiles had come to be at the heart of the city's economy.[76] The abundant deposits of millstone grit to the north of the city acted as a natural filter so that the streams feeding into the River Aire have soft water ideal for the washing and scouring of wool.[77] By the time of the English Civil Wars when the Parliamentarians won control of the city, the wealth of Leeds rested on wool—a fact which made a particularly memorable impression on one famous visitor.[78] Writing in 1720, Daniel Defoe found Leeds to be 'a large, wealthy and populous town' and hailed the local cloth market as being 'indeed a Prodigy of its kind and is not to be equalled in the world' before going on to describe how cloth made in Leeds was bought by merchants from Holland and Hamburg and was even exported in large quantities to the colonies in America.[79]

As a result of the importance of textiles to the local economy, the Department of Textiles at the University of Leeds was the largest department in the university, as well as being one of the biggest in the country. Between 1917–1919 research in this department was supported by WIRA,[80] but in 1920, WIRA acquired a four-acre site on Headingley Lane, which lies between the University of Leeds and the world-famous Test cricket ground. Here, a grand old Victorian house called 'Torridon' was converted into a research institute. The drawing room became a chemistry laboratory, while physics was to be done in the billiard room and the wine cellar used for work involving controlled temperature and humidity.[81]

With these facilities at their disposal, WIRA could now conduct its own research independently of the University.

Though Martin later confessed to Synge that he was not particularly fond of Leeds, he said that he had remained there because the work with which he was involved was excellent.[82] He was not the first to have expressed this sentiment. In 1909, the physicist and future Nobel laureate William Bragg had arrived in Leeds from Adelaide to take up the post of Cavendish Professor of Physics and probably regretted his decision. Bragg had once described life in Adelaide as having been 'like sunshine and fresh invigorating air'—Leeds, by contrast, could hardly have been more different.[83] The city may well have been a one of the world's first industrial powerhouses, but it had paid a price for this success and when the Cornish businessman and diarist Barclay Fox (1817–1855) visited Leeds, he was far from impressed:

> This amongst all others of its species [manufacturing towns] is the vilest of the vile. At a mile distant from the town we came under a vast dingy canopy formed by the impure exhalation of a hundred furnaces. It sits on the town like an everlasting incubus, shutting out the light of heaven & the breath of summer. I pity the poor denizens. London is a joke to it. Our inn was consistent with its locality; one doesn't look for a clean floor in a colliery or a decent hotel in Leeds.[84]

By the time that Bragg—and later, Martin—had arrived, little seemed to have changed. The sunshine was obscured by clouds of smoke from the factories, and as a result, the air was neither fresh nor invigorating. In a letter Bragg tried to be diplomatic about his new home, but his true feelings were hard to disguise:

> The Leeds people are really very nice: all those I met anyway. The place itself is grimy, even the suburbs; but you can get out into beautiful country to the North.[85]

Like Bragg, Martin also found solace in the countryside to the north of the city. The stunning views from his lodgings at Quarry Farm in Pool-in-Wharfedale, just north of Leeds, quickly dispelled any notions of the north being grim, but Martin wanted to experience the Yorkshire countryside in a very novel way. In his correspondence with Synge, he talked not only about their ongoing work but also mentioned that he had now acquired a car, which made the pursuit of his new found hobby of gliding off Sutton Bank much easier.[86]

When Synge had completed construction of the second counter-current flow machine (christened 'Roger') Martin came down to Cambridge and carefully attached it to the running board of his new car before setting off to drive the 170 miles back up 'The Great North Road' to Leeds at a gentle 20 mph so as not to damage his precious cargo.[87]

Shortly afterwards, having completed his PhD in 1939, Synge also moved north to Leeds where he took up the post of a biochemist at WIRA. His colleague Hedley Marston, who had been instrumental in securing IWS funding for Synge's research at Cambridge, advised him to consider the post in Leeds as only a temporary excursion into industrial chemistry from which he would soon return to more serious work at Cambridge. Marston's concern was that, with his passion for tackling fundamental problems in biochemistry, Synge would find the industrial environment of WIRA dull and uninspiring, saying, 'I don't consider the work which you have undertaken will be favourably influenced by the practical outlook at Torridon'.[88] Much of the scientific research in Leeds was intimately bound up with its history as a textile town—a fact that had made a striking impression on another young academic, who had also once been a resident of Headingley before going on to find fame elsewhere:

> … there was such a huge group crowded that I could hardly count the young scholars that wanted to learn how people here, in deep vats with strange smells, boil and dye hides and want to weave them into soft, pretty cloth; or to burn coal and not yet pollute the air; and a great crowd sought mathematics, and fragrant chemistry and cunning medicine.[89]

This parody of *The Canterbury Tales* was written by the young J. R. R. Tolkien, who had lived in Headingley and the Leeds suburb of West Park from 1920—1925 before taking up a professorial chair at Oxford. Tolkien's description was particularly apt for Martin and Synge's work, but their chemistry, while certainly fragrant, was far from pleasant. Working in a building that had been converted from a former stable and hayloft into a laboratory (Figure 23), they began using the counter-current flow machine that Martin had brought up from Cambridge.

First, they prepared the sample by treating the wool with strong acid to break apart the bonds between individual amino acids in the protein chains. Then, the samples were loaded into the machine to begin separation of the amino acids. As Martin recalled, this was neither a pleasant nor easy operation to carry out:

> It was a fiendish piece of apparatus, we had to sit by it for a week for one separation; it had 39 theoretical plates and filled the room with chloroform vapour. We used to watch it in 4-hour shifts. We had constantly to adjust small silver baffles to keep the apparatus working properly. One of the effects of four hours of chloroform intoxication was that when our partner arrived to take the next shift he was invariably sworn at by the one who had been watching the machine.[90]

Synge joked that drinking in a lungful of the 'fresh' air of Leeds as they emerged from the lab at the end of a shift was most welcome, but intoxication from the solvents they were using was not the only danger they faced in the course of their work. Synge recalled a rather cavalier approach on Martin's part towards electrical safety:[91]

> At a lower but much more dangerous voltage, Archer had a 6000V DC generator under a bench driven by an AC motor. It was used for electrophoresis experiments and the leads from it were of ordinary domestic flex lying on the bench, on which also sometimes lay puddles. Archer is a skilled experimenter and has survived much but I was really scared, and not just for myself.[92]

Figure 23 *Longfield Stables on Headingley Lane, Leeds as it looked when it had been converted into the laboratory in which Martin and Synge worked (a) and as it looks today (b). Along with a lone gatepost (c) which today bears a commemorative plaque to Martin and Synge unveiled by the Leeds Philosophical and Literary Society to celebrate their bicentenary in 2019, this is all that remains of the original WIRA research centre.*

Credit: With kind thanks to Mr. Matthew Synge for photograph (a). Photograph (b) and (c) taken by K.T. Hall.

The most serious accident that Synge suffered however, was not due to Martin's cavalier attitude, but his own misfortune. While cooling a flask of ether at the sink, it exploded, sending shards of glass in all directions. Luckily, none of the shards hit his face, but some became deeply embedded in his hands. It was only many years later that they were finally removed when a colleague at the Rowett Institute in Aberdeen was able to locate them by means of an X-ray image.[93]

In 1941, Martin and Synge became housemates. Following difficulties with his lodgings, Martin was invited by Synge to lodge with him and Ann Stephens, whom Synge married in 1943, at their home on Cardigan Road in Headingley. Life with Martin proved to be anything but dull. In a draft memoir written shortly before his death, Synge described snippets of his domestic life with Martin that could easily have been written by the fictional scientist Dr. Leonard Hofstadter about his eccentric flatmate and fellow physicist Dr. Sheldon Cooper in the hit TV comedy 'Big Bang Theory'.

Mealtimes were particularly interesting, thanks to what Synge tactfully described as Martin's dietary 'peculiarities':

> Eggs, meat, ham and bacon for choice, but no fish. No vegetables other than potatoes and no fruit other than apples. Not many cereals other than bread ... nutritionally a good all round diet—just a bit difficult to provide on war-time rations.[94]

Synge and Ann went out of their way to accommodate the culinary challenges presented by their new lodger. Out of respect for Martin's vehement loathing of onions, they were forced to cook with them only when they knew for certain that Martin would be out of the house. When Martin expressed some concerns that the silver spoon set Ann used to serve her apple stew had the faint whiff of onion, he was given his own special 'apple' spoon to use exclusively for this particular dessert.[95]

Martin's rigid and particular dietary regime may have been due to a gastric ulcer, which he treated by drinking milk containing Aluminium hydroxide—an unorthodox nutritional supplement that one WIRA colleague alleged had been purchased from the company BDH at Torridon's expense.[96] However, despite his best efforts at self-medication, Martin was eventually admitted to Leeds General Infirmary for treatment. But, like his fictional counterpart Sheldon, Martin didn't let the inconvenience of a hospital stay get in the way of doing science. In addition to the more common toothbrush and pyjamas, Martin also brought a few sheets of Perspex, a fret saw, and some chloroform. What Synge failed to record was the response of the hospital staff when, at the end of four days, Martin was discharged from hospital having used the tools and materials he had brought with him to construct what Synge described as 'a mechanical computer incorporating point ratchet wheels as a prototype towards solving the differential equations for chromatograms with curved distribution isotherms'.[97]

Martin also recognized the importance of relaxation. But, as Synge recalled, his particular method of relaxation sometimes caused alarm to local residents:

> They were given to nude sunbathing on the balcony outside our kitchen. One Sunday morning there was a loud knock on our door, which I answered. There was a policeman. He said 'I was called by an old lady on Headingley Lane (¼ mile up the hill). She said she could see with her opera glasses a naked couple on your balcony. She handed the glasses to me, and I looked, and said I thought they were wearing bathing gear, but I'd go down and check. So could you please tell them they had better be?' The police had long periods during the War without much on their hands.[98]

Martin's partner-in-crime was Judith Bagenal, a nursery teacher and friend of Ann's who had called by one day. Sufficiently charmed by Martin, she not only joined him sunbathing on the balcony, but also later married him at a Leeds registry office in January 1943.[99]

Despite Martin's eccentricities, Synge said that 'our household rubbed along alright' and the only time that the two of them ever quarrelled was when they were suffering from inhalation of chloroform vapour after a long shift at the machine.[100]

It probably came as a great relief to both of them, therefore, when Martin had an epiphany. He realized that the whole process could be greatly simplified by dispensing with the entire contraption. There was a much easier way of achieving their aim. The separation of the amino acids could be accomplished just as easily if one of the solvents, in this case the water, was held immobile on silica gel while the other solvent, containing the amino acid mixture flowed over it. Taking some silica gel from an old packing case in the laboratory, Martin and Synge soaked it in water before stuffing it into a glass column. They then loaded the mixture of amino acids dissolved in chloroform onto the top of the silica gel column and watched in utter delight as a red marker dye began to move down the length of the column, indicating that the different amino acids were separating from each other according to their solubility.

Better still, with this new innovation, they avoided the lengthy vigil at the machine, breathing in chloroform:

> ... we were in business straight away ... We had a most remarkably simple piece of apparatus. It was a better device than the machine which we built over a couple of years, and had to sit and watch for a week at a time for an analysis.[101]

The principle behind this remarkably simple piece of apparatus soon had a new name.

Since it was based on separating—or 'partitioning'—solutes according to their differing solubility in two immiscible liquids (water and chloroform) it became known as 'partition chromatography'.[102] This distinguished it from Tsvett's earlier chromatography procedure that relied on separating a solute from a mixture by adsorbing it to a solid matrix (Figure 24).

Over the years, technology usually evolves by becoming more complex, but the development of partition chromatography bucked this trend. Here, although technology improved by evolving from complex to simple, even the new format had its problems. Certain types of amino acid, such as those with two basic, or hydroxy, groups, were simply adsorbed by the silica gel and could not be separated using this method.

The problem was solved by further simplification—with an innovation familiar to many primary school teachers. Key Stage 3 of the national science curriculum teaches the principle of chromatography by showing how water travelling up a piece of filter paper separates out different coloured felt pen pigments. Martin knew that chemists in the dyeing industry had successfully used filter paper to check the purity of dyes. He considered whether this tactic would work for partition chromatography, instead of silica-gel:[103]

> ... all the apparatus that was required was a sheet of filter paper and a closed box where saturated atmospheres of water vapour or of organic solvents could be maintained. The original box was part of a drain pipe! On the filter paper, the various amino acids were persuaded to run a race and after a short time they became spaced out as spots on the paper, much as horses appear in a race.[104]

Using filter paper brought three major advantages. Firstly, they avoided the chemical conversion of amino-acids into their acetyl derivatives beforehand, as well as substantial losses during acetylation and subsequent extraction; it further solved the problem that certain amino-acids could not be satisfactorily separated when acetylated. Secondly, the method enabled the separation of a wider range of amino acids. Finally—and perhaps most importantly—it allowed their quantification at levels that were previously undetectable.

But how were they to detect the spots of amino acids on the paper in the first place? Unlike the spots of felt pen pigments in a primary school experiment, the spots of amino acids don't show up in bright bold colours. After a careful search through Beilstein's *Handbuch der Organischen Chemie*, Hugh Gordon, a PhD student under Synge's supervision, found a solution. By drying the filter paper and then spraying it with the reagent ninhydrin, the positions of the amino acids on the paper were revealed as distinct blobs of purple colour.[105]

When Gordon's colleague Raphael Consden' saw these purple blobs forming for the first time on one of the early chromatograms, he was unable to contain his excitement and is said to have exclaimed in sheer delight, 'Behold the Lord Cometh!'[106] Consden's response stood in sharp contrast, however, to the wider biochemical community. When Martin, Gordon, and Consden first presented this improvement of their method to a meeting of the Biochemical Society, Martin later recalled that it 'raised not a flicker of interest'.[107,108,109,110] It mirrored the underwhelming response they received when he and Synge had first unveiled their new method of partition chromatography at an earlier meeting of the Biochemical Society:

> We were rather disturbed to find that when we gave the paper to the biochemical society that it provoked no discussion at all; nobody appeared to be in the slightest interest.[111]

However, there was one at the meeting who was definitely interested. Astbury worked at the University of Leeds just down the road from WIRA, and had high

(a)

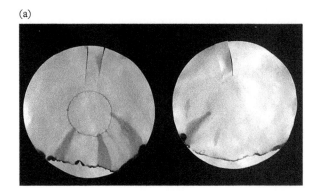

(b)

Glass sheet

Trough

Drain-pipe

Paper

Water and solvent

Lead tray

(c)

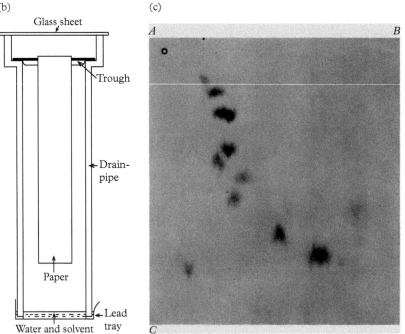

hopes that Martin's and Synge's work on the chemical composition of wool might bolster the X-ray analyses that his own research assistant Ian MacArthur, was making of wool fibres.[112]

Martin and Synge often strolled down Headingley Lane from their laboratories to visit Astbury in the Department of Textiles at the University of Leeds and to share their findings with him. Astbury saw this as the beginning of a beautiful relationship and on one occasion wrote to Sir Charles Martin saying that he and Synge were 'suddenly rushing together in a mutually helpful manner', and which left him feeling that 'we are on the verge of something epoch-making in protein studies'.[113,114] But although Astbury looked forward to Martin and Synge's visits, he was barred from making reciprocal trips up the road to WIRA. Astbury had filed patents on a process of producing textile fibres from monkey nuts, which could result in a conflict of commercial interests with WIRA.[115,116,117] How well this ban was observed is unclear, for Astbury once remarked in a letter to his colleague Kenneth Bailey that he had 'had a look at Martin and Synge's partition chromatography machine and it thrills me a lot'.[118]

Astbury was certain that Martin's and Synge's method would offer irrefutable proof that amino acids in proteins were arranged in neat numerical patterns, but others still expressed scepticism. Astbury had collaborated closely with Albert C. Chibnall, Professor of Biochemistry at Imperial College, London in the attempt to develop textile fibres from monkey nut proteins. In 1942, Chibnall warned that Bergmann and Niemann's methods of measuring amino acids were unreliable. Chibnall dismissed their conclusions, and those of Astbury—that amino acids

---

Figure 24 a) 'A most faithful servant' - this was how Synge once described chromatography, adding that it was '(sometimes more like a genie out of the Arabian nights) both for revealing and isolating known substances and for analyzing new ones'; *Synge in J. Chromatography Library 1979; p.450.* The concept of partition chromatography invented by Martin and Synge is a simple one and is an elaboration of the principle that a mixture of substances can be separated according to their different degrees of solubility in a particular solvent. This can be demonstrated very easily at home by making spots with different coloured felt pens on a piece of coffee filter paper and placing it in water. Martin and Synge realised that a much more effective separation could be achieved by passing the mixture over a solid matrix that had been soaked in a second solvent. Photograph by K.T. Hall.
(b) A diagram of their apparatus using for performing partition chromatography using filter paper to separate amino acids from a mixture.
(c) An example of a chromatogram obtained using the apparatus in (b) in which a mixture of amino acids from the breakdown of wool proteins has been applied to the filter paper at the position marked by the circle and have then been separated by partition chromatography in two different directions, first with the solvent collidine, and then with phenol. Each each different type of amino acid moves a different distance in the respective solvent and appears as a distinct black spot.

Credit: Taken from Figure 3, Consden, Gordon and Martin, 'Qualitative Analysis of Proteins: A Partition Chromatographic Method Using Paper' The Biochemical Journal 38 (1944): 224–232. Reproduced with permission of Portland Press.
Credit: Taken from Plate 1, Consden, Gordon and Martin, 'Qualitative Analysis of Proteins: A Partition Chromatographic Method Using Paper' The Biochemical Journal 38 (1944): 224–232. Reproduced with permission of Portland Press.

were arranged in regular mathematical patterns as being 'nothing more than the hypnotic power of numerology'.[119,120]

Martin and Synge had an idea of how this hypnotic power might be broken. Using the necklace analogy, Bergmann and Niemann had drawn their grand conclusions after having snapped the 'string' between each individual 'bead' and then measuring the relative amounts of each colour, shape, and size of 'bead'. However, Martin and Synge spotted the flaw in this approach. Here, an alternative analogy of considering the protein as a building might be helpful. For example, we can list all the different building materials—bricks, timber, tiles, etc.—from which the building is made and then measure their relative amount: this was essentially the approach used by Bergmann, Niemann, and Astbury in their analysis of proteins. But this approach tells us nothing about how these different materials are arranged spatially in relation to each other to actually make a house. This understanding is crucial if we want to grasp how the building is constructed.

In 1941, Martin and Synge published their first papers describing the method of partition chromatography, providing the scientific world with a powerful new tool for the chemical analysis of proteins.[121,122] But a single modest sentence in the introduction to one of these papers hinted their method might go much further than providing a new means for the analysis of wool fibres. It suggested that the insights offered by their method might reveal the secret of how the chemical composition of proteins, e.g. haemoglobin or insulin, determined their properties:

> ... the distinctive properties of proteins reside in the arrangement rather than the nature of the amino-acid residues.[123]

In other words, the ability of haemoglobin to carry oxygen, or insulin to lower blood glucose levels, was a result not just of the various different amino acids from which those proteins were composed. It was also due to the specific *order* in which they were joined together with each other.[124] Martin and Synge then went on to propose a method by which this specific order of amino acids in a protein chain might be determined:

> ... the only direct evidence in favour of a hypothesis concerning the order of amino-acid residues in a peptide chain can come from the isolation and identification of fragments of this chain containing at least two amino acids.[125]

The WIRA team discovered a very simple way of testing that amino acids were arranged with neat mathematical regularity. Martin himself chose the analogy of trying to solve a jigsaw puzzle.[126] But whether solving a jigsaw or working out the linear arrangement of amino acids in a protein chain, the principle is the same—the task of solving the final complete structure is far easier to achieve using *partially assembled portions*, rather than *individual pieces*.

Martin and Synge realized that this approach offered an easy way to test the hypothesis. If protein chains were indeed made up of a small number of amino acids arranged in simple repeating patterns (such as XYZXYZXYZ)—as proposed by Astbury, Bergmann, and Niemann—then breaking the protein down into fragments, known as dipeptides, which contained only two amino acids, should generate only a limited number of permutations: XY, YZ, and ZX. However, when they began to break proteins such as edestin and gelatine into smaller fragments, a much more complex picture began to emerge. They concluded tactfully that:

> It is not easy to reconcile the picture of protein structure presented by the present work with the Bergmann–Niemann hypothesis ... While the present work does not by any means form a rigid refutation of the Bergmann–Niemann hypothesis, it seems to suggest for proteins a considerably more complicated structure.[127,128]

Wool provided the answer, for according to Astbury, the chains of the keratin proteins in wool contained a simple arrangement of alternating polar (water-loving) and non-polar (water-hating) amino acids.[129] Breaking wool proteins down into dipeptides should therefore give only fragments containing a polar amino acid bonded to a non-polar one. But when Martin put this to the test using partition chromatography, he found something else (Figure 25).

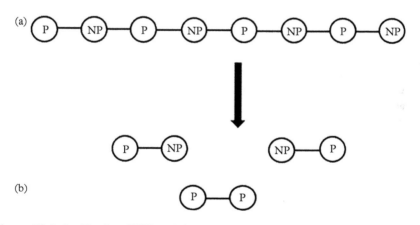

Figure 25 *Archer Martin vs William Astbury – dipeptides land the knockout punch. (a) According to Astbury's model, the proteins in wool were made of chains consisting of a simple alternating pattern of polar (P) and non-polar (NP) amino acids. If this model was correct, then breaking wool proteins down into fragments containing two amino acids should give rise to only two kinds of dipeptide. But when Archer Martin put this to the test (b) using partition chromatography, he found dipeptides containing two polar amino acids – which defied Astbury's predictions. Diagram by K.T. Hall.*

When he presented his findings at a meeting of textile scientists held in Leeds in 1946, he offered a clear challenge to Astbury and his acolytes:

... the structure of wool is more complicated than has been suggested by Astbury and does not conform to the Bergmann—Niemann hypotheses. Astbury has looked at the structure of proteins 'constructively and with the eye of stoichiometric faith'. Lacking this eye and approaching the problem analytically, the present author is unable to find the same degree of order and simplicity.[130]

In response, Astbury praised Martin as an 'outstanding experimentalist', but added that it saddened him that Martin did not share his faith:

This is an age of little faith, and it would be a pity were the depression to spread to the proteins just when hopes were beginning to run so high. One thing we can be sure of is that the proteins are not a hodge-podge, but are constructed according to ordered plans, and it is essential to keep this point steadfastly in mind.[131]

Although this might sound like a compliment to Martin, 'Big Bang Theory' fans may immediately grasp why it was backhanded. Astbury's words could have come straight from Sheldon, a purely theoretical physicist who scorns the methods of his flat mate—and fellow scientist—Leonard. Where Sheldon spends his days pondering over equations scrawled on a whiteboard, Leonard performs experiments and gathers data in a lab.

The comic tension between Sheldon and Leonard highlights an important and ongoing theme in the philosophy of science: which is more important—theory or experiment? And while Astbury was certainly not a purely theoretical scientist, his apparent compliment to Martin betrays a hint of the same condescension that comes from Sheldon.[132] As a physicist, Astbury felt that he occupied the highest rung on the scientific ladder—not only gathering data through experiments but also fitting that data into theoretical frameworks for interpretation. He disdained those for whom science was simply about experiments and instrumentation—chemists like Martin and Synge, who spent their days tinkering with machinery and working with chloroform.

A dispute over the molecular structure of wool fibres might seem of little importance to anyone outside the textile industry. But Martin and Synge had done much more than simply invent a new lab technique for separating amino acids. These two 'experimentalists' had also come up with a novel idea—one so powerful that it would transform our understanding of life at the molecular level. They believed it was possible not only to separate the individual amino acids in a wool protein, but also to work out their precise linear order—or, as Martin and Synge called it, their *'sequence'*.[133]

In a 1955 lecture called 'In Praise of Wool', Astbury reflected that, when he had first begun studies on wool in the late 1920s, it had been dismissed as being 'thoroughly dead, unbelievably dull and unprofitable scientifically, and altogether the kind of protein ... that no respectable, aspiring biochemist would touch with a barge pole'.[134] Yet, by the time that Astbury gave this lecture, things had changed dramatically. From its humble origins rooted in work on wool, Martin and Synge's idea of 'sequence' was helping to set in motion a revolution in molecular biology.[135] Astbury lauded Martin's and Synge's work as being an 'episode to the glory of wool that catches the imagination'.[136]

Nor was Astbury the only person to be singing the praises of wool. When he gave the Banting lecture in 1952, Charles Best also hailed Martin and Synge's work as 'an invaluable tool'. It revealed that, far from being 'thick brown muck', insulin was a protein in which the amino acids were ordered in a precise chemical arrangement.[137]

But there remained one major obstacle to overcome on the road from wool to insulin. Although Martin and Synge had used partition chromatography to separate the amino acids in wool proteins, they could not yet use it to determine their linear order. The proteins in wool were simply too large.[138] What they needed was a test subject made of amino acids but that was much smaller. The rage of the Second World War provided the ideal candidate.

8

A Boastful Undertaking

As bombs began to fall on Britain, Synge found himself becoming restless. Wool might be of great economic importance, but he was desperate to apply his skill and knowledge more directly to the war effort. In October 1941 he wrote to the research manager of Imperial Chemical Industries (Fertilizer and Synthetic Products) (ICI) at Billingham-on-Tees, asking whether the company might be able to offer him 'work more relevant to war-time production than that on which I am presently engaged'.[1] In particular, he stressed that he would like to gain some experience of chemical engineering in the hope that he might learn how his work on proteins could be made more applicable to medicine.[2]

Within only a few days, Synge received a response, and it was not what he had hoped. In his reply, Dr. M. P. Applebey, Delegate Director and Research Manager at ICI said that, although he would be willing to discuss a position for Synge at the company, he could not guarantee that the work would have any bearing on the war effort. Instead, he strongly advised Synge to remain in Leeds, reassuring him that the work he was doing at WIRA was 'of sufficient national importance to justify your continuing it'.[3] Applebey's words may have come as a disappointment to Synge at the time, but they proved prophetic.

In due course, the war came to WIRA, albeit indirectly. Around the same time as Synge was writing to ICI, he and Martin received a visit from B. C. J. G. Knight, a Lister Institute biochemist, who, together with the scientist Marjorie Macfarlane, had made a landmark discovery in bacteriology.[4] Knight and Macfarlane had isolated the toxin of the bacterium *Clostridium welchii*, which had been one of the main causes of gas gangrene in infected wounds during the First World War, and shown that it was an enzyme.[5] This was the first demonstration that a bacterial toxin could act as an enzyme and raised the possibility of developing anti-toxins that might block its action.

The discovery could not have come at a more opportune time. In the trenches of the First World War, disease had been as deadly a killer as enemy action and with the outbreak of war, the development of new anti-bacterial agents had taken on a fresh urgency. While the Spitfires and Hurricanes of the RAF repelled the Luftwaffe, Howard Florey (1898–1968) and Ernst Chain (1906–1979) at the Oxford Dunn School of Pathology were developing a new secret weapon to be deployed against hordes of microscopic—but no less dangerous—invaders. Having isolated the antibiotic penicillin and demonstrated its effectiveness as a chemotherapeutic agent, Florey and Chain were now attempting to scale up its production. But in addition to penicillin, there was also a growing interest in other types of antibiotic, and it was this that Knight wanted to discuss with Martin and Synge.

Knight was particularly interested in an antibiotic called gramicidin and this finally gave Synge the chance to apply his skills to the war effort. It also offered much more, for gramicidin was made up of only a few amino acids. Therefore, it was the ideal candidate on which to test whether partition chromatography could not only separate and identify these amino acids but also determine their precise arrangement. To obtain some of this material, Synge sent some reprints of his early papers on partition chromatography to a research group at the Rockefeller Hospital, New York City who were working with gramicidin, assuring them that the 'use of this for a study of the arrangement of the amino-acid residues in gramicidin seems to us decidedly promising'.[6]

By summer 1943 the samples had arrived, and Martin and Synge set to work analysing them using first silica-gel and then partition chromatography on paper.[7] Somewhat surprisingly, this early analysis appeared to actually confirm the hypothesis championed by Astbury, Bergmann, and Niemann that the amino acids were arranged with a mathematical regularity based on multiples of 2 and 3.[8] But this was soon to change—and spectacularly so. In 1943, Synge left Leeds to continue his work on antibiotic research at the Lister Institute, but he remained in close contact with Martin and Gordon at WIRA and expanded his collaboration with them to work on yet another type of antibiotic, called 'gramicidin S'. Thanks to its discovery in a soil bacterium by Soviet biologist Georgy F. Gauze, this was also known as 'Soviet Gramicidin' and the Red Army were using it to treat infected wounds on the battlefield.

When Howard Florey returned from a visit to the USSR, he brought with him some samples of gramicidin S, a few of which found their way to Synge.[9,10,11,12] Through his contacts with the Soviet Red Cross and Red Crescent Societies in London, Synge was able to procure more samples of gramicidin S and began using them to iron out a major shortcoming of the paper chromatography method.[13,14] Although it was highly effective in separating different amino acids from a mixture, paper chromatography could not be used to prepare them in any appreciable amounts for further work. As ever in the evolution of partition chromatography, the solution was simple; in this particular instance, it came in the form of the humble potato.

With his colleague Sydney Elsden, Synge found that using raw potato starch instead of paper as a solid medium for partition chromatography allowed the purification of amino acids in higher quantities. But Synge had far more grand plans in mind for gramicidin S than simply using it to tweak and improve existing laboratory protocols.[15,16]

Reflecting on the formidable challenge of unravelling protein structure, the biochemist William Stein observed that he and his fellow protein chemists were 'roaming unknown country without benefit of map or compass'.[17] Synge, however, was confident that gramicidin S provided both. To do this, he attacked the problem on two fronts. The first, what Synge fondly described as a 'boastful undertaking' by Martin and the team at WIRA in Leeds,[18] aimed to use partition chromatography not just to separate the different amino acids in gramicidin S, but also to work out the precise linear order in which they were arranged. In a letter to the scientists in the Soviet Union from whom he had received the material, Synge even coined a new term for this precise order of amino acids, calling it a 'sequence'.[19,20]

More than just a new technical term, the idea of amino acids being arranged in a specific one-dimensional *sequence* would come to define the core of molecular biology. It would eventually extend to include DNA, explaining how it carried hereditary information. At the same time, in Leeds Martin was wrestling with more immediately practical challenges that were worrying.[21] When certain solvents, such as phenol or butanol, were used in paper chromatography, treatment with ninhydrin revealed some amino acids not as a distinct purple blob, but instead a messy pink smear. Archer's colleague Hugh Gordon later admitted that he was tempted to just overlook this as a minor technical glitch and forge on with the experimental work.[22] But for Martin, this was more than just a question of cosmetics. In a letter to Synge he called it 'an urgent problem' and felt that if it were not solved then the usefulness of their method would be severely hindered.[23,24]

While Martin racked his brain to find a solution to the problem, he faced even more formidable challenges in other areas of his life. In a letter to Synge, he confessed that he found the gruelling regime of endless feeds for his new baby daughter to be 'a great nuisance'—presumably because they distracted him from paper chromatography.[25] And this was by no means the biggest of Martin's personal difficulties. Sleepless nights and the many other rigours of new parenthood are a challenge at the best of times, but when coupled with war-time air raids, the situation was taking its toll on his wife, Judith. For although Martin described her as being merely 'rather in the dumps', Judith's condition was much more serious.[26] In one letter, Martin asked whether Synge's wife Ann might be able to provide the details of a female psychologist whom she had once recommended as Judith was seeking admission to hospital for what he described as 'a neurotic state'.[27,28]

Amidst this domestic turmoil, Martin tried to focus on the problems in the lab, and admitted to Synge that, with Judith away visiting relatives for a week, 'I find it easier to think with wife + daughter away'.[29] After many weeks of frustration, his resolve finally paid off. Reasoning that the pink smears might arise due to an oxidation reaction with the air, Martin carried out the chromatography process in the absence of oxygen by replacing it instead with ammonia. But although this made the annoying pink smears disappear, it brought a new problem. In place of pink smears, Martin observed the rapid formation of black patches on the filter paper.

A search of the literature revealed the culprit. When phenol was used as a solvent in the chromatography, it became oxidized to form the black spots, and the catalyst for this oxidation was the presence of copper. Although Martin knew that the filter-paper he was using probably did contain traces of copper, he realized that a far more likely source of contamination was the lab atmosphere itself. Under war-time conditions, lab work was a case of simply having to make do with what-ever equipment happened to be at hand, regardless of its condition. Very often this required a degree of technical improvisation, and in one such instance, Mar-tin and his team had to use an old fan in a drying cabinet to evaporate away the solvent from their strips of paper at the end of a chromatography run. The com-mutator of the fan was made of copper and, as it sparked badly, the atmosphere in the cabinet was laden with copper. Having identified the source of the prob-lem, Martin found a simple solution. By carrying out the whole procedure in the presence of a chemical which absorbed the copper, the pink spots vanished. He could finally begin working out the order of amino acids in gramicidin S.

Synge, meanwhile, had opened a second line of attack on the problem of pro-tein structure: to determine how the linear chain of amino acids might be folded up in three-dimensions. Did the linear chain of amino acids close to form a circle? Or was it perhaps folded into some even more complex three-dimensional struc-ture? And might this in turn provide some clues as to how the polypeptide chains of proteins fold up in three-dimensions? To answer these questions, he enlisted the help of Dorothy Crowfoot Hodgkin (1910–1994), a chemist based in Oxford whom he had first met during sessions in The Bun Shop pub while on her visits to Cambridge (Figure 26).[30]

Figure 26 *Dorothy Crowfoot Hodgkin (1910–1994).*
Credit: Ramsey & Muspratt bromide print, circa 1937.

Like Astbury at Leeds, Hodgkin was pioneering the use of X-ray crystal-lography to solve the structure of complex molecules found in living systems, although her working conditions were hardly what might be called state-of-the art. Hodgkin's lab was housed in a dingy corner of the University Museum on South Parks Road, Oxford, where, in order to obtain an AC electrical power supply for her equipment, holes had to be drilled through the two-metre-thick wall of the museum into the physics department next door.[31] Synge likened the romantic character of Hodgkin's laboratory to that of Goethe's scholar Faust, who seeks to gain knowledge and power through a pact with the Devil. For others, such as the crystallographer Max Perutz (1914–2002), it was a far from inspiring location:[32]

> Once arrived, I made for Ruskin's Cathedral of Science—the University Museum—walked past the skeletons of extinct species populating its nave to its darkest corner and descended the stairs to Dorothy's crypt-like office where she laboured on the structure of life in a place that was, but for her vitality, quite dead.[33]

Like Archer Martin, Hodgkin's fascination with chemistry had begun at an early age. Hiding away in the attic of her home she and her sister had set up their own chemistry lab. When she had once suffered a nosebleed as a girl, her first thought had been that it was 'a pity all this good blood should go to waste so I collected it in a test-tube and used it to make haematoporphyrin'.[34]

But there was one field of chemistry that held a particular fascination for her. Having dissolved copper sulphate and then watched the steady growth of crystals over the next few days, she later recalled how this had left her 'captured for life' by this phenomenon.[35]

Her passion was encouraged by her mother, Grace Mary, who bought her daughter a couple of published lectures for children that had been given by the distinguished physicist Sir William Bragg. It was whilst he had been Cavendish Professor of Physics at the University of Leeds in 1912, that William Bragg and his son Lawrence had first learned of a discovery made in Germany by physicist Max von Laue (1879–1960) that an X-ray beam could be scattered as it passed through a crystal.

William and Lawrence quickly realized that von Laue's results meant that X-rays could reveal far more than just broken bones. Lawrence discovered that, with the application of a certain mathematical treatment, the precise spatial arrangement of atoms and molecules in the crystal could be deduced from the patterns made by the scattered X-rays on a photographic film. Meanwhile, in Leeds, his father had built a new instrument called the X-ray spectrometer, which allowed the measurement of the angle and intensity of the scattered X-rays. Coming together, the Braggs spent the next year working feverishly using what they called 'X-ray crystallography' to reveal the atomic structures of crystals such as common salt and ice. Their efforts earned them the 1915 Nobel Prize in Physics and, in the introduction to his children's lecture, Bragg said that, with X-ray crystallography:

... we have, so to speak, been given new eyes ... We can now understand so many things that were dim before; and we see a wonderful new world opening out before us, waiting to be explored. I do not think it is very difficult to reach it or to walk about in it ... We will try to take the first steps into the new country, so that we may share in the knowledge that has already come, and comes in faster every day ... How far our new powers will carry us, we do not yet know; but it is certain they will take us far.[36]

Thanks to X-ray crystallography Hodgkin became one of the most prominent pioneers in the exploration of Bragg's 'new country'. However, the journey was far from easy for her. At school she had to fight to be allowed to study chemistry rather than do needlework, as was expected of girls, but her tenacity paid off. After completing her undergraduate studies in Chemistry at Somerville College, Oxford, she moved to Cambridge to carry out doctoral research with John Desmond Bernal (1901–1971). Like William Astbury, Bernal was another of the young rising stars of X-ray crystallography who had joined William Bragg after he had moved from Leeds to the Royal Institution. Bernal now hoped to show that this method could be used to reveal the structure of biological molecules.

Bernal's first test subject was the sterols, a group of compounds, including cholesterol, vitamin D, and the sex hormones, and he had shown that X-ray studies could give invaluable insights into the physical arrangement of atoms and molecules that allowed chemists to differentiate between possible structures. Hodgkin threw herself into this work, but in 1934, Bernal invited her to join him in taking up another challenge. While Astbury was using X-rays to determine the structure of insoluble fibrous proteins in wool, in Cambridge Bernal focused on another type of protein group—in which the polypeptide chains were folded up, not into insoluble fibres, but into complex globular structures.

One such protein was the digestive enzyme pepsin, and early in the summer of 1934, Bernal obtained some pepsin crystals. When dried, they gave only a vague messy blur on X-ray analysis; but when hydrated they gave a beautiful pattern of spots. At the time, Hodgkin was already starting to suffer from the painful rheumatoid arthritis that would plague her for the rest of her life but, ignoring medical advice to rest, she threw herself into working on the X-ray patterns generated by the pepsin crystals. From these patterns, she and Bernal were able to calculate the most basic arrangement, which, when repeated in three-dimensions, gave rise to the entire crystal as well as calculating an estimate for the molecular weight of the protein. What they could not do, however, is determine the complete position of every atom. As proteins were made up of hundreds of atoms, it seemed that their structures might simply be too complex to be resolved in atomic detail.

Later that year, Hodgkin was offered a post back in Oxford and returned to Somerville where, in addition to teaching duties, she was able to establish her own research group. One of the first tasks that she set herself was to continue the work that she had been doing for her PhD at Cambridge using X-ray crystallography to solve the structure of cholesterol. Her aim was to show that X-ray crystallography could be used not just to determine the positions of atoms in a simple inorganic crystal (as the Braggs had done for common salt), but also to do the same for a complex biological molecule.

There was, however, a formidable challenge. Scattered rays of visible light can be focused by a lens to form an image, but because X-rays have a much shorter wavelength there is no material that can act as a lens to focus them. Instead, this process must be carried out using a mathematical operation called a Fourier Transform, named after the French mathematician and physicist, Jean-Baptiste Joseph Fourier (1768–1830). To work out the position of atoms in the unit cell (The simplest 3D arrangement of atoms or molecules that, when repeated, gives rise to the crystal structure) of a crystal, a series of Fourier calculations are required that use information about the position and intensity of spots made on a photographic film by diffracted X-rays, as well as their phase. In other words, has the spot on the film been made by the peak or a trough of a diffracted wave? This was known as 'the phase problem' and for very simple compounds such as inorganic crystals, it can easily be solved by first performing the Fourier calculations assuming the wave to be at a peak, and then repeating them using the corresponding value for a trough. In this way, a map of the variation in the density of electrons across the unit cell, with higher density areas corresponding to individual atoms, can be constructed.

But as the molecule becomes more complex, the number of such calculations becomes so increasingly large that, for structures as huge as proteins, the task appeared to be impossible. In 1933, however, crystallographer Arthur Lindo Patterson (1902–1966) found a way to deal with this problem. It was far from ideal, and although it could hardly be called a solution, it did at least offer a useful rule of thumb. Patterson showed that, by simply using intensities and not phases, a contour map could be calculated that, while it did not show the precise location of atoms in a structure, did give an indication of the relative position of pairs of atoms. It was far from perfect and prone to being misleading but, at the time, it was better than nothing and could perhaps give some indication of structural features. Moreover, by comparing the differences between Patterson maps for a molecule and a derivative containing a significantly heavier atom then, at least in small biological molecules, it might be possible to determine the precise position of atoms.

Hodgkin's work on cholesterol proved this point. Starting in 1941, Hodgkin's PhD student Harry Carlisle began collecting X-ray diffraction data from cholesterol that had been modified to contain Iodine—a relatively heavy atom in comparison to the carbon atoms that make up most of the molecule. By calculating Patterson maps from this data and comparing them with corresponding maps derived from unmodified cholesterol, they were able to pinpoint the locations of the atoms in the molecule. Carlisle recalled that many of the laborious calculations required for this work were carried out while sitting in freezing cold railway carriages during the blackout as he commuted between Oxford and Princes Risborough, where he had been enlisted for war-related research.[37]

From Carlisle's meticulous calculations, he and Hodgkin produced not only two-dimensional Patterson maps, but an actual three-dimensional wire model of what the molecule looked like.

Hodgkin's biographer Georgina Ferry has called this moment a 'milestone'—for it was the first successful demonstration that X-ray crystallography could be used to determine the three-dimensional structure of a biological molecule.[38] But there was no time for Hodgkin to rest on her laurels, for already a new challenge was looming. When Hodgkin and Carlisle had begun their work on cholesterol, its chemical formula was already well known. Would the X-ray methods that they had pioneered be as successful with a biological compound for which the chemical composition was still unclear?

The particular compound in question was penicillin which, thanks to its potent antibiotic properties and its potential for the treatment of battlefield casualties, was the subject of intense research. The main priority of this work was to establish large-scale production of penicillin, either by scaling up existing fermentation methods or by finding some way of artificially synthesizing penicillin. However, before attempting to synthesize penicillin by artificial methods, its chemical composition had first to be determined. The Oxford Dyson Perrins Laboratory was one of the main centres for this work and the chemists there were eager to involve Hodgkin with their work. What frustrated Hodgkin, however, was that no one had yet managed to grow crystals of penicillin—all she had to work with were crystals of products derived from its chemical breakdown. Nevertheless, she and her research student Barbara Low were able to offer useful independent confirmation of molecular weights obtained by the chemists.

In 1943, the chemical composition of penicillin was finally established. Attention turned to the question of the particular spatial arrangement by which these atoms bonded with each other to form the molecule. This was a subject over which the chemists involved in this work were deeply—and passionately—divided. While there was a consensus that the atoms in penicillin must be bonded to form some kind of cyclic shape, opinion was sharply divided into two camps about the exact form that such a structure might take. Ernst Chain, who was working with Howard Florey on the isolation of penicillin, was convinced that penicillin must consist of a five-membered ring joined to one containing four atoms—a structure known as beta-lactam. However, Professor Robert Robinson, head of the Dyson Perrins Laboratory, favoured a particular arrangement known as the thiazolidine-oxazolone structure, which consisted of two five-membered rings joined together. Robinson seemed to be furious that anyone could think otherwise, and his younger colleague John Cornforth (1917–2013) announced that if the beta-lactam structure turned out to be correct, he would give up science and grow mushrooms instead.

That same year a development occurred that would have a crucial bearing on Cornforth's proposed change of career. Chemists in the United States succeeded in the crystallization of penicillin and Hodgkin seized this opportunity to obtain some samples for her work. In a letter to Richard Synge thanking him for some samples of gramicidin, she said she was now working 'day and night' to solve the structure of penicillin.[39] Having suffered a miscarriage in the previous year, she battled her way through this tragedy to present the results of her research to a meeting of the Penicillin Committee in February 1945.

Hodgkin's X-ray data on penicillin ought to have given John Cornforth good reason to start contemplating a future in fungi, for her studies hinted that the beta-lactam structure might indeed be correct. Hodgkin's structures were still only projections of the molecule in two-dimensions—but what she really needed was to determine the precise location of each atom in a three-dimensional structure. To do this would require a formidable amount of Fourier calculations, and although by no means impossible, would be fiendishly complex. There was, however, a way to do it. In 1937, Hodgkin had first learned of a device called a Hollerith machine which was a mechanical computer. By feeding it with data on punched cards, the Hollerith machine could be used to crunch the numbers in the required Fourier transforms necessary to compute the three-dimensional structure of penicillin. Working with George Hey from the Scientific Computing Service, Barbara Low wrote a program that would carry out the necessary calculations, and which was then run at night on an American Hollerith machine at Cirencester that, by day, was being used for tracking ships' cargoes.[40]

Thanks to the perseverance of Hodgkin and Low in using the Hollerith machine, the structure of penicillin was solved and the beta-lactam ring hypothesis proved. John Cornforth, as it happened, went back on his word, remained in science, and went on to win the 1975 Nobel Prize in Chemistry. Although news of Hodgkin's success was first confined to the Committee for Penicillin Synthesis, word soon leaked out.

Having been kept up to date on Hodgkin's work by his daughter Maureen, who was one of her students at Somerville, Astbury offered her his hearty congratulations:

> I want to be one of the first to congratulate you and to say how terribly bucked I am that you have pulled it off. It's simply great, and not a little pleasure is that you have kept it in the Old Country! ... it is a triumph for (a) crystallography; (b) women, (c) Oxford and (d) Somerville.[41]

When the Royal Society held a meeting in 1945 to celebrate the fiftieth anniversary of the discovery of X-rays by the German physicist Wilhelm Röntgen, Hodgkin was looking forward to presenting her discovery of the structure of penicillin. But she was annoyed to be told by the Committee for Penicillin Synthesis that, as her work was still considered to be a war-time secret, she would not be allowed to do so. As a result of this decision she was unable to make her work public until 1949.

Hodgkin's discovery of the structure of penicillin did not directly aid chemists in developing a means by which the drug could be mass-produced. However, it did enable the synthesis of a wide range of modified versions of the molecule, from which many of the different antibiotics in use today have been derived. This in itself was a triumph—but another was to follow.

In 1948, scientists at the company Merck in the United States, and Glaxo in the United Kingdom had successfully isolated and crystallized vitamin B_{12}—deficiency of which causes pernicious anaemia, a condition that can be fatal. In the wake of her success in solving the structure of penicillin, Hodgkin was asked by Lester Smith, one of the Glaxo scientists, whether she would carry out X-ray studies on crystals of vitamin B_{12}.

In addition to its obvious clinical value, vitamin B_{12} held a particular allure for biochemists, as its size was roughly midway between that of penicillin and the proteins. If the structure of vitamin B_{12} could be solved by X-ray crystallography, then the task of doing the same for the proteins might yet prove possible. Furthermore, chemical analysis of its composition had already shown that vitamin B_{12} contained an atom of the metal cobalt, which would make interpreting the Patterson maps much easier.

The complete chemical composition of vitamin B_{12} was yet to be determined, and Hodgkin began a collaboration with the chemist Alexander Todd (1907–1997). While Todd's group in Cambridge attempted a precise determination of its chemical composition, Hodgkin's group in Oxford carried out X-ray studies of crystals of vitamin B_{12} to unravel its structure; in July 1955, *The New York Times* announced their success. The article 'Britons Discover Structure of B-12' should have been cause for celebration, but the opening quickly caused annoyance:

> The solution of one of nature's most formidable puzzles—the exact chemical structure of Vitamin B-12—has been achieved by a team of British chemists at Cambridge University ... The team that won the race and with it one of the greatest prizes in chemistry, was headed by Sir Alexander Todd of Cambridge University chemical laboratory.[42]

After having spent long hours slaving over a Hollerith machine to compute a structure, Hodgkin's assistant Jenny Pickworth (now Professor Jenny Glusker) was understandably put out that the article omitted to mention the work of Hodgkin's team. Todd himself was also very annoyed that the only reference to Hodgkin's team was at the very end of the article when they were described as simply having 'assisted' in the work. To rectify this gross error, Todd wrote to the editor of *The New York Times*:

> The main credit for the progress which has been made should be given to Dr. Dorothy Hodgkin of Oxford and her colleagues Miss J. Pickworth and Mr. J. H. Robertson ... I feel strongly that Dr. Hodgkin's outstanding contribution to the structural elucidation of vitamin B12 should be recognized.[43]

Yet, when Hodgkin attended at a scientific meeting held in Brussels that August, she was surprised to find herself still being asked for her thoughts on what was described as *Todd's* structure for vitamin B_{12}.[44] However, writing in

1962, Lawrence Bragg compared the monumental challenge of using X-ray crystallography to solve the structure of complex biological compounds with attempts in aviation to break the sound barrier. Hodgkin and her work on vitamin B_{12}, said Bragg, had successfully smashed through this barrier 'with a loud report'.[45] Nor was Bragg the only person to recognize the importance of Hodgkin's achievement. At the end of October 1964, the news was announced that the Nobel Prize in Chemistry for that year would be awarded to Hodgkin for her discovery of the structure of penicillin and vitamin B_{12}.

Hodgkin remains the only British female scientist to have been awarded a Nobel Prize and it seems that her gender was of as much interest to the national newspapers as was her science. *The Daily Mail* announced the news of Hodgkin's success with the headline 'Nobel Prize for British Wife' while, alongside a photograph of Hodgkin, *The Observer* told its readers that, 'On Thursday the affable-looking housewife below won Britain's fifty-eighth Nobel Prize for a thoroughly un-housewifely skill'.[46]

Hodgkin's success with vitamin B_{12} gave fresh hope that X-ray crystallography might yet reveal the structure of protein molecules in atomic detail and the Nobel Prize was not the only distinguished accolade that she received for this achievement. She also became the very first woman to have her portrait hung at the Royal Society. The portrait was rather unconventional as it showed not Hodgkin's face but only her hands and been drawn by the artist Henry Moore whom Hodgkin had known since the 1930s. Moore wanted to convey a sense of the incredible skill and finesse manifest in Hodgkin's hands that had enabled her to prepare samples of delicate crystals for analysis by X-ray crystallography. But he also had another reason for his particular choice of subject: her crippling rheumatoid arthritis, which flared up particularly at times of stress. Describing Moore's work in an essay on the depiction of Hodgkin by artists, the historian Patricia Fara has said that, while other artists might seek to conceal this aspect of Hodgkin's life, Moore chose deliberately to highlight it and, by doing so, he evoked a greater sense of admiration for her.[47]

Another artist who chose Hodgkin as a subject was Maggi Hambling, whose painting in the National Portrait Gallery shows Hodgkin, aged 75, sitting at a desk in her study that is strewn with manuscripts, notes, files, and a three-dimensional model of a complex molecule (Figure 27). Amidst the pile of papers on her desk lies a half-eaten sandwich, suggesting that Hodgkin is still so immersed in her work that she has forgotten it. But in her essay, Patricia Fara, suggests another meaning:

> At the centre, the half-eaten sandwich is like a modern *memento mori*: already stale, it is destined to disintegrate, unlike the engineered replica of an organic molecule which will, like the scientific knowledge it represents, endure forever.[48]

This 'engineered replica of an organic molecule' to which Fara refers was probably the achievement of which Hodgkin was most proud. For although Hodgkin had received the highest accolade in science for her work on penicillin and vitamin

B_{12}, her real ambition remained unfulfilled. She still wanted to solve the complete structure of a protein. And she had one particular candidate in mind.

Figure 27 *Dorothy Hodgkin by Maggi Hambling oil on canvas, 1985.*

Credit: © National Portrait Gallery, London. Reproduced under license by National Portrait Gallery.

Hodgkin had first read about insulin in a book that she had chosen as a school prize, in which Sir Henry Dale described the isolation of pancreatic extract by the team in Toronto as 'a development so rapid and romantic and so fruitful in practical therapeutic results that it has struck the imagination of the public in all civilized countries with a force without parallel in the history of technological discovery'.[49]

In 1934, Robert Robinson, the Waynflete Professor of Chemistry (1886–1975), had obtained from the Boots Pure Drug company a sample of crystalline insulin, which was given to Hodgkin to study. When she first examined the crystals under the microscope, she saw immediately that they would be too small to use in X-ray crystallography experiments. She then set about trying to grow larger crystals using a method developed by the Canadian chemist David Scott, which involved cooling the solution of insulin slowly down from 60°C over several days.[50]

By early 1935, she had finally obtained decent crystals and when she developed the photograph at ten o'clock one evening after a ten hour X-ray exposure, she experienced what she described as 'probably the most exciting moment of my life'.[51,52] Euphoric, she spent the rest of the night wandering around Oxford in a state of ecstasy that was interrupted only when a concerned police officer asked her what she was doing out alone in the middle of the night.

The pattern of X-ray reflections from the insulin crystals were not yet detailed enough to give a precise molecular structure, but they allowed Hodgkin to estimate the weight of the molecule and so confirm measurements obtained by Theodor Svedberg from his study of insulin by ultracentrifugation. Hodgkin's short article on insulin in the journal *Nature* was her first piece of work published as a single author and it quickly brought her to the attention of the scientific community, as well as establishing her as one of the leading lights of X-ray crystallography.[53]

At around the same time as Hodgkin was developing her photos in Oxford, the benefits of understanding of the physical and chemical nature of insulin were becoming abundantly clear. The steady advances in the purification of insulin meant that patients were no longer erupting in abscesses due to impurities. But the advances had come at a cost. The purified insulin was now absorbed so rapidly into the blood from the site of injection that its time of action was only short. This meant that when its effects wore off after only a couple of hours, the patient might experience a sudden rapid rise in blood sugar levels. Of even greater concern was the risk that blood sugar levels might also suddenly drop during the night, plunging the patient into a potentially fatal hypoglycaemic coma whilst asleep. In an attempt to prevent such violent oscillations of blood sugar levels, patients needed to inject themselves before each meal, as well as at three o'clock in the morning.[54] The situation was far from ideal—but what if patients were able to inject themselves with a single dose of insulin that lasted for 24 hours?

Hans Christian Hagedorn (1888–1971), a researcher at Nordisk Insulin-laboratorium in Copenhagen, grasped the essence of the problem. Hagedorn's supervisor, the Nobel laureate August Krogh (1874–1949), had been in the United States on a visiting fellowship in 1922 when he had first heard about the work being carried out in Toronto. As Krogh's wife was herself a diabetic patient, Krogh was eager to learn more.[55] He wrote to Macleod asking whether it might be possible to collaborate in setting up insulin research in Denmark and whether he could come to Toronto to discuss the matter further.[56,57] When he arrived in Toronto, Krogh was impressed at what he found—so much so, that he later nominated Banting and Macleod for the Nobel Prize.[58] Just under a year after his return to Denmark, Krogh was delighted to inform Macleod that large-scale production of insulin was going well and that he had plans to build a factory to produce 500,000 clinical units a week.[59] As Denmark made strides in scaling up the production of insulin and increasing yields by using frozen material, Hagedorn, meanwhile, solved a more fundamental problem.[60]

As far as Hagedorn was concerned, multiple daily injections of purified insulin were but a 'poor imitation of nature's own mechanism', which could rapidly detect changes in blood sugar levels and fine tune the release of insulin from the pancreas in response to them.[61] Hagedorn had realized that what was needed was a means of slowing down the absorption of insulin from injection sites so that, instead of being released into the blood all at once, it did so over a much longer period of time. This might be achieved if a way could be found to reduce the solubility of the injected insulin.

A means of doing this came thanks to both the physical chemistry of insulin, and the somewhat unlikely help of the bulging testes of the rainbow trout. Hagedorn had experimented with a number of different chemical additives which, when combined with insulin, altered its solubility. However, it was his co-worker Norman Jensen who suggested the use of protamine, a basic compound found in the nuclei of sperm cells, and which was particularly abundant in those obtained from the rainbow trout.[62] When bound as a complex with protamine, insulin formed a precipitate with minimal solubility at physiological pH. This meant that when injected, it ought to be absorbed much more slowly, and therefore have a longer time of action.[63] When the Danish preparation was tested upon patients in Boston, it did not disappoint. Its effect lasted twice as long, not only allowing patients to maintain stable blood glucose through the night but also meaning that they could now take their entire daily dosage of insulin in one single injection. Clinicians were stunned at this success. Fred Banting described Hagedorn's innovation as 'the greatest advance in the treatment of diabetes since the discovery of insulin', while Howard Root, the clinician in charge of the trial, declared that 'a new revolution in the treatment of diabetes must follow'.[64,65]

Thanks to her own work on the structure of insulin, Hodgkin soon came to the attention of the Danish researchers—and the rest of the international scientific community.[66] Invitations to attend prestigious scientific meetings began to arrive—as did other approaches. Someone who was particularly interested in Hodgkin's work on insulin was Dorothy Wrinch (1894–1976), one of her former undergraduate lecturers at Somerville College (Figure 28). Wrinch's admiration for Hodgkin had begun even before her work on insulin was published. When Hodgkin had first returned to Oxford from Cambridge, she had found a bouquet of red roses at the door of her room, sent by Wrinch.[67] However, the relationship between the two was not always to be smooth or straightforward.

Wrinch was the first woman to hold a university lectureship in mathematics at Oxford and, as a result of having published an impressive 42 papers by 1929, she was also the first woman to be awarded a DSc degree there. As a result, she now no longer had to be known by her double-barrelled married name, but

Figure 28 *Dorothy Wrinch (1894–1976) examining a model of a cyclol with Nobel laureate Irving Langmuir (1881–1957).*

Credit: Courtesy of Special Collections, Smith College.

could call herself simply Dr. Wrinch.[68,69] Although she was a mathematician by training—and a highly successful one at that—Wrinch refused to be confined to one particular intellectual discipline. In her younger days she had been an acolyte of the philosopher Bertrand Russell and had written to him when he was imprisoned for his pacifist stance during the First World War. She also wrote papers on problems in physics such as general relativity, the philosophy of science, and, in 1930, even ventured into sociology with a book called *The Retreat from Parenthood.* Written under the pseudonym Jean Ayling, the book explored concerns that the professional classes such as doctors, lawyers, scientists, and civil servants appeared to be having fewer children. In her view this was a worrying development that threatened the future of the British Empire. The solution that she proposed was quite straightforward: 'In the case of many people the greatest—perhaps the only—service they can do mankind is to remain barren'.[70]

The book also advocated radical changes in social policy such as the establishment of national child-rearing centres. These measures, it was argued, would allow women to negotiate the balancing act of raising a family while pursuing of a professional career. This idea was personally relevant to Wrinch herself. Following years of her husband's heavy drinking, Wrinch's marriage had begun to unravel, leaving her to juggle the formidable demands of being a single mother with a young daughter while at the same time facing the challenges of her professional career.

Another of Wrinch's growing interests was biology, particularly at the molecular level. In the summer of 1932, she became a founding member of a group known as the Biotheoretical Gathering, which sought to apply mathematics to biology.[71] Together with Astbury, Bergmann, and Niemann, Wrinch tried to find mathematical order beneath the apparent messiness of the living world. But unlike the others, who sought neat, regular repeating one-dimensional patterns, Wrinch looked for specific two-dimensional geometry. In a paper published in 1936, she argued that, despite their differences in chemical composition, there was sufficient uniformity among the proteins to suggest a 'simple geometric plan in the arrangement of the amino-acid residues, characteristic of proteins in general'.[72] The simple geometric plan that she had in mind was that the amino acid chains from which proteins were made might be folded into hexagonal units, or 'cyclols', to form flat sheets (Figure 29).[73]

In this paper, Wrinch was cautious to make clear that her proposed cyclol theory was 'to be regarded as offering for consideration, a simple working hypothesis, for which no finality is claimed'.[74] But the following year, she became sufficiently confident to extend her ideas into three dimensions, proposing that the planar hexagonal sheets of amino acids were folded up to form regular polyhedral structures.

Although Wrinch was a theoretician, the data from the handful of proteins she had studied so far appeared to be on her side. Wrinch claimed that the regular patterns of amino acids found by Astbury, Bergmann, and Niemann in proteins like gelatine and albumin fitted perfectly with her proposed cyclol structures, as did measurements of the molecular weight of proteins made by Theodor Svedberg using ultracentrifugation. Of this handful of proteins, there was one in particular that Wrinch seized upon as proof for her argument. In a paper entitled 'On the Structure of Insulin' published in *Science* in 1937, Wrinch presented insulin as the standard bearer for her cyclol theory.[75] Not only did the number of amino acids in insulin and its molecular weight fit with her proposed structure, but there was now a third piece of evidence in her favour—Dorothy Hodgkin's X-ray crystallographic studies.

According to Wrinch, X-ray crystallography was 'the most stringent test of any proposed structure' and she was pleased to report that Hodgkin's work supported her argument.[76] Hodgkin's studies had shown that the molecule was arranged

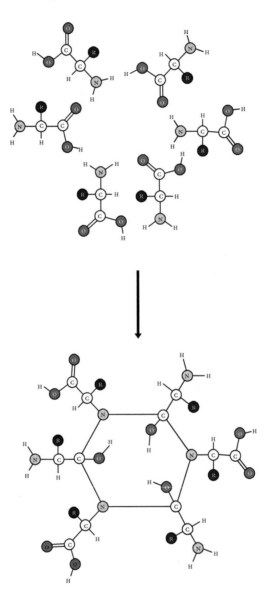

Figure 29 *Dorothy Wrinch's cyclol hypothesis. Wrinch proposed that, instead of individual amino acids linking up to form a long chain, they might instead bond together to form a ring shaped arrangement which could then combine with each other to form repeating cyclical structures, or as she called them 'cyclols'. Diagram by K.T. Hall.*

with a three-fold symmetry and Wrinch argued that this was consistent with its protein chain being folded into a cyclol polyhedron. In triumph, she declared that her proposed polyhedral structure 'fits adequately with the measurements given by the X-ray analysis of the lattices of insulin'.[77,78] But what Hodgkin's X-ray studies could not yet reveal was the exact position of individual atoms in the insulin molecule. Until a solution to 'the phase problem' was found, the molecule's atomic structure would remain stubbornly elusive.

In 1938, Hodgkin published Patterson maps for insulin and, presumably with Wrinch in mind, pointed out diplomatically that 'the patterns calculated do not appear to have any direct relation either to the cyclol or to the various chain structures put forward for the globular proteins'.[79] Wrinch, however, did not share Hodgkin's caution and, in collaboration with the Nobel laureate Irving Langmuir (1882–1957), produced her own calculations of a Patterson map for insulin that, perhaps unsurprisingly, showed that it fitted the structure of a cyclol polyhedron perfectly.[80]

Langmuir wasn't the only person whom Wrinch had sought to enlist in promoting her cyclol theory of protein structure. Denis Riley, one of Hodgkin's research students, recalled how, before she teamed up with Langmuir, Wrinch had tried to lure him away from Hodgkin. Acutely aware that her own theoretical work needed good experimental evidence and recognizing that Riley showed promise as an X-ray crystallographer, Wrinch had hoped that the offer of a research grant made over tea and sherry might entice him to her team.[81] When Riley declined her offer, Wrinch asked whether he would allow her access to his experimental data— a request which must have left him feeling rather awkward, to say the least. He had not yet attempted to interpret the data himself, but perhaps more importantly, to comply with Wrinch's request would surely amount to a betrayal of Hodgkin's trust in him.

Riley must have breathed a sigh of relief when Wrinch left Britain in 1939 to live in the USA. Having both failed to recruit Riley and to gain access to his data, Wrinch had now resorted to asking Hodgkin directly for her experimental results. Hodgkin's response was a masterful piece of diplomacy:

> My present lot of photographs are most beautiful ... I'm working them up in a rather careful way so that I observe most reflections twice, but there are hundreds of reflections (2 or 3,000 at least) and sometimes I gasp at the thought of what my series will be like ... It would be silly to send you the list of intensities I used last summer—the present ones would be so much more reliable. But I don't calculate to be through with collecting them till the end of the summer.[82]

But if Hodgkin was hoping that her tactful reply might deter further requests, she had underestimated Wrinch's tenacity—or her inability to pick up a hint. When Wrinch continued with her requests, Hodgkin had to be a little more forthright:

> I have every hope now that we are going to get something out of the wet insulin data that will bring us information much nearer to atomic dimensions. But I'm

afraid I'm not going to part with my intensity data to anyone yet ... I'm doing preliminary calculations as we go along—but I don't want to let anything out till I'm reasonably certain of it.[83]

Even without full access to all of Hodgkin's data, Wrinch's star continued to rise. Her cyclol theory had established her as a luminary in the field of protein chemistry—and beyond. Newspaper headlines like 'Woman Einstein—Dr. Wrinch Solves Biological Problems Mathematically' appeared alongside photographs of her holding models of her cyclol polyhedra built by her young daughter, Pamela.[84]

Irving Langmuir, with whom Wrinch had collaborated on her latest insulin paper, was so impressed that he nominated her for the 1939 Nobel Prize in Chemistry. But not everyone was so overwhelmed. X-ray crystallography was still a relatively new method and although it had been used successfully to solve the atomic structures of simple materials— diamond, ice, and table salt—those of huge biological compounds like proteins had so far remained elusive. Some chemists were even sceptical that structures as complex as proteins could ever be mapped to the same level of atomic detail as more simple compounds. Fearing that the young discipline which he had helped to found might be dragged into disrepute by Wrinch's grand claims, Lawrence Bragg urged for caution:

> In a protein, the assumptions must as yet be vague and provisional. Exaggerated claims ... are only too likely, at this stage, the bring discredit upon the patient work which has placed the analysis of simpler structures upon a sure foundation.[85]

In the same issue of *Nature*, John Desmond Bernal argued that Wrinch's calculation of a Patterson map for insulin made far too many arbitrary assumptions and that a cyclol model was only one of several possible structures that could correspond to such a map. In view of this, he concluded that 'the cyclol hypothesis fails completely to account for the X-ray evidence for insulin'.[86]

But Bragg and Bernal's criticisms were tame in comparison to that of the US chemist Linus Pauling (1901–1994). By applying quantum mechanics to the problem of chemical bonding, Pauling had formulated a set of rules governing how atoms could link with each other to form molecules. A string of papers on this subject had established him as a scientist of international renown and, aged only 30 years old, he was awarded a full professorship and recipient of an award from the American Chemical Society for 'the most noteworthy work in pure science done by a man under 30 years of age'.[87] He discussed his work on chemical bonding with such luminaries as Albert Einstein and was hailed by *The New York Times* as being a possible Nobel Prize winner. Pauling went on to become one of the few people to win two.

By the mid-1930s, Pauling had turned his attention to biology and was considering how his ideas about chemical bonding might be applied to protein structure. Now that Wrinch was also being hailed as an international celebrity in this field, she was eager to meet Pauling and present her ideas to him. After failing to arrange a meeting with Pauling during a visit to the United States, she wrote to him saying, 'I was very much disappointed that it did not prove possible to visit you during my time in the States. I am, however, still extremely anxious to sit at your feet'.[88]

Wrinch may have held Pauling in awe, but unfortunately for her, the feeling was not mutual. Only two days after he received Wrinch's letter declaring her admiration for him, Pauling was contacted by Warren Weaver, director of the Rockefeller Foundation. In 1935, the Rockefeller Foundation had awarded Wrinch a generous five-year research grant for her work on protein structure and Weaver wanted to know what Pauling thought of Wrinch's research.[89] In his reply, Pauling tactfully expressed some caution:

> I have the impression that she is a very clever person, and I am sympathetic to the type of speculative consideration which she is carrying on now. Without doubt there is a great deal of truth in her general picture. This picture is, however still very far from definite—she suggests various alternatives and does not make any definite predictions. I have felt that the definite suggestion which she did make regarding protein structure, dealing with a type of polypeptide condensation involving hexagonal rings, is incorrect ... I feel that Dr. Wrinch's work suffers a little bit from being similarly too speculative and from being based too largely on the assumption that nicely symmetrical structures are the right ones.[90]

A year later, having met with Wrinch when she was on a lecture tour in the US, Pauling dropped the tact and in his correspondence with Weaver, he was brutally honest:

> I had two extended talks with Dr. Wrinch, who also spoke at our seminar at Cornell. I was greatly disappointed in her. She did not present as good a case for her theory as do her published papers. I shall send you my report about her soon.[91]

True to his word, Pauling sent his report—and it has been alleged that, as a result of his comments, Wrinch's research grant from the Rockefeller Foundation was not renewed.[92] He also made more public criticisms of Wrinch's work. In the *Journal of the American Chemical Society*, he and Carl Niemann cautioned that:

> Because of the great and widespread interest in the structure of proteins, it is important that this claim that insulin has been proved to have the cyclol structure be investigated thoroughly.[93]

In the pages that followed they undertook such a thorough investigation and, after reviewing evidence from X-ray crystallography and calculations about the stability of the cyclol bond, they came to a very blunt conclusion:

> ... there exists no evidence whatever in support of this hypothesis and that instead strong evidence can be advanced in support of the contention that bonds of the cyclol type do not occur at all in any protein ... It is concluded from a critical examination of the X-ray evidence and other arguments which have been proposed in

support of the cyclol hypothesis of the structure of proteins that these arguments have little force.[94]

Wrinch took Pauling's attack badly. Having submitted to *The Journal of Chemical Physics* a manuscript for publication in which she defended her work against Pauling's attacks, she poured out her feelings to the British mathematician Eric H. Neville (1890–1961), whom she described as being 'one of the few people who know how desperately miserable and lonely I have been all my life':[95]

> This new Pauling business gets me down. He is a most dangerous fellow. Even decent people hesitate to stand up to LP [Linus Pauling]. He is bright and quick and mercyless [sic] in repartee when he likes and I think people are just afraid of him ... The big paper on bond strengths and lengths as come back from JCPhys [Journal of Chemical Physics] with reports from six referees. They all seize upon my comments which apply to P [Pauling] and want them deleted. They are cowards.[96]

Even Wrinch's thirteen-year-old daughter became acutely aware of the toll that these criticisms were taking on her mother. Putting pen to paper, Pamela decided to appeal to Pauling's better nature:

> Dear Dr. Pauling, Your attacks on my mother have been made rather too frequently. If you both think each other to be wrong it is best to prove it instead of writing disagreeable things about each other in papers. There are many quarrels in the world. Alas!! Don't please let yours be one [sic] it is these things that help to make the world a kingdom of misery!![97]

The letter was never sent, but Wrinch must have been touched by this display of loyalty from her teenage daughter. Nevertheless, her confidence was badly shaken. She even considered leaving science altogether but lamented that, while fields such as music or politics might once have offered the prospect of a successful alternative career, it was now too late in the day to explore such options:

> I just can't seem to bear it and everything else alone. Sometimes I take poor lovely Pam into my confidence and for days on end treat her as an adult friend but it is a strain on her and I then have to go back to be her mother and protector which leaves me quite alone to stand against the world. Really and truly my life has been a complete washout, without any happiness except Pam and proteins.[98]

The tide had well and truly turned against Wrinch. As more proteins were studied it became apparent that her proposed cyclol structure did not fit with the emerging data. But as far as Wrinch was concerned, the cyclol hypothesis was far from dead and she clung on despite the mounting experimental evidence against it. As was evident in a letter to Hodgkin written in 1939, Wrinch felt that some of the motivation for the attacks against her might not be for entirely scientific reasons:

As to the controversy and polemics ... I my dear have not polemicized. I am out only for the truth about protein structure. Accordingly I attempted to interpret your data. Misstatements concerning this attempt and my other works have to be corrected. Desmond etc were of course not the only attackers since as I expect you have pointed out to me organic chemists and others of course hold fast to traditional ideas, even in the face of plain statements and deductions by physicists, x ray crystallographers and mathematicians. Our chromosome count of course does not tend to weaken the desire to (sic) others to attack us. When I am driven to make statements, I do it with an eye to the feelings of say Pamela in 10 years or Luke in 20 years on reading the literature. I am of course delighted to collaborate with you.[99]

Over time the correspondence between them dwindled and in 1944, Wrinch wrote to Hodgkin saying, 'It seems a long time since I heard from you ... '. She was, apparently, quite oblivious to the possibility that Hodgkin might well be deliberately trying to put some distance between them.[100]

But when her second husband, Otto Glaser, died in 1951, Wrinch used this as an opportunity to rekindle her relationship with Hodgkin. Ever the mathematician, Wrinch described her distraught emotional state as being 'pretty well at minus infinity' and one of her chief concerns was to clear the air over a disturbing rumour that had reached her:

Your letter, Dorothy, is so kind and it looks as if my fears that you gave up being friends with me long, long years ago are not justified. I cannot tell you how much I have worried, specially since I have been so far away, about your apparent distrust of me. There was the quite terrible occasion when Booth (WHY do people always repeat things to me?) said you had said I had stolen your data ... I would do almost anything, Dorothy, to get these things straight between us.[101,102]

In her reply, Hodgkin was keen to set the record straight:

I am so very sorry you should have been unhappy about my attitude over insulin. I do assure you that I am not in the least distrustful [sic], nor do I regard the data as having been given to you otherwise than entirely by my own free will.

It is true that at first I did not give them very willingly. I regard this as an imperfection in my character, but one which is perhaps understandable. It was in the middle of the war, I was working on penicillin and had deliberately put insulin aside as a subject that I ought to defer. Yet I was fond of the stuff and wanted some day to have the pleasure of working on it again ...

[...]

The thing that inhibited my talking to you much about insulin and which has probably led to this feeling of suspicion, is quite different, and perhaps it would be best if I said what it was. I was only worried for your sake since it seemed to me that you cared a great deal that the cyclol idea should be right and that it would make you most unhappy if it should be proved wrong ... I may, of course, be quite wrong about cyclols and proteins but I just felt our attitudes to the problem were so different that we should probably only hurt one another by discussing it—and that therefore, perhaps it was best not to.[103]

Wrinch admitted that, in her persistent requests for Hodgkin's data, she had perhaps overstepped a mark in scientific etiquette. But in the 1950s it had been discovered that, despite having once been dismissed as too unstable, the type of chemical bond she had proposed between amino acids in her theoretical cyclol structures did indeed exist in both nature and the laboratory. Emboldened by this new discovery, she had no intention of giving up on her beloved cyclol theory so easily:[104]

> I am not fixated on the little cyclols: I am fixated on the protein: and I can distinguish between the two. I did I must admit long very greatly indeed for data which would make it possible to see if they were roughly correct or even on the right lines and I see now rather more clearly what a big sacrifice it was for you to let me see the intensities when the war made it impossible to spend time on them yourself. Again my deepest thanks for this: I appreciate now fully what it meant ... I do realise that no one has any right to expect that data which have taken years to assemble should be made available to a theoretical worker ...[105]

When Wrinch died in 1976, shortly after her only daughter Pamela was killed tragically in a house fire, Hodgkin described her as having been 'enthusiastic and adventurous, courageous in the face of much misfortune, and very kind', but in a letter to Dr. Caroline Cohen of Brandeis University the following year, she acknowledged that their relationship had been complex:[106]

> I have, as you realise, very mixed emotions about Dorothy Wrinch. When I first knew her she was the mathematics tutor at Somerville and to her pupils who were my friends, a very exciting teacher to have. Her turning towards other worlds, and particularly the cyclol hypothesis occurred a little later ... I'm sure there are worthwhile things to be salvaged from much of the work she did.[107]

Wrinch was not the only person who had high hopes riding on Hodgkin's X-ray studies of insulin. Astbury had done some chemical analysis of insulin and although he confessed to Hodgkin that this work was only in preliminary stages, he was confident that it too would eventually conform to the laws proposed by Bergmann and Niemann:

> Such chemical data as I have been able to collect for insulin look pretty scrappy ... I am not worrying any more about getting strict 2's and 3's for whole molecules or structures. The constituent units are built on 2's and 3's though – of that I have no doubt.[108,109]

Astbury was convinced that the twin approaches of chemical analysis combined with X-ray studies of structure were converging towards a momentous breakthrough in the understanding of proteins. And at the heart of it all was insulin:

> ... obviously both X-rays and chemistry are here converging on something very fundamental in protein structure. And that something is undoubtedly the same

thing as has shown itself from time to time in relation to the 2n3m rule and throughout the whole of this stoichiometric discussion—we refer to the building up of proteins from subunits whose make-up must be comparatively simple, and whose stoichiometry may conceivably be always a matter of 2's and 3's ... In the meantime X-ray progress on insulin will be followed with a keenness as great as any in protein studies.[110]

Both Astbury and Wrinch were convinced that the apparent complexity of proteins could be reduced to elegant mathematical structures. But with hindsight and ever more data, this idea became untenable. But Synge recalled that even in the 1930s, one of his Cambridge colleagues had mocked the naivety of Astbury and Wrinch's blind faith in finding mathematical order in biology with the quip:[111] 'With Astbury and Wrinch, protein chemistry's a cinch'.[112]

Yet Synge also harboured hopes that a combination of chemical analysis and X-ray studies would solve the mystery of protein structure. Where his approach differed from that of Astbury, however, was that he placed his hopes not in insulin, but a much smaller compound. For although gramicidin S proved to be too toxic to have any value as an antibiotic, Synge was convinced that it would be invaluable in unlocking the mystery of protein structure:[113]

> Dr. A. J. P. Martin, Dr. Crowfoot and I have been for some time searching for a protein-like compound of molecular weight of the order of a few thousand which (1) should exist in the form of crystals giving extensive X-ray diffraction data and (2) should offer a prospect of determination of structure by partial hydrolysis and other organic chemical methods, employing analytical techniques recently developed ... and ionophoretic techniques now being developed at Leeds. In this way we might hope to obtain a complete physical picture of the structure, which would be of general value for the progress of knowledge about proteins.[114]

If Synge and Hodgkin had hoped that X-ray studies of a simple compound such as gramicidin S would shed important light on larger more complex structures such as insulin, they were to be disappointed. Hodgkin continued her work on gramicidin S until well into the 1950s, but its structure proved to be stubbornly elusive and was not solved until 1978.[115]

Not that this mattered. By the early 1950s, Hodgkin had turned her attention back to insulin and her Patterson map for the structure of insulin was even used as a design for wallpaper and carpets at the 1951 Festival of Great Britain.[116] Working with her at this time was research student Beryl Oughton (later Rimmer), who recalled how, after a drive to Cambridge, she and Hodgkin became some of the first people to see James Watson's and Francis Crick's new double-helical model of DNA. Oughton herself went on to become a distinguished scientist in the field of structural biology, as did a number of Hodgkin's other female proteges. One of these was Barbara Low, who had worked with Hodgkin to carry out computations on the Hollerith machine for the structural studies of penicillin in the 1940s before going on to work with Linus Pauling at Caltech. Low then went on to take up associate Professorships at Harvard and Columbia, although her prestige in the field did not prevent her on one occasion from being barred entry to an academic gathering after being mistaken for the tea lady.[117]

This was a mistake that no one would ever have dared to make with the most famous of Hodgkin's proteges, however. The student in question had helped Hodgkin with the early work on gramicidin S, but unlike Oughton and Low, she went on to distinguish herself in a very different field to that of structural biology. It has even been suggested that it was the frustration of trying to prepare derivatives of gramicidin S that drove this particular young woman to turn her back on science and opt instead first for a career in law, and then politics.[118] Yet had she known at the time that gramicidin S would yet play a crucial role in unravelling the chemical mystery that was insulin, she might well have opted to remain in science, in which case Britain—and perhaps even the world—might have been a very different place indeed.

9

The Blobs That Won a Nobel Prize

In a 1977 interview, when Archer Martin was asked whether a scientist could ever make a good social or political leader, his reply was blunt: 'I think it's improbable. The scientist normally has too much concern for truth. He will be an inefficient politician'.[1] Yet within only two years of making this confident assertion, one of Dorothy Hodgkin's former students was to prove Martin spectacularly wrong on two counts. Firstly, for his assertion that a scientist could not be an efficient politician and secondly for his implicit assumption that a politician must be male.

This particular student had also made a striking impression on Richard Synge who remembered her in a collection of anecdotes written four years before his death:

> Among the numerous acolytes girls in that basement was Margaret Roberts (later Thatcher) doing her 4th year Chemistry thesis which involved some fetching and carrying of gramicidins. I've often thought Dorothy's mailed-fist-in-velvet-glove approach got passed on to her pupil (who in later life virtually discarded the velvet glove)[2]

By the time that Margaret Thatcher (1925–2013) had become the first female Prime Minister of the United Kingdom on May 4, 1979, her days of toiling as an undergraduate student in Hodgkin's laboratory to prepare samples of gramicidin S were long behind her. But she never forgot her scientific mentor and kept in touch over the years, despite a stark difference in their political views. 'We just took a different view', Thatcher once told Hodgkin's biographer, Georgina Ferry, 'and we both knew it; she couldn't dissuade me and I couldn't dissuade her'.[3]

Unlike her protégé, Hodgkin was firmly on the left of the political spectrum. She believed in the socialist goals of decent healthcare and education for all, and when the University of Bristol, of which she was then the Chancellor, was faced with severe budget cuts she petitioned both the Education Secretary Sir Keith Joseph, and the Prime Minister to reconsider their policies.[4] She was also a firm believer in internationalism and was keen to promote the cause of world peace. To this end she campaigned against the Vietnam war and wrote to Henry Kissinger condemning the bombing of North Vietnam.[5] As an active member of organizations such Pugwash and the Campaign for Nuclear Disarmament, she tried to persuade Thatcher that dialogue with the Soviet Union would ultimately be in everyone's best interests.[6,7]

The PM certainly listened—although she did not necessarily always agree. 'Yes, she was idealistic', said Thatcher of her former mentor, 'but no amount of idealism could conceal the tyranny of the Soviet regime'.[8] Despite this, however, it was clear that Hodgkin and Thatcher held each other in great respect. When a delegation from the USSR Academy of Sciences attended a reception at 10 Downing Street in 1988, Boris Vainshtein, Director of the Academy's Institute of Crystallography who had known Hodgkin since the early 1950s, was impressed to see a watercolour portrait of her hanging above the writing desk in Thatcher's study.[9]

Like Hodgkin, Synge was also firmly on the political left, but he does not seem to have had the same respect that she did for her former pupil. When Synge received an invitation from the Prime Minister in 1989 to attend a lunch to be held at 10 Downing Street in honour of distinguished British scientists, he politely declined. In his reply, Synge said that his reasons for not attending were the same as those given by fellow scientist Sir John Cornforth who, in his own response to the invitation, had been somewhat more forthright than Synge:

> My wife (a scientist) and I think it possible that your recently discovered concern for the environment has led you to assemble the surviving members of an endangered species. Certainly you, and a sequence of inept or malevolent ministers chosen by you have already done your worst to ensure its extinction. We feel that by accepting your invitation we would be betraying all those—schoolchildren, students, researchers, teachers—whom <u>you</u> have betrayed.[10]

Throughout his life Synge was an active supporter of causes on the political left. The fact that he actually noted the date of Thatcher's resignation in a diary indicates the strength of his feelings towards her and her politics. On November 22, 1990, Synge had been travelling from Leeds to Harrogate to go walking in the local countryside when he heard the news that the PM was leaving Downing Street for the last time. On reaching the top of Almscliffe Crag, an impressive geological feature in the lower Wharfe valley, he wrote that he had enjoyed a 'good view'.[11] Given his ideological convictions, this may not have been simply a literal reference to the stunning vista of Lower Wharfedale, but also a metaphor for his hopes of a new political outlook for the whole country.

Of all the political causes to which Synge gave his support, the one that probably mattered to him most of all was nuclear disarmament. In 1989 he was asked to add his signature to a list of eminent scientists who had signed a public statement issued by Greenpeace expressing scepticism about claims made by British Nuclear Fuels that nuclear energy might solve the problem of the greenhouse effect. But in response to this request, Synge gave a very guarded reply:

I feel that this whole issue is being taken as scientifically established while in fact still controversial (as to effects). Improved photosynthesis and extension of agriculture into lands now too cold for it might well more than compensate for any rise in sea level ... It is also worth remembering that the environmental changes between 12,000 and 5,000 years ago were far more dramatic than anything now predicted (read H. Lamb's 'Climate'). Yet, 'civilised' or not, all the ancestors of all of us survived to reproductive age nor, despite the tale of Noah, is there serious archaeological evidence of human catastrophe during that period.[12]

As far as Synge was concerned, humanity faced a much more imminent threat. The greenhouse effect was, he wrote, a red herring 'set in motion to distract public attention from the far more serious threat from nuclear weapons'.[13] At the time that Synge wrote this, a thaw in the Cold War was already well underway and only five months later, the Berlin Wall fell. But with hindsight, it is now clear that Synge's fears about nuclear war at the time were much more than just shrill alarmism—especially during the 1980s. The final decade of the Cold War had also seen its most dangerous episode. In November 1983, NATO had carried out a war games exercise that was mistaken by the Soviet Union as preparation for a genuine attack. The chilling revelation that Soviet forces had actually prepared to respond with a pre-emptive nuclear strike on the West suggests that, far from being mere 'red herrings' themselves, Synge's concerns at the time were well founded.[14]

As a scientist, Synge's position on the left was far from unusual for the time. The idea that history was a process of social and political progress towards a more equal world had an inherent appeal to many scientists, as it seemed to be a natural extension of the way in which science advances over time. Moreover, with its vision of a society ordered along scientific lines, Marxism and its claim to have revealed the very laws by which such social progress would unfold, had a particular appeal to the scientific imagination.

Although the political left was the orthodox position for most scientists at that time, Archer Martin seems to have been an exception. Certainly, by the time that he was being asked his thoughts on whether scientists could make good politicians, his own political stance had diverged sharply from that of Synge. While Synge remained firmly on the political left all his life, Martin despaired at the Miners' strike of 1973–1974 and the election of a Labour government. Torn between loyalty to his country and a growing sense that Britain was in a state of terminal political decline, he accepted a professorship at the University of Houston:

I felt that I had a patriotic duty to stay in England to try to help the English scene. Now, the situation has changed: A, in England the situation has become quite appalling and the political direction is such that I can only forsee [sic] disaster, and B, Houston offered me a post without a retiring age.[15]

Despite maintaining that scientists should stay out of politics, in Houston Martin spoke candidly about some of his own political views. One of these was education, the decline of which was contributing to the growing sense of disaster that he felt looming over Britain:

The English system of education at the moment is being altered, under political pressures, to be very much more similar to the American system. And, in my opinion, it's being ruined. The selection of pupils into groups of ability is being largely abolished: very wide ranges of ability are being taught in the same class which I think is a basically most undesirable method. The result is that the poorest student usually learns nothing at all, and the best students don't learn nearly as much as they should, and get profoundly bored. Whereas, if you have proper segregation you can make sure that all those capable of learning to read do so, and can do simple arithmetic. Whereas those who are capable of going further can move reasonably fast and their interest in school is maintained.[16]

It was a view that stood in stark contrast to the egalitarian politics of his old friend and colleague, Synge. And while Synge believed that nuclear war presented the greatest threat to humankind, Martin had very different fears for the future—and how to solve them:

I think overwhelmingly the most important problem in the world at the moment is the human population. It seems to be that, like every other species, we have to limit our population ... we shall be exceedingly uncomfortable if we leave it to natural forces. We should take in hand our own population control, and it would seem to be wholly ridiculous just to reduce the fertility of everybody. We should select the better members to survive. The method that I would suggest would be to offer a voluntary bonus to anybody except those below a certain age to get themselves sterilized. I think in this way it would have a rapid effect in removing those who found life difficult in various ways. The bonus would be more attractive.[17]

According to his entry in the *Oxford Dictionary of National Biography (ODNB)*, Martin made the extreme suggestion that, rather than being incarcerated in prison, criminals should instead be exposed to a dose of radiation that would cause them to age by the same number of years as a prison sentence.[18] Unsurprisingly, such extreme views did not go unchallenged. According to Martin's *ODNB* entry, this and other controversial pronouncements resulted in student riots and almost certainly contributed to the termination of his contract with the University of Houston in 1979.

Synge's politics also landed him in trouble—but for other reasons. In the immediate aftermath of the Second World War, Synge wrote to Georgy Gauze in Moscow (from whom he had obtained some of the samples of gramicidin S) and expressed optimism for the future:

I must say how grateful I am to yourself and Professor Sergiev and the Soviet authorities generally for making materials and information available for this work. I very much hope that this sort of collaboration begun in war-time may continue and strengthen the bonds of friendship between our two countries in times of peace.[19]

But Synge's optimism was sadly out of step with the political realities of the post-war world. Winston Churchill, in his bleak declaration made in Fulton, Missouri the following year, stated that 'From Stettin in the Baltic to Trieste in the Adriatic, an iron curtain has descended across the continent'. The USA and the USSR, once former allies, now viewed each other with suspicion and outright hostility. The Cold War had officially begun, and Synge was about to become embroiled in its politics.

For although partition chromatography had met initially with a muted response from the biochemical community when it was first presented in 1941, things had now changed. As the impact of Martin's and Synge's method began to gain recognition, Synge was invited to present a paper at a major international conference held in 1949 at Cold Spring Harbor, on Long Island.[20,21] Founded in 1890, Cold Spring Harbor has today been home to eight Nobel prize winners and was already renowned as an international centre of research at the cutting edge of biology. But in order to make the trip to the United States, Synge first needed to apply for an entry visa and attend an interview at the U. S. Consulate. It did not go well.

During the interview, Mr. A. T. Fliflet, the U. S. Vice-Consul, had leafed through the documents Synge had brought with him and been particularly interested to come across a Polish visa. On finding this document, Synge recalled that Fliflet's 'demeanour changed'.[22] Fliflet then asked Synge about the purpose of this visit to Poland, to which Synge answered that it had been to attend a peace conference. When Synge went on to explain that he had attended this conference as an individual and not as the delegate of any particular organization, Fliflet asked whether he was a member of the Communist Party.

Synge replied that he was not, but Fliflet kept pushing. He then listed a number of organizations, including the Science Section of the Society for Cultural Relations with the Soviet Union, of which Synge admitted that he was a member. Now Fliflet had truly scented blood and he must have been delighted with Synge's answer to his next question. For when Fliflet asked 'Have you any communist sympathies?', Synge replied:

> Yes, I am in agreement with some points of policy put forward by communists ...
> Proposals to outlaw atomic and biological warfare. Furthermore, I consider the
> use of threats to employ such methods as politically misguided in the extreme.[23]

Fliflet's response gave little away. He replied simply, 'So do we'.[24]

When Synge was then asked by Fliflet whether he wished to overthrow the Constitution of the United States by violence, he (perhaps unsurprisingly) denied that this was his intention and conversation turned to the Marshall Plan. This was the grand investment by the USA in rebuilding the infrastructure and economy of a Europe that had been devastated by war. But those with sympathies towards the Soviet Union might have interpreted it as a strategic move calculated to prevent a continent exhausted by war from falling into the political orbit of the USSR. When Fliflet asked what Synge thought of the Marshall Plan, he simply replied 'I should like to reserve judgement for a year or so …', so Fliflet tried another approach and asked how Synge would feel about seeing a Communist government in France. 'I must protest at being asked this question,' replied Synge, 'I think it is not a matter for me, but for the French.'[25]

Again, Fliflet gave little away. He just said, 'So do we', and with that the interview was concluded.

A couple of months after the interview, Synge received a letter from the American Consul informing him that, due to the 'circumstances of this application' it had been passed on to the Department of State for approval.[26] Frustrated at this delay, Synge wrote to the Director of Cold Spring Harbor asking whether he might be able to look into the matter. Synge complained that the interview had involved a series of intrusive questions about his politics 'that we in this country would regard as highly improper, but to which, nevertheless, I submitted'.[27]

In a letter to M. Demerec, the Director of Cold Spring Harbor, Synge admitted that 'Evidently some of the answers I gave were unsatisfactory'.[28] He was not wrong. At the start of June he was informed that the U. S. State Department had cancelled his visa. There would be no trip to Cold Spring Harbor.

Synge's reaction was one of surprise. He had never considered himself to be so sufficiently politically active that it might result in such serious measures:

> … in recent years I have been so preoccupied with scientific work and with domestic responsibilities and have paid scant attention to politics. Nevertheless, like so many others I have felt deeply the frustration of war-time hopes and the diversion of so much scientific effort to the mass destruction of human beings.[29]

The response of the scientific community, meanwhile, was one of indignation. Seventy-nine scientists signed a letter in support of Synge, and it was published in *Science*. In the letter, they denounced Synge's treatment by the U. S. authorities as having 'deprived the symposium of the scientific judgement of an exceedingly able worker, and thus has done disservice to the progress of science in this country'.[30] Writing to Synge, the American biochemist William Stein spoke for all of them:

> All of the American group was both angry and embarrassed at the treatment accorded to you by the authorities. It is difficult to convey in a letter the sense of outrage caused by this affair. There is, of course, no excuse for such behaviour. It is a reflection of the hysteria abroad in this country right now. One can only hope that sanity will return before long, and bend one's efforts towards hastening the arrival of that day.[31]

This public show of support was a great source of consolation to Synge, as was a letter that he received from Hodgkin and several other delegates who sent 'our greetings and our great regret at not having you with us here. Your presence has been universally missed and we offer you our sympathy'.[32,33] In his absence at the conference, Synge had asked Hodgkin to present his paper but due to an error in communication with the organizers, in the end it was given by someone else;[34] far more important was the impact that Synge's work was to have on the audience—and on biology itself.[35]

The reception of Synge's paper at Cold Spring Harbor was very different from the muted response that he and Martin had encountered when they had first presented partition chromatography only a few years earlier. Back then, they had just been two industrial chemists whose work on wool fibres had barely raised an eyebrow, even within the Biochemical Society. But within a very short space of time, their names would come to be known well outside of scientific circles.

Martin himself went on to receive an accolade that, with the exception of Professor Stephen Hawking, has been shared by very few scientists. For in an episode of the hit TV series, *The Simpsons*, Martin is mentioned by name as a scientific hero for his invention of yet another method of chemical separation, known as gas chromatography.[36]

This was done after Martin had left Leeds and moved to the National Institute for Medical Research in London where, with his colleague Tony James, he had begun to address a question raised in his first paper with Synge on partition chromatography. This was whether highly refined separations of chemicals might also be obtained using partition chromatography if the mobile phase was a gas, instead of being a liquid.[37] In 1952 he and James published a paper showing that this method could indeed be used to achieve effective separation of lipids—fatty compounds that had previously 'been painfully slow and little more than a greasy sticky confusion of beaker chemistry'.[38,39,40] Known as gas-liquid chromatography (abbreviated to GC), this powerful new piece of technology was quickly adopted in medicine and manufacturing, particularly in the petrochemical industry, where it enabled the rapid and accurate analysis of petrol and Martin even said that he had once used it to test the quality of petrol in his own car.[41] It is perhaps more than a little ironic that, in addition to the petrochemical industry, another great beneficiary of GC was the environmental movement. Sir James Lovelock, who found fame as the proponent of the 'Gaia' hypothesis wrote in an obituary to Martin:

> [T]he environmental awareness that Rachel Carson gave us would never have solidified as it did without the evidence of global change measured by GC. This instrumental method provided accurate evidence about the ubiquity of pesticides and pollutants and later made us aware of the growing accumulation in the atmosphere of chlorinated fluorocarbons, nitrous oxide and other ozone-depleting chemicals.[42]

Long before Martin had found fame on *The Simpsons*, Synge had also made a television appearance thanks to his own rather unusual claim to fame. As a boy, Synge had been staying at a hotel near Loch Ness when, glancing out of the window he spotted a large object moving in the murky waters of the Loch. Quickly he begged his mother to drive him along the road that ran near the shore of the

Loch so that, by using the speed and angle of the car, he might calculate how fast the mysterious marine object was moving.[43]

In May 1958, Synge recalled his boyhood encounter with the most famous—and notoriously elusive—inhabitant of Loch Ness in a BBC programme called *Legends of the Loch* presented by Raymond Baxter.[44,45] But by the time of this TV appearance, Synge's name, together with that of Martin were already very well known for a somewhat more conventional though no less prestigious honour.

On November 5, 1952, the headlines in *The New York Times* reported that, in addition to the election of Dwight Eisenhower as President of the United States, the Nobel Prize in Chemistry for that year had been awarded to Martin and Synge for their development of partition chromatography.[46] On the other side of the Atlantic, meanwhile, *The Yorkshire Post* seized the opportunity to fly the flag of regional pride and proudly announced the award of a 'Nobel Prize for Leeds Invention', although the headline in the *News Chronicle* a couple of days later was far more memorable and lively—'Blobs Win Nobel Prize For Two Young Researchers'—a reference to the purple spots of amino acids on a paper chromatogram (Figure 30).[47]

(a)   (b)

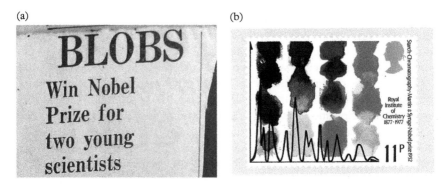

Figure 30 *a) A newspaper headline announcing the award of the 1952 Nobel Prize in Chemistry to Martin and Synge for their development of partition chromatography.*
b) Commemorative stamp released by the Post Office to mark the centenary of the Royal Society of Chemistry in 1977. The stamp depicts a paper chromatogram showing the 'blobs' which earned Martin and Synge their Nobel Prize. (Although the stamp actually incorrectly describes their method as 'starch chromatography', which prompted Synge to complain to the Post Office (Richard Synge to D. Burn, Post Office Marketing Department, 26th November 1976, the papers and correspondence of Richard Laurence Millington Synge, A13, Trinity College Archives, GBR/0016/SYNG).

Credit: The papers and correspondence of Richard Laurence Millington Synge. Trinity College Library, Cambridge. GBR/0016/SYNG.
Credit: Stamp Design © Royal Mail Group Limited

In the days that followed, Martin and Synge were inundated by letters of congratulations from colleagues around the world. Stanford Moore, who had presented Synge's paper on gramicidin S at the Cold Spring Harbor meeting, wrote saying 'I am sure that this was one time when the Nobel Committee had to ponder not who, but when. Yours is a perfect award . . '.[48] A. C. Chibnall, who, like Synge had challenged the ideas of Bergmann, Niemann, and Astbury, claimed that he could think 'of no other case within my active life in which the award of the Nobel Prize has given me so much satisfaction … '.[49]

Sydney Fox of Iowa State College said that the popular image 'of a chemist peering at a test tube held aloft is now replaced by a picture in which instead of a test tube there is a paper strip'.[50] Even Astbury, who had seen his cherished conviction that the sequence of amino acids in proteins was dictated by mathematics demolished by partition chromatography, offered Martin and Synge his 'heartiest congratulations' on their achievement.[51]

Moore said that Martin and Synge had 'set the world on fire'—a point that was also made, though with slightly more sobriety, by the Swedish scientist and Nobel Laureate Arne Tiselius, with whom Synge had worked in 1946. Tiselius was particularly impressed that such a powerful development could have been made with relatively simple resources, for as he pointed out—'almost any schoolboy with ordinary lab equipment could demonstrate the procedure … British scientists seem to have a special gift for making great discoveries with small resources'.[52] But this was in no way meant to detract from Martin and Synge's achievement for, as Tiselius went on to say, 'the method has given us the key to solve a number of very important problems of great current interest in chemistry, biology and medicine'.

Some were more reserved in their praise. B. H Wilsden, who had been Director of Research at WIRA when Martin and Synge were working there, wrote to ask Synge for a copy of the speeches that he and Martin had given to the Nobel Institute. In his letter, Wilsden was muted in his praise, referring only to Martin and Synge's 'recent distinction'.[53] When Synge duly sent him a copy of the speeches and Wilsden read them, he wrote back to complain that Synge had not 'included some instance of the application of paper chromatography to provide knowledge about wool' and that the support of their work by WIRA 'is not over-emphasized'.[54]

But while Wilsden was grumbling about wool, it had already become quite clear that the impact of partition chromatography would be much more profound than simply offering new insights into the structure of textiles. When Melvin Calvin (1911–1997) of the University of California at Berkeley wrote to offer them his congratulations, he had particular reason to be thankful to Martin and Synge. From 1946 onwards, Calvin had been investigating the biochemical pathways involved in photosynthesis, the process by which plants use photons of energy in sunlight to convert atmospheric carbon dioxide and water vapour into carbohydrate compounds—and upon which all life on Earth depends. To do this, Calvin had fed algae with radioactively labelled carbon dioxide with the aim of tracing the metabolic route by which this carbon ultimately became assimilated into sugar.[55] His plan was to identify the various intermediate compounds involved in this process, and in so doing 'to draw a map of the path of carbon as it flows into the plant in the form of carbon dioxide and distributes itself among all the plant constituents'.[56]

In 1961, Calvin was awarded the Nobel Prize in Chemistry for this work, which Synge once described as being 'the most striking available example of the use of autoradiography with paper chromatograms'.[57] For both the silica gel and paper partition chromatography that Martin and Synge had developed had been vital in allowing Calvin to separate and analyse the different intermediate compounds at various stages as carbon dioxide was converted to sugar in the plant. From this he discovered a cyclical set of metabolic reactions by which plants convert carbon dioxide into sugar, and which today bears his name.

As a result, when Martin and Synge were awarded the Nobel Prize, Calvin wrote to Synge offering not only his congratulations, but also his thanks.[58] And he was not the only researcher who had been set on the road to a Nobel Prize thanks to partition chromatography. For Synge also received a letter of thanks from young biochemist Fred Sanger (1918–2013), who congratulated Martin and Synge on what he felt to be 'a well-deserved award, especially as I have been making such extensive use of your methods, and all of us who are interested in proteins realize the enormous contribution of partition chromatography to the subject (Figure 31)'.[59]

Figure 31 *Fred Sanger (1918–2013).*
Reproduced with kind permission of MRC Laboratory of Molecular Biology.

In Sanger's case, this contribution was to be enormous. His name had been on the list of delegates from the 1949 Cold Spring Harbor meeting who had written to Synge offering their support as well as signing the public statement in *Science* condemning his treatment by the U. S. government. He had begun working in collaboration with Synge a few years earlier when, along with Hodgkin, he had been enlisted by Synge to help with the work on solving the structure of gramicidin S. But like Hodgkin, Sanger also had a growing interest in insulin at this time, and it was thanks to this that he would eventually achieve scientific immortality.

He was born in 1918, to parents who both had strong connections with Quakerism, which was to shape much of his life's outlook. While studying Natural Sciences as an undergraduate at Cambridge, he signed the 'Peace Pledge Union' and joined the Cambridge Scientists Anti-War Group, which changed his life in a way he never anticipated. Since the group opposed going to war, Sanger was given the task of organizing a report about the economic consequences of rearmament.[60] By his own admission he had little clue how to go about this and was therefore delighted to be introduced to Joan Howe, a fellow member of the group who happened to be an economist. In a short space of time, however, Fred and Joan found that they had a lot more in common than columns of figures showing the production costs of tanks and fighters. As a result, they were married in 1940.

With the outbreak of the Second World War, Fred registered as a conscientious objector and attended a Quaker Relief Training centre. Here, he learned subjects like agriculture and building and, after his training was complete, he went on to work as an orderly in a hospital near Bristol.[61] As the first casualties began to arrive from the evacuation of Dunkirk, Fred said that the experience 'brought me into contact with real life'—as did cleaning the hospital's toilets and floors.[62,63]

Although Fred had hoped that his work as a conscientious objector might 'help to save lives, instead of taking lives', he had also realized by this time that his talents would be better put to use in scientific research.[64] Modest and quiet by nature (at school he had earned the nickname 'Mouse'), he had never been one to overestimate his academic ability.[65] At the end of his undergraduate study at Cambridge he concluded that:

> I had really not decided that I was that good. I hadn't been getting first classes in my Part I exams, and so I really didn't have the confidence that I could do research. Usually you have to get a first to go into research. In this Part II Biochemistry there were all these brainy people who had got their firsts in their Part I. They all seemed very clever compared to me. So at the end of the year I took the exams and I sort of went off and didn't think too much about what I was going to do.[66]

But by now Fred had definite thoughts about what he wanted to do with his life. He wrote to Professor Gowland Hopkins, head of the biochemistry laboratory at the University of Cambridge, asking whether it might be possible to return to undertake research for a PhD.

The subject of Fred's doctoral research was the metabolism of the amino acid lysine but after he completed his thesis in 1943, a change of leadership in the department brought him a much more exciting subject to tackle. For although Fred found Hopkins inspirational, he was also getting on in years and this was taking its toll. As Hopkins tottered around the lab grumbling about his poor eyesight and hearing, certain members of the department (whom Sanger diplomatically described as having only 'various degrees of competence') took advantage of the situation for their own financial ends.[67]

Order was gradually restored after the main administrator had been sent to prison for having embezzled lab funds and a new pair of hands arrived at the helm to lead the department.[68] A.C. Chibnall had been Professor at Imperial College London and when he arrived in Cambridge to take over as head of department, he brought with him the perfect research problem for Sanger to tackle.

Before his move to Cambridge, Chibnall had already begun to investigate the chemistry of insulin. Working with some samples that he had obtained from the company Boots, Chibnall had found a feature of insulin that intrigued him. At the end of its protein chains were free amine groups—a chemical moiety composed of nitrogen bonded with two hydrogen atoms. As these groups are highly reactive, it might be possible to link them with a chemical tag, such as a coloured dye, and in so doing to identify the terminal amino acid in the chain. This was the task that Chibnall set before Sanger. But before he could even begin, he needed to find a suitable chemical tag that could be used to label the end of the protein chain. Thanks to a conversation over a glass of beer with a colleague in the chemistry department, he found exactly what he needed.

Fluorodinitrobenzene (FDNB), a highly reactive compound, had been synthesized as part of the war effort—although Sanger said that he was never clear whether it had been developed with the intention to use it as a poison gas or in the course of research on antidotes to chemical weapons.[69] When FDNB reacted with the free amine group of whatever amino acid was at one end of the protein chain, it formed a bright, yellow-coloured derivative that could then be easily split off from the protein and separated using the silica-gel method of partition chromatography. In an interview given many years later, Sanger paid tribute to Martin and Synge for having first come up with this innovation:

> Nearly all these fractionation methods were invented by them, particularly Martin. He was the great brain behind this. He was a very bright, original person and a most inspiring person to talk to. I can't say I talked to him. I listened to him, really. I think he was as near to a genius as anyone I have ever met ... he was the brains behind it, a rather eccentric person, and I don't think he did much by himself ... Synge was a very good complement to him because he was a very practical, down-to-earth person, and a very efficient experimentalist.[70]

Despite his accolades for Martin, Sanger actually had a much closer working relationship with Synge. After learning that Sanger was using FDNB to label the terminal amino acids in insulin, Synge had recruited both him and Hodgkin to help with the studies of gramicidin S.

Sanger's task was to use his labelling method to identify terminal amino groups in gramicidin and gramicidin S, and to send labelled derivatives of these compounds to Hodgkin for X-ray analysis.[71,72,73,74,75] As the work progressed, a friendship developed between Sanger and Synge. When Sanger later worked for a short period in Tiselius's laboratory in Uppsala, Sweden, he thanked Synge, who was also a visiting researcher there, for 'helping to make my stay in Sweden such a pleasant one both in the lab and out of it'.[76,77,78]

Within a short space of time, Sanger's work on gramicidin S had shown it not to be a linear chain, but a cyclical molecule,[79] but insulin would not give up its secrets quite so easily. Once he had treated the free amino groups at the end of the insulin chains with FDNB, Sanger broke apart the links in the protein chain with acid to release a mixture of labelled and non-labelled amino acids. As the labelled amino acids now had a bright yellow colour due to the FDNB, they could then be easily separated using silica-gel partition chromatography, just as Martin and Synge had first used for wool (Figure 32).[80]

By 1945, Sanger had already made a crucial discovery about insulin using this method. He had found that insulin was composed not just of a single protein chain, but rather of at least two distinct chains of amino acids. One, known as the 'A' chain, had glycine at its terminus—the same amino acid found in deep space by the Rosetta probe. The other chain, designated the 'B' chain, had the amino acid phenylalanine at its N-terminus. Sanger was not content to just stop there. He wanted to forge on and see whether he could push beyond identifying the terminal amino acids of each of these two chains and work out the linear order in which their immediate neighbours were arranged.

To do this, he used the same two-stage approach that Martin and Synge had developed for their work on wool proteins. The first stage was to subject insulin to only a gentle treatment with acid, which meant that the protein chain was not broken at every link, but rather only in specific places to release peptide fragments of a certain size. These fragments were then labelled with FDNB and separated using partition chromatography on a silica-gel column. The second stage of the process was to then subject each of these peptide fragments to a harsh treatment with acid that broke the bonds between individual amino acids. The resulting mixture of amino acids was then analysed using paper chromatography.[81] Put another way, the first stage was to dismantle a Lego model into big chunks, and the second stage to break apart each chunk into its constituent bricks.

It proved to be laborious and tedious work. In one letter, Sanger confessed to Synge, 'I am still struggling on with insulin—it's being a bit elusive'.[82] Ultimately, his tenacity and determination paid off. Like Synge, he, too, had been invited to give a paper on his work to the Cold Spring Harbor symposium in 1949. When planning his trip, he hoped that he and Synge would be able to travel there together but with no visa for Synge, Sanger found himself making the journey alone.[83]

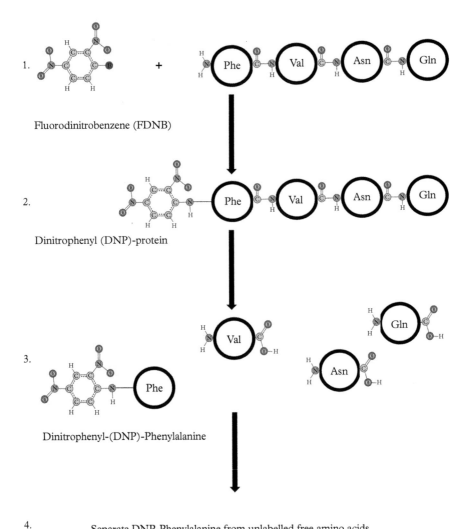

1. Fluorodinitrobenzene (FDNB)

2. Dinitrophenyl (DNP)-protein

3. Dinitrophenyl-(DNP)-Phenylalanine

4. Separate DNP-Phenylalanine from unlabelled free amino acids using silica gel partition chromatography

Figure 32 *Sanger's strategy to determine the terminal amino acids at the A and B chains of insulin (for simplicity only the four endmost amino acids of the B chain are shown). Step 1. Insulin is reacted with the compound Dinitrofluorobenzene (DFNB) Step 2. DFNB reacts with the amino (-NH2) of the terminal amino acid to produce a dinitrophenyl (DNP) derivative of the protein. Step 3. Treat the DNP-protein with acid to break the bonds between individual amino acids. This produces a mixture of free unlabelled amino acids and the terminal amino acid which is labelled with DNP. Step 4. Separate the DNP-labelled amino acid from the mixture using silica gel partition chromatography and identify it by comparison with the chromatographic movement of known DNP-amino acids. Diagram by K.T. Hall.*

At the meeting, Sanger presented his initial work on the amino acids found at the very end of the two chains of insulin, but he admitted in his introduction that 'Very little is known, however, about the order in which these amino-acids are arranged to form the whole protein molecule'.[84] Sounding a note of optimism, however, he went on to point out that the newly developed methods of chromatography gave cause for hope. And even as he spoke these words, Sanger knew that his optimism was well founded. For by now, Synge, in collaboration with Martin and the team had WIRA, had at last succeeded in their 'boastful undertaking'— thanks to partition chromatography they had not only analysed which amino acids were found in gramicidin S, but had also determined the precise order in which they were arranged (Figure 33).

Valine   Ornithine   Leucine   Phenylalanine   Proline

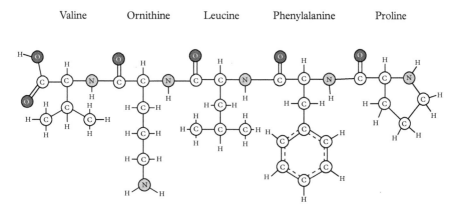

Figure 33 *'A boastful undertaking'... Using partition chromatography, Martin, Synge, Gordon and Consden were able to determine the complete amino acid sequence of gramicidin S. And if it could be done successfully for a small antibiotic like gramicidin S, Fred Sanger saw no reason why the same could not be done with insulin... Diagram by K.T. Hall.*

This achievement left them confident that partition chromatography could now be used 'for studies of the sequence of amino-acid residues in peptide structures generally'—and Sanger shared their confidence.[85] He wondered if what the WIRA team had achieved with gramicidin S could also be now done for insulin. Could he push beyond working out the sequence of just a few amino acids at the end of the chain, and instead aim for the entire protein? It seemed worth a try:

> The results with the terminal peptides suggested that both fractions A and B of the oxidized insulin consist of only one polypeptide chain containing about 20 and 30 amino-acid residues respectively. It was thus considered that it might be possible to determine the complete peptide sequence of these chains from an investigation of the lower peptides produced on partial hydrolysis by the chromatographic method of Consden, Gordon and Martin (Biochem. J., 41, 590).[86]

Over the course of the next few years the amino acid sequence of each of the two component chains began to emerge, thanks to the painstaking effort of Sanger and his colleagues Hans Tuppy and Ted Thompson. The approach was essentially the same that Sanger used to work out the sequence of the terminal amino acids—partially break the protein chain down into specific fragments, which were then treated with FDNB and separated. These fragments were then broken down further into their component amino acids, which were separated using paper chromatography.[87] In this way, Sanger, Tuppy, and Thompson began to slowly amass a set of fragments of defined sequence—rather like trying to assemble a jigsaw puzzle by collecting many partially assembled fragments.

However, the peptide fragments produced by partially breaking down the protein chain with mild acid were too short, which meant that, as Sanger began to build up the sequence of insulin, there were gaps. Thankfully, nature had provided a means of generating a set of much larger fragments. Using naturally occurring digestive enzymes that act at specific sites along the protein chain, Sanger and his co-workers were able to fill in the gaps in their sequence. But the work was slow and plodding. When, in 1948, immunologist Brigitte Askonas joined Sanger's lab as a PhD student, she recalled how the work on insulin was proceeding at a snail's pace:

> When one would ask him how his work was going, he would say very little. 'Oh, I've got another peptide'. Then, at a lab meeting he would bring a stack of cards showing overlapping short sequences, and slowly, diffidently, build up his latest segment of the molecule.[88]

Nevertheless, Sanger's slow, methodical approach paid off. In the spring of 1951, he wrote to Synge in triumph with a diagram showing the complete order of amino acids in the B chain. When he first published this work in a journal, Sanger acknowledged that the success of Consden, Gordon, Martin, and Synge in having used partition chromatography to work out the amino acid sequence of gramicidin S had been of key importance, and two years later he followed this up with the sequence of the A chain.[89,90,91] Sanger had now been working on insulin for nearly a decade, producing a publication every other year. This was a rate of progress that would certainly not be acceptable in the metric-obsessed research culture of contemporary academia, and even at the time, there were members of Sanger's own department who felt that it was not good enough. Askonas recalled, 'some had wanted to kick him out'.[92] Even Francis Crick is said to have once complained, 'The trouble with Fred is that he has developed all these powerful methods, but we can't persuade him to do anything interesting with them'.[93]

In stark contrast to Sanger's disgruntled naysayers in Cambridge, the Nobel Committee in Stockholm felt very differently. In October 1958, Sanger was awarded the Nobel Prize in Chemistry for his determination of the complete amino acid sequence of bovine insulin. Reporting on the award, *The Times* compared Sanger's achievement with Sir Roger Bannister's success in running the four-minute mile, but, with a characteristic and endearing modesty, Sanger refused to bask in the limelight.[94] Many years later, he turned down a knighthood, saying 'I don't want to be different', and when his work on insulin earned him the Nobel Prize, he was quick to point out that 'with the insulin chains we were again lucky in that the method of paper chromatography had just been developed by Martin and his colleagues'.[95,96]

Sanger's sequencing of insulin was a landmark in protein chemistry and was achieved thanks to what he described as '[p]robably the greatest advance that has been made recently in this field'—partition chromatography.[97] This was the first ever time that the complete sequence of amino acids in a protein had been determined and its consequences for our understanding of biology would be profound (Figure 34). The most immediate of these was that it demolished once and for all the conviction that the order of amino acids in a protein was dictated according to some mathematical law.[98]

But while figures such as Astbury, Bergmann, and Niemann saw their long-cherished hypotheses demolished by Sanger's work, other scientists found cause to celebrate. At the end of her Nobel speech, Hodgkin had, with characteristic modesty, remarked that 'I seem to have spent much more of my life not solving structures than solving them'.[99] And of all these unsolved structures, there was one in particular that had nagged away at her over the years. According to her biographer, Georgina Ferry, 'there was only one mountain left to climb for Dorothy, and that was insulin. Insulin was unfinished business'.[100] For although penicillin and vitamin B_{12} had earned Hodgkin her Nobel Prize, her colleague Jia-Huai Wang recognized that 'insulin had long been Dorothy's favorite molecule'.[101]

Fred Sanger's success had rekindled Hodgkin's interest in insulin.[102] In 1951, Sanger had written to Hodgkin and sent her an early part of his sequence, hoping that she would 'be able to make something out of it'.[103] She did not disappoint him. In December 1954, she told the American chemist Robert Corey that she was trying to build a model of the structure of insulin based on Sanger's results. Corey responded to this news with excitement:[104]

> … insulin is your crystal, and I should very much like to see you rewarded for all your efforts by coming up with the detailed structure. It would give me a bigger thrill than the B12 structure, though goodness knows, that should be big enough for anyone.[105]

One crucial piece of information that helped Hodgkin with her model was a preprint of a paper by Sanger, which she received in 1955 from his colleague Max Perutz.

Sanger had discovered that the two polypeptide chains of insulin were physically connected by bonds formed between sulphur atoms at specific positions on each chain.[106] Hodgkin described this as being 'terribly important' for her model, but Perutz had much more to offer.[107]

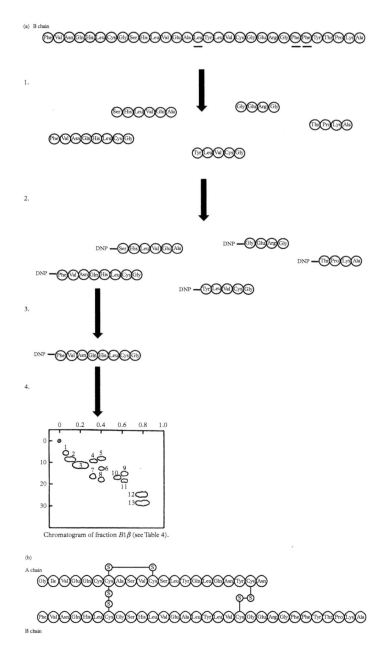

Chromatogram of fraction $B1\beta$ (see Table 4).

Perutz had left his native Austria in 1936 to come to Cambridge where he had begun a PhD with the X-ray crystallographer John Desmond Bernal, but his timing had been unfortunate. When the Second World War started, he found himself in an internment camp outside Liverpool before being transferred to one in Canada. Thankfully, the British Government realized that this was a waste of his invaluable expertise and he was brought back to Britain to put his talents to use in the war effort. Perutz now spent some time involved with a top-secret project that attempted to construct aircraft carriers from a mixture of ice and wood pulp. When it (unsurprisingly) came to nothing, he was at last able to return to X-ray crystallography.

Turning his attention to the 'phase problem' that was causing Dorothy Hodgkin a major headache in her own X-ray studies of insulin, Perutz believed that he had at last found a solution. The challenge was to determine the particular phase of each reflection made by diffracted X-rays. By modifying the blood proteins haemoglobin and myoglobin with atoms of heavy metals, Perutz and his colleague John Kendrew were able to obtain not only the intensity and position of spots in the pattern made by diffracted X-rays, but also their respective phases. Armed with this extra information, they were now able to deduce molecular structures for haemoglobin and myoglobin. Although the resolution of these structures was only low, Perutz and Kendrew had shown that, contrary to the scepticism of many chemists, even the structure of biological molecules as complex as proteins could now be revealed by X-ray crystallography.

For this achievement, Perutz and Kendrew were awarded the 1962 Nobel prize in Chemistry and the insights they gained were not lost on Hodgkin. Only a few years earlier, she had received a sample of porcine insulin crystals from Jorgen Schlichtkrull, a Danish chemist who worked for the company Novo Nordisk.

Figure 34 *a) Simplified diagram showing Sanger's strategy for determining the amino acid sequence of the A and B chains of insulin. For simplicity, only the B chain is shown. 1) Partially digest the B chain with mild acid treatment to break the chain down into 5 peptides. 2) React these peptides with DFNB to label the terminal amino acids as DNP-derivatives. 3) Separate the DNP-labelled peptides using methods based on adsorption or electrical charge 4) Once each DNP-labelled peptide (only one of the five is shown) has been separated from the mixture, break down with acid and separate and identify the resulting amino acids using paper partition chromatography. Note that breaking up the B chain with acid still did not produce peptides containing the Leucine residue at position 15, or Phenylalanine at positions 24 and 25 (all underlined). For this reason, Sanger had to use a second way of breaking up the chain using protease enzymes. Diagram by K.T. Hall. Figure of the chromatogram is for peptide B1gamma, taken from Fig. 6, p.469, Sanger, F., & Tuppy, H., 'The Amino-acid Sequence in the Phenylalanyl Chain of Insulin,' The Biochemical Journal 49, 1951; 463-480.)*
b) The complete amino acid sequence of the A and B chains of insulin as determined by Sanger and his co-workers. Based on sequence shown in Sanger, F., Thompson, E.O.P., Kitai, R., 'The Amide Groups of Insulin,' The Biochemical Journal 59, (1954): 509-518.

Schlichtkrull's interest in insulin was more than just professional. Having a diabetic daughter, he was keen to understand more about the structure of insulin and how this knowledge might be used to improve the management of diabetes. With her success in using X-ray crystallography to solve the structure of biological molecules, Hodgkin was the ideal person to approach.

Hodgkin's work had shown that insulin was actually made up of six individual protein molecules to form a unit called a hexamer, which also contained two atoms of zinc. The discovery that these zinc atoms could be removed raised the possibility that they could be replaced by atoms of a heavier metal such as cadmium or lead.[108] This would then allow the phases of diffracted X-rays to be calculated, which might, in turn, allow the calculation of electron density maps to reveal the precise locations of all the atoms in the protein chains of insulin. The task of computing Fourier transformations and electron density maps was laborious but as S. Ramaseshan, a visiting researcher from India recalled, the task was made less onerous by lab games of cricket in which discarded pages of calculations were put to good use as an improvised bat and ball.[109]

At the close of her Nobel speech, Hodgkin outlined the challenges of trying to solve the structure of insulin, admitting that the electron density maps were difficult to interpret. A few years later, she confessed in a letter: 'I am feeling a little down about the late insulin work'.[110,111] The lack of progress was in no way due to a lack of determination from her team. When M. Vijayan, another visiting researcher, realized that his research on insulin would be interrupted by his having to return to India to get married, to sidestep this minor distraction, he arranged for his future wife to come to Oxford on a fellowship so that their wedding ceremony could be conducted without hindering progress in the lab.[112] The honeymoon that followed was unconventional: Vijayan and his bride chose to savour the thrill of seeing the first three-dimensional structures emerge from the electron density maps that he had helped to prepare.

In the end, their dogged tenacity paid off. Only a few weeks after Neil Armstrong had stepped onto the moon's surface, Hodgkin and her team made one giant leap of their own:

> I used to say the evening that I developed the first X-ray photograph I took of insulin in 1935 was the most exciting moment of my life. But the Saturday afternoon in late July 1969, when we realized that the insulin electron density map was interpretable, runs that moment very close.[113]

Perutz at the Cambridge Laboratory of Molecular Biology was one of the first people to hear of Hodgkin's success. Having received a phone call from Guy Dodson, one of Hodgkin's team, inviting him to come to Oxford and join the celebration, Perutz pinned up a hastily written poster that declared, 'Late night news from Dorothy Hodgkin: INSULIN IS SOLVED'.[114] He then jumped into his car and dashed over to Oxford, taking with him an unopened bottle of champagne he had received as a gift when he had won the Nobel prize seven years earlier. While the champagne had gone somewhat flat, the mood was certainly not.[115]

For the wider world, however, all the excitement of that summer was focused on the moon landings. The plenary lecture of the International Union of Crystallography Congress held at the State University of New York at Stony Brook, Long Island in August 1969 had originally been scheduled as a presentation of analyses of moon rock brought back by the Apollo 11 astronauts. But after concerns that the rocks had not passed biological safety tests, the lecture had been cancelled and Hodgkin was offered the vacant slot in the schedule to talk about her latest work on insulin. Rather than give the lecture herself, however, she chose the youngest member of her team, Tom Blundell to officially unveil the structure of insulin to the scientific world (Figure 35).[116]

Figure 35 *Photograph of Dorothy Hodgkin taken in 1989, twenty years after her discovery of the crystal structure of insulin, shows her with a 3D model of the molecule in the background.*
Credit: CORBIN O'GRADY STUDIO/SCIENCE PHOTO LIBRARY.

Word quickly reached the press. *The Times* hailed the achievement as being 'the fruit of more than 30 years' work' for Hodgkin.[117] In some quarters, however, there was a more reserved response. Although he had never doubted that Hodgkin would be successful in her quest to solve the structure of insulin, fellow crystallographer Jack Dunitz said that 'it did not bowl us over'.[118] As far as Dunitz was concerned, Hodgkin's greatest achievement had been to solve the structure of vitamin B_{12}. By doing this, she had demonstrated to the scientific world that the structure of complex biological molecules could be revealed by X-ray crystallography. In the wake of this accomplishment, solving the structure of insulin, impressive as it was, was anticlimactic.

Elsewhere, there were hopes that Hodgkin's success in solving the structure of insulin might lead to more than a scientific triumph. In Hodgkin's papers in the Bodleian Library, Oxford, is a copy of the letter that Ion Pavel sent in 1969 to Charles Best as part of his campaign to win recognition for Nicolai Paulescu.[119] Perhaps Pavel was hoping that some of Hodgkin's prestige in the field of insulin research might now rub off upon his own crusade to restore Paulescu to his rightful place. Hodgkin was certainly aware of Paulescu and his work. At around the same time that Pavel was stepping up his campaign, Hodgkin received a report on the discovery of insulin from Norman Lazarus, a colleague at the Wellcome Foundation: 'I think you will agree from what has been said here that Paulescu I think should have shared in the general back slapping or else he must have missed it by a whisker'.[120] Hodgkin seemed inclined to agree with Lazarus. In 1978, when she gave the Presidential Address to a meeting in Bath of the British Association for the Advancement of Science, she drew attention to the respective contributions of both Banting and Paulescu in the discovery of insulin:

> Successful extracts were obtained almost simultaneously in very different circumstances, in Roumania by Paulescu and Banting and Best in Toronto. Paulescu, an experienced medical scientist was returning to an old interest after interruption by the war: his results were not immediately recognised.[121]

Pavel must have been delighted. But in stark contrast to Paulescu, Hodgkin's work on insulin was most definitely being recognized. And while insulin was by no means the first protein structure to be solved by X-ray crystallography, Hodgkin's work held the promise of offering new insights into how the molecule functioned. This, in turn, raised the exciting possibility that new, improved forms of insulin might now be deliberately engineered. One of the main drawbacks of having to manage diabetes with porcine or bovine insulin was that they did not mimic the exact physiological response of the human pancreas in releasing insulin. Now that the structure of insulin was known, it might be possible to understand not only how the molecule worked, but also how to improve its efficiency in the management of diabetes.

Even before Hodgkin's announcement of the structure of insulin, some had already taken the first steps in this direction. Using chemical treatment, independent researchers at the University of Toronto and in China had prised apart the A and B chains of the insulin molecule by breaking the bonds that held the sulphur atoms. Having separated the two chains, they had then shown that it was possible to recombine them to reconstitute a molecule with full biological activity.[122] In the conclusion to their paper, the Canadian researchers speculated on a tantalizing possibility:

... if chemically synthesized A and B chains were available ... it should be possible to obtain insulin by the above method and thereby provide the terminal step in the total synthesis of a protein with biological activity.[123]

Since the two chains of insulin contained a combined total of only 51 amino acids, the possibility of synthesizing it artificially did not seem too outlandish. The chemical methods for forming bonds between individual amino acids to join them were well known, and biochemist Panayotis Katsoyannis (1924–2019) at the University of Pittsburgh seized upon the challenge of making this a reality. Using the information from Sanger's work, he and his team began by chemically stitching together individual amino acids to form small polypeptide fragments that corresponded to those found at the extreme ends of both chains of insulin. They were clear that their ultimate goal was nothing less than 'the total synthesis of this protein'.[124] From 1961 onwards, Katsoyannis and his team ground out their results in a steady succession of papers, each of which described the synthesis of short peptide fragments of insulin. The first of these described the synthesis of a single fragment that contained the last five amino acids from one end of the A chain from sheep insulin. Then, having established this first foothold, they extended this to include ten amino acids.[125,126] Like a team of climbers slowly inching their way towards a distant summit, each successive paper gave them yet one more precious foothold on their gruelling ascent.

It was a slow, laborious approach that took over five years and two hundred individual steps. After successfully synthesizing each of the two individual chains of insulin, Katsoyannis announced in 1963 that his team had successfully combined them to produce a functional molecule.[127,128] The work was presented at a scientific conference, but the news soon went much further. By May of 1964, Katsoyannis had made the pages of *Life* magazine, which likened his achievement to being 'roughly equivalent to working a dozen jigsaw puzzles simultaneously while blindfolded'.[129]

When Katsoyannis published this work in a scientific journal, he made clear in his introduction that this was nothing short of a scientific landmark.[130] For it was the first time that a natural protein had ever been synthesized completely from scratch. Or so he thought.

For Katsoyannis and his team had not been the only ones struggling to towards the summit of artificially synthesized insulin. Upon reaching their goal, they discovered that someone else had planted their flag there, too. A team of German researchers working at the German Institute for Wool Research in Aachen announced that they had not only succeeded in synthesizing the two chains of insulin but had also recombined them to produce a biologically active protein.[131,132,133]

The German team was led by Helmut Zahn (1916–2004) who, like Martin and Synge, had a background in textile science and, like them, the impact of his expertise in this field had gone far beyond the study of wool fibres. It was thanks to

Zahn's expertise in forging chemical cross-links between protein chains—a technique that he had learned from the study of wool proteins—that his team had been able to successfully recombine the two artificial insulin chains to make a functional protein.[134] As a result, when *Life* magazine reported on the synthesis of artificial insulin, Katsoyannis found his photograph alongside one of Zahn.

What the article failed to mention, however, was that there had actually been yet another team aiming for the summit. For Chinese scientists, Mao Zedong's 1958 announcement of the 'Great Leap Forward' seemed to offer the ideal opportunity to propose grand plans of research. In response to Mao's vision of transforming China from an agrarian system that had endured for centuries into an advanced industrialized economy, scientists began to put ideas forward that they hoped would prove to the world that their country was capable of research greatness. Such suggestions included the determination of the amino acid sequence of the blood protein myoglobin, or cytochrome c, which is crucial to the release of metabolic energy from respiration—all of which seemed to be suitably grand. But while none of these proposals lacked in ambition, they fell short on ideological grounds. For although the successful determination of the amino acid sequence of a protein might well have been of interest to a few Chinese biochemists, it was deemed unlikely to grip the imagination of the wider population.

But why devote time and money to sequencing the amino acids in yet another protein, when Sanger had already done this for insulin? Rather than determine the sequence of a protein, why not raise the stakes, and synthesize an entire protein completely from scratch? For young scientists like Tang Youqi, it was a thrilling prospect. Tang, who acted as an interpreter when Dorothy Hodgkin gave a lecture in Beijing on her studies of vitamin B_{12} and later became a Professor of Biophysics, had one particular protein in mind as the target:

> In 1958 the young students were asked what they wanted to do. They didn't know much, but they wanted to do something great, so they chose to synthesize insulin. In fact our government spent a lot of money on that. The solvent used in the insulin synthesis could fill a small swimming pool. The amino acids also cost a lot of money. In China we cannot go to the moon, but I think for a country of 800 million people to spend that much money was something of an experiment.[135]

As a relatively small protein for which the entire amino acid sequence was now known, insulin was the perfect candidate. In 1958, the Shanghai Institute of Biochemistry proposed that a nationwide collaboration be established and dedicated towards its complete synthesis. The project also had the advantage that it ticked all the ideological boxes of 'The Great Leap Forward'. Not only was it of demonstrable practical value, but it was perceived as having the blessing of one of the founders of Communism himself. Friedrich Engels had once remarked that life was nothing but 'the mode of existence of albumen [protein]' and with an endorsement like this, the proposal to artificially synthesize an entire protein could hardly fail to win the approval of the political authorities.[136]

Hodgkin first learned of this work while she was in Beijing to give her lecture on vitamin B_{12}, and Tang later reflected that insulin was to be the start of an 'intimate and lasting friendship' between her and the Chinese scientists.[137] The task was divided so that while researchers at the Shanghai Institute of Organic Chemistry and the Chemistry Department of Beijing University focused on synthesis of the A-chain, other groups at the Shanghai Institute of Biochemistry concentrated on synthesis of the B-chain and its recombination with the A-chain. Due to the upheavals of the 'Great Leap Forward' no scientific journals had been published for some time. In 1961, when the journal *Scientia Sinica* resumed publication in 1961, the scientists began to present their results, first with the announcement that they too had succeeded in recombining the natural A and B chains of insulin to reconstitute a functional protein.[138] Having secured this foothold, they went on to produce a functional hybrid insulin protein consisting of the natural A-chain combined with the synthetic B-chain.[139,140,141]

While the political regime repressed many other areas of scientific work at this time, they recognized the value of allowing research on insulin to continue. Although the scientists involved in this work had only minimal contact with scientists outside China and were not allowed to publish in Western journals, they did have a loyal ally in Hodgkin. When she paid another visit to China in 1965, Hodgkin was delighted to find that they had been successful and, knowing that while the Chinese scientists themselves would be unable to attend a forthcoming conference to be held on insulin in the USA, she hoped that she at least might be able to bring word of their triumph to the scientific community.[142] She also encouraged them to do what neither Katsoyannis nor Zahn had yet managed to do—to obtain crystals of their synthetic material that could be used for X-ray analysis in order to compare the structures of both natural and artificial insulin.[143]

Within only a year of Hodgkin's visit, the Chinese teams announced that they had successfully crystallized synthetic insulin and, in the opening line of their paper, reminded readers that, 'The first successful total synthesis of a protein was accomplished in 1965 in the People's Republic of China'.[144] Thanks to the work of Katsoyannis and Zahn, this claim is not quite true, but the authors of this paper had very good reasons for putting on such an overt display of patriotism and stressing that their success belonged to the whole nation: China was heading towards a political and social disaster.

When the Nobel laureate Sir John Kendrew visited China that year, he told Chen-Lu Tsou, who had led the team working on recombination of the two chains, that as a result of having been featured on the BBC news, the synthesis of insulin was 'definitely the best Chinese scientific achievement known to the public in the UK'.[145] But Kendrew's visit would be one of the last made by a Westerner for a number of years. Concerned that Soviet Communism had lost its way, Mao was determined that China should avoid a similar fate and in 1966 he launched 'The Cultural Revolution' to set the country back on the straight and narrow. The human cost of his quest to restore ideological purity was horrific. Millions died, were forced from their homes, or interned in 're-education' camps.

Presumably to avoid a similar fate, the authors of the paper on synthetic insulin preceded the account of their scientific work with the following announcement:

> Holding aloft the great red banner of Chairman Mao Tse-tung's thinking and manifesting the superiority of the socialist system, we have achieved under the correct leadership of our Party, the total synthesis of bovine insulin.[146]

And in addition to giving precise details of the reagents and experimental methods used as is usual in a scientific publication, the authors felt it also necessary to add that:

> Throughout the various stages of our investigation, we followed closely the teachings of Chairman Mao Tse-tung: eliminating superstitions, analysing contradictions, paying respect to practice, and frequently summing up experiences.[147]

To dispel any doubts about the strength of their ideological conviction, they cited Engels and stressed that the ultimate aim of their work was more than just synthesizing insulin. It was, they declared, nothing less than 'an important step forward in the long pursuit to synthesize life from inorganic compounds'.[148]

Amidst this chaos, universities were closed down and basic scientific research was prohibited—presumably on the grounds that it was insufficiently practical. One casualty was Tsou, who, despite being a leading figure in the synthesis of insulin, had now fallen so far from favour with the political authorities that by the time Tiselius visited China in 1966, Tsou was forbidden to go anywhere near him. When asked to comment on another relatively recent Chinese technological achievement—the successful detonation of their first atomic bomb—Tiselius is said to have replied 'the making of an atom bomb you can learn from textbooks but not the synthesis of insulin'.[149] Tiselius clearly recognized that the synthesis of insulin was a monumental achievement, and one for which Tsou and his colleagues deserved to be contenders for the Nobel Prize, but the authorities had other ideas. As Tsou recalled, 'People who had very little to do with the effort [to synthesize insulin] either elbowed themselves, or were pushed by the authorities, to the forefront, but the Prize was not to be'.[150]

As chaos erupted on the streets and millions of people were evicted from their homes or deported to camps for 're-education', the scientists struggled on with their studies of insulin. Working in isolation from the rest of the scientific world, however, they faced huge problems, including lacking the computing power to carry out the necessary calculations, and writing their software in machine code rather than high-level programming languages like FORTRAN. Under the dictates of the Cultural Revolution, they were not allowed any contact with Western scientists and the only available scientific journals were those that had arrived in the library a couple of years after publication. In addition, they were expected to break off from their scientific work to attend political activities.

Nevertheless, they persisted, and in 1972 were finally allowed to report their work in the *People's Daily* and a domestic journal, taking care to include the required dutiful references to Marx, Engels, and Chairman Mao. It was thanks to Hodgkin that the rest of the world first came to hear of this achievement. While travelling to the International Congress of Crystallography in Japan that same year, Hodgkin had decided to pay a visit to her colleagues in Beijing. There, sitting on the floor of a hotel just west of Tiananmen Square, she and the young Chinese researchers compared the structures that they had computed for insulin. At first things did not look good—the structures did not correspond with each other. But the room soon burst into cheers when someone realized that one set of coordinates simply needed to be rotated through 180 degrees for both structures to make a match. 'It was', said Jia-Huai Wang, 'an unforgettable moment. Dorothy jumped up from the floor!'[151] This was the first time that two groups of crystallographers had obtained the same result and Hodgkin announced the result at the International Congress of Crystallography, before bringing the tireless efforts of the Chinese researchers to the attention of the world through an article in *Nature*. In her conclusion she wrote, 'It will be splendid if we can some day soon all meet and talk over the very interesting observations that are accumulating, East and West, on the structure and function of insulin'.[152] As a lifelong adherent of progressive political views, Hodgkin felt that the pursuit of science transcended national and ideological boundaries and her work on insulin offered the chance to build a welcome bridge between East and West during the dangerous years of the Cold War. Sadly, however, idealism stood in stark contrast to the Communist authorities in Beijing. After the Chinese scientists had announced their success in solving the structure of synthetic insulin, their group was disbanded with several members being dispatched for 're-education' or to work on farms.[153]

The work of Katsoyannis, Zahn, and the Chinese researchers was an impressive scientific achievement, but was it also a medical one? With diabetic patients already able to manage their condition using bovine or porcine insulin, was the complete synthesis of human insulin from individual amino acids just a scientific vanity project? Writing in 1964, Katsoyannis was clear that it was not, and that 'the anticipated steady increase in the demand for insulin, coupled with the fact that the supply from natural sources is not unlimited, justifies the speculation that synthetic insulin may soon be needed'.[154]

Moreover, since porcine and bovine insulin were foreign proteins, when injected, they elicited an immune response resulting in allergic reaction in some patients. With pure human insulin this problem should disappear, thus strengthening Katsoyannis's case. However, there was one huge obstacle: complete artificial synthesis of human insulin at the lab bench was expensive—in terms of both time and money. When the company Ciba-Geigy produced some insulin by artificial synthesis in 1974 and tested it in a very small-scale trial, they concluded that the production costs would be prohibitive for what was only a small amount of material.[155,156]

Thankfully however, nature had already provided the tools for an alternative method of making insulin. When Martin and Synge had first published their new method of partition chromatography in 1941, they had casually suggested 'that the possible field of usefulness of the new chromatogram is by no means confined to protein chemistry'.[157] This would prove to be a memorable understatement, for although partition chromatography had originally been developed for the study of wool, it would also offer the first hint at how another white, stringy fibre—DNA—carried the instructions not only for making insulin, but for life itself.

10

The Prophet in the Labyrinth

On February 28, 1953, drinkers enjoying a quiet pint at Cambridge's The Eagle pub in the city centre are said to have been treated to a memorable sight. Flinging open the door, two young men burst in and declared to anyone within earshot that they had just discovered the secret of life.[1] Today, in honour of this event, The Eagle serves a specially brewed beer called 'DNA' and in one corner of the pub hangs a plaque which bears the following words (Figure 36):

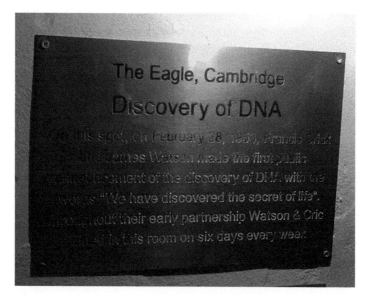

Figure 36 *Commemorative plaque on the wall of 'The Eagle' pub in Cambridge which incorrectly ascribes the discovery of DNA to James Watson and Francis Crick. The inscription reads:* Discovery of DNA: On this spot on 28th February 1953 Francis Crick and James Watson made the first public announcement of the discovery of DNA with the words, "We have discovered the secret of life". Throughout their early partnership, Watson and Crick dined in this room on six days every week. *Photograph by K.T. Hall.*

> On this spot, on February 28, 1953, Francis Crick and James Watson made the first public announcement of the discovery of DNA with the words 'We have discovered the secret of life'. Throughout their early partnership Watson & Crick dined in this room on six days every week.

But the tourists who cluster around the plaque snapping photos on their phones might well be disappointed to learn that it commemorates something that never actually happened.

For in his memoir *What Mad Pursuit*, Crick said that he has no recollection of this event and has ascribed it more to the poetic license of James Watson, who has since admitted that he embellished his memories of his incident for dramatic effect.[2,3,4] And while the beer certainly tastes pleasant, its name is something of a misnomer. This is because, although Watson and Crick discovered the molecular structure of DNA, the actual substance itself had already been discovered nearly a century before they are said to have gone bursting into the pub.

The actual discovery of DNA was far less dramatic and was first announced in an 1871 paper published with the riveting title of 'On the Chemical Composition of Pus Cells'. It was the work of Johann Friedrich Miescher (1844–1895), a young Swiss physiologist and is a landmark in biology. Moreover, it owes its existence to what has been described as 'a louse that, in its modest way, changed the course of science'.[5] As a result of contracting typhus louse bite as a child, Miescher developed a hearing impairment. Although he had trained to become a doctor, he feared that he would be unable to use a stethoscope due to his impaired hearing and opted instead to follow a career in research. But the question was, which field of research should he choose? On this matter, the advice of his uncle, the distinguished physician Wilhelm His, proved to be invaluable.

His directed Miescher towards an emerging and exciting new field of research that he believed held great promise for a young scientist seeking to carve out a career. This was the field of 'physiological chemistry', or what is better known today as biochemistry—the chemistry of living systems. One of its founding figures was Professor Felix Hoppe-Seyler (1825–1895) who, at the University of Tübingen, had established the very first laboratory in Europe devoted to this subject. Hoping to learn from the best, Miescher made his way to Tübingen where Hoppe-Seyler set him the task of studying the chemical composition of the cell nucleus.

Miescher's plan was to use white blood cells as his experimental material since they were widely found throughout the body and thought to be a good general model of the cell. But obtaining them was not a task for the faint hearted or those easily put off their dinner. Knowing that pus was a rich source of white blood cells, Miescher used his initiative and procured a regular supply of discarded surgical bandages from a local hospital.[6] He then carefully washed the pus off the bandages before assessing whether the material was sufficiently fresh for use in his experiments on the basis of its appearance and aroma.

Miescher had a ferocious work ethic and a formidable dedication to his research. When he found himself to be short of laboratory glassware, he simply resorted to using his own china dinnerware from home. Later, one of his students recalled how, on Miescher's own wedding day, a group of friends had been forced to drag him away from the laboratory bench in order that he might attend the ceremony.[7] This stoicism and focus served him well when he began his research working in the freezing cold former kitchen of Tübingen castle, now converted into laboratories for the University.

Modern readers of his paper who expect to find some seismic moment of revelation transforming the landscape of biology are likely to be disappointed. With its detailed accounts of the chemical methods employed, it does rather resemble a glorified cookbook.[8] For while experimenting with his pus extracts, Miescher detected a substance that, unlike proteins, could be precipitated with acid but which redissolved in alkali and, moreover, was rich in phosphorus. Because this new substance was localized in the cell nucleus, Miescher dubbed it 'nuclein' and he was confident that, rather than being due to contamination, he had isolated 'a specific chemical entity or a mixture of closely related entities'.[9]

While acknowledging that this was at best only a preliminary study, Miescher was also confident that this chemical entity would prove to be of some great significance:

> … it now seems to me likely that we will find an entire family of such phosphorus-containing bodies, varying slightly from one another, [and that this] group of nucleic bodies [might prove equal in importance to] the proteins.[10]

Moreover, he was sure that other researchers would now be inspired to investigate the properties of this substance, which would lead to discovering the internal processes of cell growth.[11]

But for the moment such inspiration would have to wait. Any research scientist, whether PhD student, post doc, or professor, will empathize with frustrations that result from publications being delayed. Although Miescher had completed his experiments in 1869, his work was not published until 1871, when it appeared in a volume compiled by Hoppe-Seyler, who was very cautious about Miescher's claim. His caution was understandable. Only few years earlier, another of his students, Oscar Liebrich, had made a similar claim with the discovery of a new cellular substance that he named 'protagon'. But when Liebrich's work had come under fire from eminent scientists in the field, Hoppe-Seyler's reputation had taken a blow. Wary from this experience, Hoppe-Seyler only allowed Miescher's work to be published after he had repeated it himself to his own satisfaction.

When Miescher died of tuberculosis at fifty-one, he was haunted by a sense of frustration and opportunities missed. He once described himself as going to bed every evening feeling like a schoolboy who has not done his homework. One of his students likened him to a ship that sails into port laden with treasure, only to sink as it enters the harbour.[12] Even as a young man, Miescher he wrote to Hoppe-Seyler, expressing his frustration that other physical scientists were not turning their attention to the mysteries of the cell:

> I believe that the physiologist does science a great service when he arrives at precise, pure chemical problems ... Unfortunately few physicists and chemists are so inclined to sacrifice themselves for the accomplishment of tasks, the significance of which is to be found in applied, rather than pure disciplines.[13]

Miescher was frustrated at what he perceived to be the lack of interest shown by physicists and chemists in tackling the chemical mysteries of the cell. However, while the plaque in The Eagle incorrectly ascribes the discovery of DNA to Watson and Crick, it may have offered Miescher some consolation to know that physicists had finally turned their attention to his 'nuclein'.

Crick was just one of a number of physicists who had steadily been becoming interested in living systems and the problem of the gene. One of the most famous of these was Erwin Schrödinger (1887–1961), one of the founders of quantum mechanics. Better known for his thought experiment involving cats, Schrödinger published *What is Life?* in 1944. While its title was deceptively simple, the answer to it was not. In wrestling with a possible response, Schrödinger proposed that hereditary traits such as the Austro-Hungarian Habsburg's family lip must somehow be manifest in a molecular carrier that was passed on from one generation to the next. Such a carrier would have to have two key properties: it would have to be very large, and its structure would have to encompass a large amount of variation.

There was one type of biological molecule that met both these two criteria—and it wasn't DNA. Thanks to the success of blockbuster films like *Jurassic Park* and *The X-Men*, most people today grasp that DNA plays a central role in passing on genetic information, and its double-helical structure has become something of a cultural icon. But when Schrödinger's book appeared, the situation was very different: proteins were considered to be the most likely carrier of hereditary traits, not DNA. As proteins were both giant molecules and composed of up to 20 different types of amino acids, they were capable of an enormous amount of structural variation and so easily fulfilled the criteria that Schrödinger had identified as being essential to carry genetic information.

In comparison, DNA seemed to be a poor candidate as the carrier of genetic material. Far from showing sophisticated structural variation, it appeared to be a rather dull molecule—a continuous repetition of the same four nucleotide bases along the length of the entire chain. According to this model, which became known as the tetranucleotide theory, it was difficult to envisage how a molecule with such a simple chemical structure could encompass sufficient variation in its structure to carry the genetic message. DNA was relegated to being simply a structural support or, as the writer Horace Judson poetically said, 'the wooden stretcher behind the Rembrandt'.[14]

But in the same year that Schrödinger's book was published, a paper appeared that challenged this view. Oswald Avery (1877–1955), a shy, retiring microbiologist (Figure 37) and his co-workers MacLyn McCarty and Colin MacLeod at the Rockefeller Hospital in New York City published a paper showing that DNA—and not protein—could confer the ability to cause disease onto strains of pneumococcus bacteria that had previously been non-virulent.[15]

Figure 37 *Oswald T. Avery (1877–1955) celebrating Christmas, but not, sadly, a Nobel Prize.*
Credit: © Oswald T. Avery Collection, US National Library of Medicine.

This experiment has since been compared by one Nobel laureate to the D-Day landings of that same year in terms of its historical impact—and hailed by one modern undergraduate textbook of molecular biology as 'Avery's Bombshell'.[16,17]

At the time of its publication, however, the response to Avery's paper was mixed. One of Avery's Rockefeller colleagues, Alfred Mirsky, argued that the DNA used in the experiments could contain traces of protein contaminants that caused the observed effects. Mirsky asked how Avery could be absolutely sure that he had completely removed all traces of protein from his preparations. Yet, other researchers, such as the microbiologists Andre Boivin in France and Hattie

Alexander at the Columbia University College of Physicians and Surgeons, New York, were sufficiently intrigued by Avery's results to see whether they could be reproduced in other bacterial systems such as *E. coli* and *H. influenzae*, and his work on transformation was soon being discussed at a number of international scientific meetings.[18,19]

Avery himself would probably have been surprised that his work could have such a dramatic effect on anyone. He was not one for accolades and never attended scientific meetings, preferring instead to keep a low profile and just focus on doing his lab work.[20] When he was awarded the Paul Ehrlich Gold medal in 1933 in recognition of his research into the polysaccharides on the surface of bacterial cells, he did not travel to Germany for the ceremony. Nor did he make the trip to England when the University of Cambridge conferred an honorary degree upon him in 1944, or to Sweden, when he was awarded the Pasteur Gold Medal by the Swedish Medical Society in 1950.[21] Even the award of the prestigious Copley Medal by the Royal Society in 1945 could not entice Avery away from his beloved laboratory.[22] Instead, Banting's and Best's champion Sir Henry Dale, who was then the President of the Royal Society, decided to bring the medal in person to Avery while visiting the Rockefeller Institute. As Dale approached the laboratory, he found Avery so completely absorbed in pipetting bacterial cultures at his workbench that he was completely unaware of his visitor. Retreating quietly without disturbing Avery, Dale turned to his colleague Edgar Todd who was with him and said simply, 'Now I understand everything'.[23]

Avery preferred to keep a low profile and would much rather be at the lab bench rather than engage with the circus of scientific conferences. His colleague Rene Dubos praised Avery for not being 'a person who makes himself or his work obvious to international committees', but his reclusive tendencies did not mean that the scientific establishment was oblivious to him.[24] In the course of his career he was nominated forty-four times for the Nobel Prize.[25] Most of these nominations were actually for the earlier work that he had done on polysaccharides found on the surface of bacteria and their interaction with the immune response. But from 1946 onwards he began also to receive nominations for his demonstration that nucleic acid could transform bacteria.[26]

According to the Danish biochemist Hermann Kalckar, Avery's work 'was really worth two Nobel Prizes', yet despite his numerous nominations, it went unrecognized by the Nobel Committee.[27] This might seem to have been a monumental oversight given that Avery's work was the first real evidence that the genetic message was carried by DNA and not protein. But as the earlier example of the 1923 Nobel Prize to Banting and Macleod shows, the history of this particular accolade is rarely without those who feel they have not received due recognition.

It has been suggested that perhaps the Nobel committee were all too aware of Avery's reclusive nature and feared that, were he to be given the award, he would embarrass them by not bothering to show up at the ceremony in Stockholm.[28] Another problem for Avery was that the criteria for the award of the Nobel Prize in Medicine and Physiology state that it should be for work that has 'conferred the greatest benefit on mankind' and at the time his discovery fell well short of this requirement.[29] It was still possible that the effects he had observed might only be confined to bacteria, and, in addition, nothing was yet known about the mechanism by which transformation took place.[30] Perhaps nucleic acids were not the hereditary substance themselves, but simply mutagens that somehow altered the genetic material, causing it to confer the property of virulence? Whatever the reasons behind their decision, the failure of the Nobel Foundation to recognize Avery's work left his colleagues baffled and disappointed. For Dubos, it was 'a matter of painful surprise', while Alvin Coburn, another microbiologist, reflected:[31]

> Throughout my years of communion with this extraordinarily selfless person, I had just one regret. It was the failure of the men in Stockholm to understand the incalculable worth of Fess [a name for Avery used by those close to him].[32]

Some of those 'men in Stockholm' were actually willing to admit that they may have made a huge mistake. And for some of them, it would not be the last time that they had to do so. For by the time that Professor Arne Tiselius, who had become Vice President of the Nobel Foundation in 1947, had conceded that Nicolai Paulescu should have at least had some share in the 1923 Nobel Prize for the discovery of insulin, he had already described the failure of Avery to receive this award as 'the most conspicuous omission' in the history of the prize'.[33]

Avery's attitude to being denied the Nobel Prize stood in stark contrast to a figure such as Charles Best. Whereas Best strove to 'convince the world' of his place in the discovery of insulin, Avery felt no such compulsion about his own work on DNA. In place of excessive speculation and attention, he preferred caution and sobriety:

> ... with mechanisms I am not now concerned. One step at a time and the first step is, what is the chemical nature of the transforming principle? Someone else can work out the rest.[34]

Alongside Schrödinger and Crick, another physicist who had turned their attention to biology was William Astbury. Even before he was sparring with Archer Martin over the arrangement of amino acids in proteins, Astbury had become interested in DNA and was so excited by Avery's work that he called it 'one of the most remarkable discoveries of our time'.[35] Writing to Avery in January 1945, Astbury said he felt 'extremely thrilled' and asked whether Avery would be willing to send a sample of DNA for X-ray analysis as this was 'a wonderful chance here to make an important step forward'.[36]

A physicist by training, Astbury had become passionate in his conviction that the complexity of life was best studied using the tools of physics. Well before the publication of Schrödinger's book, Astbury had pioneered the application of physics to biological systems.

This work put Astbury on the map, but it was his research assistant Florence Bell (1913–2000) who expanded this work to include biological fibres other than just proteins (Figure 38).

Figure 38 *Florence Bell (1913–2000).*
Reproduced with kind thanks of Mr. Chris Sawyer.

With chapters describing the analysis of protein fibres found in jellyfish, shark's fin, and rare diseases of hair, Bell's published 1939 PhD thesis might seem to be an unlikely scientific landmark.[37] But one chapter stands out. It describes X-ray crystallographic studies of what was then called thymonucleic acid, or DNA. It contains the very first successful X-ray image of DNA, from which Astbury and Bell proposed an early model of what the molecule might look like. While several key features of this model are wrong, it is important for two reasons. Firstly, it showed that the bases stack up on top of each other in a manner described by Astbury and Bell as being like 'a pile of plates' and allowed them to measure the distance between adjacent bases—and it is this that provided Watson and Crick with a vital clue when they began their own work.[38,39,40] Secondly, it showed that the regular, ordered structure of DNA could be revealed by X-ray crystallography.

Despite these achievements, when Bell caught the attention of the local press, it was not for her scientific work. When the Institute of Physics held a conference in Leeds in 1939, Bell's presentation of her work was reported by the Yorkshire Evening News with the stunned headline 'Woman Scientist Explains', giving readers the impression that some new zoological specimen, previously unknown to science, had been discovered (Figure 39).[41]

Figure 39 *Newspaper headline from 'The Yorkshire Evening Press' newspaper, 23rd March 1939 reporting on a presentation given by Bell at a conference held in Leeds by the Institute of Physics. MS419 A.1, Astbury Papers, Special Collections, Brotherton Library, University of Leeds.*

Reproduced with permission of Special Collections, Brotherton Library, University of Leeds.

While Bell was described by the press as a 'slim 25-year old Cambridge University graduate', it was for her work on DNA that she would more likely wish to be remembered.[42] For by showing that the structure of DNA could be revealed by X-ray crystallography, she had helped to lay the foundations for the work of another more well-known heroine in the story of DNA.

Thanks to several biographies, an award-winning West End play starring Nicole Kidman, a Mars rover bearing her name, and a commemorative coin released by the Royal Mint to mark what would have been her 100th birthday, Rosalind Franklin (1920–1958) can no longer be cast as the unsung heroine of DNA that she once was. For many years her crucial contribution to Watson's and Crick's discovery was overlooked. Working at King's College, London in 1952, Franklin and her PhD student Raymond Gosling took an X-ray image of DNA that is described on a commemorative plaque outside King's College, London as being 'one of the world's most important photographs'.[43]

Labelled by Franklin and Gosling simply as 'Photo 51', it showed the striking pattern of a black cross formed by the diffracted X-rays. When James Watson first saw this image, its effect on him was electric: 'The instant I saw the picture my mouth fell open and my pulse began to race'.[44] He knew that the distinctive cross-shaped pattern was the characteristic X-ray signature of a molecule coiled into a helical shape. Although this clue alone was not enough to allow him and Crick to solve the structure of DNA, it had a profound psychological effect upon them.[45] As Watson explained in a 2001 address given at the inauguration of the Harvard Center for Genomic Research: 'the Franklin photograph was the key event. It was, psychologically, it mobilized us'.[46]

Armed with this clue, together with other vital X-ray data contained in a report that Franklin had written for the Medical Research Council, Watson and Crick were able to do what Astbury and Bell had not: solve the structure of DNA. In a paper published in *Nature* in April 1953, they proposed their now-famous double-helical model of DNA, in which two chains of sugar and phosphorus are entwined around each other in a double-helix rather like a spiral staircase. Jutting out at right angles along the length of each chain were the nucleotide bases, of which there are four different types: adenine (A), guanine (G), cytosine (C), and thymine (T). In Watson's and Crick's model, the bases on one strand pair up with those on the opposite strand in a very specific way—A with T, and G with C. This specific pattern of pairing forms the rungs of the DNA ladder that hold the two strands of the molecule together. And as Watson and Crick explained, in what must rate as one of the greatest understatements in the history of science, it also does an awful lot more:

> It has not escaped our notice that the specific base pairing we have postulated immediately suggests a possible copying mechanism for the genetic material.[47]

The statement was to have profound consequences, for if DNA really *was* the genetic material, then it would have to have some means of copying itself. By showing that the bases on one strand paired up in a very precise way with those on the other, Watson and Crick's model gave an elegant explanation of how the molecule might be able to replicate (Figure 40).

The inspiration for this had come while Crick was sitting in The Bun Shop, a pub close to the Cavendish laboratory in Cambridge. Having just attended a lecture on 'The Perfect Cosmological Principle', he wondered whether there

(a)   (b)

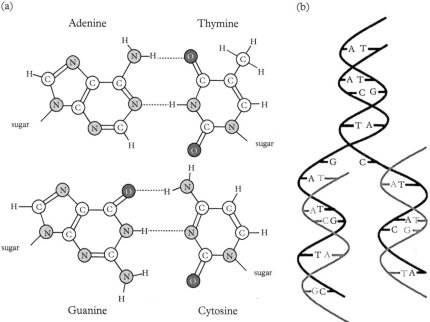

Figure 40 *a) The specific pairing between bases in DNA. Adenine on one strand of the DNA double helix forms hydrogen bonds (dotted lines) with thymine on the opposite strand, while guanine forms bonds with cytosine. This pairing explains the ratios discovered by Erwin Chargaff, holds the two strands of the double-helix together.*
b) Base-pairing also explains how the DNA molecule can make copies of itself to pass on genetic information. When a molecule of DNA replicates, such as during cell division, the two strands of the double helix unwind temporarily breaking the bonds between the bases. Thanks to the specific pattern of pairing between bases on opposite strands, each single chain from the parent molecule (shown in black) then acts as a template directing the synthesis of a new daughter strand (shown in grey). Diagram by K.T. Hall.

might also be a perfect biological principle—namely, the ability of DNA to make copies of itself. Over a pint with theoretical chemist John Griffith, Crick wondered whether similar bases might be able to physically attract each other. If this was indeed possible, then all the adenine bases jutting out of a single chain of DNA might be able to attract free adenines to slot between them and, in this way, a new strand of DNA might be synthesized. As Griffith knew a lot more about chemistry, Crick suggested that he try to calculate whether this might actually be possible.[48]

Griffith was ahead of Crick in his thinking and had actually already devised some possible schemes of attraction between bases. When Crick spotted him in the queue of the tea room at the Cavendish laboratory some time afterwards, Griffith had some interesting news. His calculations had shown that bases *could* attract each other—but not in the way that Crick had expected. Crick had envisaged an arrangement in which the attraction between similar bases occurred between their flat surfaces and so resulted in them slotting on top of each other. But Griffiths had not been clear on what Crick meant, so he had instead gone away and carried out calculations on the assumption that the interaction between bases occurred between their edges, not their flat surfaces.

It would prove to be a very fortuitous misunderstanding. What Griffiths found was that, if bases were indeed attracted to each other in this way then, adenine would always pair with thymine, while cytosine would always pair with guanine. Crick suddenly realized that this opened up an intriguing possibility. Until now, he had assumed that each of the two chains in a DNA molecule made an exact copy of itself during replication, but what if this were not the case? Instead, what if each chain directed the synthesis of a copy that had a complementary sequence of bases to its own? Griffiths' calculations suggested a way this could happen, for according to his pairing scheme, adenine on the parent chain would be copied as thymine on a daughter chain, while guanine was copied as cytosine.[49]

Initially, this was just a theoretical possibility, and then came a fortuitous meeting. In May 1952, Watson and Crick were introduced to biochemist Erwin Chargaff (1905–2002), who was visiting Cambridge from the United States (Figure 41).

Figure 41 *Portrait of biochemist Erwin Chargaff (1905–2002); Photographed in 1970.*

Credit: SCIENCE PHOTO LIBRARY. Reproduced with kind permission of the American Philosophical Society.

Chargaff had been born in 1905 in what is now Ukraine and after studying chemistry at the University of Vienna had eventually gone on to spend most of his career at the University of Columbia in New York—despite having once confessed to feeling 'afraid of going to a country that was younger than most of Vienna's toilets'.[50] When he first arrived at Columbia, Chargaff's work had focused on understanding the biochemical mechanism of blood coagulation, but in 1944, he, like Astbury, had been one of the few scientists to experience an epiphany about the role of DNA, thanks to Avery's experiments on pneumococcus.

Chargaff had nominated Avery for the Nobel Prize,[51] and had no doubts at all that Avery had been short-changed in being denied the award. Years later, he recalled the moment when he first learned of Avery's work and its implications began to sink in.

> I saw before me in dark contours the beginning of a grammar of biology…. Avery gave us the first text of a new language, or rather he showed us where to look for it. I resolved to search for this text.[52]

Chargaff had a hunch that this text had something to do with the bases that Bell and Astbury had shown were stacked up on top of each other along the length of a DNA molecule.[53] But he was also convinced that this text could not be revealed merely by inspecting the physical shape of the DNA molecule using X-rays as Bell and Astbury had done. What was needed was a rigorous analysis of 'the nature and the proportions of their constituents, on the sequence in which these constituents are arranged in the molecule, and on the type of linkages which hold them together'.[54] 'What was to be done was clear to me', said Chargaff, 'but not at all how to do it'. For he was aware that he faced a formidable technical problem:[55]

> Even for the characterization of the basic components [of the nucleic acids], enormous amounts were required; a quantitative analysis was out of the question. If my assumptions about the nucleic acids were to be proved correct, it was evident that extremely accurate quantitative methods had to be discovered. Moreover, these had to be applicable to minute quantities of nucleic acids, because several organs of many different species and also relatively inaccessible microorganisms were to be compared.[56]

Chargaff needed some means of separating and quantifying the four different bases in DNA, but there was currently no method of separation with sufficient precision and accuracy to do this. His grand plan to decipher the new grammar of biology revealed by Avery seemed doomed to fail before it began.

At least it seemed that way until Archer Martin arrived at Columbia to give a seminar, in which he showed a paper chromatogram like the ones he had developed during his wool work with Synge at WIRA. Sitting in the seminar was Ernst Vischer, one of Chargaff's post-doctoral fellows, who realized that Martin had provided the answer to their problem.

In one of his first papers on partition chromatography, Martin had noted that 'paper chromatograms are by no means limited to the separation of amino-acids'.[57] Chargaff and Vischer realized that, although partition chromatography had been used to separate and quantify amino acids in the protein chains of wool, it might also be adapted to the analysis of nucleotides from a chain of DNA. In his classic book *The Path to the Double Helix*, historian Robert Olby underlined the importance of this insight, saying 'the introduction of chromatography by Martin and Synge and its application to nucleic acids by Chargaff is therefore of great importance in the history of our subject'. [58] Meanwhile, with his characteristic gift for understatement, Synge suggested in his Nobel Lecture that 'it is evident that partition chromatography has considerable application in the study of nucleic acids'.[59]

In his excitement, Vischer tore off a piece of Martin's chromatogram to check that they had the correct type of paper in the laboratory before dashing back to tell Chargaff that he may have found the answer to their problem. He and Chargaff then spent the next two years adapting partition chromatography to the analysis of nucleic acids in the hope of discovering how they might spell out a 'grammar of biology'.[60,61] In 1949 Chargaff gave a series of lectures in which he described both published and unpublished results of his analyses. Then, in 1950, he published a summary of this work in the journal *Experientia*, in which he acknowledged the invaluable contribution of Martin and Synge, stressing that 'The basis of the procedure is the partition chromatography on filter paper'.[62]

When Chargaff was first introduced to Watson and Crick during a visit to Cambridge in 1952, his chromatographic analysis of the bases in DNA quickly became the main topic of conversation. His first impression of Watson and Crick on meeting them was not a particularly good one. He described Crick as having 'the looks of a faded racing tout' who spoke in 'an incessant falsetto, with occasional nuggets glittering in the turbid stream of prattle'. Watson, meanwhile, struck Chargaff as being 'quite undeveloped at twenty-three, a grin, more sly than sheepish; saying little, nothing of consequence'.[63] As Chargaff later recalled, this underwhelming first impression 'was not improved by the many farcical elements that enlivened the ensuing conversation'. One such source of irritation to Chargaff was what he felt to be their apparent obsession with fitting DNA into a helical model and working out the height of one complete turn. The pair seemed to have become so fixated on this particular parameter—known as the 'pitch'—to the exclusion of everything else that Chargaff later dismissed them as 'two pitchmen in search of a helix'.[64] If the *Oxford English Dictionary* definition of 'pitchman' is anything to go by, it was hardly a compliment.[65]

Chargaff was also appalled that they appeared to be 'unencumbered by any knowledge of the chemistry involved, to fit DNA into a helix'.[66] The grand scale of their ambition was 'coupled with an almost complete ignorance of, and contempt for, chemistry, that most real of exact sciences'.[67] As a result of this ignorance, both Watson and Crick were unfamiliar with Chargaff's work—a situation which Chargaff quickly sought to rectify.

In a couple of papers published in 1949, Chargaff and his co-workers described how they had used partition chromatography to successfully separate and measure the amounts of the four bases in DNA isolated from bovine thymus and spleen and a couple of micro-organisms.[68,69] In the course of this work they had made a curious observation. In each different organism, the proportion of each of the four bases showed 'certain striking, but perhaps meaningless, regularities'.[70] In a series of lectures given that same year, Chargaff elaborated slightly on these striking regularities, saying that, in samples from all the organisms he had examined so far—which now also included human sperm—the relative proportion 'of adenine to thymine and of guanine to cytosine, were not far from 1'.[71]

This meant that the amounts of the bases A and T must be equal, as must also be the amounts of G and C. Chargaff described this correspondence as 'note-worthy', but he also cautioned against over-interpretation, saying 'whether this is more than accidental, cannot yet be said'.[72] But for Crick, Chargaff's ratios were anything but a coincidence. They answered his question of how bases in DNA might form pairs, and he described the effect upon him as being 'electric ... I suddenly thought, "Why, my God, if you have complementary pairing, you are bound to get a one-to-one ratio'.[73] In a letter to his son Michael, Crick explained the significance of this revelation:

> the exciting thing is that while these are 4 different bases, we find that we can only put certain pairs of them together ... we find that the pairs we can make—which have one base from one chain joined with one base from another are only A with T and G with C. Now on one chain, as far as we can see, one can have the bases in any order, but if their order is fixed, then the order on the other chain is also fixed ... It is like a code. If you are given one set of letters you can write down the others. You can now see how Nature makes copies of the genes. Because if the two chains unwind into two separate chains, and if each chain makes another come together on it, then because A always goes with T, and G with C, we shall get two copies where we had one before ... In other words, we think we have found the basic mechanism by which life comes from life.[74]

When Chargaff published his autobiography twenty-six years after his enco-unter with Watson and Crick, he was in no doubt as to the historical significance of their brief meeting, writing 'I believe that the double-stranded model of DNA came about as a consequence of our conversation'.[75] But he also felt that Wat-son and Crick had not given him sufficient credit for his contribution to their discovery.[76] A year before meeting them in Cambridge, Chargaff had actually speculated about the possible significance of these ratios:

> As the number of examples of such regularity increases, the question will become pertinent whether it is merely accidental or whether it is an expression of certain structural principles that are shared by many desoxypentose nucleic acids ... It is believed that the time has not yet come to attempt an answer.[77]

Writing many years later, it had become painfully apparent to him that he had failed to grasp their full significance at the time. He lamented: 'I seem to have missed a shiver of recognition of a historical moment ... a change in the rhythm of the heartbeats of biology'.[78]

Watson and Crick had a very different understanding of what Chargaff's contri-bution to their work had been. Writing in 1978, Chargaff spoke of 'the regularities that I then used to call the complementarity relationships and that are now known as base-pairing'.[79] But in his first papers written at the end of the 1940s, Chargaff never actually made any reference to complementarity between bases—he simply presented the numbers and made no attempt to interpret what they might mean. For, unlike Watson and Crick, he had no access to structural studies on DNA, such as those made by Franklin, that suggested a double-helical structure.

Nearly every biology textbook today hails the specific pairing of complementary bases as 'Chargaff's Rule', which may have offered him scant consolation at missing out on a Nobel Prize, but even his claim to have a 'rule' named after him has been disputed. In a review of Chargaff's autobiography, chemist Jerry Donohue—Watson's colleague in Cambridge—noted that the degree of variation in some of Chargaff's data on base ratios was hardly suggestive of a rule.[80] Donohue went on to point out that the so-called 'rule' now lauded in textbooks had been constructed with the benefit of hindsight, saying, 'When the final model of DNA was discovered – more or less by accident – it wasn't Chargaff's rule that made the model, but the model that made the rule.[81]

Watson and Crick's proposed model for the structure of DNA, first published in April 1953, is without doubt a landmark in biology, alongside Darwin and Mendel. But the scientist and historian Matthew Cobb has argued that, only six weeks later, they published a second paper of equal significance. For while the specific pairing of bases between the two strands of the double-helical structure explained how the molecule could replicate and make copies of the genetic message, it did not explain how it carried that message in the first place.[82] It was in this second paper that they addressed this question, saying 'it therefore seems likely that the precise sequence of the bases is the code which carries the genetical information'.[83] Cobb describes this insight by Watson and Crick as their 'brilliant suggestion'—that is, the bases didn't just hold the DNA molecule together or enable it to make copies of itself by complementary pairing—they also *carried the genetic message*.[84] Brilliant as this insight was, it, too, owed its origins to Chargaff.

In 1949, Chargaff had first used partition chromatography to analyse the chemical composition of DNA from bovine thymus and spleen and noticed that the amounts of the four bases were 'not in accord' with what would be expected if DNA was just a monotonous repetition of the same four bases, as many scientists believed to be the case.[85] He had lectured on his findings that same year, [86] and in 1950, explained further that he had found what he'd hoped for:

> ... desoxypentose nucleic acids from different species differ in their chemical composition ... and I think that there will be no objection to the statement that, as far as chemical possibilities go, they could very well serve as one of the agents, or possibly as the agent, concerned with the transmission of inherited properties.[87]

Chargaff had not only dealt a fatal blow to the dominant idea that DNA was little more than a monotonous repeat of tetranucleotides, but he had also offered the first hint that the genetic message was somehow carried by the bases. The question was how did this happen? In July 1953, shortly after the publication of their proposed double-helical structure for DNA, Watson and Crick received a curious letter from someone who thought he had the answer.[88]

George Gamow (1904–1968) was a cosmologist whose other interests besides the Big Bang and nuclear physics included card tricks, whisky, and how DNA might function as the genetic material. Born in Odessa, Gamow studied physics at the University of Leningrad before moving to Göttingen and then Copenhagen, where he pioneered the application of quantum physics to nuclear phenomena, such as the release of alpha particles from unstable nuclei.

It was with some reluctance that Gamow had eventually returned to the Soviet Union. In spring 1932, he and his wife, Lyubov Vokhminzeva, who went by the nickname 'Rho' and was also a physicist, attempted to escape to the West by paddling across the Black Sea from the Crimea to Turkey. When this failed due to a storm, they considered travelling by sleigh from the Khibiny Mountains into Finland but gave up on this idea as they feared that any prospective sleigh drivers might turn them over to the Soviet authorities. The following August they investigated the possibility of travelling by motor-boat from a marine station in Murmansk to Norway but quickly realized that the rapid expansion of the Soviet navy in these waters would make this impossible.[89]

Gamow's eventual defection turned out to be much less dramatic. In 1933, he received an invitation to speak at the prestigious Solvay Conference which was to be held in Brussels. When applying for authorization to travel, Gamow insisted that he would not make the trip unless Rho was allowed to come with him as he had promised her a shopping trip in the West. Amused by this explanation, the authorities granted him permission to travel to Belgium and so, in mid-October he and Rho boarded a train bound for Helsinki—and freedom.

Having successfully escaped from the Soviet Union, Gamow and Rho made their way to the USA where he began to think about how physics might be applied to the problems of biology. His particular interest was in how genes might work and when he first read Watson and Crick's papers on the structure of DNA, he quickly put pen to paper.

The letter was to be beginning of a cordial friendship between Watson, Crick, and Gamow, for as Crick recalled:

> ... he was what is called good company was Gamow. I wouldn't quite say a buffoon, but—yes, a bit of that, in the nicest possible way. You always knew, if you were going to spend the evening with Gamow you would have a 'jolly time'.[90]

Such 'jolly times' were usually accompanied by copious amounts of whisky and when Watson met him for the first time in 1954, he was struck by how 'the very tall thinness of his Russian youth had given way to a middle-age girth, accentuated by more alcohol than generally compatible with high-powered manipulation of mathematical symbols', while Crick later described a short visit by Gamow as having been 'rather exhausting as I do not live on whisky'.[91,92]

Gamow's very first letter to Watson and Crick, written on July 8, 1953, just over a month after their second paper on DNA, appeared written while under the influence of his favourite tipple. Described by Watson as a 'zany communication', it was written in Gamow's own hand, littered with words crossed out and amorphous paragraphs that defied the convention of margins.[93] In places, the letter read almost like that from a giddy teenage fan and Watson confessed that it 'had so many whimsical qualities that we did not know how serious he [Gamow] might be'. However, the somewhat chaotic presentation, coupled with the almost stream-of-consciousness flow of the contents, were actually evidence of the seriousness of Gamow's proposal.

He began by praising Watson and Crick, saying that their model for the structure of DNA had now brought 'biology over into the group of "exact" sciences' and how he was 'very much excited' by their *Nature* paper suggesting that bases in DNA might carry the genetic information.[94] He continued by saying he had an idea how this might happen.

He proposed that each different organism might be represented by a specific number that could be derived from the order of bases in their DNA. The following year he elaborated on this idea with a paper in *Nature*. Since it was now becoming accepted that genes were made of DNA, and that genes directed the synthesis of proteins, which were, in turn, made of amino acids, it seemed logical that there must be some relationship between the sequence of bases in DNA and that of amino acids in proteins. The problem was that, while protein chains were known to be made of up to twenty different types of amino acid, the chains of DNA contained only four different types of base. How could only four bases represent as many as up to twenty different types of amino acid?

Gamow suggested that each different type of amino acid was specified not by a single base, but by a particular combination of them. If so, then a combination of two bases would allow sixteen possible permutations, which would still be insufficient, but combinations of three would extend this to sixty-four—plenty to encompass the twenty different amino acids found in proteins. In his *Nature* paper, he outlined a model in which bases on opposite chains of DNA formed a spatial arrangement, at the centre of which was a diamond-shaped hole into which a specific amino acid could fit.[95] In this way, specific combinations of four bases winding around the outside of the DNA molecule could direct the assembly of a chain of amino acids to form a complete protein (Figure 42).

Gamow proposed that amino acids were assembled into protein chains directly on the surface of DNA, but Watson persuaded him to consider an alternative. In his model, Gamow likened the synthesis of proteins by the cell to the operation of a factory. Unlike bacteria, the DNA of eukaryotic cells resides in the nucleus, which Gamow likened to the factory manager's office, containing filing cabinets of blueprints for the manufacture of the product. The sites of protein synthesis on structures known as ribosomes, which were found in the cytoplasm outside the nucleus, were the factory floor.[96] According to this analogy, Gamow's proposal that protein chains were assembled directly on the surface of DNA was as if

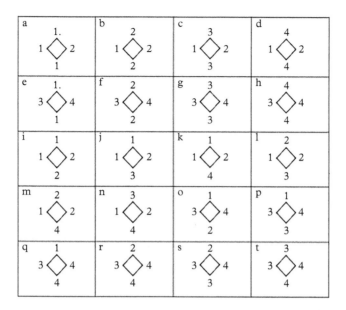

Figure 42 *George Gamow's model for how the genetic code might work. Gamow proposed that amino acids were assembled into protein chains directly on the surface of DNA. As the two strands of the double helix wound around each other, bases from opposite chains (Gamow designated the four different bases not as A,C,G, and T, but 1,2,3,4) formed groups of four with a diamond shaped hole at the centre of this arrangement. The different permutations of four bases arranged around the diamond shaped hole dictated which of the 20 amino acids (designated by Gamow as 'a' to 't') could be fitted into it. Because two amino acids of each group forming a diamond were shared by neighbouring groups, this model made two key predictions: firstly, that only certain amino acids would ever be found as neighbours, and secondly, that altering a specific base by mutation should also alter the amino acids at neighbouring positions. Both these predictions were refuted by Sanger's analysis of the amino acid sequence of insulin. Diagram by K. T. Hall based on Gamow, G., 'Possible relation between deoxyribonucleic acid and protein structure,' Nature, 173, 1954; p.318.*

the manager and the factory floor staff had all squeezed into the office and were attempting to carry out all their operations there.

But there was an alternative scenario. What if the manager remained at their office desk and communicated instructions and orders for assembly to a foreman who then took these out onto the factory floor? In molecular terms, this suggested that DNA might act through a chemical intermediate that carried the genetic information out of the nucleus and to the sites of protein assembly in the cytoplasm.

But Gamow was not the first to make this suggestion. In 1947, the French scientists Andre Boivin and Roger Vendrely had already proposed that DNA might direct the synthesis of proteins via a chemical intermediate. And they also had a particular candidate in mind.[97] In the 1940s, it had been discovered that, although DNA was confined to the nucleus, its close molecular cousin, ribonucleic acid (RNA), was found to accumulate outside the nucleus in the cytoplasm during protein synthesis.[98] Although DNA has tended to hog the limelight in popular culture, RNA plays its own vital part in the story, as reflected by an essay question once posed to finalists in Biochemistry at the University of Oxford which asked 'DNA is, while RNA does. Discuss'.[99]

For any finalist struggling to pick up some desperately needed marks, this question came gift-wrapped from the examiners.[100] For although RNA can, like DNA, be a store of genetic information as is the case in certain viruses, it is a much more versatile molecule than its more familiar cousin. Certain types of RNA can also have catalytic activity like an enzyme and in the thirty years that have passed since this particular exam question was set, small fragments of RNA have been shown to be capable of specifically inhibiting the expression of certain genes. The discovery of these siRNAs (as they are known) has given researchers a powerful new tool for studying the regulation of genes. It might also have potential to be used as a therapeutic method for the treatment of conditions like cancer, in which normal gene expression has gone awry.

Some of this versatility was already becoming evident in the 1950s when two distinct forms of RNA were discovered, both of which played crucial roles in the synthesis of proteins.[101] One of these was found to be a major structural component of ribosomes, the cytoplasmic sites at which protein synthesis occurs and so has come to be known as ribosomal RNA (rRNA). The other, known today as transfer RNA (tRNA), transports specific amino acids to the ribosomes, where they are assembled into polypeptide chains. But despite the discovery of these two forms of RNA, convincing evidence that RNA was also the chemical intermediate carrying genetic information from DNA to the sites of protein synthesis remained elusive.

Thanks to an impressive act of dietary genetic gymnastics by the humble bacterium *E. coli* however, it was becoming very clear that such a messenger must exist. In 1957 the French scientists Francois Jacob (1920–2013) and Jacques Monod (1910–1976), together with American scientist Arthur Pardee (1921–2019) visiting from the US, were investigating how bacteria respond to a change in their nutritional environment. The 'PaJaMo' experiments, as they became known, were to offer a crucial insight into how DNA and RNA work together to turn genetic information into protein.

Glucose is the usual source of energy for *E. coli*, but if it is in short supply and the sugar lactose is present, the bacterium rapidly synthesizes a suite of enzymes for the utilization of lactose instead. That these lactose metabolising enzymes are not synthesized when the sugar is absent makes perfect evolutionary sense. After all, why waste valuable energy and resources on synthesizing enzymes for the utilization of a nutrient that is not present in the environment? Bacteria with this ability to respond to the presence of lactose are called lac+ and contain a set of genes on a particular region of their chromosome that direct the synthesis of an enzyme called beta-galactosidase, which enables the metabolism of lactose. Mutants lacking these genes are unable to synthesize this enzyme and so cannot grow on lactose. Jacob, Monod, and Pardee found, however, that when genetic material from lac+ organisms was transferred into lac– mutants, the beta-galactosidase enzyme became suddenly active in the mutants, allowing them to grow on lactose. The immediacy with which this happened suggested that the DNA from the lac+ strain triggered the rapid production of an intermediate chemical messenger, which, in turn, directed the synthesis of the beta-galactosidase enzyme.

On Good Friday, 1960, Jacob, Monod, and Pardee were discussing this work with Crick and Sydney Brenner, a South African medical student, during a gathering at Brenner's room in King's College Cambridge when Brenner suddenly gave a loud yell before he and Crick both leapt to their feet.[102] Jacob then recalled how Crick and Brenner then 'Began to gestic'ulate. To argue at top speed in great agitation. A red-faced Francis. A Sydney with bristling eyebrows. The two talked at once, all but shouting'.[103]

Crick and Brenner's excitement was about viruses—or more accurately, the realization that Jacob, Monod, and Pardee's results made perfect sense in the light of some research done throughout the 1950s on the infection of bacteria by a type of virus known as a bacteriophage. A number of researchers had shown that infection of bacteria by bacteriophages resulted in the rapid production of RNA. Crucially, this RNA had a very short-half life and also bore no resemblance to the DNA sequence of the host bacterium, but rather to that of the infecting bacteriophage.

Was this new type of RNA the mysterious messenger?[104] Confident that it could be, Jacob and Monod dubbed it 'messenger RNA', or mRNA—a name now much more widely known, thanks to the plentiful information about how the Covid-19 vaccine works.[105] To understand how mRNA might work, Crick invoked a piece of technology that is now acquiring a kind of retro-chic. He likened mRNA to a magnetic tape and proposed that, in an analogous fashion to the way in which the magnetic head reads information from a tape in a cassette recorder before converting it into sound, the ribosome reads the information in the mRNA sequence and uses it to assemble amino acids into a protein chain. The challenge now was to find some experimental evidence in support of this model. After all, the increase in RNA that was observed after the infection of bacteria by bacteriophage might simply have been due to the synthesis of new ribosomal RNA.

While other guests enjoyed themselves at a party hosted by Crick and his wife later that evening, Jacob recalled how he and Brenner sat in a corner engrossed in planning how they could show that this short-lived RNA species was indeed the mysterious chemical intermediate:

> A very British evening with the cream of Cambridge, an abundance of pretty girls, various kinds of drink, and pop music. Sydney and I, however, were much too busy and excited to take an active part in the festivities … It was difficult to isolate ourselves at such a brilliant, lively gathering, with all the people crowding

around us, talking, shouting, laughing, singing, dancing. Nevertheless, squeezed up next to a little table as though on a desert island, we went on, in the rhythm of our own excitement, discussing our new model and the preparations for our experiment.[106]

In order to show that this short-lived RNA was not just newly synthesized ribosomal material, Jacob and Brenner hit on the idea of using ultracentrifugation to separate the different types of RNA in the cell. Mat Meselson and Frank Stahl at Caltech had shown that radiolabelled DNA could be separated using this method and it seemed logical to assume that it would be successful if applied to RNA, too. Brenner and Jacob took themselves off to California to make use of Meselson's expertise in this area.

But they were not the only ones hunting for mRNA. Watson had enjoyed a famous and fruitful collaboration with Crick only a few years earlier when they had discovered the double helical structure of DNA. He now threw his hat into the ring and entered into direct competition with his old colleague in the hunt for mRNA. Working with him was Walter Gilbert, a young physicist who had been lured into biology thanks to what has been described as 'a seduction skilfully orchestrated' by Watson.[107]

Thanks to his desertion of physics for molecular biology, Gilbert would later find himself as one of the main participants in another breakneck race, this time to produce human insulin. But for the moment, his attention was on mRNA. When Watson learned that Brenner, Jacob, and Meselson were on the verge of publishing their discovery of a short-lived intermediate RNA molecule that they believed was responsible for carrying the genetic message from DNA to the sites of protein synthesis, he sent Brenner a telegram asking him to delay publication of his research until he and Gilbert had submitted their own work for publication.[108] With great magnanimity Brenner honoured this request and the two papers were published together in 1961.[109,110] Thus, the race to discover mRNA appeared to have ended in a draw.[111,112]

Even before the discovery of mRNA, Gamow had turned his attention to thinking how RNA could carry a code that translated the base sequence of DNA into the amino acid sequence of proteins. To crack this code, Watson and the British biochemist Leslie Orgel proposed the formation of a group of fellow scientists who were interested in RNA. Gamow loved the idea and quickly became one of the pivotal figures in organizing what was to become known as The RNA Tie Club. Each member was to receive a tie embroidered with a strand of RNA and a tailor-made tie pin engraved with the three-letter abbreviation for a particular amino acid. As there were thought to be only twenty naturally occurring amino acids, this immediately made membership of this club a highly exclusive affair. This elite group drew on expertise from a range of academic disciplines including biologists such as Chargaff and physicists such as Edward Teller (1903–2003), inventor of the hydrogen bomb. Each member, including the bongo-playing, Nobel laureate physicist Richard Feynmann (1918–1988) signed off their letters with their particular amino acid 'codename'—Gamow was alanine. Despite the diverse range of their academic disciplines, what united them all was what has been described as 'a lunatic enthusiasm' to figure out how RNA might carry a code that translated the base sequence of DNA into the amino acid sequence of proteins.[113,114,115,116] Their discussion focused on how the code might work at the abstract level of information theory, with little regard for the actual biochemical mechanisms involved. Yet, while these purely theoretical discussions were being conducted over whisky and cigars, experimental evidence was starting to emerge from elsewhere that would shed important light on these questions.

Gamow had immersed himself in using mathematical approaches to calculate how various permutations of bases in DNA might correspond to different amino acids in his diamond code. Thanks to the help of a friend who worked at the Los Alamos laboratory, where the first atomic bomb had been developed, Gamow had been able to use the MANIAC computer there to help with this formidable task. But Crick realized that there was no need to go to these lengths to put Gamow's code to the test.

Of the four bases that made up a single 'diamond' unit in Gamow's model, one of these would be shared with the unit that immediately preceded it, and one with that which immediately followed it. From this emerged a simple prediction: the degree of overlap between adjacent diamond units placed strict restrictions on which particular amino acids could lie on either side of a particular position in the protein chain. Thanks to Sanger's analysis of the sequence of insulin, this prediction could be put to the test.

Thanks to regular updates on Sanger's work over lunch in The Eagle pub, Crick realized that the partial amino acid sequences of insulin obtained by Sanger and his colleagues, as well as ones from the peptide hormone beta corticotropin, were casting serious doubt on Gamow's model for the code. Then came the killer blow: Gamow's code predicted that if a single amino acid in a protein chain were altered by a change in the bases of its coding unit, then, due to the overlapping nature of the coding units, the amino acids on either side should also be altered. But Sanger's experimental data on insulin showed that this was not the case. In a communication to the RNA Tie Club, Crick pointed out that the complete sequences of both bovine and sheep insulin differed by only a single amino acid. This was impossible according to Gamow's model. Crick wrote: 'It is surprising how quickly, with a little thought, a scheme can be rejected. It is better to use one's head for a few minutes than a computing machine for a few days!'[117]

Even though his initial scheme had been wrong, Gamow had nevertheless made an enormous contribution: he had introduced the idea of a code that related the base sequence in DNA to the amino acid sequence in proteins. Yet, as geneticist Boris Ephrussi (1901–1979) pointed out to Crick on at least two occasions, there was still one huge problem yet to be addressed: that no one had yet obtained any experimental evidence to show that the order of bases in DNA did actually determine the sequence of amino acids in a protein.[118]

In response to Ephrussi's challenge, Crick set about trying to find an example of variation in the amino acid sequence of a protein that could be directly related to a specific genetic alteration in the sequence of DNA. A good candidate was lysozyme, an enzyme easily isolated and crystallized from the egg white of domestic fowl. Since the genetics of domestic fowl were well established this seemed liked a promising line of enquiry. But after obtaining lysozyme from the eggs of a wide range of different species, including guinea fowl, turkey, chick, duck, goose, and lesser black-backed gull, Crick had nothing to show.

His co-worker, chemist Vernon Ingram (1924–2006), came up with an alternative. In 1949, the American geneticist James Neel, who had studied the genetic effects of radiation damage on the survivors of the atomic bomb on Hiroshima, had proposed an explanation for the patterns of inheritance of sickle-cell anaemia. First described in 1917, it is a genetic disorder in which the red blood cells are misshapen, giving rise to debilitating effects. Neel proposed that the patterns of inheritance of this condition could be best explained if it was caused by a mutation in a single gene that was carried in two copies by sufferers.[119]

Later that year, Linus Pauling (chemist and nemesis of Dorothy Wrinch) took the first steps towards offering a molecular explanation as to why this might be the case. Pauling observed that the normal form of haemoglobin and that found in patients suffering from sickle-cell anaemia showed significantly different mobility when moving through an electric field.[120] This occurred because, unlike normal haemoglobin, the sickle-cell form of the protein carried a net positive charge. What remained unclear was whether this was because the polypeptide chain of the sickle-cell protein was mis-folded in such a way as to mask one or more of the negatively charged amino acids, or because the polypeptide chain of the sickle-cell protein contained a positively charged amino acid that was absent from the normal form.

Ingram realized that these two different forms of haemoglobin were exactly what he and Crick needed to show that a genetic change led to an alteration in the amino acid sequence of a protein. Haemoglobin was too large, however, for a complete analysis of its amino acids to be possible so Ingram tackled the problem by digesting the polypeptide chain of both the normal and sickle-cell proteins into fragments using trypsin. The fragments resulting from the partial digest were applied to a piece of cellulose filter paper on which they were separated first by electrophoresis in one direction, and then partition chromatography in the other. For both the normal and sickle-cell forms of the protein, this treatment generated a pattern of thirty spots on the resulting chromatogram, which Ingram called a 'finger print'.[121] Each spot corresponded to a particular peptide fragment, of which twenty-nine were in identical positions on the chromatograms for both the normal and sickle-cell form of the protein. But the position of the spot corresponding to peptide fragment number 4 showed a marked difference between the two forms of the protein, which suggested to Ingram that whatever the difference between the sickle-cell and normal forms of haemoglobin might be, it was located in this particular peptide fragment.

Using partition chromatography on paper, he now determined the amino acid composition of this one fragment, and by subjecting it to partial hydrolysis with acid, he was able to determine its amino acid sequence. Using this approach, he found that the sequence of normal and sickle-cell haemoglobin did indeed differ by only one amino acid—a glutamic acid residue carrying an overall negative electrical charge in the normal protein was substituted by the electrically neutral valine in the sickle-cell protein. This alteration of a single amino acid, arising from a genetic mutation, is enough to cause the chains of haemoglobin to misfold in such a way that the mutant protein is less soluble. As a result, it forms crystals that deform the red blood cells, making them less flexible and prone to clumping together. The result is that they can then clog up blood vessels, giving rise to the debilitating pathology of sickle-cell anaemia. All this, from the simple substitution of one type of amino acid for another.[122]

Ingram's discovery that a single genetic mutation in the DNA sequence of haemoglobin gave rise to the substitution of glutamic acid for valine at a specific position in the protein chain led him to draw a bold and far more general conclusion:[123]

> ... the sequence of base-pairs along the chain of nucleic acid provides the information which determines the sequence of amino-acids in the polypeptide chain for which the particular gene, or length of nucleic acid is responsible. A substitution in the nucleic acids leads to a substitution in the polypeptide.[124]

The Times hailed Ingram's discovery as nothing less than 'a landmark in genetics' and when he gave a lecture entitled 'On Protein Synthesis' at University College, London, to the Society for Experimental Biology that same year, Crick declared that 'Ingram's result is just what I expected'.[125,126]

Crick's lecture was a turning point in the history of biology. In it he proposed an idea that was simple, but one that, according to Judson, nevertheless 'permanently altered the logic of biology'.[127] Dubbed by Crick 'The Sequence Hypothesis', it has come to be the foundation stone of molecular biology:

> ... the specificity of a piece of nucleic acid is expressed solely by the sequence of its bases, and that this sequence is a (simple) code for the amino acid sequence of a particular protein.[128]

A picture was beginning to emerge—DNA made RNA, and this in turn directed the synthesis of proteins (Figure 43). But despite this triumph, the question of how exactly the sequence of bases might encode the order of amino acids in a protein such as haemoglobin or insulin remained elusive.

In 1961, the first hint of an answer began to emerge. Watson did not have particular high hopes for the Fifth International Congress of Biochemistry, held in Moscow during August that year, dismissing it as nothing other 'than a chance to see communism in action'.[129] As a large parade passed through Red Square celebrating the achievement of the cosmonaut Titov who had just circled the Earth,

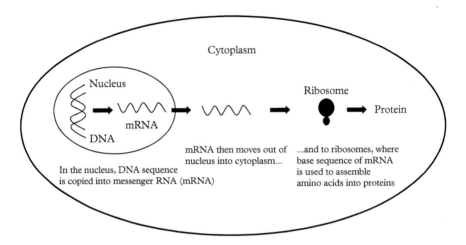

Figure 43 *DNA makes RNA makes protein. The sequence of bases in DNA is copied into single stranded messenger DNA (mRNA) that is complementary in sequence and contains the base uracil instead of thymine. In eukaryotic cells this takes place in the nucleus and the mRNA is then exported into the cytoplasm. The base sequence of messenger RNA is then read by the ribosomes in the cytoplasm and used to form a protein chain of amino acids. Diagram by K.T. Hall.*

Watson's prediction seemed to be confirmed. But then came what he described as 'an unexpected bombshell talk'.[130]

The talk was given by Marshall Nirenberg (1927–2010), a researcher at the National Institute of Health in Bethesda, whose biographer Franklin Portugal described him as 'the most famous person you have never heard of'.[131] Nirenberg did not belong to the gilded ranks of the RNA Tie Club, nor at that time was the institute at which he worked considered to be particularly prestigious. These reasons might explain why only thirty-five people bothered to attend his session (and considering there were a total of six thousand delegates at the conference, this meagre turnout left Nirenberg a little disappointed).[132] However, Mat Meselson, whose ultracentrifugation methods had been crucial in helping to discover messenger RNA, did attend. As Meselson sat listening to Nirenberg describe his work, he quickly realized that he was witnessing history in the making—'I heard the talk. And I was bowled over by it'.[133]

After the talk, Meselson quickly dashed off to find Crick and tell him about Nirenberg's results. Crick had never heard of Nirenberg but arranged for him to give a repeat performance of his talk during a session chaired by Crick on the final day of the meeting. When Watson learned of Nirenberg's work and began to discuss it with other prominent scientists in advance of this second presentation, he recalled that some of them believed he was having a practical joke,[134] but when Nirenberg got up to speak on the final day, Watson said that not only had he convinced the audience, but also he left them stunned.[135] After the talk, Meselson went up and embraced Nirenberg.[136]

For Crick, Nirenberg's work was nothing less than a 'breakthrough'.[137] Nirenberg had described how he and his post-doctoral fellow Heinrich Matthaei had artificially synthesized mRNA molecules that consisted solely of the base uracil, designated as U. When they had then added these artificial polyuracil RNA molecules to an *in vitro* experimental system containing all the cellular biochemical apparatus required for protein production, they had been delighted to observe the synthesis of a protein chain. But better still, when they analysed the chemical composition of this artificially synthesized protein, it was found to contain only repeats of a single amino acid—phenylalanine. The implication was obvious—the presence of the base uracil in mRNA somehow directed a cell to insert phenylalanine during the construction of a protein chain. For Watson, the implications went even further: 'From that moment on, it seemed likely that the genetic code would soon be completely cracked'.[138]

Not that Nirenberg could afford to stand still. For although he had given a demonstration of the experimental tools required to crack the genetic code, others were now quick to use them. When Nirenberg returned to his lab after the Moscow meeting, he found himself in intense competition with Severo Ochoa at New York University who, in 1959, had been awarded the Nobel Prize for his part in the discovery of the enzyme that allowed the synthesis of RNA chains. Ochoa was a fierce competitor renowned for taking no prisoners and he began to synthesize artificial mRNA molecules and test them in cell free systems at breakneck speed in order to discover what other combinations of bases might determine particular amino acids. Having attended a meeting at which Ochoa had presented his latest work, Nirenberg lamented, 'It floored me, that Ochoa had made such advances'.[139]

What neither Nirenberg nor Ochoa had yet established, however, was the minimal unit of bases required to determine a single amino acid. This question was also puzzling Crick and Brenner. Using mutants of a type of virus called bacteriophage T4 that infects bacteria, they showed that the genetic code must be specified by bases arranged in triplets. In what became his second landmark experiment, Nirenberg then confirmed this to be the case. Using a method developed by chemist Har Gobind Khorana (1922–2011), Nirenberg and his post-doctoral fellow Philip Leder synthesized molecules of RNA composed of only three bases and showed that a specific RNA triplet could bind to only one particular type of amino acid.

It was thanks to this work that, when Nirenberg arrived at his laboratory on the morning of October 16, 1968 he found a giant banner hanging there saying 'UUU are great' (Figure 44). It had been hung there by the rest of his team, who, sipping champagne from lab glassware, celebrated the news that, together with Robert Holley and Har Gobind Khorana, Nirenberg had just been awarded the Nobel Prize in Physiology or Medicine.[140]

By this time the base triplets coding for each of the twenty naturally occurring amino acids had been identified and it had been shown that the code was degenerate. This meant that since there were sixty-four possible triplet combinations of bases and only twenty naturally occurring amino acids, each type of amino acid

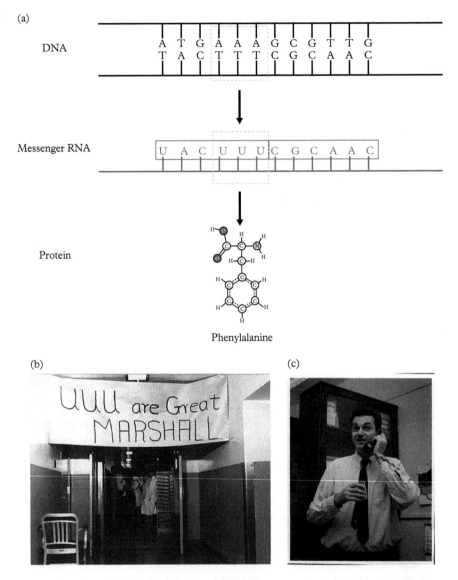

(a)

DNA

Messenger RNA

Protein

Phenylalanine

(b)

(c)

Figure 44 *a) How the genetic code works. Each different type of amino acid is specified by a triplet of three bases in messenger RNA called a codon. The amino acid phenylalanine for example is specified by the triplet AAA in DNA which is copied into UUU in messenger RNA (RNA uses uracil 'U' instead of thymine 'T'). b) 'UUU are great Marshall' – a banner hung up in the lab by Nirenberg's team to celebrate him winning the 1968 Nobel Prize for his part in helping to crack the genetic code by showing that the amino acid phenylalanine was specified by the codon UUU in messenger RNA. c) Marshall Nirenberg (1927–2010) receives a telephone call congratulating him on winning the Nobel Prize for this discovery and celebrates with some champagne in a lab beaker. Diagram by K.T. Hall.*

could be specified by more than one type of base triplet. Intriguingly, the code appeared to be universal throughout living systems and Crick had high hopes for what this might ultimately reveal:

> The genetic code may perhaps give us some hints about that speculative and difficult problem, the Origin of Life ... The great biochemical uniformity of living things certainly that they had a common origin, and it is not impossible that certain features of the present system actually contain, frozen into them, the early history of its development.[141]

A few years later, having reflected some more on this universality of the code, Crick made an even more grand proposal. Together with the chemist Leslie Orgel (1927–2007), Crick suggested that life on Earth might well have been seeded by extra-terrestrial micro-organisms.[142] The idea that life on Earth might have its origins in outer space was not new, and had first been suggested by the Swedish scientist Arrhenius in 1908, who called it the 'Panspermia hypothesis'. Crick and Orgel now pushed this idea even further with a suggestion that they called 'Directed Panspermia'—a bold proposition:[143]

> Could life have started on Earth as a result of infection by microorganisms sent here deliberately by a technological society on another planet, by means of a special long-range unmanned spaceship?[144]

Whatever the origins of the genetic code, Crick felt that its deciphering marked the end of an era. This had begun with the discovery by himself and Watson of the double-helical structure of DNA and had ended with the work of Nirenberg and Matthaei.[145] But for others, such as the historian Horace Judson, the work of Nirenberg and Matthaei was the climax of a journey that had begun not with the double helix, but rather with Avery's discovery that DNA could pass on the property of virulence in bacteria. For it was this discovery that Chargaff said had first hinted 'a new grammar' of biology.[146] Inspired by Avery's work, Chargaff had taken the first tentative steps on the road to finding this new grammar, and now, thanks to Nirenberg's work, its syntax had at last been discovered.

Judson went so far as to say that Chargaff's analysis of DNA was nothing less than 'the discovery that made molecular biology possible'.[147] But far from being recognised for his discovery that DNA was not merely a repeat of tetranucleotides, Chargaff had not received the accolades that Judson felt were owed him and had been honoured with only 'the parsimonious justice of a ten-per-cent tip'.[148] Chargaff himself resented that he had never been awarded the Nobel Prize for his work on DNA, but hearing himself hailed by Judson as one of the architects of molecular biology would have brought him little consolation.

For despite being instrumental in the emergence of molecular biology, Chargaff felt no sense of pride at this achievement. In his autobiography, Chargaff recalled that, as a young man, he had first been attracted to biochemistry, 'partly through the romantic notion that the natural sciences had something to do with nature'.[149] Instead, now he felt only fear, dread, and despair for the future.

One reason for his disillusionment was that he felt the aims of science had changed. It was no longer an attempt to understand nature by uncovering its truths, and in so doing to learn wisdom—but instead simply to 'outsmart' it.[150] It was not a development he welcomed:

> I would say that most of the great scientists of the past could not have arisen, that, in fact most sciences could not have been founded, if the present utility-drunk and goal-directed attitude had prevailed.[151]

Of all the sciences, Chargaff saw molecular biology as the nadir of what he once described as this 'strip-mining of nature'.[152] He famously despised the term 'molecular biology', deriding it as 'the practise of biochemistry without a licence'.[153] But it wasn't just the nomenclature that offended him. Rather, he felt the whole structure was built on two flawed philosophical foundations. The first of these was that the complexity of life could be understood simply by breaking it down into its component molecules. He lamented that 'life is what's lost in the test tube'.[154] Seeking to illustrate the philosophical shortcomings of a purely reductionist view, he asked: 'Who could understand music only from an analysis of the composition of the instruments of an orchestra? The news that all trombones are made of brass is trivial when measured against the immensity of the musical universe'.[155] Moreover, Chargaff felt that this was not only a flawed philosophy, but also one that had detrimental consequences as it diminished other spheres of intellectual activity, and he joked, darkly: 'Soon we shall have molecular sociology, molecular history, and a little later perhaps molecular theology'.[156]

He also objected to the way in which DNA was now revered as offering a single unifying explanatory concept for the diversity and complexity of life: 'I hear the start of a new apocryphal gospel with DNA as the logos of our times … the double-helix, quite apart from its many undeniable scientific merits, has become a mighty symbol; it has replaced the cross as the signature of the biological alphabet'.[157,158]

His other major gripe with molecular biology was that it had become institutionalized. In this culture, scientists became ever-more specialized, delving into the minutiae of one particular gene or protein, but at the risk of developing a kind of myopia about any other field of intellectual endeavour. Molecular biology also had become an industrialized process, dedicated to churning out new facts simply so that new papers and grant applications could be written for the perpetuation of careers. The yardstick of success was no longer through novel insights into nature. Although the grant writers would seek to justify their research on the grounds that it might lead to cures for disease, Chargaff found himself very sceptical about the whole endeavour:

> Let us assume that there is a National Music Foundation in Washington, and that this NMF makes large grants many schools. These schools erect large buildings

and fill them with pianos or electric caterwaulers or whatever is needed to supply hundreds of graduate composers per year. Each of these students is now supposed to write a symphony every year of his fellowship support. What then? There is, of course, no audience for all this music; but since it exists, it has to be crammed down the ears of humanity. And after some time the musical Moloch will demand a suitable supply of young people so as to have something to do. Anything said against this becomes an attack on the noblest aspirations of mankind, for who can be against music?[159]

This, said Chargaff, was the situation in which young molecular biologists now found themselves, as their science became institutionalized. Judson put it another way, as he lamented that what was once the 'golden age' of molecular biology had now become an 'age of brass':[160]

The remarkable small self-governing democracies of the classical era, represented by the phage group, the garret at the Institut Pasteur, the RNA Tie Club, the Medical Research Council unit at the Cavendish—few such collaborations are now to be found ... They are steadily being superseded by a different sort of laboratory life and ethos: laboratories that are large and rigidly hierarchical, and an ethos driven by careerism, in which doing science is in large part a way to secure more grants, to get more promotions, to gain power. Perhaps to get rich.[161]

Chargaff himself turned to classical mythology in order to expose how this kind of institutionalized science perpetuated itself. In his final dialogue, he pictured it as a ravenous minotaur sitting in the darkness of its labyrinth:

Ah, a new bunch of graduate students, unspoiled, enthusiastic, bringing new forces to exploring the old labyrinth. How I love to work with these young minds. There is so much to do, and the labyrinth grows all the time. No sooner have we explored one spiral than we find it has branched out into a hundred new ones.[162]

The never-ending spirals of Chargaff's eternally growing labyrinth were not a set of experiments leading to a whole new set of questions. Rather, they were the more cynical process of generating questions simply to publish more papers and generate more research grants.

To illustrate his point, he imagined the discussion between an older biochemist and a bright-eyed, eager young molecular biologist, who declares in triumph that the complexities of biology can all be reduced to an understanding in terms of DNA:

I am not at all discouraged; quite the contrary. I have only to think of the unprece-dented possibilities that are opening up before us ... we shall be able to look up the nucleotide sequence of every DNA; and each purine and pyrimidine will have a number; and we shall know what happens when we change it. And boy, will we change it![163]

But on hearing this, the older biochemist feels only despair:

> I cannot stomach people who claim that they have understood and explained
> Hamlet by telling me how often the word 'and' appears in the first act ... Look
> at the enormous variety in the shapes of organisms, organs, even cellular com-
> ponents; where is the biochemistry of specific shape? Where is the biochemistry
> of cell differentiation? Is there a separate nucleotide code for your fingerprints,
> which are different from mine? Is it a pairing error in position seventy-nine which
> has produced the visions of Blake? Above all, it is against this shabby mechaniza-
> tion of our scientific imagination, which kills all ability to notice the foreseen that
> I protest.[164]

Far from condemning these young molecular biologists, however, Chargaff's
fictional older biochemist felt pity for them. He believed that young PhD and
post-doc scientists entering molecular biology were being sold a lie. But he also
feared them. For, now that the grammar of biology had been deciphered, might it
not also be deliberately rewritten? It was a prospect which left Chargaff terrified.

Chargaff confessed that he had already lost his faith in science long before
these new developments in biology—in fact, he could recall the exact moment
that disillusionment set in. Walking along the beach one evening in August 1945
he met a stranger who told him that the Japanese city of Hiroshima had been
incinerated by the dropping of a new kind of bomb with terrifying destructive
capability.[165] Chargaff's cherished faith that science was a noble profession was
now 'shattered' and the effect upon him was profound:[166]

> The double horror of two Japanese city names grew for me into another kind of
> double-horror ... an estranging awareness of what the United States was capable
> of, the country that five years before had given me its citizenship; a nauseating
> terror at the direction in which the natural sciences were going. Never far from an
> apocalyptic vision of the world, I saw the end of the essence of mankind; an end
> brought nearer, or even made possible, by the profession to which I belonged[167]

Chargaff's greatest fears, however, were reserved not for the destructive powers
unleashed by physics, but rather by what biology might unleash in the not-
too-distant future. He pointed out that, while nuclear weapons were capable of
destruction on a horrific scale, they could at least be regulated by arms control
treaties and, given sufficient political will, perhaps even abolished and nuclear
research brought to a stop. The results of genetic engineering, by contrast, were
much more difficult to control: 'You cannot recall a new form of life'.[168]

Towards the end of their discussion, Chargaff's fictional young molecular
biologist declares in triumph:

> When we know the Universal Code, we shall soon learn how to interfere with cer-
> tain nucleotide sequences in DNA, how to change them specifically and thereby
> to produce desirable genetic changes.[169]

But before he departs, the older biochemist urges caution:

Once you can alter the chromosomes at will, you will be able to tailor the Average Consumer, the predictable user of a given soap, the reliable imbiber of a certain poison gas. You will have given humanity a present compared with which the Hiroshima bomb was a friendly Easter egg. You will indeed have touched the ecology of death. I shudder to think in whose image this new man will be made.[170]

The older biochemist is quite clearly Chargaff, and it is tempting to dismiss his jeremiad tone as the classic case of an old man who is adamant that everything was better in the old days. Perhaps Chargaff's caustic views were simply the result of suddenly finding himself forced into early retirement on a substantially reduced pension as a result of having showed up for work one day only to find himself locked out of his laboratory by the university authorities.[171,172] Chargaff, not wishing to compromise his own integrity had even stopped taking on graduate students.[173] He had no wish to be complicit in sending them into the labyrinth and feared that their faith in DNA to act as some kind of guiding thread through its tunnels was dangerously misplaced.[174] Likening himself to a 'minor apocryphal prophet', Chargaff made dire predictions about what horrors might emerge from the dark, twisting tunnels of this labyrinth in which the younger generation were losing themselves.[175]

But the new generation of younger scientists who were carving out careers in molecular biology felt very differently. For them, the possibility of deliberately rewriting the grammar of life was an exciting opportunity to understand more about how DNA worked—and to build a glittering career in the process. They were determined to prove Chargaff and wrong, and the secret weapon with which they hoped to achieve this victory was insulin.

11

The Clone 'Wars'

Herb Boyer (Figure 45) was one of those eager young molecular biologists for whom Erwin Chargaff reserved his pity. But on this particular day, even Boyer felt his optimism and enthusiasm draining from him. Because from where he was sitting in the conference auditorium, the future for molecular biology was starting to look decidedly bleak.

Figure 45 *Herb Boyer sitting in the auditorium at the Asilomar conference and pondering what the future holds for recombinant DNA technology…*

Credit: US National Library of Medicine. Courtesy of US National Library of Medicine.

Normally a scientific conference would have had Boyer on the edge of his seat with excitement, but this one held at Asilomar in California in February 1975 had him feeling depressed. Boyer later described those few days as 'an absolutely disgusting week'—nothing less than 'a nightmare'.[1] It was a sentiment shared by his co-worker Stanley Cohen, who found the meeting to be 'one of the most depressing that I can remember' (Figure 46).[2] As the stress began to take its toll, Boyer found himself unable to sleep, while Cohen resorted to popping antacid tablets in an attempt to calm his gastric reflux.[3]

Figure 46 *Stanley N. Cohen photographed in 1980, having been awarded the Lasker Prize.*
Credit: Courtesy of US National Library of Medicine.

The Asilomar meeting had the potential to kill their future careers in science. The issue under discussion by the 150 scientists who had gathered was the creation of recombinant DNA—hybrid DNA molecules containing the genetic information from two or more unrelated organisms. Boyer and Cohen were both key figures in pioneering this new technology and, with their supporters, they argued that not only did it have the potential to open up our understanding of how genes work, but also—and perhaps more importantly—to allow the creation of new medicines. In response, its detractors, such as Chargaff, pointed out the dangers— what if, for example, a gene known to cause cancer in humans was inserted into the DNA of a common bacterium?

Thirty years earlier, physicists had been forced to confront the moral dilemmas raised by their work when the first atomic bomb had been exploded in the deserts of New Mexico. Now biologists faced the new ethical dilemmas posed by their work to ensure that they did not result in catastrophe. The consensus among many of the senior scientists attending the Asilomar meeting was the need to draft a set of guidelines that imposed severe restrictions on working with recombinant DNA.

But Boyer and Cohen feared that these regulations might be so restrictive that this new field of research would be halted before it began. Normally, Boyer loved nothing more than to talk openly at scientific meetings about his work, but as the discussions at Asilomar became increasingly heated, he may have regretted his previous frankness. For it was thanks to his very own enthusiasm for sharing his work that he now found himself in the midst of a debate that might destroy it.

Two years before Asilomar, Boyer had given a presentation at the prestigious Gordon conference, named in honour of Dr. Neil E. Gordon (1886–1949) chemistry professor at Johns Hopkins University. In 1931, Gordon had organized a week long meeting of prominent scientific minds at a remote island in Chesapeake Bay in response to his concerns that science had become 'too big, too hurried, too specialized, too chopped up, and too cooped up to encourage speculative thought'.[4] This meeting proved to be so successful that it became an annual event, at which the dress code was strictly one of 'no jacket, no necktie, shorts if you want to wear them'.[5] With this relaxed dress code, along with a schedule of sessions on the golf course or in the swimming pool, Gordon aimed to promote an informal environment in which researchers could feel at ease to talk openly and freely about work in progress.

In keeping with the spirit of its founder, Boyer got up and talked eagerly about an experiment that he and Cohen had just completed, but he never anticipated the consequences. His collaboration with Cohen had begun over sandwiches and beer in a delicatessen near Waikiki beach in Honolulu,[6] and it seemed like an unlikely partnership. When the science writer Janet Hopson first interviewed Boyer, he defied all her expectations of what an academic scientist would look like:

> I expected to meet a distinguished man in his middle fifties, irritated by my tardiness, and as stiff and colorless as a lab report. But the fellow who opened the door was wearing gym shorts and munching an apple. Boyer's curly hair and mustache were still damp from jogging and showering. He looked to be about 35, was tanned and talked easily between bites of the apple, blending phrases like 'EcoRI endonuclease recognition sequence' and 'Wow! That's far out!' in the same sentence.[7]

By contrast, Cohen was quiet and more reflective. Rather than gym shorts or leather jackets, his spectacles and sports jacket gave him what historian Sally Smith-Hughes has described as 'the quintessential image of the professor, the solid citizen, and serious intellectual'.[8]

Boyer and Cohen had both come to Hawaii to attend a joint Japanese–American conference on the subject of plasmids—circles of DNA that are able to replicate in bacterial cells and be transferred easily between them. Whilst this might seem a somewhat esoteric subject for study, it had serious health implications. In 1955, doctors in Japan had treated a patient with a serious case of dysentery caused by the Shigella bacterium. What baffled the clinicians was that this particular strain of Shigella did not respond to four common anti-bacterial drugs. Four years later, researchers discovered that the Shigella bacterium had somehow acquired its drug resistance from the population of *E. coli* that are normally resident in the human gut. Because their work showed that the trait of drug resistance was not carried on the bacterial chromosome, they inferred that it must therefore be passed on by some other cellular factor.

In the 1960s, these mysterious extra-chromosomal genetic factors were found to be plasmids—circular DNA molecules passed from one bacterium to another and that carry genes conferring traits like resistance to antibiotics, heavy metals, and ultraviolet (UV) light. This made plasmids an attractive subject for Cohen to study, for although he had graduated in clinical medicine, he was also interested in basic scientific research. To pursue these interests, Cohen accepted a post at the Department of Medicine at Stanford University in 1968, but soon found that his colleagues did not share his interest in plasmids. Hoping to find more enthusiasm in the Biochemistry Department, he approached its chair, Nobel laureate Arthur Kornberg, but was told by him that plasmid research was of little interest.[9]

Despite his dismissal of Cohen's proposed research, Kornberg had unknowingly actually started to carve out the path for him. For only a short while earlier, Kornberg had warned another young scientist against heading into the unknown. This was Paul Berg, one of Kornberg's own protégés, with whom he had worked on DNA replication and gene expression in the *E. coli* bacterium (Figure 47).

Figure 47 *Paul Berg.*

Credit: Paul Berg Papers, US National Library of Medicine. Courtesy of US National Library of Medicine.

Kornberg was keen that Berg should continue to build his career on the tried and tested biochemistry of *E. coli*, but when Berg announced that he now felt the time had come to strike out on his own in a new direction, his mentor was not happy:

> Arthur was furious. I won't say we ever came to blows, but there were times when I was so furious with him because he was so critical. He more or less said, 'You're wasting your talent. You're destroying your career. You have so much of a gift for doing enzyme research. The only true path to knowledge is E. coli', and so on and so forth … [10]

Berg's pioneering spirit and his refusal to listen to his mentor's admonitions helped to blaze the trail for Boyer's and Cohen's own work. By the end of the 1960s, much was known about how DNA was regulated and expressed to form proteins in bacteria, but far less was known about how this occurred in more complex eukaryotic cells. Having studied the expression of genes in bacteria together with Kornberg, Berg wanted to tackle the more ambitious—and formidable—challenge of understanding how genes in mammalian cells are controlled—with particular focus on how loss of this control can give rise to tumours.

Berg was proposing to do this by adopting a trick that had been used to study the function of genes in bacteria. This method relied on the fact that bacteria are susceptible to infection by a type of virus called a bacteriophage, or phage. When a phage infects a host bacterium, it replicates its DNA and packages it up in a protein coat to form new virus particles that can then go on to infect new host cells. However, sometimes during this process of replication, the phage packages not only its own DNA into a new infectious particle, but also fragments of the bacterial host's DNA. These discrete chunks of bacterial DNA are then carried by the phage into its new host and if they contain genes coding for a readily observable trait, they can then be detected.

Berg wondered whether a virus could be used to similarly transport discrete chunks of DNA from higher organisms. Might it be possible to physically insert a defined sequence of foreign DNA into the genome of a virus? If so, the resulting hybrid virus could be used as a vehicle to transport the foreign DNA into a mammalian cell, where its activity could be studied.

Berg had one particular virus in mind that he thought would make an ideal delivery vehicle for the transport of foreign DNA. This was Simian Virus 40 (SV40) which, after entry into a cell can insert its genetic material into the genome of the host cell. This meant that the viral DNA—along with any foreign DNA that it was carrying—would be passed on during cell division to daughter cells in a stable manner providing researchers a reliable long-term experimental system in which to study gene expression. But alongside this advantage, SV40 also came with one significant snag. Its genome was a closed circle and so, before any foreign DNA could be inserted into it, this circle would somehow have to be cut and opened up.

Boyer came to the rescue. At the time, he was based in the Department of Microbiology at the University of California at San Francisco (UCSF), where he was encountering a couple of difficulties. The first of these was that, on arrival, he found that the lab space that he had been promised had not materialized. The second problem was that his post at UCSF also included some teaching, which not only distracted from his research, but was also hampered by his own lack of enthusiasm for the lecture hall: 'I was a horrible teacher ... teaching, particularly teaching medical students—God, I dreaded that'.[11]

As a result, Boyer found himself spending far more time hanging around the Department of Biochemistry, where he found the intellectual environment much more stimulating. The subject of his research was an obscure set of bacterial enzymes known as restriction endonucleases and although this was a source of fascination for Boyer, his father, a railway worker in Pennsylvania, was left baffled. Boyer recalled that, when explaining something called 'restriction endonuclease modification', his father demanded to know, 'Well, what good is it? What are you going to do with that?'[12] The best answer that Boyer could offer was, 'I don't know—cure the common cold?'[13]

Restriction endonucleases may not have cured the common cold, but they did earn their discoverers Werner Arber, Hamilton Smith and Dan Nathans the 1978 Nobel Prize in Physiology or Medicine. In the early 1960s, Arbers and his colleague Daisy Dussoix at the University of Geneva had been studying the infection of bacteria by bacteriophages. They were particularly intrigued by the observation that, in certain strains of bacteria, these bacteriophages showed only limited, or 'restricted' growth.[14] Having observed that the DNA of these infecting phages was being degraded, Arber and Dussoix proposed that this might be due to the action of a specific set of bacterial enzymes.[15] Within only a few years, strong evidence had emerged that they were correct. Mat Meselson and Robert Yuan at the University Harvard, and Hamilton Smith at Johns Hopkins University Medical School, had isolated and purified restriction enzymes from *E. coli* and *H. influenzae*.[16]

Restriction enzymes appeared to have evolved in bacteria as a defence against viral infection rather like a very primitive immune system. Later, Daniel Nathans went on to discover that restriction enzymes could cut not only the DNA of bacterial viruses, but also of the SV40 virus that Berg was planning to use.[17] Moreover, these cuts were made at very specific sites along the length of the DNA molecule. It was for this discovery that Nathans was awarded a share in the Nobel Prize with Arber and Smith, and it would go on to revolutionize the life sciences.

The wheels of this revolution were set in motion when Boyer gave Berg a sample of restriction enzyme that one of his graduate students, Robert Yashamori, had isolated and purified from *E. coli*. Known as Escherichia coli Restriction Endonuclease I—or EcoRI—it was one of the enzymes that Nathans had shown could cut the circular DNA of SV40 at one single location. This was exactly what Berg needed, as this particular pair of molecular scissors could snip open the circular DNA of SV40 into a linear molecule.

Using the same set of molecular scissors, Berg's group now snipped open the genome of a type of bacteriophage called lambda to obtain a discrete fragment of DNA. This contained a set of bacterial genes involved in metabolism of the sugar galactose that had been picked up from bacterial hosts during infection by the phage and provided the ideal test material for Berg's experiment to see whether a recombinant DNA molecule could be made using genetic material from two different organisms. The next step was to find a way of physically joining these two linearized fragments of DNA: one from the animal virus and the other from the bacteriophage.

As part of his PhD research, graduate student Pete Lobban had studied an enzyme called terminal transferase, and found that it could add a short single stranded overhang composed of only one type of base, such as adenine, to the end of a DNA fragment.[18] If an overhang like this was attached to the ends of the linear SV40 fragment, and a second overhang made of a complementary base added to the bacterial DNA fragment, then pairing up between the bases on opposite overhangs would allow the two DNA fragments to be joined together.

But the process of preparing the two DNA fragments in this way was tedious and required a cocktail of different enzymes. First, the DNA had to be cut with EcoRI, and then Lobban's overhanging strands had to be added to form what became known as 'sticky ends'. These sticky ends allowed the two DNA fragments to be annealed and bonded together by an enzyme called a ligase. Finally, any gaps in the joint were filled in by yet another enzyme, known as DNA polymerase. Each of these different types of enzyme had first to be isolated and purified—a process that was time consuming and laborious. Finally, in 1972, Berg and his team announced that they had made the very first hybrid DNA molecule using genetic material from two unrelated organisms (Figure 48).[19]

Building on their success, they now planned to show that recombinant SV40 could carry foreign DNA into mammalian cells. If it worked, it could help study the activity of discrete chunks of human DNA, such as those thought to be involved in tumourigenesis. Berg hoped that this might give new insights into our understanding of cancer, but before he could get any further, the work hit a major—unscientific—problem.

During the work to make the hybrid virus with SV40 and bacterial DNA, Janet Mertz, one of Berg's graduate students, had attended a course on viruses at Cold Spring Harbor, Long Island. Here, Mertz had mentioned a second experiment that Berg was planning to carry out with his new hybrid. This would involve introducing the hybrid SV40 back into *E. coli* cells in order to see whether the fragment of bacterial DNA that it carried was still active in making protein.

For some who were present at the meeting, this proposal set alarm bells ringing. On the afternoon of June 28, 1971, microbiologist Robert Pollack rang Berg to express his concern at what he had heard from Mertz at the meeting. The conversation that followed has since been described as 'a telephone call that would

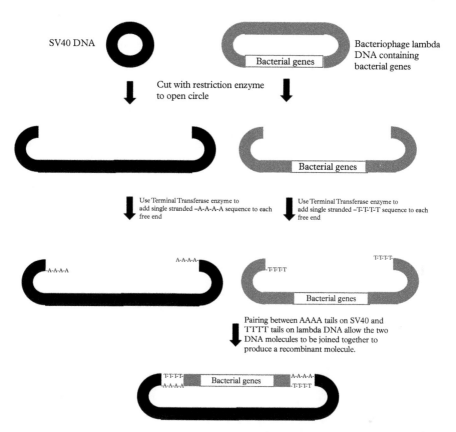

Figure 48 *Paul Berg's proposed experiment which involved cutting and splicing together DNA from the genome of the mammalian SV40 virus and a section of the genome from bacteriophage lambda containing bacterial genes involved in the metabolism of lactose. A crucial step was the innovation to add single stranded tails of polyA and poly T to the two DNA fragments which enabled them to be joined together. Diagram by K.T. Hall.*

fundamentally change the relationship of American science to the democratic society that shelters it'.[20]

Pollack's concerns were based on the discovery that when batches of polio vaccine intended for use in humans had first been tested on hamsters, some of the animals had been found to develop tumours. The culprit was identified as the SV40 virus lurking within the cultures of Rhesus monkey cells used to grow the polio virus. Although SV40 had never been shown to cause cancer in humans, it was known that when human cells grown in tissue culture were infected with it, they underwent a process known as 'transformation', in which they took on the physiological and morphological properties of tumour cells. Writing in *Science*, Pollack had already expressed concern about research involving transforming viruses like SV40: 'We're in a pre-Hiroshima situation … It would be a real disaster if one of the agents now being handled in research should in fact be a real human cancer agent'.[21]

Pollack was horrified by Berg's proposal to deliberately introduce genetic material from such a virus into a common bacterium known to colonize the human gut. The virologist Wallace Rowe of the National Institute of Allergy and Infectious Diseases echoed Pollack's concern and suggested that even Berg himself was not without his own doubts about the wisdom of his plan. Interviewed in *Science*, he warned that 'The Berg experiment scares the pants off a lot of people, including him'.[22]

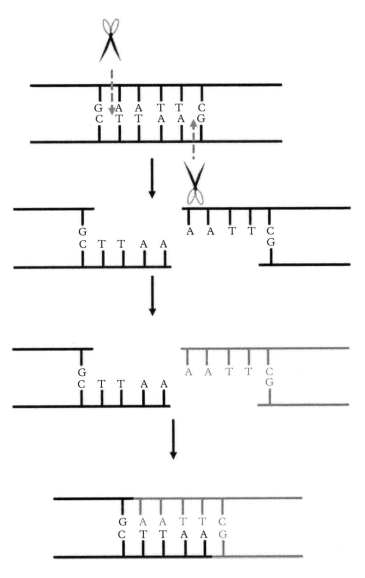

Wallace's insight proved to be astute. Acknowledging the possible danger, Berg decided not to do the next part of the experiment. Not that this mattered, because Berg had shown that it was indeed possible to construct a hybrid DNA molecule using the genetic material from two unrelated sources.

In recognition of this achievement, Berg went on to win the 1980 Nobel Prize in Chemistry, but the victory had been a tough one. Today, the kind of enzymatic reagents that Berg and his team used are commercially available and any PhD student or post-doc can easily scroll through the online catalogue of a supplier and order a battery of such enzymes at the click of a mouse. But in the early 1970s, only a handful of researchers had the necessary skills required to isolate, purify, and use the enzymes needed to cut, trim, weld, and re-seal the DNA. The isolation, purification, and characterization of one single restriction enzyme was such an impressive feat of specialist craftsmanship that this achievement alone might be enough to earn a doctorate. Moreover, the tiny amounts in which these enzymes were obtained made them a precious commodity. But luckily for Berg, Stanford's Biochemistry Department was an epicentre of expertise in the field: he had ready access not only to these precious reagents, but also to people like Boyer, who knew how to make and use them.

The difficulty in preparing and purifying the battery of enzymes required made it very unlikely that other labs were going to start producing recombinant DNA molecules any time soon. However, around the time Berg published his work, Mertz discovered something that was to change everything.[23] Boyer and his colleagues had now determined the precise sequence of bases that was recognized by EcoRI as a target when it cut DNA. This corresponded to a short run of six-bases (GAATTC). However, Mertz found that, when EcoRI cleaved double-stranded DNA containing this sequence, it made a cut that was not blunt, but staggered, with one strand of the double helix overhanging the other by a few bases.[24] Put simply, cleavage with EcoRI resulted in DNA fragments with naturally occurring sticky ends, meaning Berg's original process was now unnecessary (Figure 49).

Figure 49 *'The Servant With the Scissors' was how ten year old Silvia Arbers described her father's discovery of what became more widely known by the less imaginative name of 'Restriction enzymes'. These are bacterial enzymes which limit or 'restrict' the growth of an invading bacteriophage thanks to their ability to recognise and cut specific sequences of DNA. Or, as Arbers' daughter put it somewhat more poetically –* 'My father has discovered a servant who serves as a pair of scissors. If a foreign king invades a bacterium, this servant can cut him in small fragments, but he does not do any harm to his own king. Clever people use the servant with the scissors to find out the secrets of the kings.' *(Arber, S. cited in (Konforti 2000)). The enzyme Eco RI, for example, recognises the sequence GAATTC and makes a staggered cut across the double-helix, leaving a single stranded overhang. This means that if another fragment of DNA (shown in grey) has also been cut with Eco RI, it will have the same overhangs and, thanks to the specific pairing of bases on opposite strands which emerged from Erwin Chargaff's work and the two fragments can now be physically joined to each other to produce a recombinant DNA molecule. Diagram by K.T. Hall.*

This meant that making hybrid DNA molecules was available to all, or, as Berg was quick to point out—'Now anybody could do it. You could buy an enzyme, take two DNAs, cut them, mix them, tie them together and, presto, a recombinant DNA'.[25]

Cohen was quick to seize on the implications of this discovery. He was isolating plasmids and trying to identify specific regions of DNA involved in controlling replication. To do this, he had developed a method for breaking up the plasmid DNA by shearing it in a blender, before re-introducing it into *E. coli* in the hope of studying its properties. However, not only was the process slow and inefficient, but also the shearing was a random process that produced a large amount of DNA fragments of variable length. Restriction enzymes that generated sticky-ended DNA fragments promised to change all this. Not only would Cohen be able to cut plasmid DNA at a specific site, generating fragments of uniform size, but it might also be possible to take a defined fragment from one plasmid and physically insert it into a second complete plasmid in which the region controlling replication was intact. The resulting hybrid molecule could then be introduced into *E. coli* where it would be replicated. In this way, the bacterium could be used as a factory to churn out identical copies of the hybrid plasmid containing the inserted foreign fragment.

This procedure of inserting a known DNA fragment into a plasmid for the purpose of making identical copies—or cloning—was the hot subject of discussion between Boyer and Cohen in the Honolulu deli. Boyer had only been present at the conference as a result of having received Cohen's last-minute invitation, and plasmids were not really his field of expertise. However, walking away from the deli that night, he nevertheless described himself as feeling 'jazzed'.[26]

The following January, the two labs began their collaboration. Cohen's group focused on preparing plasmids, while Boyer's group carried out the restriction enzyme work. Until now, isolating and purifying fragments of DNA had been a painstaking task that involved centrifugation through a density gradient. But in a stroke of luck, on his way home from Hawaii, Boyer had called in to visit some colleagues who showed him an easier method involving nothing more than pipetting a mixture of DNA fragments onto a slab of agarose gel and applying an electric field.[27] Since the DNA molecules all carried a negative electric charge, they would begin to move through the gel slab towards the positive end, but the fibres of the gel acted as a molecular sieve, slowing down the progress of the larger fragments and allowing smaller fragments to pass through. In this way, a mixture of DNA fragments could easily be separated according to their size and then visualized as a pattern of bands by staining the gel with ethidium bromide, a dye that makes DNA glow under UV light. Specific fragments of interest could then be cut out of the gel and purified. Known as gel electrophoresis, this same method also allowed confirmation that a desired clone had been obtained. Plasmid DNA could be isolated from a bacterial culture, cut with restriction enzymes, and the resulting fragments separated and analysed to see whether the molecule contained the desired insert.

This method of carrying out the separation of DNA fragments by gel electrophoresis is today such a routine task for legions of technicians, PhD students, and post-docs that most of them could probably do it in their sleep. But for Boyer, the moment he first looked at a gel under UV light and saw the glowing bands of DNA fragments was one he would never forget:

> The [DNA] bands were lined up [on the gel] and you could just look at them and you knew ... I was just ecstatic ... I remember going home and showing a photograph [of the gel] to my wife ... You know, I looked at that thing until early in the morning... I knew that you could do just about anything ... I was really moved by it. I had tears welling up in my eyes because it was sort of a cloudy vision of what was to come'.[28]

Boyer simply could not contain his excitement. In June 1973 he attended the Gordon conference in New Hampshire. At this point, he and Cohen were in the process of writing up their work for submission to a journal and had made an agreement that they would not talk about their work in public until it was published.[29] But knowing that one of the rules of the Gordon conference was that all discussions should be treated in confidence, Boyer figured that he was on safe ground. Moreover, he was a firm believer in the free flow of scientific ideas and he was buzzing with excitement.

There was actually very little immediate discussion following Boyer's talk beyond some technical details. But later that morning, another speaker, William Sugden, made a casual quip in reference to Boyer's earlier presentation: 'Well, now we can put together any DNA we want to'. It was then that the real implications of Boyer and Cohen's work began to dawn.[30]

Maxine Singer, a scientist at the National Institute for Allergy and Infectious Diseases in Bethesda, Maryland, was one of the first people to recognize that Boyer's and Cohen's work might have monumental consequences. When she chaired the final session the following day, she acknowledged that 'We all share the excitement and enthusiasm of yesterday morning's speaker who pointed out that the scientific developments reported then would permit interesting experiments involving the linking together of a variety of DNA molecules',[31] but went on to express concern that 'such experiments raise moral and ethical issues because of the potential hazards such molecules may engender'.

Her concerns were echoed by Edward Ziff, an American scientist who was a visiting scholar at the Medical Research Council Lab of Molecular Biology in Cambridge, UK. Ziff had been among those who had helped to dissuade Berg from pressing ahead with his plan to introduce a hybrid SV40 carrying bacterial DNA into *E. coli*. Reporting on the Gordon conference for the magazine *New Scientist*, Ziff conceded that 'Many sorts of DNA hybrids are likely to prove entirely innocuous', but expressed concern that Boyer's and Cohen's methods might allow 'the linkage of oncogenic DNA to DNA which could permit its replication in a bacterium associated with man'.[32]

In response to these concerns, Singer, together with Professor Dieter Soll of Yale University (with whom she was co-chairing the final session of the conference), proposed to the delegates that they vote on whether a letter should be sent to the Presidents of both the National Academy of Sciences (NAS) and the National Institute for Health (NIH) to express their concerns about this work. The motion was carried and in September 1973, their letter appeared in the journal *Science*. While it made clear that 'These experiments offer exciting and interesting potential both for advancing knowledge of fundamental biological processes and for alleviation of human health problems', the letter went on to warn that certain hybrid DNA molecules might also be dangerous and cautioned that 'Although no hazard has yet been established, prudence suggests that the potential hazard be seriously considered'.[33]

The letter concluded with the call for the establishment of a committee to 'consider this problem and to recommend specific actions or guidelines, should that seem appropriate'. But while Singer and Soll were voicing their concerns in *Science*, Boyer and Cohen had already completed the next stage of their work. Cloning a fragment of bacterial DNA in a plasmid was one thing—but could the same feat be achieved with a fragment of DNA from a more complex—eukaryotic—organism? Boyer had considered this question while at the Gordon conference and, during one of the technical discussions following his paper, he was made an offer that he simply could not refuse.[34]

John Morrow, one of Berg's graduate students, told Boyer that he had a sample of DNA from the South African toad *Xenopus laevis* that had been cut with the enzyme EcoRI. Morrow suggested that it should therefore be possible to slot the *Xenopus* DNA into plasmid DNA that had been cut with EcoRI. It certainly seemed worth a try and Boyer and Cohen began the work in July.

Once again, they were successful—cloning was now not limited only to bacterial DNA but could also be done with the DNA from complex, multicellular organisms, too.[35] But their experiment showed something else as well. Not only was the *Xenopus* DNA replicated in the bacterial plasmid, but also it was also still functional—it was being transcribed into RNA: the DNA was being *actively expressed*.

In May 1974, Boyer and Cohen published their work in the journal *Proceedings of the National Academy of Sciences of the United States of America*, but news of their success did not remain confined to scientific circles for long. On May 20, *The New York Times* reported on Boyer's and Cohen's work with the headline 'Gene Transplants Seen Helping Farmers and Doctors'.[36] The article went on to speculate about how recombinant DNA technology might solve major problems that faced the world, such as that of feeding a growing global population. One suggestion was that this might be done by genetically altering plants so that they could use atmospheric nitrogen. As well as being one of the main gases in the Earth's atmosphere, nitrogen is also an essential nutrient for plant growth. But as cereal crops lack the ability to absorb and fix atmospheric nitrogen, they must instead be fed with large amounts of nitrogen-based fertilizer. At the time this was proving to be increasingly costly since the fertilizers were derived from petroleum and, due to political events in the Middle East, the price of crude oil was rising sharply. What if, however, the genes from certain bacteria that were known to be able to fix atmospheric nitrogen could be inserted into plasmids and transferred into cereal crops? Aside from the economic benefits, the increased efficiency of nitrogen fixation would allow a massive increase in cereal crop production.

Buried within the article was another suggested use for recombinant DNA technology. The proposal was confined to a single sentence, but it was one laden with possibility:

> The same ability to create what amounts to a new species of bacteria, according to Dr. Cohen, could lead to colonies of Escherichia coli, equipped with the gene-carrying plasmids, growing large supplies of insulin for diabetics, who now depend on supplies obtained from beef and pork pancreases.[37]

The significance of this sentence and what it might mean for the potential of the new technology was not lost on Neils Reimer of Stanford's Office of Technology and Licensing (OTL). Along with most other biological scientists there, Cohen was probably unaware that the OTL even existed and had little reason to concern themselves with the likes of Reimers. This changed when Reimers called Cohen to invite him for a chat about the possibility of filing a patent on the methods that he and Boyer had used.

This was not something about which Cohen felt particularly comfortable. His cloning work with Boyer had been done using a plasmid that Cohen himself had constructed—and named pSC101 (Stanley Cohen 101). When the work with Boyer was published, this was the only plasmid capable of being used for cloning and so Cohen now found himself inundated with requests from other laboratories asking whether he might send them a sample for use in their own experiments. Observing the etiquette in science that, once a paper has been published, the authors are expected to make any reagents and samples that they have used freely available to other researchers, Cohen duly granted these requests—albeit on certain conditions. For along with each sample of pSC101, Cohen would also send a letter stating that the plasmid was being sent on the understanding that it would not be used to introduce novel antibiotic resistance genes into bacteria.

Like many other scientists, Cohen felt that knowledge should be shared openly. He was concerned that if a patent was filed on his work, then he would no longer be legally able to share pSC101 with fellow researchers. Reimers took a different view entirely. He believed that Boyer and Cohen had just developed what might well become the basic tool of an entire industry and, far from wanting to restrict access by other researchers to this tool, he wanted to put in place a legal framework to enable this. His view was that once Boyer's and Cohen's method was protected by a patent, it could be used under license by commercial companies in return for a royalty fee, which would generate funds for Stanford and the University of

California.[38] In an interview given in 1997, Reimers recalled how he persuaded Cohen to change his mind:

> Through a patent we might be able to get an exclusive license to a company to develop a recombinant insulin and so on. You can't get drugs developed today without some kind of proprietary protection, because they require an investment of a couple of hundred million dollars for R & D. So through the mechanism of the license we would be able to do that sort of thing.[39]

Once Cohen had been persuaded, he made a decision that some might see as noble, and others as naive. Although Stanford had a policy of awarding one-third of royalties on a patent to an inventor, Cohen told Reimers that he would renounce his share.[40] His motivation for this was to provide Stanford with an income, and it would also prove to be a politically expedient move. As Cohen became a prominent voice defending recombinant DNA research against its opponents in the years that followed, his decision to renounce his share of the royalties made him immune to any charge that his support was based on personal gain.[41]

But time was against them. Under US law, the application for a patent must be made within one year of the first public disclosure of the invention, and since Cohen's and Boyer's paper on the cloning method had been published in November 1973, they had to act fast. On November 4, 1974, with only a week to go before the deadline, Reimers submitted the application by Stanford and the University of California to file a patent called 'Process and Composition for Biologically Functional Molecular Chimeras'.[42]

The filing of the patent application was a watershed moment for the life sciences. While this had long been a practice in the physical sciences, the idea of patenting techniques in the life sciences was novel and it took academic scientists into uncharted waters. There were concerns that if this practice became commonplace, it would taint the purity of academic research and restrict researchers' freedom. Research would be driven not by intellectual curiosity, but by the market. Moreover, not only it would threaten the cherished ideal that science was about the sharing of ideas, but it also raised troubling questions. For example, was it acceptable to file patents on publicly funded work with a view to procuring commercial gain? And how broad should such patents be allowed to be? The opening sentence of the Cohen–Boyer patent stated that a 'method and compositions are provided for replication and expression of exogenous genes in microorganisms'. Writing many years later, Berg and Mertz offered a damning verdict on the Cohen–Boyer patent, describing the breadth of its claims as being 'dubious, presumptuous, and hubristic'.[43]

In addition to concerns about the ethics of ownership, the Cohen–Boyer patent faced far bigger challenges. The original article in *The New York Times* reporting on Boyer's and Cohen's work had given the impression that recombinant DNA technology offered countless benefits. Soon, the media began paying much less attention to these potential benefits of this new technology, and focused instead on the dangers it might pose.

Cohen felt that scientists—himself included—were, to some extent, to blame for this change. In response to the concerns about recombinant DNA research raised by Singer and Soll after Boyer's presentation at the Gordon Conference, the NAS had formed a committee to explore how this field of research should be regulated.

The committee was chaired by Berg and in July 1974, published its conclusions in a letter to *Science* with the title 'Potential Biohazards of Recombinant DNA Molecules'.[44] Signed by ten prominent pioneers in this field, including Boyer and Cohen, the letter gave a summary of the technical developments that allowed the creation of hybrid DNA and went on to say that:

> Although such experiments are likely to facilitate the solution of important the-
> oretical and practical biological problems, they would also result in the creation
> of novel types of infectious DNA elements whose biological properties cannot
> be completely predicted in advance. There is serious concern that some of these
> artificial recombinant DNA molecules could prove biologically hazardous.[45]

The authors and signatories of the 'Berg letter' recognized that, because this research 'creates new recombinant DNA molecules whose biological properties cannot be predicted with certainty, such experiments should not be undertaken lightly'.[46] But it identified three particular types of experiment that gave special cause for concern. The first of these, designated 'type 1', involved the introduction of DNA sequences conferring antibiotic resistance or the capacity to make toxins into bacterial strains that did not normally have these abilities. 'Type 2' experiments involved the linkage of genes from oncogenic or other animal viruses to plasmid DNA. The third category, known as 'shotgun' cloning, was one in which fragments of DNA spanning the entire genome of an animal were inserted into a plasmid and introduced into a bacterium.

Berg and his co-signatories called upon the Director of the National Institutes of Health (NIH) to establish an advisory committee that would evaluate the potential biological and ecological risks of this work. Furthermore, this committee would develop procedures to prevent the spread of recombinant molecules and establish a set of guidelines to be followed by all researchers working in this field.

Until such a set of guidelines were in place, the co-signatories of the Berg letter called for all scientists to observe a voluntary moratorium on any further work involving recombinant DNA.[47] During this period, scientists were asked not to initiate experiments of type 1 or 2, and to refrain from cloning animal genes into bacteria, 'until attempts have been made to evaluate the hazards and some resolution of the outstanding questions has been achieved'.[48]

In the wake of the publication of the Berg letter, the media were now alerted to the possibility that recombinant DNA might bring dangers as well as benefits.

The New York Times told its readers that 'Possible Danger Halts Gene Tests', while *The Washington Post* raised a very important—and awkward question:

> Ever since Hiroshima, scientists have been concerned that probing the secrets of nature without caution and moral restraint might open a Pandora's box of ills, if not disasters, that could do terrible damage. But who is to be the keeper of the keys?[49]

In anticipation of this very question, the Berg letter had made one final proposal. This was to hold an international meeting the following year at which scientists involved in recombinant DNA research could debate its future and how, if at all, it should be regulated. In an interview many years later, Berg described how he felt at the time:

> All of us were of a like liberal mind, and we felt that ethics and responsibility in science were important. The nuclear weapons program was exploding. All of us felt an obligation not to do something that was involved with germ warfare and things of that sort. So there was a sense of wanting to do the right thing. And not doing the wrong thing because you were selfish, because it was your experiment, your idea, and you were going to pursue it hellbent; no matter what anybody said, you were going to do your experiment. It was much more a matter of a social thing: We ought to talk this out; we ought to think this through.[50]

Acting on Berg's imperative, around 150 scientists from all around the world gathered on February 24, 1975 at the Asilomar Conference Centre, a converted former chapel in Pacific Grove on the Monterey Peninsula, California. For the next three days they reflected, discussed, and argued about how they should 'do the right thing'—all under the watchful eye of the press. Like the bacteria that they studied down their microscopes, the scientists now found themselves the subject of intense scrutiny and not everyone was happy about it. Scientists were not used to this level of attention from the press—many of them were uncomfortable with it. The journalist Michael Rogers, writing an article on the Asilomar meeting for *Rolling Stone* magazine, reported that when one eminent young scientist (most likely Stanley Cohen) was approached by a press photographer, 'he retreats, face covered, like a newly busted bigtime mobster hiding behind his fedora on the steps of a precinct house'.[51]

The stated aim of the Asilomar meeting was to 'review scientific progress in research on recombinant DNA molecules and to discuss appropriate ways to deal with the potential biohazards of this work'. [52] David Baltimore, one of the signatories on Berg's letter, stated: 'If we come out of here split and unhappy then we will have failed in the mission before us'.[53] This, however, was going to prove far from easy. As one young molecular biologist lamented—'Here we are sitting in a chapel, next to the ocean, huddled around a forbidden tree, trying to create some new Commandments—and there's no goddamn Moses in sight'.[54] Opinions were sharply divided. At one end of the spectrum was a handful of scientists such as James Watson, for whom talk of hazards was overblown and calls for regulation were simply a hindrance to progress. Occupying the centre ground were figures like Berg, who were tearing their hair out trying to solve the unenviable task of assessing the risks and then implementing practical measures to control them. Finally, in opposition to Watson were the likes of Chargaff, whose analysis of the chemical composition of DNA with partition chromatography had led to the discovery of the base-pairing rules by which fragments of DNA from different organisms could now be joined together. But although his work had laid the foundations for this new method, Chargaff was deeply suspicious of this new technology and its consequences:

> You can stop splitting the atom; you can stop visiting the moon; you can stop using aerosols; you may even decide not to kill entire populations by the use of a few bombs. But you cannot recall a new form of life. Once you have constructed a viable *E. coli* cell carrying a plasmid DNA into which a piece of eukaryotic DNA has been spliced, it will survive you and your children and your children's children. An irreversible attack on the biosphere is something so unheard of, so unthinkable to previous generations, that I could only wish mine had not been guilty of it.[55]

The discussions were charged with emotion and tempers flared. As Berg tried to keep the discussion focused on the question of how the risks involved in this work might be quantified, Watson found himself becoming increasingly frustrated. How, he demanded to know, could anyone consider subjecting this work to restrictive regulations (Figure 50a, b) when, as he colourfully put it, 'We can't even measure the fucking risks!'[56]

Watson's scepticism about how this work could be regulated was shared by Joshua Lederberg, another Nobel Laureate, who was already aware that a spectre was rising over the entire proceedings. Lederberg feared that any guidelines and regulations drawn up by the scientists gathered at Asilomar might very quickly be in danger of 'crystallizing into legislation'.[57] In a statement prepared to be given at the conference, Lederberg raised an issue that, at the start of the 21st century has become all too relevant:

> Research on Recombinant DNA is, in my opinion, the CENTRAL WAY in which molecular genetics can contribute to the solution of important medical problems … In assessing the need to continue vigorous research on the molecular biology of viruses, for which DNA recombination is an invaluable tool, I believe that most people are OVEROPTIMISTIC with respect to the means we have available to forfend global epidemics … We have no guarantee that the natural evolutionary competition of viruses with the human species will always find ourselves the winner.[58]

After an exhausting and sometimes acrimonious three days, a consensus was reached. In article written to *Science*, Berg and his fellow organizers of the meeting

(a)   (b)

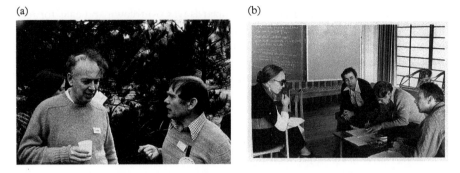

Figure 50 *The Asilomar Meeting a) James Watson (left) and Sydney Brenner (right) enjoy a breather from often heated discussions at the Asilomar conference in 1975 about the future of recombinant DNA research. b) (Left to right) Maxine Singer, Norton Zinder, Sydney Brenner and Paul Berg try to work out a set of guidelines at the Asilomar conference for how this research should be regulated.*

Credit: © *Maxine Singer Papers. US National Library of Medicine. Courtesy of US National Library of Medicine.*

summed up their achievements. The voluntary moratorium on recombinant DNA research for which they had called eight months earlier would be suspended and workers could proceed providing that they adhered to appropriate safeguards.[59] Experiments were to be divided into four categories—minimal, low, moderate, and high—based on the perceived level of risk involved and physical containment measures were specified that reflected each increasing level of risk. Experiments deemed to be of minimal risk could be carried out at the laboratory bench simply by following standard protocols already employed for good microbiological practice such as no drinking, eating, or smoking in the laboratory and the wearing of lab coats. Experiments deemed to be of high risk, meanwhile, could only be performed in laboratories equipped with air locks, showers, full body protective clothing, and negative pressure to prevent the escape of aerosols.

One measure glaringly obvious to those with a background in microbiology was to avoid pipetting by mouth so as to prevent the possible ingestion of pathogens. Yet, much of the new research in molecular biology was being done by biochemists who were far less experienced at handling cultures of bacteria and for whom pipetting of reagents such as enzymes by mouth was routine practice. Berg's report on the meeting reminded all principal investigators that they had a responsibility to ensure that all their staff were informed about potential biohazards and were given adequate training in containment measures and emergency procedures.

In addition to physical containment measures, there was another means by which any potential hazard from recombinant DNA work might be confined to the laboratory. With his 'bushy eyebrows, gleaming eyes and nonstop animation that blend to an impression midway between leprechaun and gnome', Sydney Brenner, who had worked with Francis Crick on deciphering the genetic code, was described by Rolling Stone magazine as 'the most forceful presence at Asilomar'.[60] Chairing a session, Brenner warned the delegates what was at stake—'In some countries, this would be done by the government, and once guidelines were set and you broke them there would be no question of peer censure—the police would simply come out and arrest you'.[61]

But for Brenner, the Asilomar meeting was a chance for scientists to win the trust of the wider public by showing that they were willing to take on the responsibility for regulating their work—'to reject the attitude that we'll go along and pretend there's no biohazard and hope we can arrive at a compromise that won't affect my own grants and be appointed to the National Academy and all the other things that scientists seem to be interested in'.[62] It was imperative to find ways of minimizing the risk from recombinant DNA experiments and the session chaired by Brenner was devoted to a novel and potentially effective means by which this might be achieved. This involved biological containment—the use of bacterial hosts that were disabled in such a way that they could not survive and propagate outside the laboratory.

In addition to specifying physical containment measures, the guidelines that emerged from the Asilomar meeting also adopted this idea of only using disabled bacterial hosts for work on recombinant DNA, but the authors of this summary report were modest and sober about what they had achieved:

> Research in this area will develop very quickly and the methods will be applied to many different biological problems. At any given time it is impossible to foresee the entire range of all potential experiments and make judgements on them. Therefore it is essential to undertake a continuing re-assessment of the problems in the light of new scientific knowledge.[63]

Even before the Asilomar meeting, the National Institutes for Health (NIH) formed the Recombinant DNA Advisory Committee, which began the task of taking the very general points that emerged from the Asilomar meeting and casting them into a set of concrete guidelines. The task was a formidable. On the one hand, the NIH did not want to make the guidelines so restrictive that scientists, eager to make progress in this field, were tempted to carry out their experiments in secret. At the same time, they needed to produce something that showed the general public and media that they were taking their concerns seriously.

After a meeting at Woods Hole in July 1975, a first draft of guidelines was proposed. Building on the discussions at Asilomar, these classified four different levels of physical containment (P1 to P4), with P4 facilities being the kind used for handling of the most dangerous types of pathogens and requiring airlocks, full protective clothing, labs under negative pressure, and showering on exit. The report also specified three levels of biological containment—EK1–EK3—each of which was based on use of a strain of *E. coli* called K12, which most microbiologists believed was unable to colonize the human bowel. EK1 containment used the standard strain of K12, while EK2 used a strain that had been genetically altered so that it had only a one in 100 million chance of surviving outside the laboratory. EK3 experiments also used this strain but with the added measure of having demonstrated that the bacterium could not survive in plants, animals, and other external environments.

This early set of guidelines was not met with enthusiasm by many of the scientists who had attended the Asilomar meeting.[64] The consensus was that they were too lax and did not 'adequately reflect the tone of caution implicit in the Asilomar conference's recommendations'.[65] One of the most serious criticisms was the recommendation that cloning of mammalian DNA could be carried out under P3 levels of containment, when many of the researchers felt that this work should only done under the highest level. In response to these criticisms, the NIH went back to the drawing board while researchers waited in frustration for the outcome.[66] This brought its own pressure, for as one researcher pointed out—'If you keep everyone waiting, there is going to be stuff done on a Saturday night'.[67]

By end of the year a new—third—set of guidelines had emerged.[68] When these were finally made public on June 23, 1976, they were considerably more strict than those drawn up at the Asilomar meeting.[69] Nevertheless, many of those who had attended the meeting in California felt that they had won a moral victory. In only four days, said one commentator, the scientists at Asilomar had made 'a unique moral decision for the future of science' while another lauded the scientists as having 'performed a service of practical and symbolic importance to themselves, to science, and to society'.[70,71] Rather than simply forging ahead with their research, the scientists had demonstrated to the world that they were willing to listen to public concerns and take responsibility for the social consequences of their research. However, this was only the beginning of the battle.

Unfortunately, as soon as the guidelines published, they came under fire—from both sides of the debate. For Watson they were too restrictive and based on what he dismissed as 'an imaginary monster'.[72] He was candid in his view that everyday household chemicals in modern life presented other, far greater, risks than recombinant DNA technology: 'I don't like PCBs [polychlorinated biphenols] ... I don't like hair dyes—they're a real story that would scare the shit out of you ... But all that the civilised English majors want to talk about is recombinant DNA'.[73]

But Watson was wrong that concerns over recombinant DNA were confined only to arts majors who might be ignorant of science. Robert Sinsheimer, chair of the biology division at Caltech (and who would later go on to play a crucial role in initiating the Human Genome Project), argued that the main danger from recombinant DNA work was not just the possibility of inserting oncogenic DNA sequences into common pathogens, but simply that the very act of splicing together eukaryotic and prokaryotic DNA violated a barrier that had remained unbroken over the course of millions of years of evolution. Sinsheimer was concerned that we simply could not anticipate what the consequences of tearing down this barrier might be and warned that 'if the guidelines are adopted and nothing untoward happens, we will owe this success far more to good fortune than to human wisdom'.[74] Speaking before a Senate sub-committee on health, Sinsheimer told Senator Edward Kennedy that, in his view, recombinant DNA technology was 'an accomplishment as significant as the splitting of the atom'.[75] Sinsheimer accepted that imposing restrictions on any field of research 'is especially bitter for the scientist whose life is one of inquiry' but, in the case of recombinant DNA, he was happy to make an exception: 'Science has become too potent ... It is no longer enough to wave the flag of Galileo'.[76]

Writing in *Science*, the Rockefeller cell biologist Phillip Siekevitz echoed Sinsheimer's concerns and articulated the key question at the heart of the debate: 'Should the acquisition of new knowledge, for whatever purpose be the ultimate and sole criterion in the pursuits of biological research?'[77] For Siekevitz, the answer was no'. He argued that scientific research must be 'modified by ethical and social consideration of the society in which we live' and warned his fellow biologists that they should tread very carefully, for 'Like the physicists before us, we have entered the realm of the Faustian bargain'.[78]

Meanwhile, the environmental organization Friends of the Earth dismissed the NIH guidelines as little more than being 'window dressing' that gave researchers the *appearance* of being concerned while at the same time allowing them to forge ahead 'with as little impediment as possible'.[79] Chargaff also shared this concern, and expressed these views with his characteristic poetic flair. One of Chargaff's immediate fears was that *E. coli* was a common micro-organism known to colonize the human gut. He therefore proposed 'a complete prohibition of the use of bacterial hosts that are indigenous to man' and, furthermore, that all such work should only be carried out in one single location, for example, the Department of Defence facility at Fort Detrick in Maryland. But even such extreme measures would still not be sufficient to sooth Chargaff's deep philosophical reservations about recombinant DNA:

> Are we wise in getting ready to mix up what nature has kept apart, namely the genomes of prokaryotic and eukaryotic cells? ... Have we the right to counteract, irreversibly, the evolutionary wisdom of millions of years, in order to satisfy the ambition and the curiosity of a few scientists? The world is given to us on loan. We come and we go: and after a time we leave earth and air and water to others who come after us. My generation, or perhaps the one preceding mine, has been the first to engage, under the leadership of the exact sciences, in a destructive colonial warfare against nature. The future will curse us for it.[80]

Unsurprisingly, Chargaff's damning verdict on recombinant DNA did not go unchallenged by its supporters. In a letter to *Science*, Berg and Singer said they felt:

> deeply disturbed by the distortions, derision, and pessimism that permeate Chargaff's comments. He appears to see science as a curse on our time, and men as feeble. In our view it is knowledge and understanding derived from science and scholarship that lead men to rationality and wisdom.[81]

Cohen took particular exception to Chargaff's claim that research in this area was a violation of 'evolutionary wisdom'. If research into recombinant DNA was indeed an act of 'colonial warfare against nature', then it was an entirely just war. For as Cohen pointed out, evolution was neither as wise nor benevolent as Chargaff claimed:

> It is this so-called evolutionary wisdom that gave us the gene combinations for bubonic plague, smallpox, yellow fever, typhoid, polio, diabetes, and cancer. It is this wisdom that continues to give us uncontrollable diseases such as Lassa fever, Marburg virus, and very recently the Marburg-related hemorrhagic fever virus, which has resulted in nearly 100 percent mortality in infected individuals in Zaire and the Sudan.[82]

As far as Cohen was concerned, if medical research could find cures for such diseases, then he was quite happy to wage 'an intentional and continuing assault on evolutionary wisdom'.[83] Scientists spectating from other disciplines, such as the eminent physicist Freeman Dyson, lent their support to this view:

> I do not deny or belittle the dangers. I say only, let us not leave the starving millions of humanity out of account when we balance the dangers against the benefits. It is perhaps not irresponsible, but rather an act of enlightened courage, to expose ourselves to an unknown risk of disastrous epidemics in order to give ourselves a change of lifting some hundreds of millions of our fellow humans out of the degradation of poverty.[84]

In a private letter to Berg written after this article had appeared, Dyson urged him to stress the benefits that recombinant DNA technology might bring:

> No professional biologist can allow himself to say, 'I work for the good of humanity'. Such a claim smells too strongly of hypocrisy and self-serving. It is much more comfortable, and sounds more honest, to say 'I do my work because it is fun'. I noticed when I was at the Salk Institute recently that even people working on tumor viruses are quick to say 'Of course, we are not primarily interested in curing cancer, only in understanding it', as if the desire to cure cancer would somehow compromise their scientific respectability. So I think it is this reluctance to pose as benefactors of humanity that has caused you and your colleagues to omit from your statements any discussion of the human and social costs of delaying your work. Nevertheless, however unwilling you are to say so in public, it may turn out to be true that the fate of the starving millions of India may hinge upon the timely availability of some biological technique from your experiments … When I gave my talk in Madrid last year, I was forcefully reminded that poverty is a more serious threat to human health even than recombinant DNA.[85]

To give a perspective on the risks involved with recombinant DNA, Cohen tried to reframe the debate by pointing out that millions of people were willing to be treated with vaccines for diseases. Could it be guaranteed that these vaccines might not cause some, as yet unrealized, future health problem? No, said Cohen. And yet, the proven benefit of receiving such a vaccine in protecting against disease outweighed any such speculative and as yet unknown risk:

> The statement that potential hazards could result from certain experiments involving recombinant DNA techniques is akin to the statement that a vaccine injected today into millions of people could lead to infectious cancer in 20 years … We have no reason to expect that any of these things will happen, but we are unable to say for certain that they will not happen … Can we in fact point to one major area of human activity where one can say for certain that there is zero risk? … we must distinguish fear of the unknown from fear that has some basis in fact; this appears to be the crux of the controversy surrounding recombinant DNA.[86]

It is somewhat ironic that Cohen's point resonates in our own time with ongoing debates about the risk/benefit ratio of vaccines against Covid-19 that are themselves the products of recombinant DNA technology. Meanwhile, Chargaff's claim that the construction of recombinant DNA violated a rigid evolutionary boundary between the genetic material of bacteria and higher organisms also came under fire. Harvard Medical School's Bernard Davis argued that, having already colonized the mammalian gut for the several million years, *E. coli* would already have had prolonged exposure to fragments of foreign DNA present in its host.[87] There was, therefore, a high probability that, during this time, some of this host DNA had already become integrated into the genome of *E. coli* giving rise to naturally occurring new combinations of hybrid mammalian and bacterial genetic material. If this were the case, and new combinations of hybrid DNA could indeed arise purely by natural processes—and not just as a result of artificial manipulation in the lab—then this would be a powerful counter-argument against those such as Sinsheimer and Chargaff, who claimed that the new technology was a breach of some inviolable evolutionary barrier. The challenge was therefore to find some empirical evidence that this was indeed the case, and it was to this task that Cohen turned his efforts.

However, the greatest challenge came from outside the scientific community. In a statement given to the Senate Subcommittee on Health and Scientific Research, NIH Director Donald Frederickson said that the advent of recombinant DNA technology marked a watershed moment in the relationship between science and wider society:

> Biomedical research is entering a new era in its relationship to society. It is passing from an extended period of relative privacy and autonomy to an engagement with new ethical, legal and social imperatives under concerned public scrutiny.[88]

And there was no greater champion of public scrutiny for this new technology than Alfred Vellucci, Mayor of Cambridge City, Massachusetts. When Vellucci summoned his fellow local politicians to gather for a city council meeting on June

23, 1976—the very same day that the NIH published their guidelines—it was not to discuss usual matters like tax, rubbish disposal, and rent control. Vellucci was outraged to have discovered that Harvard University had submitted plans to construct P3 containment facilities for work on recombinant DNA without first informing the local council. Vellucci was determined to haul the scientists responsible over the coals—and he intended to do it in full public view.

'The Cambridge Incident' began when Mark Ptashne, a molecular biologist at Harvard, applied for $500,000 in federal funding from the National Cancer Institute to refurbish three laboratories on the fourth floor of the Bio Labs building. One of these was to have P3 containment facilities and, as required by law, the University of Harvard had published its intention in a legal notice in *The Boston Globe* and other local newspapers, along with an invitation for members of the public to attend a meeting on February 26, where the plan would be discussed. However, it seemed that even if any members of the public had actually read the notice, they had no concerns about the plans. As Ptashne later recalled, although he and a few University administrators were present to answer questions, no one came to the meeting.[89]

Lack of people at the meeting, however, did not mean the plan was unopposed. Senior members within Ptashne's own faculty were very uneasy about the plan. Among these, two of the most vocal were Ruth Hubbard and her husband, Nobel Laureate George Wald. Hubbard was not reassured by the arguments of advocates like Dyson and Cohen that recombinant DNA research was justified on the grounds that it might offer new cures for diseases. She argued that the significant reduction in child mortality due to infectious diseases had been achieved, not by novel scientific research, but rather by adopting social and public health measures such as improved nutrition, better housing, and clean water. For her, the biggest problem was that the risks were simply unknown. Citing the recent outbreak of a mysterious disease that had killed twenty-nine attendees at a conference of the American Legion in Philadelphia (later classified as Legionnaire's Disease) she pleaded 'that it would be good to face up to ignorance in areas in which we have no experience, instead of engaging in facile speculation'.[90]

Hubbard had grave reservations about conducting recombinant DNA experiments in a building where, earlier, it had proved impossible to control an infestation of ants. Her biggest concern was that, largely thanks to Watson's account in *The Double Helix*, molecular biology had become hypercompetitive to the point of becoming downright vicious. Hubbard feared that in this cut-throat, 'devil take the hindmost' culture, concerns about safety would hardly be top of the priority list. The resulting debate polarized the department into two opposing camps, with junior members choosing to remain silent rather than jeopardize their careers by taking sides. Eventually, Hubbard made a decision that she later described as 'the sin'.[91] Picking up the telephone early one morning around mid-May, she called Barbara Ackermann, a former mayor of Cambridge who now sat on the city council. Hubbard knew Ackermann from protesting against the Vietnam War, and now they came together to protest a new cause.

Hubbard explained her concerns and invited Ackermann to attend a meeting called by Dean Henry Rosovsky at Harvard on May 28 to engage in a campus-wide discussion of the proposed work, but if Rosovsky hoped that this might enable Harvard to resolve this problem internally, he was to be gravely disappointed. At the end of the meeting, Ackermann stood up and announced that the city council was keeping a careful eye on developments, and at the same time, a reporter from the *Boston Phoenix* newspaper was sitting in the audience quietly scribbling notes.

Far from bringing the issue of the P3 lab to a quiet conclusion, the meeting had quite the opposite effect. When Mayor Al Vellucci picked up the June 8 edition of the *Boston Phoenix* and read a detailed report on the proposed new lab at Harvard, he was livid. He was particularly incensed that 'no representative of the public has participated in a decision which some Harvard officials concede may affect the health and safety of the general public'.[92] Until now he had known nothing about the planned research, and he was furious that Harvard had intended to go ahead without first bringing the matter to the attention of the council. When Rosovsky announced on June 13 that construction of the P3 lab would go ahead, the public reassurances that he offered in *The New York Times* that the strictest safety measures would be followed did nothing to assuage Vellucci's fear—nor his rage. To find out more, he met with Hubbard and her husband before announcing in his own colourful way that the city council would hold a public hearing at which the scientists involved in this work could explain what they were up to—in full view of Vellucci's constituents. He raged:

It's about time the scientists began to throw their goddamn shit right out on the table so that we can discuss it ... Who the hell do the scientists think that they are that they can take federal tax dollars that are coming out of our tax returns and do research work that we cannot then come in and question?[93]

Vellucci took no half-measures in ensuring that the public were given full involvement in the meeting. The proceedings were opened by a high school choir that sang 'This Land is Your Land', and the TV crews filming the event had camera-ready spectacles of protestors gathered, with banners that gave a nod to 1776 showing the slogan, 'No Cloning without Representation'. It was, according to one onlooker, nothing less than a 'a circus'.[94] As far as Vellucci was concerned, he was defending his mostly blue-collar electorate against a bunch of arrogant intellectuals—a task he performed with the utmost zeal and pleasure. Graham Chedd, who was covering the hearings for *New Scientist* magazine, observed:

Vellucci was in his element ... the scientists appearing before him were not. The consequences were sometimes farcical, sometimes painful and on occasion—such as when Vellucci demanded that everyone present in favour of the new laboratory stand up and identify themselves—positively frightening.[95]

The Cambridge hearing was the first major confrontation between scientists involved in recombinant DNA work and the general public and it did not go particularly well for the former. Ptashne came under particularly heavy fire. His visit to North Vietnam in 1970–1971 had left him impressed by the claims of the socialist government there to support science. Indeed, most scientists at that time considered themselves to be on the political left—they believed in social progress and that this could be driven by government intervention—particularly through harnessing science as an engine of change. Sadly, they had not reckoned with politicians like Vellucci.

In his book 'Invisible Frontiers', Stephen Hall gives a vivid account of the exchange that followed. Vellucci's questions to Ptashne were structured in such a way as to manoeuvre his opponent into an impossible situation. Why, demanded Vellucci, was it even necessary to construct a special P3 facility in the first place if—as the molecular biologists claimed—there was no significant risk with recombinant DNA work? Moreover, could Ptashne give 'an absolute, one hundred per cent guarantee that there is no possible risk which might arise from this experimentation. Is there zero risk of danger?'[96] Would the experiments be safer if they were done in a P4 facility in an isolated, non-populated area of the country, or a P3 lab in a densely populated, urban area? In the whole history of science, have mistakes ever been made? Do scientists ever show lack of judgment?

By forcing Ptashne to answer questions that demanded a simple response of 'yes' or 'no', Vellucci sought to back him into a corner where he would have to admit that there was some level of risk with these experiments. Trying to bring some perspective on the understanding of risk, molecular biologist Walter Gilbert pointed out that 'Fire too, is a vague hazard', but it was Vellucci who emerged as the victor from this battle.[97]

Across the nation, the mood of the public towards science had changed. Perhaps this was part of a larger cultural change akin to the challenge of authority during the anti-Vietnam protests and the growing environmental movement. The growing sense of scepticism towards the idea that scientific discoveries guaranteed material, social, and political progress was also being reflected in popular culture. By the end of the 1960s and the start of the 1970s, the optimistic liberal outlook of the TV sci-fi show *Star Trek*, with its faith that, thanks to science, reason, and progress, humanity would ultimately transcend bigotry and tribalism to bring liberal democracy to the stars, were being superseded by far more bleak visions of the future, for example, *The Omega Man*, *Soylent Green*, and *Planet of the Apes*. Ackermann even later acknowledged that her feelings about recombinant DNA might have been coloured to some extent by having watched the sci-fi thriller *The Andromeda Strain* the evening before Ruth Hubbard called her.[98]

When Vellucci invoked the image of Baron Frankenstein, Ptashne could not resist a smile. This was not a wise move. 'This is a deadly matter, sir', said Vellucci in response to Ptashne's apparent flippancy, '... If worse comes to worse, we could have a major disaster on our hands. I guarantee everyone in this room that if that happens no one will be laughing then'.[99]

The entire nation was watching what happened in Cambridge—because what unfolded there potentially could dictate the future course of research in the field of recombinant DNA. Three Nobel laureates—Lederberg, Howard Temin, and Kornberg—all wrote letters to Vellucci to defend research into recombinant DNA. Kornberg pleaded: 'I implore you again not to suppress the serious and responsible search for new knowledge ... If scientific inquiry is stifled in Cambridge, it will be done in Waltham, Palo Alto, or Moscow. In 1976, please do not squander your most precious human resources'.[100]

But despite these petitions, Vellucci would not soften his stance. Scenting victory, Vellucci concluded proceedings with the proposal of a moratorium in which no recombinant DNA work could be done for two years and declared that the hearing would resume on July 7. Of the twenty-two witnesses who testified at this second meeting, one was Gilbert, now so revered as a pioneer in molecular biology that one of his graduate students, Forrest Fuller had nicknamed him 'the Sorcerer' (Figure 51).[101]

Figure 51 *Walter Gilbert. US National Library of Medicine. Courtesy of US National Library of Medicine.*

The nickname would prove to be well deserved—not only for Gilbert's skills at the lab bench, but also for his political intuitions in reading the current situation. For as he gave his own testimony at the hearing, Gilbert conjured up something that he hoped might work a little magic on the sceptical public:

> One of the experiments we're interested in is putting the gene for human insulin into a bacterium ... The purpose of that experiment is to make human insulin, a specific hormone, available ... We are making something which is, in a sense, beyond price ... It is not that we are making cheaper medication. We are making something which we cannot get by any other means.[102]

By proposing that recombinant DNA technology might be used to produce human insulin, Gilbert sought to steer the debate away from sci-fi horror stories and back to potential medical benefits.[103] He was also drawing attention to a growing concern, of which he had first become aware when he was shown a graph during a presentation at a conference sponsored by Eli Lilly that May. It was a simple graph, showing just two lines—one represented the predicted increase in cases of diabetes in the US over the coming years, while the other showed the projected supply of insulin extracted from bovine and porcine pancreatic tissue. Because this material was recovered from abattoirs, insulin was simply a by-product of the meat industry, and its supply was therefore vulnerable to a host of factors, for example, if a spell of poor weather resulted in diminished quality of animal feed.

This made the supply of insulin unreliable and, according to the graph, this was set only to get worse in the near future. For while the line on the graph showing demand continued to rise, the second one, showing supply, looked set to level off by the early 1980s. As Irving Johnson, the vice-president of Lilly explained starkly, the implication was obvious—and serious—if cases of diabetes continued to increase as predicted, then the demand for insulin would soon outstrip the supply from animal pancreas:

> ... we were concerned about being able to produce enough insulin to treat all the diabetics ... Even if we could collect all the pancreases from all the pigs in China, you couldn't produce enough insulin to treat all of the people diabetics [sic] in the world.[104]

Elsewhere, others had also turned their attention to this same problem. In the 1950s, Archer Martin had used his Nobel Prize money to buy an old house in the countryside, where he converted most of the rooms into laboratories, workshops, and an office, relegating his family to the attic—and its dry rot.[105,106] Tired of the shackles of life in industry and academia, Martin had hoped to boost his career working as a freelance technical consultant for clients like Esso. Later, Synge recalled that one of Martin's other interests at the time was 'messing around with pancreases' in the hope of finding more effective ways of extracting insulin.[107] Having first become interested in this problem during his stint working at the Boots laboratories in Nottingham, he was now so isolated and immersed in his own research that he appeared to have no interest in the revolution his earlier work with Synge had set in motion thanks to Chargaff and Sanger, nor awareness of the drama that was unfolding on the other side of the Atlantic. Instead, he became increasingly detached and inward-looking, concerned only with the minutiae of technical details. One of Synge's daughters recalled that, when she visited Martin on one occasion, she found the house strewn with pig guts—from which he was trying to extract insulin.[108]

A few years later, Martin was diagnosed with Alzheimer's disease. When he died in 2002, *The Independent* newspaper hailed Martin's development of partition chromatography as having set in motion 'a revolution over the last 50 years ... a revolution that is still taking place as biochemists learn how to master larger and larger molecules that play key roles in living processes'.[109] And although Martin had been tragically oblivious to this revolution, Eli Lilly were quick to recognize its potential for transforming the production of insulin.

In 1976, sales of insulin accounted for eighty to eighty-five per cent of the US market, totalling $160 million a year, and Eli Lilly was keen to tackle this problem of supply. An alternative needed to be found, and Johnson was confident that the solution lay with molecular biology and recombinant DNA technology:

> Lilly's interest in developing human insulin as a fermentation product, in its simplest terms, was one of supply and demand. The number of insulin-using diabetic patients had been growing rapidly ... However, animal insulins were derived exclusively from the pancreas gland, which were a by-product of the meat industry. The animals were not raised just for their pancreases, and demands for meat, or climatic or economic factors that influenced meat production affected the availability of the glands. One possible approach to addressing the problem of supplying enough insulin in the future was to use the emerging recombinant DNA (rDNA) technology; that is, to manufacture human insulin through fermentation of bacteria expressing DNA that are coded for human insulin.[110]

The problem was where to begin? Where among the three billion base pairs found in the human genome should they start to look for the specific DNA sequence that encodes the insulin protein? The task was not so much akin to looking for a needle in a haystack but rather a needle in a field of haystacks.[111] At that time the complete DNA sequence of the human insulin gene had not yet even been determined. And, when the complete sequence was eventually published in 1980, the sequences of the insulin gene that code for the protein were found to be interrupted by regions known as 'introns' that do not code for any amino acids.[112] This was found to be a common feature of eukaryotic genes and means that when the insulin gene is copied into mRNA, these introns are transcribed along with the coding regions before being snipped out to generate an mRNA that contains only the coding sequence.

But 'the sorcerer' had a solution to this particular problem. In a series of papers published in the journal *Cell*, Argiris Efstratiadis, one of Gilbert's post-doctoral workers, had described how he and a team of colleagues had successfully isolated the specific rabbit mRNA that carried the information needed to synthesize globin, the protein component of haemoglobin.[113] The next stage had been to use a neat biochemical trick carried out by a recently discovered enzyme called reverse transcriptase. Isolated from a virus, this enzyme has the ability to take a sequence of single-stranded RNA and copy it into double-stranded DNA—a process that the writer Stephen Hall has likened to taking a photographic print into the darkroom and, through a series of manipulations, emerge with the original negative.[114]

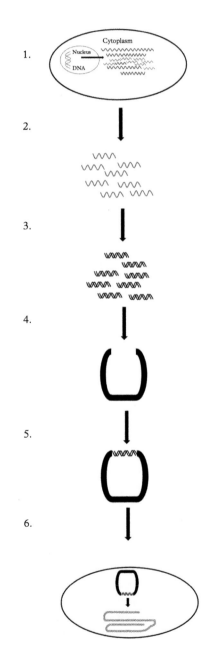

Gilbert reasoned that it should be possible to do exactly the same for human insulin. By taking the specific mRNA for human insulin out of cellular extracts and then incubating this with reverse transcriptase, a DNA copy of the coding sequence could be synthesized. Using restriction enzymes this could then be slotted into a bacterial plasmid where (hopefully) it would make the active insulin protein (Figure 52).

The first major technical hurdle was how to obtain the mRNA for insulin in sufficient quantities. The beta cells that produce insulin make up only about one per cent of the total pancreatic tissue and mRNA is both short lived and synthesized in only tiny amounts. This meant that vast quantities of pancreatic material and an awful lot of onerous preparative work would be required to obtain sufficient mRNA. To add to these difficulties, RNA is notoriously unstable and much more susceptible to degradation than its molecular cousin. The slightest contamination by RNA degrading enzymes from the surface of workers' fingers, for example, could easily ruin a sample and result in the waste of an awful lot of work.

Gilbert was not the only scientist facing these challenges. Howard Goodman, a molecular biologist at UCSF had also been present at the Lilly conference and he was also thinking about insulin. The previous year, Axel Ullrich, a German postdoc had arrived in Goodman's lab with the ambitious plan to isolate the mRNA for human insulin and clone the gene. But shortly after his arrival, Ullrich had learned that he had serious competition. What dismayed him was that this competition came from Bill Rutter, his very own head of department, who was planning to do the very same experiment.

Ullrich's boss Goodman did not always see eye to eye with Rutter, but when news reached the department at UCSF that Gilbert was planning to clone insulin, they put their differences aside. Joining forces in response to the competition from Gilbert, they formed what was described by one member of the department as 'a marriage of distress'.[115]

Figure 52 *Walter Gilbert's strategy to clone and express human insulin in bacteria. (1) An insulinoma cell line produces large amounts of insulin messenger RNA (mRNA) (2) Total mRNA is extracted from the cells and the particular mRNA encoding insulin is isolated and purified (3) Once isolated and purified, insulin mRNA is copied into double-helical DNA (cDNA) by the enzyme reverse transcriptase (4) The insulin cDNA is inserted into a bacterial plasmid which is then introduced into a bacterium (5). The insulin cDNA is expressed by the bacterium into mRNA which is translated into preproinsulin, the single chain precursor of insulin. (A similar approach was adopted by Bill Rutter and Howard Goodman's team at UCSF but unlike Gilbert, the UCSF team did not have the advantage of starting with an insulinoma cell line that produced a large amount of insulin mRNA. Instead, they had to isolate insulin mRNA from rat pancreatic tissue.) Diagram by K.T. Hall based on information found in Villa-Komaroff, L, A Efstratiadis, S Broome, P Lomedico, R Tizard, SP Naber, WL Chick, and W Gilbert. 1978. "A Bacterial Clone Synthesizing Proinsulin."* Proceedings of the National Academy of Sciences *1978 (75): 3727–31.*

With Gilbert having stepped into the ring, Rutter suddenly felt the thrill of the chase, later saying in an interview with historian Sally Smith-Hughes that it 'dialled up my competitive spirit',[116] The race to clone human insulin was on, and the starting pistol was fired during William Chick's presentation at the Eli Lilly conference in Indianapolis. Chick, from the University of Harvard, reported that he had successfully grown tumour cells from a type of cancer that affects the insulin-producing beta cells of the rat pancreas. Known as insulinoma cells, these churned out vast quantities of insulin and were therefore a rich source of the mRNA coding for insulin that both labs so desperately needed.

The first stage in the race was to prove that insulin could be cloned from a simpler organism like a rat and, for both teams, Chick's insulinoma cells were a biochemical holy grail. These cells produced insulin mRNA in large quantities and so would save massive amounts of time and effort in preparing material. As a result, Chick soon found himself being courted by both teams and, at the end of the conference, the UCSF team headed back to California confident that they had persuaded him to provide them with some of his precious cells. Shortly afterwards, however, Chick informed them that unfortunately there was insufficient material available for him to share. It was a frustrating turn of events that Rutter suspected had less to do with low yields during preparation of the cells, and more to do with Gilbert's influence.[117]

Gilbert, meanwhile, had succeeded in obtaining some of Chick's insulinoma cell line and now had a significant head start over his West Coast competitors. However, this didn't mean that when he returned from the Lilly conference, he could start work immediately.

The two-year moratorium proposed by Mayor Al Vellucci on any recombinant DNA work had now been watered down to a matter of months, but it was a ban nonetheless, and it was now in effect.[118] Following his public hearings, Vellucci had formed The Cambridge Experimentation Review Board, a citizens group to monitor and assess any future recombinant DNA work being done on their doorstep and ensure that it was being carried out in compliance with NIH regulations. This was the first time that a group of lay people had been given the power to have a say over scientific research conducted in academia and Vellucci's conviction that such a body was necessary was evident from a letter that he wrote in 1977 to NAS President Phillip Handler:

> In today's edition of the *Boston Herald American*, a Hearst Publication, there are two reports which concern me greatly. In Dover, MA, a 'strange, orange-eyed creature' was sighted and in Hollis, New Hampshire, a man and his two sons were confronted by a 'hairy, nine foot creature'. I would respectfully ask that your prestigious institution investigate these findings. I would hope as well that you might check to see whether or not these "strange creatures," (should they in fact exist) are in any way connected to recombinant DNA experiments taking place in the New England area.[119]

It was only when the ban was lifted in February 1977 that Gilbert's team could begin their work in earnest and they did so with the ferocious intensity that was characteristic of life in his lab. This was not a 'nine-to-five' culture. It was not uncommon for graduate students and post-doctoral fellows to be still at their benches well into the early hours of the morning, chopping, purifying, and mixing fragments of DNA to the sounds of Joni Mitchell or the Rolling Stones blasting out of a stereo.[120]

Besides the formidable task of isolating the specific mRNA for insulin and copying it back into DNA, there was another major technical hurdle. Inserting the newly synthesized DNA fragment into a plasmid and introducing it into a bacterium was not enough: the bacterium had to actually be able to transcribe the DNA sequence into mRNA, which could then be used to produce the functional insulin protein. In order to activate copying of the DNA sequence into mRNA, another fragment of DNA, known as the 'promoter', had to be slotted in *alongside* the fragment that coded for the protein. This sequence did not code for protein, but instead acted as a switch to activate transcription of the coding region by allowing the binding of enzymes and factors necessary for the synthesis of mRNA.

Meanwhile in California, Rutter and Goodman were determined to win. With no access to the insulinoma line that Gilbert was using, they had to look around for an alternate source. Having discarded an initial idea to use fish from an aquarium, they were forced to resort to obtaining insulin mRNA by the more conventional— and laborious—method of extracting it from pancreatic tissue.[121] Luckily, John Chirgwin, one of Rutter's post-docs, had found a way to deactivate the cellular enzyme which degrades RNA, but this still meant that they faced the onerous task of making pancreatic extracts from two hundred rats per week.[122] This was an unenviable task for many reasons. Each pancreas had to be surgically removed and islet cells separated from the other cell types by centrifugation before being treated with chemicals to break open the cells and extract the mRNA, all the while taking careful steps to ensure that it was not degraded in the process. The miniscule amounts of RNA were then spun by further centrifugation into a pellet yielding only a few micrograms of mRNA. Meanwhile, rat carcasses piled up in boxes outside the lab and the air stank of rodent faeces and urine.[123] Ullrich described this as one of the most stressful experiments that he ever conducted and in an interview given much later, he was clear to emphasize that it had not been done frivolously with regard to the animals' welfare.[124] Sometimes, when he found himself unable to sleep due to the pressure and stress, he would get up at two in the morning and go to the lab.[125] A further source of stress was that Ullrich did not find Goodman to be particularly encouraging about the work. Goodman seemed so convinced from the outset that the project was doomed to fail that when he left to take a sabbatical, Ullrich found himself feeling rather relieved at his supervisor's departure.[126]

Once enough mRNA was obtained, Ullrich began the task of copying it back into DNA using reverse transcriptase. By January 1977, he was ready for the next stage—the preparation of the DNA fragment and its insertion into a plasmid, which could then be introduced into bacteria for expression of the protein. The question, however, was which specific plasmid to use? Different types of plasmid have different properties—some are easier than others to introduce into bacteria, others replicate more quickly, and some express foreign proteins more efficiently. As luck would have it, Herb Boyer's lab had made what Ullrich considered to be the perfect candidate for the experiment.

pBR322 was a plasmid made by Paco Bolivar and Mary Betlach, two of Boyer's team, and for Ullrich it was a gift. In only a matter of weeks, he had successfully cloned the cDNA of rat insulin into the plasmid and in so doing catapulted the UCSF group ahead of their rivals in Gilbert's lab. But Ullrich's triumph proved to be premature. For although pBR322 ticked all the boxes from a scientific point of view, it had one major flaw. The guidelines issued by NIH stated that all plasmids must go through a two-stage process of validation before they could be used. First, they had to be designated as having been officially 'approved' for use, and then they had to be signed off as having been 'certified' for use by the Director of the NIH. Rutter has since said that, at the time, this distinction between the two stages was not sufficiently clear, and most researchers therefore assumed that 'approved' and 'certified' were interchangeable terms.[127] As a result, when Ullrich had initially requested a sample of pBR322 from Boyer's lab and been told that it was approved for use, he had assumed that this gave him a green light to go ahead with his work. By his own admission, 'I was no expert in that plasmid stuff, so I just took the word of the guys from Herb Boyer's lab with whom we communicated frequently'.[128]

The confusion over the meaning of 'approved' and 'certified' would prove to be much more than just a matter of mere semantics. When Rutter and Ullrich realized that they had used a plasmid that had not yet been officially certified for use, Rutter made an informal call to NIH's Deputy Director of Science Hans DeWitt Stetten to ask his advice about what they should do. Rutter posed the question in very broad, hypothetical terms, asking Stetten what the consequences might be, were a researcher—purely for argument's sake—to have used a plasmid that was not yet certified. Stetten's response was simple—it would be a disaster.[129]

Stetten was not being melodramatic. Given the fearful mood of the public and the media that had been evident at the Cambridge hearings, his concern was for the political fallout should it come to light that an uncertified plasmid had been used in a recombinant DNA experiment: 'My God … if this happens, we are certain to have legislation which would restrict the development of the whole field. This would be a unique thing in science and would be terrible for the United States'.[130]

Two years earlier, Lederberg had already anticipated Stetten's fears that the regulations drawn up by consensus at the Asilomar meeting might become legislation:

> The main danger is that tentative questions will be incorporated by some political imperative into ironclad regulations that will be with us long after anyone has forgotten why they were instituted. The particular field of DNA-splicing research,

far from being an idle scientific toy, or the basis of expensive and specialized aid to a few lives, promises some of the most pervasive benefits for the public health since the discovery and promulgation of antibiotics.[131]

Now those fears were looking increasingly like becoming a reality. In the wake of Vellucci's hearings to scrutinize recombinant DNA research in Cambridge, other city councils across the country had followed suit in setting up similar committees.[132] By now, recombinant DNA technology had come to the attention of the highest power in the land. In July, Senator Edward Kennedy wrote to President Gerald Ford urging him to place recombinant DNA research not just under the control of local committees of citizens, but of federal law.[133] Kennedy's intention was not to ban such research but to 'assure compliance with the guidelines in all sectors of the research community'.[134] Scientists looked on nervously as bills began to be introduced into the Senate and House of Representatives which sought to change the NIH guidelines into federal law that would limit and constrain recombinant DNA work.[135] There was talk that violation of such legislation might result in $10,000 fines, NIH grants being withdrawn, or even imprisonment, prompting some critics to draw comparisons with how the Soviet regime had commandeered the work of the biologist Trofim Lysenko in order to make science subservient to political ends and crush dissent.[136] When Kennedy addressed a hearing of the Senate sub-committee on recombinant DNA and the NIH guidelines, however, he made clear that it was not his intention to stifle such research or punish workers in this area, but to ask important questions about its aims and consequences:

> The plain fact is that genetic engineering has the capacity to change our society. How do we want it changed? What uses can we make of this knowledge? What degree of change is desirable and at what rate? What kind of society do we want to become?[137]

Although Gilbert complained that molecular biologists 'are being hassled out of existence for no reason at all',[138] their careers were, for the moment at least, still safe. But the same could not be said for the precious clones of rat insulin cDNA made by Ullrich.

With restrictive legislation looking ever more likely, the scientists could not afford to put a single step wrong. And for this reason, Stetten's advice to Rutter about what to do with Ullrich's clones was simple, but painful. All the clones of rat insulin cDNA that Ullrich had made must be destroyed.

Rutter actually considered Ullrich's clones to be such a historic achievement that he proposed holding a small ceremony to commemorate their destruction but, on March 19, 1977, Ullrich quietly and unceremoniously destroyed them himself using acid.[139] Yet even then, all was not lost. For Ullrich still had a small amount of rat insulin cDNA that he had not inserted into the plasmid, stored in his freezer. Determined to 'meet with triumph and disaster, and treat those two impostors just the same', he began the cloning experiment again—but this time inserting the rat DNA into a plasmid that was legal.[140] His original plan had been to clone the insulin gene into pCR1, a plasmid that was both approved *and* certified, but which had the drawback of simply not being as efficient as pBR322. Yet when he now tried to insert the rat gene into this plasmid, nothing worked. This, however, proved to be only a minor setback, because in April, the plasmid pMB9 made by Boyer's technician Mary Betlach received official certification. Like pCR1, it was not as efficient as pBR322, but at least it had the obvious and crucial advantage of legitimacy.

Ullrich quickly wrote up his work and submitted it for publication in *Science*. Publication in a peer-reviewed journal was the conventional channel for scientists to announce their results but Rutter decided that as the stakes were so high with this work, it was time to break with tradition. On May 23, 1977, the UCSF team held a press conference where, in front of TV cameras, they announced to the world their success in cloning the coding sequence for rat insulin.[141]

The researchers were breaking new ground—and not just scientifically. A couple of weeks later, UCSF filed a patent on the process which named Goodman, Rutter, Ullrich, and three other post-docs, Shine, Chirgwin, and Pictet as inventors.[142] Rutter's decision to announce the results in a press conference ahead of their publication in *Science* meanwhile was a historic break from the established practice of presenting research by submitting a manuscript to a peer-reviewed journal. And where Rutter had blazed the trail in breaking with tradition, others would soon follow. Charles Best must have been nodding in approval to see that scientists had taken on board his lesson of the need to 'convince the world' and were harnessing the full power of the mass media to do so.

But this new trend did not go without criticism. Writing only a few years later, one correspondent to the *New England Journal of Medicine* complained that:

> In recent months, however, we have begun to witness a reversal unheard of in the annals of scientific communication: the phenomenon of scientists publishing research data by press conference. It is not entirely clear what is causing this departure from the established norms. However, there is evidence that competition and the increasing involvement of academic scientists in the field of commercial application may be part of the problem.[143]

The author of the article went on to express concerns that the rush towards 'gene cloning by press conference' would bypass the critical evaluation of data that takes place during peer review of papers. The result might well be that thoughtful analysis and discussion would be replaced by media-friendly soundbites, hyperbole, and overinflated claims for potential benefits. These fears appeared to be vindicated by the tone used by the national newspapers that covered Ullrich's success. In the same breath as they acknowledged that Ullrich's results were still speculative, newspaper articles went on to make grand predictions that a cure for diabetes might well be in sight. If bacteria could be induced to produce human insulin, then the feared world shortage might be averted thanks to a 'virtually limitless supply of the vital hormone'.[144] Moreover, they raised the possibility that if Ullrich's success in cloning the gene for rat insulin could be repeated with human DNA, then this might eventually allow scientists to find a means of reactivating the production of insulin in a damaged pancreas.[145]

But Rutter and Ullrich soon had far more to worry about than these criticisms. For while Rutter had happily courted the media to publicize Ullrich's triumph in cloning rat insulin, he very quickly found himself the unwanted subject of their attention for other reasons.

At the start of September, Rutter received a phone call which left him feeling very uneasy. The call was from Nicholas Wade, a writer at *Science*, who sensed that behind the triumph of Ullrich's success in cloning the rat insulin gene there lay a much bigger story. In the course of his investigations, Wade had spoken with a number of other researchers at UCSF, one of whom joked that Wade seemed to be so well acquainted with the internal politics of the department that he must have had a ten-page dossier on the various feuds and rivalries that existed.[146] The most likely explanation for Wade's intimate knowledge was simple but unsettling—that he was receiving his information from a source within the department. Sifting through what was mostly water-cooler gossip, Wade found one sparkling nugget of information—that Ullrich's success had been built on a violation of the NIH rules.

Wade called Rutter to ask for clarification on the matter and to enquire whether the rumour that the experiment had first been performed using an uncertified plasmid was true—and if so, why had this error not been made public to the NIH when the lab announced their success in cloning the rat insulin gene? Rutter could see where the questions were heading and in attempt to prevent what could well be a catastrophic piece of PR, not only for his own lab but for recombinant DNA research in general, he called the editor of *Science* to argue that publishing Wade's article would serve no useful purpose and be counterproductive.

His efforts were in vain. At the end of September, Wade's article appeared in *Science* under the headline—'Recombinant DNA: NIH Rules Broken in Insulin Gene Project'.[147] Rutter now found himself having to answer some awkward questions—such as why had the UCSF team believed that pBR322 was sanctioned for use? Boyer, in whose lab pBR322 had been made, maintained that he had made it clear that the plasmid had not yet been certified.[148] In his defence, Rutter argued that he had been unaware that 'approval' and 'certification' were two distinct stages and that when he first heard that pBR322 had been approved by NIH on January 15, he had therefore assumed there would be no problem in using it.

Then there was the question of why the Biosafety Committee at UCSF had not been informed about the use of pBR322 until May, when the actual experiment had been done a few months earlier. In response, Rutter pointed out that, prior to beginning any experiments, they had filed a memorandum with the Biosafety Committee that gave a general description of their intended plan to clone the rat insulin gene using pCR1 and any other plasmids which were 'approved' by NIH.

This general statement given in the memorandum also helped to answer another awkward question. All cloning work had to be done in a P3 containment facility and workers using this laboratory had to sign a logbook. While there was an entry in the logbook for April when Ullrich had successfully cloned the rat insulin DNA into pMB9, there were no signatures for the earlier work with the pBR322 plasmid. There was, however, an entry in February for work with pCR1 and, as the logbook had to be signed for each overall experiment and not for each individual use of the P3 lab, Rutter felt that this entry had been sufficiently general to include the use of pBR322.

Although Wade's article certainly brought unwelcome attention to Rutter's group, it was no trial-by-media condemnation of Rutter and Ullrich. If anything, it highlighted the challenges and confusion that researchers faced in trying to interpret—and comply with—rules that were perceived to be ambiguous, unclear, and still undergoing development. Wade was clear to point out that this was the only incident in which NIH guidelines had been broken and reassured readers that 'it is clear that the experiment posed no issue of public health since it was performed in the required type of laboratory—a "P3" facility—and with a vector which has now been certified as safe'.[149]

But the publication of the article could not have come at a worse time. A few months earlier, an article written by journalist Janet Hopson called 'Recombinant DNA Lab and My 95 days In It' had appeared in *The Smithsonian*. In it, she described her experience of spending three months working in Herb Boyer's lab at UCSF. Although the article vividly evoked the energy and enthusiasm of daily life among the postgraduate students and post-doctoral fellows in the lab, Boyer might have been alarmed to read some of Hopson's other observations of daily life:

> After last week's luncheon I brushed up on the NIH guidelines and watched day-to-day lab procedure more closely … The workers rarely wear white laboratory coats (suggested by the guidelines) and they smoke, eat and drink in the lab while experiments are in progress (expressly) discouraged. I also watched people sucking recombinant organisms into glass pipettes with their mouths instead of the recommended rubber bulbs, and saw work surfaces decontaminated only sporadically—instead of daily, as required. Half of the researchers here follow the guidelines fastidiously; others seem to care little … Among the young graduate students and postdoctorates it seemed almost chic not to know the NIH rules[150]

Hopson's account of the cavalier attitude by some researchers towards the NIH guidelines, followed shortly afterwards by the news that the UCSF team had breached them, was hardly likely to offer any reassurance to a general public who were becoming more vocal in their opposition to recombinant DNA research. In March 1977, the NAS held an open forum in Washington D.C. at which members of the public could contribute to the debate about recombinant DNA technology. When Irving Johnson of Eli Lilly gave a presentation to explain that the current supply of insulin from bovine and porcine pancreatic tissue was unreliable and that recombinant DNA technology could offer a stable supply of human insulin, he may have expected these benefits to be immediately obvious to his audience. But if so, he was in for a shock:[151]

> It was at this meeting that I first came into contact with the intensity of the opposition to rDNA technology, both on safety and ethical grounds. As the only industrial speaker on the programme, my presentation was interrupted by a group of opponents who attempted to physically take over the meeting, waving banners, comparing proponents to Hitler and so on.[152]

Not that Johnson was the sole target for criticism. Protestors draped a banner in front of the panel of speakers while others waved placards that read 'I will not be cloned' (Figure 53).[153]

Figure 53 *Protestors campaigning against recombinant DNA research hold a sign quoting Adolf Hitler, "We will create the perfect race" in front of a panel (which included Maxine Singer, Paul Berg, and NIH Director Donald Fredrickson) at a meeting of the National Academy of Sciences held in Washington D.C. in 1977.*

Credit: *Courtesy of US National Library of Medicine.*

Amidst this ongoing furore, the NIH had reviewed its guidelines and was proposing to release an updated and more relaxed version of them. For the various environmental groups who had now joined the fight against recombinant DNA work, such as Friends of the Earth, this was a step too far, while many scientists working in this field still considered these new proposed guidelines to be too strict.

Johnson felt them to be overblown, but acknowledged that they were of value in giving 'the public more confidence that scientists were behaving appropriately'.[154]

With over a dozen bills and amendments introduced before the Senate and House of Representatives from January 1977 to March 1978, it was looking increasingly likely that these guidelines drafted by NIH would become federal law.[155] Penalties for the violation of these laws ranged from fines of anywhere between \$10,000–50,000 per day—with the result that universities might be unable to get insurance, and even result in imprisonment for up to a year.[156] The seriousness of the situation became all too clear to Bill Rutter in November 1978, when he was summoned to give an account of the incident involving pBR322 and the breach of NIH guidelines before a Senate committee on commerce, science and transportation. Alongside him was Boyer, who, although he had not been directly involved in the experiment to clone rat insulin, had been responsible for making the plasmid pBR322. In the hours that followed, Senator Adlai Stevenson III, who was chairing the committee, grilled Rutter and Boyer with the forensic precision of a prosecuting lawyer, which Rutter admitted had left him and Boyer feeling 'totally outmatched'.[157] Stevenson's approach stood in sharp contrast to that of the cultural anthropologist Margaret Mead, who had made a memorable entrance to the proceedings 'wearing a huge long robe and with a long shepherd's staff'.[158] Rutter was left distinctly unimpressed:

> Here was a social anthropologist with her shepherd's staff giving her advice on molecular, microbiological, and physiological science. [laughter] She said something like, "You're going to hear today from these scientists that this is not dangerous. I'm here to tell you it is dangerous." After every significant statement she would pound the floor with her staff for emphasis. "These people are here, telling you it's safe. I'm telling you it's not safe." Boom! Boom! "These people are here telling you that they have only interests in promulgating the truth. Well, I'm telling you here, this is not the truth." Boom! Boom! Her entourage were there clapping after her significant remarks. Obviously it was a total setup, an amazing setup.[159]

Together, they endured a thorough grilling that Rutter said made them feel like 'schoolboys'.[160] He later described the experience as an inquisition—particularly when Senator Stevenson issued a damning reprimand to them both: 'You say you don't want legislation. If there is legislation, you gentlemen would be the authors of it'.[161]

Although Wade's article in *Science* had concluded that the breach of guidelines by Rutter and Ullrich was an innocent mistake, it did raise another awkward issue. When Ullrich's paper appeared in *Science* announcing that he had successfully cloned the rat insulin gene into pMB9, some researchers in the field were sceptical about this claim. Only three weeks had passed between Ullrich destroying all the pBR322 clones and submitting the manuscript to *Science* in which he described cloning the cDNA for insulin into the certified pMB9 plasmid. To these critics, it all sounded just too easy. There were murmurings of dissent that no one could have carried out a successful cloning experiment from scratch in such a short space of time. Suspicions began to ferment that not all of the pBR322 clones had been destroyed and that the actual experiments might not have been carried out as described in the *Science* paper. Molecular biology had become an intensely competitive field in which the perception was that there were no prizes to be had for coming in second place. Wade's article raised the possibility that, even if Ullrich was innocent, other researchers might be tempted to take short cuts around the guidelines to stay ahead of their rivals.

At Harvard, meanwhile, Gilbert's group were all too aware of the intense competition that they faced from their West Coast rivals, and the news that Rutter's team had held a press conference to announce the successful cloning of rat insulin came as a blow. This did not, however, mean that the race was lost. For while Rutter and Ullrich had succeeded in inserting the specific fragment of DNA that carries the sequence for the rat insulin protein into a bacterium, they had still not shown that this sequence could be expressed in bacteria to produce a functional protein. Beyond this lay the ultimate goal—the task of cloning—and expressing human insulin. The race was therefore, still very much on.

But although the insulinoma tumour cell line had given Gilbert's team a head start, they had hit problems of their own. After months of laborious and painstaking work extracting mRNA from the tumour cells and copying it into DNA, Gilbert's post-doc Efstratiadis had synthesized one microgram of cDNA encoding rat insulin, which he now gave to Gilbert's graduate student Forrest Fuller for further enzymatic preparation before cloning it into a bacterial plasmid. To comply with what had been agreed at the Asilomar meeting, Fuller was carrying out all this work using a strain of *E. coli* called K12 chi 1776 that had been distributed to researchers by the University of Alabama microbiologist Roy Curtiss. This strain of bacterium had been deliberately enfeebled by Curtiss so that it could not replicate without a host of specific growth supplements that would only be available in a laboratory environment. While this meant that it couldn't replicate outside the lab and therefore satisfied requirements for biological containment, it also didn't grow too well in the lab either—as Fuller found to his frustration when he tried to use it for his cloning work.

Gilbert was keen for results and, as Harvard still did not have the required P3 facility in which the final cloning experiment could be done, he had arranged that he and Fuller travel to Basel in Switzerland where such a laboratory was available for them to use. But Fuller was still battling against a host of technical problems and tensions began to grow between them. Feeling the pressure from Gilbert, Fuller took to working at night—and then disaster struck.[162] During a routine preparative procedure, Fuller inserted a plastic tube containing the precious insulin cDNA into a vacuum drier to remove some of the solvent used in its preparation. Only too late did he realize, to his horror, that he had not sealed the plastic tube containing the precious sample before placing it under a vacuum. In a matter of seconds, much of the cDNA which had taken months of careful preparation was lost.

Fuller later described the incident as being 'probably the worst day of my life'—worse even than the news that Rutter's team at UCSF had successfully cloned the rat insulin gene.[163] Thankfully however, there was a glimmer of hope for it turned out that despite the vacuum drier disaster, there was just enough material left to snatch victory from the jaws of defeat. Taking what little remained of the sample, Gilbert and Fuller headed for Switzerland and the final stage of the experiment. When checking in for the flight at the airport, Gilbert and Fuller deliberately asked to be seated in different sections of the aircraft.[164]

The entries in Fuller's lab notebook such as 'Negative', 'Aagh', and, ultimately, 'Shit!' all suggest that the gods of molecular biology were no kinder to him in Switzerland than they had been in Harvard.[165] Later in the summer, he returned to Harvard with only ten clones, none of which when analysed contained the rat insulin gene. That autumn, he began quietly attempting to make his own insulin cDNA, aware that this would put him in direct competition with Efstratiadis, but if he hoped that this display of determination would win the admiration of Gilbert, he had gravely misjudged his mentor's mood. Fuller was a well-liked member of the lab, but his colleagues all recognized that his cards were now marked and likened his situation to watching an animal dying a slow death in the middle of a close-knit herd.[166] When Fuller returned to work after the Christmas break at the start of January 1978, he was informed by Gilbert that he had one month in which to find another lab at Harvard in which he could complete his PhD.[167] The announcement sent shock waves through the other members of Gilbert's lab, for each of them realized that, if they failed to come up with results, Fuller's fate might easily become their own too.

When looking for someone to replace Fuller on the insulin project, Efstratiadis suggested that Gilbert approach Lydia Villa-Komaroff, a former colleague who had been forced by Vellucci's moratorium to leave Harvard and move to Cold Spring Harbor in order to continue her cloning work. According to family legend, the life of Villa-Komaroff's grandfather been spared during the Mexican Revolution by Pancho Villa himself. The revolutionary leader's grace would prove to be a blessing not only for the fortunes of Villa's family but also for those of the Gilbert lab.[168]

Villa-Komaroff had done her graduate studies at MIT and so was able to use her contacts there to allow Gilbert's team access to a much-coveted resource. For unlike Harvard, MIT now had a P3 containment lab. This meant there was no need for any further costly and time-consuming flights to Switzerland—the cloning could be done right on their doorstep. At the start of February, as Boston lay under twenty-five inches of snow with the airport and roads closed, Villa-Komaroff battled her way to the lab through the arctic drifts.[169] Here, working in an environment that was only marginally more comfortable than that outside, she began to insert the plasmids carrying insulin DNA that she and Efstratiadis had made into bacteria. In addition to the formidable technical challenge of preparing the clones, the work required meticulous attention to ensure that no contaminating material either entered or left the lab. The lab itself was small—eight by twelve feet—into which were crammed a refrigerator, centrifuges, an incubator, benches, a tissue culture hood, and the all-important radio—designated to provide a steady background of rock and classical music. Every time Villa-Komaroff entered the lab, it required donning gloves, a gown, and plastic overshoes in order to minimize the risks of carrying any contaminants with her. The air pressure within the lab was kept lower than that outside to ensure air could only flow into the lab and not out of it, preventing the escape of any airborne recombinant bacteria. Finally, every piece of waste material—pipette tips, plastic test tubes, tissue culture flasks—had to go into plastic bags that were sealed and sterilised in an autoclave. As Stephen Hall said when recounting Villa-Komaroff's labours in his book *Invisible Frontiers*, 'Nothing left the lab alive, save the researchers'.[170]

By her own admission, Villa-Komaroff's time in Cold Spring Harbor had been a fallow period. But while the endless unsuccessful attempts at cloning might have left other graduate students demoralized and despondent, Villa-Komaroff was a fighter—as James Watson found out when, during one of the traditional Cold Spring Harbor food fights, she tipped half a bottle of wine over his head.[171] Rather than lose heart from her unsuccessful experiments, she drew from them the lesson that failure can sometimes be a scientist's best teacher, and her tenacity and determination paid off. On Valentine's Day 1978 she counted over 2,300 bacterial colonies that looked as if they might well have taken up the plasmid. After weeks of laborious screening to analyse their DNA, Villa-Komaroff showed that forty-nine of these were indeed carrying the gene for insulin. Although it was certainly an impressive show of stamina by Villa-Komaroff, this alone was not a novel achievement—it was what Ullrich had achieved a year earlier. But Villa-Komaroff noticed something odd about some of these colonies, which suggested that she had done far more than just clone the insulin DNA.

As part of the cloning procedure, the fragment of DNA which encodes insulin had been inserted into a region of the plasmid that contained the gene for penicillinase, an enzyme which degrades penicillin. Inserting the insulin DNA into this region should knock out expression of this enzyme, meaning that bacteria were no longer able to grow in the presence of penicillin. Yet, when Villa-Komaroff looked at her agar plates, what she saw surprised her. Some of the bacteria were indeed growing on penicillin, albeit very slowly. From this she inferred that these bacteria must still be synthesizing penicillinase—and if that was the case, maybe they were also synthesizing the insulin protein, too. Tests with radioactive antibodies that specifically bound to the insulin protein confirmed her hunch. The Harvard group had not only cloned the DNA for rat insulin, but they had also managed to switch the DNA into an active state so that its information was read by the bacteria to make the insulin protein.[172]

In June that year, while receiving an honorary degree from the University of Chicago, Gilbert gave a seminar in which he made the first public announcement of their success. A science reporter from the *Chicago Tribune* who was present took up the story and within days the Wall Street Journal was declaring it to be 'a major step toward producing human insulin for diabetics by bacteria'.[173]

Leafing through the latest issue of *Nature* as she prepared a draft of her Presidential Address to the British Association for the Advancement of Science, Dorothy Hodgkin saw the news of the Gilbert team's triumph. When Hodgkin gave her speech, she praised the work as 'an extraordinarily intricate achievement that could not have been conceived when first I began research forty-five years ago ... It is also oddly a parable of curious hopefulness'.[174]

Thanks to Villa-Komaroff's work, Gilbert's team had, in one single step, somersaulted over Ullrich's triumph of the year before in cloning the rat insulin gene and taken the lead. Now it was time to sprint through the home stretch towards cloning human insulin, but between them and victory there still remained one towering hurdle.

The NIH regulations specified that any recombinant work using human DNA had to be done in a laboratory with the maximum level of P4 containment. Facilities of this kind require all workers to wear full body protective suits, enter via an airlock, shower on exit, and are used for research into lethal pathogens such as the Ebola virus. The only such laboratory in the United States at that time was at Fort Detrick, where it had been used by the Department of Defence to conduct research into biological warfare. With the Cold War still very much on, the chances of the military jeopardizing national security by opening their facility to a bunch of eager molecular biologists were pretty slim.

Therefore, Gilbert's team had to look further afield. Fortunately, the British Ministry of Defence were far more receptive to helping blaze the trail in molecular biology. After passing the required security checks, Gilbert and his team were allowed four weeks at the British military Microbiological Research Establishment at Porton Down, UK, in which to carry out their work. This was their one shot at success, and they took no chances. Knowing how vulnerable enzymes and the preparative stages of the DNA were to even the tiniest contaminants, Villa-Komaroff prepared all the necessary reagents from scratch and packed them for the trip, rather than have to make them on arrival in Britain. Equipment such as petri dishes and vital reagents such as enzymes were all packed, as were some Cuban cigars that were to be given to Gilbert in anticipation of the celebrations that would erupt when the experiment was successfully completed.[175]

Before starting their work, all members of the team were fitted with gas masks—and from there, things only got more uncomfortable and difficult. Prior to entering the P4 lab, the workers had to remove all clothing and wear plastic boots, gloves, protective robes, and headgear. To enter the lab, they had to walk through a formaldehyde wash and on exiting it were required to shower. Every item being brought into the lab had to be washed in formaldehyde, which meant that written instructions of the various protocols had to be placed within plastic wallets to protect them. Equipment had to be sterilized in autoclaves and the task of merely obtaining equipment was not always easy. Efstratiadis recalled how, having walked almost a mile through the various buildings in order to get a rotor for a centrifuge, he was told that he would first have to sign for a key to open the locker in which the rotor was kept. Having duly signed and obtained the locker key, he was then informed that he would have to sign for another key in order to open the room where the locker was.[176]

It wasn't just the lab environment that proved to be uncomfortable. When Gilbert and his team decided to go for lunch in the officers' mess, they caused a few raised eyebrows among the regular diners.[177] Women had only recently been admitted and even then, only if they were wearing a skirt. Dressed in jeans and T-shirts, Gilbert's lab crew made this their first—and last visit.

Despite the frustrations and difficulties of working in the P4 facilities, the cloning experiment seemed to be going well. Having cloned human insulin DNA into a plasmid, the team had introduced it into bacteria and obtained colonies. All that was needed now was to screen them to confirm that they did indeed human insulin DNA and that this was producing protein. Confident of impending success, Gilbert held a celebratory dinner and lit up one of his prized Cuban cigars.[178]

The celebration proved to be spectacularly premature. A phone call that evening from Villa-Komaroff's husband brought them crashing down to earth. Back in the USA, TV stations and the late editions of newspapers were already breaking the news that Gilbert had been beaten in the race to clone and express human insulin. This was bad enough, but there was worse to follow. As the Harvard team began to analyse the DNA of their clones, they made a shocking discovery. Sure enough, the clones contained the insulin gene—but it was not human DNA. Somehow, somewhere in the course of carefully preparing all the reagents and samples, a tiny amount of rat insulin cDNA had found its way into one of the solutions and contaminated the experiment. The result was that, having flown halfway around the globe to battle their way with airlocks, gas masks, and formaldehyde washes, the Gilbert lab had simply repeated the cloning of the rat insulin gene. Earlier in the year, this achievement had brought euphoria—now it left a very different feeling.

Gilbert described the trip to Porton Down as a 'total disaster'—and to add insult to injury, when they arrived back at Harvard, someone had forgotten to take all Villa-Komaroff's original 2335 bacterial colonies out of the incubator.[179] But the failure of Gilbert's team did not mean that Rutter and Ullrich over on the West Coast were popping champagne corks. Their mood was also far from celebratory.

After his initial success in cloning the rat insulin gene, Ullrich had tried to show that the insulin protein could be expressed in bacteria, but his results were ambiguous at best. By now, Goodman had returned from his sabbatical in Japan and there was a feeling among some of the post-docs that he was taking far too much credit for Ullrich's initial success, which created a festering toxicity in the lab.[180]

While Ullrich's work on insulin was mired in technical niggles and workplace politics, his colleague and fellow German Pete Seeburg was having much more success. When Seeburg managed to clone and express production of another protein with potential therapeutic applications—growth hormone—Ullrich was spurred into action: it was time to clone human insulin. But he also faced the lack of a P4 facility, and so, like Gilbert and his team, Ullrich also found himself boarding a flight to Europe.

His destination was not Porton Down, however, but Strasbourg. After Ullrich's success in cloning rat insulin, Eli Lilly had become interested in offering support for the attempts to clone insulin using recombinant DNA.[181] Having signed a contract with Lilly at the end of the summer, the Rutter-Goodman team were now able to use the company's P3 lab in Strasbourg which, under French regulations—unlike those in the USA—could be used for recombinant work with human DNA.

In the late summer and autumn of 1978, while Gilbert's team were donning moon suits and stepping into airlocks at Porton Down, Ullrich was flying back and forth between San Francisco and Strasbourg. One day in October, the director of the Strasbourg facility called Ullrich into his office saying that he had some news for him.

The news was that Ullrich, like Gilbert, had been beaten in the race to clone and express human insulin.

This certainly came as a disappointment for Ullrich, but it did not crush him. What came as an even greater kick in the teeth, however, was to learn on the back of this news that the patent which UCSF had filed on the cloning of insulin following his success a year earlier had just been redrafted and his name, along with those of the other three post-docs involved, had now been removed to leave Goodman and Rutter as the sole inventors. But perhaps the most galling knowledge of all must have been that, only a few months earlier, Ullrich had been offered a chance to join the very same team of researchers who had now claimed victory over both him and Gilbert.

This offer had come not from a scientist but from Bob Swanson, a venture capitalist. Taking Ullrich out for a beer and a two-player match on an early arcade game, Swanson had tried to entice Ullrich into joining a new venture that he was starting. Although initially interested, after going home and reflecting on Swanson's offer, Ullrich declined. Swanson's partner in this new venture was none other than Boyer, Ullrich's UCSF colleague. Boyer had not been idle since that day he sat fretting in the auditorium at Asilomar: and as a result he had beaten both the Harvard and UCSF labs to cloning and expressing human insulin. Better still, he had done so without ever having needed to step into an airlock, don a protective suit, or even buy an airline ticket to Europe.

12

Wall Street Gold

When Herb Boyer's phone rang, it usually meant only one thing. A call from yet one more researcher who, keen to start working on recombinant DNA, was begging Boyer for a sample of his precious restriction enzymes. But the call that Boyer received early in January 1976 would change his life.

The caller introduced himself as Bob Swanson who, having studied chemistry at MIT, had sufficient knowledge of science to understand the enormous potential of Boyer's work. But it was his business acumen, not his knowledge of organic synthesis, that he hoped would grab Boyer's attention. For having also majored in business, Swanson had gone on to forge a career in finance, specializing in venture capital and it was his expertise in this area that he believed could be invaluable to Boyer.[1] Flush with the success of Silicon Valley, San Francisco had become a crucible for new companies requiring venture capital, and Swanson was on the look-out for the next big thing.[2] With recombinant DNA technology, he believed he had found just that.

While working for the San Francisco-based venture capital company Kleiner and Perkins, Swanson had already made contact with Cetus, a company that had developed an automated screening system for micro-organisms. When Swanson had first approached Cetus about the potential of recombinant DNA technology, they assured him that, although the area showed long term promise, 'it's not going to happen for a long time'.[3]

Swanson resolved to prove them wrong. At the end of 1975, shortly after his failure to win the interest of Cetus, Swanson had parted company from Kleiner and Perkins and now found himself unemployed. After paying the rent on his apartment and the lease on his Datsun 240Z, what little money that he had left stretched only to 'peanut butter sandwiches and an occasional movie'.[4] But Swanson still burned with the ambition to start a novel commercial venture of his own based on recombinant DNA. So, he began cold calling his way through a list of the delegates who had attended the Asilomar meeting.

Many of them had the same response as Cetus. Undeterred, Swanson continued working through the list in no particular order and eventually reached Boyer's name. Boyer said that although he was very busy, he could probably spare Swanson ten minutes for a quick chat if he called by the lab at the end of the week.

When Swanson called by at the end of the week, he explained that, with his background in venture capital, he would have access to funds with which they could establish a new company using the recombinant DNA methods that Boyer had pioneered. For Boyer, it was an attractive proposition. He already had the security of his academic post at UCSF, and the funding promised by Swanson would be a very welcome source of extra financial support for the workers in his lab, thereby reducing his need to rely on research grants.

The discussion quickly moved to a nearby bar and, after few hours and several beers, Boyer and Swanson had formed a partnership. Now they needed to identify a product and Swanson already had one in mind:

> Diabetics were getting pig or cow insulin that was extracted from the pancreas glands of cattle and pigs that were slaughtered ... Once you had succeeded in overcoming the technical hurdles, it should be pretty obvious that recombinant insulin would be better than pig or cow insulin ... we believed people would rather have human insulin than pig insulin. So we said 'Okay. As best we can tell today, there's a big need'. Lilly was selling $400 million worth of insulin a year. It's a small molecule; it was probably somewhat technically feasible based on what we knew. The economics looked pretty good.[5]

From a business perspective, insulin was a very attractive choice because it was a product for which there was already a well-established and sizeable demand. Rather than go to the trouble of creating a new market, Swanson had simply to offer an improved version of what was already available. Making this case, Swanson presented a business plan in March 1976 to his former employers at Kleiner and Perkins, who were sufficiently impressed to offer an initial $100,000 in the venture. Having secured this funding, Boyer and Swanson dissolved their partnership in April and signed the required documents to form a company. But what to call this new venture? Skilled though he may have been in drumming up funding for the new company, when it came to dreaming up a name for the new venture, Swanson was no Don Draper.[6] In honour of its two founders, he proposed that the new company be named 'Herbob'.[7] Boyer tactfully offered an alternative suggestion.[8] Since the company's products would be made using genetic engineering technology, how about calling it 'Genentech'?

That very first meeting between Boyer and Swanson is today commemorated with a sculpture that stands on the campus of Genentech's impressive corporate headquarters in San Francisco (Figure 54). But back in the spring of 1976, the company existed only as a piece of paper upon which Swanson was listed as president and treasurer, and Boyer as vice-president and secretary. They now faced the same formidable challenges that were frustrating Walter Gilbert at Harvard and the Rutter/Goodman collaboration at UCSF in their race to clone and express human insulin.

Before any work could begin, Swanson's first task was to ensure that Genentech would be able to obtain a license to use the recombinant technology protected by the patent that UCSF had filed on Boyer and Cohen's work. As scientific advisor

Figure 54 *Statue of Herb Boyer and Bob Swanson on the Genentech campus.*
Reproduced with permission of Genentech Corporate Relations.

to the company, Boyer meanwhile concentrated on how to overcome the numerous technical hurdles that lay before them. Better yet, he already had spotted an opportunity to take the lead over Gilbert's and Rutter's teams.

Based on its high demand and the established market, Swanson had made the case that insulin would be the ideal product for Genentech. Additionally, Boyer had three very good scientific reasons why it was the ideal choice. The first of these was that researchers at the University of Cambridge, UK, had by now followed the trail blazed by Fred Sanger and determined the complete sequence of amino acids in human insulin.[9] The second was the successful deciphering of the genetic code, for which Marshall Nirenberg, Robert Holley, and Har Gobind Khorana had been awarded the 1968 Nobel Prize in Physiology or Medicine. These two achievements meant that it ought now be possible to deduce the precise DNA sequence that encoded the order of amino acids in human insulin. And Boyer's third reason to be optimistic was that, since being awarded the Nobel Prize for his part in deciphering the genetic code, Har Gobind Khorana had not rested on his laurels.

Four years after being awarded the Nobel Prize, Khorana and his team at MIT announced that they had synthesized the complete DNA sequence of a gene from yeast that was involved in the synthesis of proteins.[10] In rather the same way that Panayotis Katsoyannis had painstakingly made artificial human insulin by linking together fragments of synthetic amino acids, Khorana's group had now achieved exactly the same feat with DNA. A few years later they took this achievement even further. In 1976 they not only achieved the complete synthesis of a bacterial gene involved in protein synthesis, but showed that this artificial DNA was active when introduced into a cell.[11]

If it was possible to synthesize a biologically active bacterial gene, then Boyer saw no reason why the same could not be done for the gene encoding human insulin. It was certainly an attractive alternative to the laborious and time-consuming lab methods employed by Boyer's academic rivals at UCSF and Harvard. For one thing, it would spare him the arduous and unpleasant task of having to surgically remove the pancreatic tissue from 200 rats. Nor would there be any need for painstaking months of work to isolate and purify the delicate messenger RNA for insulin, all the while taking meticulous care to avoid its contamination before copying it back into DNA (Figure 55).

Most of all, Boyer's proposed strategy of synthesizing insulin had one huge advantage over Rutter's and Gilbert's methods: if a DNA fragment encoding insulin was synthesized in the laboratory using standard organic chemistry protocols, it would be classified not as human, but as artificial DNA. According to the NIH guidelines, working with DNA that was classified as 'artificial' and not human would mean that Boyer and his team need not battle with P4 facility-level precautions: the whole experiment could be done at the bench in a regular P2-category lab right there in San Francisco.

Boyer knew just the people for this job. Art Riggs, a researcher at the City of Hope Medical Center in Duarte, Southern California, was working on how the bacterial gene encoding the enzyme beta-galactosidase (or beta-gal) could be switched off by the binding of a repressor protein to a specific target sequence of DNA, known as the lac operon. As this target sequence was only 21bp in length, Riggs's colleague Keiichi Itakura had been able to artificially synthesize it in the lab. Using Boyer's cloning methods, this artificial fragment of DNA had then been inserted into plasmids and introduced into bacteria where it was found to be functionally active.[12]

The grand plan was to show that the bacterial repressor protein could be crystallized as a complex bound to this short sequence of DNA, which would offer invaluable insights into how such genetic switches might work. Riggs admitted that the project was unsuccessful—the protein could not be crystallized bound to its target DNA. But in the long run, none of this would matter—Riggs and Itakura had shown that artificially synthesized DNA could be cloned, introduced into a bacterium, and, crucially, be biologically active.[13]

But would the same be true for an artificially synthesized piece of DNA from a higher organism? To answer this question, Riggs and Itakura submitted an application to the NIH for grant funding to synthesize the DNA encoding the human hormone somatostatin. As somatostatin is only fourteen amino acids in length, the corresponding DNA fragment is a mere 42 bases long and Riggs felt confident that this could be synthesized and cloned. But the NIH felt otherwise and rejected their grant on the grounds that it was 'just an intellectual exercise'.[14]

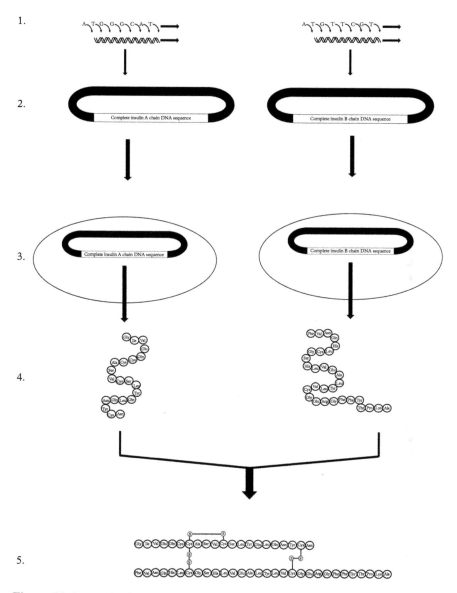

Figure 55 *Genentech adopted a radically different strategy to that of their academic rivals at UCSF and Harvard to clone and express human insulin. Knowing the complete amino acid sequence of insulin, they artificially synthesized DNA chains encoding the A and B amino acid chains of insulin (1). These two artificial DNA molecules were then inserted into bacterial plasmids (2) and introduced into bacteria (3) which expressed the A and B protein chains (4). Finally, these two separate chains were combined to produce the intact insulin protein (5). Because this method used artificial DNA, the final stage of introducing the recombinant DNA into a bacterium did not have to be carried out in top level security category P4 biosafety facility, as the UCSF and Harvard groups had to do. Diagram by K.T. Hall.*

When the news of the NIH rejection came, however, Riggs and Itakura had already received a more promising offer. Boyer had been in touch to ask whether they would be interested in bringing their expertise in the synthesis of DNA to collaborate with Genentech on the cloning of human insulin. Riggs was certainly enthusiastic about this proposal, but he suggested that Genentech should first focus on showing that the DNA encoding somatostatin could be synthesized, cloned, and expressed in bacteria, as proposed in his NIH grant application: this would prove the technology worked. Boyer agreed, but Swanson was unsure—after all, when other labs were already racing towards cloning insulin, why waste precious time on somatostatin?[15]

After being persuaded by Boyer, Swanson divided the work between the two labs. Riggs and Itakura worked on synthesizing a set of short DNA fragments that, when ligated together, would encode the complete sequence of somatostatin and the necessary sequences recognized by restriction enzymes in order to cut and insert the fragment into a plasmid. The task of assembling these fragments into a single piece of DNA and inserting this into a plasmid then fell to Boyer's team at UCSF. One of these was Herb Heynecker, a Dutch post-doc who, on first landing in San Francisco airport, had arrived carrying an ice-bucket full of precious enzymes that he had isolated and purified back in Holland. Thankfully for Heynecker and the future of recombinant DNA technology, airport security checks were not as stringent as they are today.[16] Known as ligases, Heynecker's enzymes enabled fragments of DNA to be joined together—an essential step in any cloning procedure.

Like the graduate students and post-docs in Gilbert's lab, Heynecker worked with such fury that most of the planning for his experiments was scribbled on paper towels.[17] His philosophy was a simple one:

> I see it as a race, or like playing racquetball or whatever: Lots of scientists have that competitiveness, and you need it. You have to be driven by it. If you only do science between 9:00 and 5:00, you perhaps might want to do something else.[18]

Once Heynecker had ligated the fragments of DNA and cloned them, the final stage—that of detecting the production of the hormone—was carried out by Riggs using radioactively labelled antibodies that specifically bound to somatostatin.

Expectations were high. If the DNA encoding somatostatin could be synthesized, cloned, and expressed, then theoretically, the same could surely be done for human insulin. After a few months, Riggs contacted Boyer and Swanson to say that the big moment had finally arrived: having successfully synthesized the DNA and cloned it into bacteria, the time had come to see whether the bugs were making somatostatin.

Confident of success, Swanson and Boyer took a flight down to City of Hope. If the experiment had worked—as everyone expected it would—then the somatostatin produced by the bacteria would be recognized by the radioactively labelled antibodies and give rise to black spots on a photographic film. Cautiously, Riggs began to develop the photographic films and everyone else watched and waited.

And as they continued to watch and wait, it became clear that something had gone wrong. Badly wrong. Finding himself staring at blank films that showed not the slightest hint of somatostatin production, Swanson began to feel physically ill.[19] By the next morning, he looked 'deathly white' and ended up checking himself into a local hospital.[20]

Swanson's ailments turned out to be nothing more sinister than acute indigestion. But although he later laughed this off as due to the spicy Mexican dishes he had eaten the previous evening, he also admitted that this had been an incredibly stressful experience.[21] He later described the failure to produce somatostatin as being the most frightening episode in the entire venture.[22] For if Genentech failed, then at least Riggs, Itakura, and Boyer all had the comfort of knowing that they still held academic posts. For Swanson, there was no such safety net—he was gambling everything—his professional reputation, career, and livelihood—on the success of this technology.

On paper, cloning somatostatin had all seemed so straightforward. As it was so small, somatostatin had seemed like the perfect candidate to show that this technology worked. However, it was the short length of somatostatin was the very cause of the problem. Because it was so short, somatostatin was being chewed up by bacterial enzymes that degrade fragments of polypeptide chains. Longer polypeptides fold up into complex three-dimensional configurations that can make them less susceptible to these enzymes. But somatostatin was so small that its polypeptide chain remained unfolded, leaving it vulnerable to degradation.

Riggs had actually already anticipated that these enzymes might pose a problem and had devised a solution even before the first experiment. He suggested that if somatostatin could be produced as a fusion with a larger protein such as the enzyme beta-galactosidase (beta-gal), this might protect it from degradation by the bacterial enzymes. In theory, this could be done by inserting the DNA fragment encoding somatostatin alongside a short piece of DNA that encoded a section of amino acid sequence from beta-gal. When the plasmid carrying these two sequences was introduced into bacteria, the resulting protein product would be a hybrid in which Riggs hoped the somatostatin portion might be protected from degradation. However, the blank photographic films showed otherwise.

But it wasn't Riggs's idea that was wrong. The principle was sound—it was just that the hybrid fusion product still wasn't long enough to protect somatostatin from being chewed up. Rethinking the process, Boyer and Riggs wondered whether the experiment might work if this portion of the fusion was bigger—perhaps the entire beta-gal protein. With this in mind, they repeated the experiment, this time inserting the DNA fragment for somatostatin at the very end of the DNA coding sequence for the beta-gal enzyme. A month or so later, they were ready to try again. This time it worked—an artificially synthesized piece of DNA could be introduced into bacteria to make a human protein. Riggs celebrated his victory by going along to watch the Dodgers play baseball, but he recalled very little of the game. He was too busy 'contemplating miracles'.[23]

Figure 56 *The team involved in the successful cloning and expression of somatostatin, 1977. From left to right. Back row: Arthur Riggs, Herbert Boyer, Keiichi Itakura, Roberto Crea; Front row: Lily Xi, Herbert Heynecker, Francisco Bolivar, Leonore Directo, Tadaki Hirose.*

Credit: *Photograph kindly provided by Professor Art Riggs and reproduced with permission of the City of Hope National Medical Center, Duarte, California.*

Submitting their work to *Science* for publication, Riggs and Itakura now gave a press conference at a hotel in Los Angeles on December 9, 1977,[24,25] but it wasn't only the media who were interested in their work. At the opening session of hearings before the Senate Commerce Committee's Subcommittee on Science, Technology and Space, NAS president Phillip Handler hailed their achievement as 'a scientific triumph of the first order'.[26] With Handler's announcement, politicians suddenly realized that this new technology might be ripe with commercial possibilities.[27] Swanson and Boyer now found themselves summoned to Washington D.C. to discuss their work with Senator Kennedy and argue the case as to why the NIH guidelines should not be made into federal law.[28]

Until now, Genentech had only been a paper entity with funds drummed up by Swanson supporting the lab work at UCSF and the City of Hope, but in the wake of the success with somatostatin, it became much more. Swanson decided that it was time for the company to take on a physical existence and he found some old warehouses in South San Francisco that could be used to house laboratories and offices. He also needed to find some scientists.

Enticing talented young post-doctoral researchers away from academia to work in a new commercial environment was not easy. Aware that Axel Ullrich was the lead researcher in the Rutter/Goodman team's attempt to clone and express insulin, Swanson thought he would be an ideal new recruit for Genentech. When Ullrich met with Swanson over a few beers in the Chelsea bar, a drinking hole favoured by UCSF post-docs, he recalled that some pretty unconventional methods of persuasion were used:

> ... they had these game machines at that time—very primitive compared to what's available today. There was one machine that was like a tennis game which was essentially a flat-table screen where one could move bars. A ball bounced off these bars, and you had to catch it and shoot it back and stuff. Swanson challenged me and said, 'Okay we play a game. If I win, you will join; if you win, you still have free choice'.[29]

Despite losing the game, Ullrich still refused to join. But refusing to be beaten, Swanson kept the pressure on and, early in 1978, Ullrich relented and signed a contract which offered a forty-thousand dollar a year salary and the promise of 1250 shares of Genentech stock—unheard of for an academic post-doc.[30] But even this lucrative offer was not enough to keep Ullrich's doubts at bay. Driving home, he began to panic, thinking 'God, what have I done? I've signed a contract to work at a company'.[31] When he finally arrived back at his apartment, he called Swanson to tell him that having reconsidered, the deal was off.

From the perspective of a young researcher at that time, his concerns were understandable. While patenting and the commercialization of academic work were far more common in disciplines such as engineering, it was a practice almost unheard of then in biology. Scientists working within the environment of a university had free rein to pursue their own research interests and publish the resulting work, but many doubted whether that same freedom would exist in an industrial lab. There was a widely held perception among academic scientists that commercial research would be shaped purely by the dictates of the market and that the pressure to file patents on discoveries would prevent the publication of work in academic journals. Because the career and reputation of an academic scientist is built on amassing such a collection of publications, there seemed to be little incentive to move into a research environment in which this would be stifled in the interests of protecting commercial secrets.

Some of Boyer's more senior colleagues at UCSF, who were so advanced in their careers that they no longer had to worry about publishing their work, had other concerns about Genentech—specifically, Boyer's close involvement with the venture. There were murmurs of dissent among UCSF colleagues that the whole Genentech venture was driven by greed—that by dabbling in the corporate world, Boyer was compromising the purity of his academic research and conflicts of interest would inevitably emerge.[32] Although Boyer insisted that he would remain working in his academic lab at UCSF, his involvement with Genentech made him a heretic in the eyes of some of his colleagues, and this took its toll on him:

… it was difficult to deal with the criticisms and the tension and the notoriety. It was very distracting to me … I felt ostracized. I resented the treatment that I got in the department. There were people that were supportive, so it wasn't universal … But it wasn't easy … And the way the attacks went, I felt like I was just a criminal. But I always felt that what I was doing was right. I didn't think I was doing anything unethical or immoral.[33]

Boyer described this as a 'manic-depressive time' and the criticisms stung him:

It was very difficult for me. I had a lot of anxieties and bouts of depression associated with this. Here I thought I was doing something that was valuable to society, and doing something that would make a contribution, and then to have the accusations and criticisms, it was extremely difficult.[34]

To prove his critics wrong and to allay the fears of young researchers such as Ullrich, Swanson and Boyer were determined that Genentech should be different. Swanson was clear that their company would offer young, talented scientists the opportunity not only to carry out fundamental research as they could in a university lab, but also to publish their work in top academic journals:

Everybody wanted *Nature* or *Science* or another good journal to publish their work, and so what we did had to be of a quality that would be published. So we said, 'Look, let's publish the results; let's make sure we get the patents, and we'll make the patent attorneys work overtime to get them filed before you actually get the papers out. But we'll work together on that'.[35]

Boyer, meanwhile, wanted Genentech to offer the *best* of both academia and the corporate world—there would be no selling out, no compromising of research:

… we set out with a philosophy for the company. We set out with a self-imposed mandate that employees would share in anything that came out of the company, in terms of holding stock in the company. I insisted that we have scientists publish their research in journals. Any proprietary information would have to be covered by patents. I felt this was extremely important for attracting the outstanding young scientists in the community … So we tried to set up an atmosphere which would take the best from industry and the best from the academic community, and put them together.[36]

In addition to offering young researchers all the perks of working in academia, there would be one huge extra bonus—the money would a lot better.

Eager to attract the best possible candidates for his new team, Boyer contacted Dennis Kleid at Stanford Research Institute—an expert in the field of DNA synthesis thanks to having learned his trade from the master himself, Har Gobind Khorana.[37] But despite the promise of a lucrative salary at Genentech, Kleid was cautious about the offer. At Stanford Research Institute he at least had the security of an academic post—why risk a leap into the unknown? As an undergraduate at Berkeley towards the end of the 1960s, Kleid had taken a class in molecular biology taught by one of its pioneers, Gunther Stent, who had assured him with confidence that 'molecular biology is all over. It's too bad you missed it'.[38] When Boyer and Swanson took him out to dinner at a French restaurant, they sought to persuade him otherwise.[39] Yet, when they began talking about Genentech, Kleid thought they were simply offering some welcome financial support for ongoing work on the synthesis of DNA in his lab. It was soon apparent that they proposed he leave the security of academia and become one of Genentech's very first employees.

Mulling over the offer, Kleid said that he would accept it on one condition: that he was allowed to bring his post-doc Dave Goeddel with him. Fresh out of graduate school at the University of Colorado, Goeddel had come to work with Kleid at the Stanford Research Institute on DNA synthesis and cloning although decision to move to Northern California was motivated as much by his desire to climb El Capitan in Yosemite National Park as it was to pioneer cutting edge molecular biology (Figure 57).[40]

Kleid recalled that, whether he was rock climbing or cloning DNA, Goeddel did both with a furious and formidable intensity:

> Today, every time I see Tiger Woods I think of Dave Goeddel. There's something about the way he approaches things—he's concentrating so hard, every single shot that Tiger Woods is going to take is going to be the best shot he's ever made in his life. You can kind of see that. And if it goes right, you see this huge smile, and if it doesn't go right, oh, there's a lot of anger. But when he gets to the ball, the anger is gone, even if the ball's in the woods somewhere, and the next shot is going to be the best he ever shot in his life. That's the way Dave is. An intensity that you see mostly in sports was being applied to doing experiments in the lab.[41]

Unlike Ullrich, Goeddel had no qualms about moving from academia to industry. When Kleid mentioned Swanson's and Boyer's offer to him, he jumped at the chance:

> Here was an opportunity, here was an exciting thing, we were going to do insulin. There were other people trying to do it too, and it was not going to pay to come in second. You either came first or you might as well be last.[42]

As far as Swanson was concerned, the successful cloning and expression of somatostatin indicated they were ready to try and produce human insulin. As well as pouring time and energy into attracting talented young post-docs like Goeddel, he also wanted to make sure that Riggs's group at City of Hope were still on board with this plan. Taking them out to dinner, Swanson turned to Roberto Crea, one of Riggs's post-docs whose expertise in purifying small fragments of synthetic DNA had been invaluable to the somatostatin work, and asked bluntly, 'Roberto, how long will it take to make the insulin gene?'[43] He was delighted when Crea replied that it shouldn't take more than six months.

Figure 57 *Some members of the team involved in the collaboration between Genentech and City of Hope to clone and express human insulin, circa 1978. Pictured from left to right: Keiichi Itakura, Art Riggs, Dave Goeddel and Roberto Crea.*

Credit: Photograph kindly provided by Professor Art Riggs and reproduced with permission of the City of Hope National Medical Center, Duarte, California.

But Crea's colleague Keiichi Itakura was much less optimistic.[44] Itakura recognized that, although the synthesis of the DNA for somatostatin had been straightforward, insulin presented a new challenge. Whereas somatostatin was a single short polypeptide chain of only fourteen amino acids, Fred Sanger's work had shown that a single molecule of insulin consists of two polypeptide chains, one with 21 amino acids, and the other with 30 amino acids. Almost a decade after Sanger received his Nobel Prize for this discovery, the mystery of how a single stretch of DNA can give rise to two separate protein chains had been solved. Working at the University of Chicago in 1967, the biochemist Don Steiner had shown that in a pancreatic beta cell, insulin is first synthesized as a larger precursor—proinsulin—a single polypeptide chain which is then cleaved at two specific sites to generate the separate A and B chains, which then combine to form active insulin.[45]

Crea and Itakura now began planning how to assemble fragments of DNA coding for the A and B chains of insulin from smaller blocks of bases. Once they had worked out a strategy, they began the process of synthesizing these short fragments in the lab. The task left them reeking so strongly of organic solvents that not even a hot shower would remove it from their skin.[46]

By April, they had produced a set of 29 short chunks of DNA that needed to be ligated together using enzymes to produce the two complete fragments encoding the A and B chains of insulin. To carry out this work, Kleid and Goeddel made regular trips down from San Francisco to work in a corner of Riggs's lab. Having spent his first month at Genentech occupied with the somewhat more mundane task of ordering glassware and equipment, Goeddel was delighted to be doing some real science at last, although he found the culture of Riggs's lab a little at odds with his own furious work ethic:[47]

> Our style wasn't the same as his [Riggs's] lab's. His lab was pretty easy going. They didn't work very long days in general; they made sure they had afternoon tea; took a full lunch hour, and things like that. When I went down there it was just how quickly can I get this done and get back home? We worked pretty much an around-the-clock schedule the times were down there.[48]

Once the two complete fragments of coding DNA were assembled, Goeddel and Kleid took them back to the Genentech labs in San Francisco for the next stage— to cut their ends with restriction enzymes and insert them into plasmids. The resulting recombinant plasmids carrying the insulin DNA sequences were then introduced into bacteria and vast numbers of bacterial colonies had then to be screened in order to identify those carrying the plasmid with the insulin insert.

Before this, there was one crucial task—the sequence of each of the two DNA fragments needed to be read, base by base, in order to confirm that no errors had crept in during the process of synthesis. A single mistake somewhere in the sequence would be enough to synthesize a dysfunctional protein. In the protein haemoglobin, such a single base change can have catastrophic, and tragic consequences, for example, in sickle-cell anaemia.

Confirming the fidelity of both DNA fragments was vital. Synthesis and cloning of the A chain had proved to be straightforward—and when its DNA sequence was determined, no errors were found. Unfortunately, the B chain proved to be more problematic. Because it was larger, it had to be inserted into the plasmid as two separate fragments. And when Kleid and Goeddel read the DNA sequence of the B-chain fragment, they found a mistake. Somewhere during the initial synthesis, a triplet of the bases GCA that codes for the amino acid alanine had somehow been accidentally flipped around to become ACG, the codon for the amino acid threonine.[49] An easy error, perhaps, but one with the potential to wreck the entire experiment because it meant that the wrong amino acid would be inserted into the polypeptide chain at this point.

The timing of their discovery was unfortunate. Only a few days after Goeddel and Kleid had discovered this glaring error, Swanson's former boss Tom Perkins held a dinner and reception at his mansion. Perkins was now Genentech Chairman and had invited all the scientists over to hear how work on insulin was progressing. With their fingers crossed under the table, Kleid and Goeddel told him diplomatically that they were 'making fantastic progress'. Heynecker, who was over from Holland sorting arrangements for his long-term return, later confessed that' we were all too chicken' to tell Swanson the full story.[50]

Things had to be put right—and quickly. Goeddel and Heynecker dashed back down to City of Hope where they spent the entire Memorial Weekend public holiday in the lab, taking turns sleeping.[51] Within a week, they had resynthesized the B fragment and inserted it into a plasmid. Having introduced this into bacteria and obtained positive colonies, they raced back to San Francisco to analyse the sequence of the DNA, praying for no mistakes. No doubt they were also praying that on their return, they would not see Perkins's red Ferrari Testarossa parked outside the building.[52]

Once the sequence of the newly synthesized fragment was confirmed as correct, Goeddel and Kleid worked furiously on the next stages. A typical working day involved Goeddel picking up Kleid at 6 a.m., who would finish dressing in the front seat as they drove to the lab.[53] On arrival, Goeddel dropped Kleid at the back door so that he could avoid walking past Swanson's formidable secretary Sharon Carlock, of whom he was terrified.[54] After twelve hours of solid work, they would nip home in the evening before returning to the lab for another few hours.[55] Work continued into the weekends, where *Saturday Night Live* on a small portable TV in the lab brought much-needed comic relief.

Their efforts paid off, but successful synthesis of the DNA sequences was only half the story. One major hurdle still lay ahead. Each of those two DNA sequences had to be correctly processed into the A and B protein chains, which then had to be recombined in order to produce a functional molecule of insulin. The challenge was how to demonstrate that this had taken place in the test tube. The usual way of showing the presence of functional insulin was to measure its effect on blood sugar levels. But the amount of insulin available to Goeddel was so miniscule that this would be impossible.

But help was at hand thanks to the efforts of brilliant scientist Rosalyn Yalow (1921–2011).[56] Yalow's (Figure 58) fascination with nuclear physics had begun while an undergraduate at New York's Hunter College, where she had listened enraptured to the physicist Enrico Fermi lecture on the recently discovered phenomenon of nuclear fission.

After graduating as the top student from Hunter in 1941, Yalow became the first woman since 1917 to be accepted onto a PhD program in physics at the University of Illinois at Urbana-Champaign.[57] Having achieved straight As in her graduate classes, Yalow returned to New York to work as an assistant engineer at the (IT&T) Federal Telecommunications Laboratory where she discovered that academic prowess alone was insufficient to forge a career in science. Her acceptance onto the graduate programme at Illinois had, in part, been facilitated by a lack of male applicants due to the Second World War. When men began to return after the war and look for jobs, Yalow found herself quickly surplus

Figure 58 *Rosalyn Yalow (1921–2011).*
Credit: National Library of Medicine/Science Photo Library.

to requirements. At this time, the Bronx Veterans Administration Hospital was looking for a nuclear physicist to set up a new service using radioisotopes in medicine—and Yalow came with the highest of recommendations.

Yalow began using radioisotopes to study the metabolism of substances such as iodine and albumin by injecting radio-labelled compounds into patients and monitoring the subsequent decline of the radioactivity. Working with her was clinical intern Solomon Berson, who had no previous research experience, while Yalow herself had no formal training in biology or medicine. This partnership lasted up until Berson's death in 1972.

Initially, their work focused on studying the metabolism of substances such as iodine, albumin, and thyroxine, but it was for her work on insulin that Yalow became the second woman to win the Nobel Prize in Physiology or Medicine. Although Yalow's husband Aaron had type 1 diabetes, she always insisted it wasn't a factor in inspiring her work with insulin.[58] Rather, it was thanks to Professor I. Arthur Mirsky, Chairman of Clinical Sciences at the University of Pittsburgh School of Medicine, who urged Yalow and Berson to put his new theory about the origins of type 2 diabetes to the test with their radioisotopes.

Unlike those with type 1, a type 2 diabetic patient develops the condition despite normal production of insulin by the pancreas. Mirsky proposed that this might be because the insulin produced in these patients was being degraded by the action of an enzyme. He suggested that Yalow and Berson test his hypothesis by injecting patients with insulin that had been labelled with a radioactive isotope of iodine. If Mirsky's hypothesis was correct, then the amount of radio-labelled insulin circulating in the blood should rapidly decline in patients with this form of diabetes.

But when Yalow and Berson performed the experiment, they found a surprise. In those patients who had already been receiving insulin for several months or even years as part of their clinical regime, the radio-labelled insulin actually disappeared from the circulation at a 'strikingly slower' rate than in control subjects.[59]

Yalow's and Berson's explanation for this observation was that the insulin administered to these patients as part of their clinical treatment had triggered the production of an insulin-specific antibody by their immune system. This antibody had recognized and bound the injected radio-labelled insulin to form a complex that remained in the circulation and dissociated only very slowly to release the bound insulin.

An antibody against insulin was a radical proposal—much too radical for the editors of the *Journal of Clinical Investigations*, to which Yalow and Berson submitted their paper. Orthodoxy held that insulin was far too small a protein to trigger an immune response and so the paper was rejected, only to be accepted later when Yalow and Berson made a compromise and agreed not to refer to an 'insulin antibody', but rather an 'insulin transporting antibody'.[60]

Convinced that they had made a major discovery, Yalow and Berson were unwilling to have it be obscured by mere semantics. The discovery of an immune response against insulin might well yet turn out to be of significance in understanding diabetes, but Yalow and Berson had already spotted that its immediate practical application.

Until now, the only way of assaying a patient's levels of insulin was to measure its physiological effect on blood sugar, but an antibody that specifically recognized and bound insulin might enable levels of the hormone to be measured directly. Yalow and Berson realized that if a known amount of radio-labelled insulin was bound to an antibody and introduced into the blood of a patient, then the patient's own insulin would compete with the radio-labelled material, displacing it from being bound the antibody. The resulting dilution of the amount of radioactive material bound to the antibody could then be used to calculate how much insulin had been circulating in the patient's blood.

It was a brilliant idea but would have remained as nothing more than this, had it not been for Yalow's lab guinea pigs, from which she collected antibodies raised against injections of insulin. Yalow lavished her beloved guinea pigs with affection:

> Very early in the morning, she would come with lettuce from home to supplement the guinea pig's diet. Before anyone else had arrived, she would take each guinea pig from its cage, cradle it in the crook of her left arm, and feed it with her right hand. All the while she would talk to it, calling it by its name ... cooing, soothing, entreating the animal to be happy, and cajoling it to produce the most wonderful

antiserum ... When the animals were injected with antigen, or bled for their anti-body containing plasma, she would hold them, nuzzle them, kiss the tops of their heads, whisper to them.[61]

Yalow maintained that the secret of her success lay in the affectionate way she treated her guinea pigs. She was convinced that 'it was this soft sweetness, more than the right combination of antigen injected' that coaxed the animals into raising antibodies against insulin:[62]

> Anyone can make the antigen preparation and inject it and then take blood from the animals on schedule ... but you must talk to the guinea pigs. You have to love them if they are to give you the special antiserum that you need.[63]

Whether or not it was Yalow's blend of love and lettuce that did the trick, when she injected them with bovine insulin, she found that they made antibodies that not only reacted strongly with the foreign protein, but also against human insulin and could be used to detect its presence.[64] She and Berson now had the vital tool that they needed to make their idea of an assay for insulin a reality.

Yalow and Berson called their method radioimmunoassay (RIA) and it enabled the direct detection of insulin in amounts as small as picogram quantities. Yalow's biographer Eugene Strauss has likened this to being able to detect a teaspoon of sugar dissolved in a lake that is sixty-two miles long, sixty-two miles wide, and thirty feet deep.[65] It was quickly applied to the detection of a number of other hormones and then to a wide range of other biological substances, including certain enzymes, viral antigens, tumour antigens, controlled drugs, and the rheumatoid factor.[66] It became a key part of diagnosing conditions such as thyroid deficiency and viral infection in cases of Hepatitis B. Importantly, its impact on the clinical understanding of diabetes was particularly profound. Using RIA, insulin levels in patients could now be measured directly and the results were surprising. For despite having elevated levels of blood sugar, patients with type 2 diabetes were found to have normal, or even higher—levels of insulin. This led to the important conclusion that type 2 is caused not by a lack of insulin, but rather a loss of sensitivity to insulin arising from myriad causes.[67]

In recognition RIA's power, Yalow was awarded the 1977 Nobel Prize in Physiology or Medicine. In her Nobel speech, she pointedly mentioned that her first paper had been rejected, before going on to compare RIA with the telescope and microscope in its power 'for opening new vistas in science and medicine'.[68] Five years later, she also drew attention to another new tool that would have a similar power:

> Recombinant DNA technology has already resulted in bacterial synthesis of insulin, growth hormone and interferon ... Now that the revolution in molecular biology has enabled scientists to cut, splice, and redesign hereditary material, we must consider whether it is possible to affect the transmission of genetic disease. We can hardly yet appreciate the full potential of genetic engineering.[69]

Within a year of Yalow being awarded the Nobel Prize for her invention of an assay for insulin based on antibodies, it played a vital role in the birth of recombinant DNA technology. Goeddel needed a way to show that the A and B chains of insulin synthesized by his recombinant bacteria had come together to reconstitute a single molecular complex. And Yalow's and Berson's method of RIA gave him exactly what he needed.

It all hinged on the fact that the A and B chains could only be recognized by antibodies against insulin when they had combined with each other to form a single intact molecule. Having confirmed that the bacteria were successfully churning out A and B chains by showing that these could be recombined with their radio-labelled opposite number from bovine or porcine insulin, Goeddel went to City of Hope Hospital to complete the final step of the final stage of the work. Here, Crea used his expertise to help purify the two bacterially synthe-sized protein chains, which Goeddel then labelled with radioactive sulphur. He then added some insulin-specific antibody and, after Crea had gone home at 2 a.m. to get some rest, sat up during the small hours of August 24, watching and waiting. When Riggs showed up for work the next day, he found Goeddel still there, but the nocturnal vigil had been worth it. The lab scintillation counter was lighting up, showing that the insulin antibody was bound in a complex with anoth-er protein—the one that was radioactive. The experiment had worked—they had finally recombined the A and B chains to make human insulin.[70]

On September 6, 1978, the team flew to the City of Hope Medical Center to attend a press conference organized by Swanson to announce their success to the world. Not everyone was entirely comfortable with this decision, however. Riggs was unhappy that the work was being made public before it had actually been submitted to a scientific journal and wanted Swanson to delay the press conference until then.[71] When Swanson set a date for the press conference, Riggs protested that the paper simply could not be written up in the time available and was met with the blunt response, 'Well, you better figure out how to get it done. I don't see you working on it'.[72]

Swanson had good reason to be in a hurry. The production of human insulin using recombinant DNA technology was a triumph for Genentech, but this achievement now had to be translated from a successful experiment at the lab bench to a marketable pharmaceutical product. This feat was well beyond the means of the fledgling company. Only an established pharmaceutical company had the resources and capital needed to scale up the production process and take the drug through the necessary trials required to win FDA approval. In antici-pation, Swanson had been involved in arduous negotiations with Eli Lilly. The company was keen to produce human insulin using recombinant DNA and had already signed a contract with Rutter's team at UCSF in the hope that they might be successful. With millions of dollars at their disposal, Lilly knew that they could afford to back a second potential team and had therefore begun to show an inter-est in Genentech. While Swanson was delighted at this development, he was also cautious. He wanted to avoid Lilly acquiring all Genentech's recombinant DNA technology as described in the Boyer–Cohen patent and then using it for their own projects. This would mean Genentech would lose its key piece of intellectual property upon which it had been founded.

His efforts paid off. On August 25, 1978, just over a week before the press conference was scheduled, Genentech signed an agreement with Lilly, who would provide financial support to Genentech in return for a license to use their strains of bacteria to make recombinant human insulin. A small research-based company starting from academia having a partnership with a multinational company was unheard of at the time, and in the eyes of the business community it conferred invaluable kudos upon Genentech.

All this made Swanson eager to announce their success to the world in a press conference, and that this should be done by the scientists who had actually done the work. Despite Swanson's persistent requests, Goeddel said that he was too embarrassed to speak to the media about the work and suggested instead that Kleid do the talking.

At this point in the proceedings the attention of the audience had drifted away from the production of human insulin to questions of safety and ethics surrounding recombinant DNA. Kleid's task was to draw discussion back to the main theme, but the first question that he faced from the assembled pack of reporters was 'Genentech made this insulin, and isn't that dangerous?'[73] The opening words of his response were sure to raise a few eyebrows in the audience: 'Well, if you say we made insulin in these bacteria, then that's a lie'.[74]

Kleid followed this up by trying to explain the technical details of the work— how what had actually been produced was a hybrid protein in which insulin was fused to the bacterial beta-gal enzyme to protect it from degradation and from which, the insulin could then be released by chemical cleavage before purification. But while such technical subtleties might have generated a buzz at a scientific conference, they were lost on the press. What they heard– and remembered—was Kleid's opening line, and in particular the words, 'that's a lie'.

To Swanson's horror, Kleid kept talking. In desperation, Swanson ordered Goeddel to 'get him down!' in a voice loud enough for Kleid to hear.[75] But Kleid was in full flow and Goeddel was at a loss. Thankfully, due to a fortuitous power failure, the lights went out and the room was left in darkness. But even when the power was restored, the comedy continued. When eager reporters asked to see a sample of insulin, Kleid proudly opened a fridge door to show them, but with such enthusiasm that all the vials went flying.

Although these mishaps might have suggested otherwise, the day was a success. When the Genentech team landed back in San Francisco that same day, they caught sight of the evening newspapers, where the front pages announced that Genentech had produced human insulin.[76]

There were some criticisms. Some biologists pointed out that that the Genentech team had not yet actually demonstrated that their bacterially produced insulin had any biological effect. As far as they could see, all Goeddel had done was to show that the protein synthesized by the bacteria reacted with an antibody that specifically recognized insulin.[77] None of this dented Lilly's confidence in Genentech.[78] The research agreement with Lilly had brought kudos for the new company and also much needed financial support for the next stage of the work. It also brought some new challenges.[79] On a visit to Lilly's insulin production plant in Indianapolis, the Genentech scientists were told that if their method of producing human insulin was to come anywhere even close to levels of insulin production achieved using the conventional process then the yield would have to be increased by at least fifty per cent.[80] This was no mean feat and more importantly, it was one on which the whole agreement with Lilly was dependent. The terms of the deal agreed by Swanson stated that each payment of funds from Lilly would only be released to Genentech when its scientists achieved a series of predefined benchmarks agreed with Boyer and Swanson, and which related to increasing the yield of insulin.

In order to hit these benchmark targets, Swanson needed to expand his team—and quickly. He again approached Ullrich, who this time had no reservations about leaving academia to join Genentech. The public announcement of Genentech's success in cloning and expressing human insulin in bacteria may have played a part in Ullrich's *volte-face*, but his decision was equally swayed by a growing sense of disillusionment and discontent at the way he had been treated at UCSF. This had reached its nadir when, while in Strasbourg, he had learned that, along with his fellow post-docs, Pete Seeburg and John Shine, his name had been removed from the patent filed by UCSF on cloning the insulin gene. Recognizing an opportunity, Swanson also made offers to Seeburg and Shine. Although Shine eventually returned to his native Australia, Seeburg gladly accepted.

Like Ullrich, his decision was no doubt coloured by the extent to which his working relationship with Goodman had unravelled, but unlike Ullrich, this had actually deteriorated into outright hostility. The situation had become so bad that Goodman had kicked Seeburg out of the lab and locked a chain around the freezer in which he kept the clones containing the recombinant constructs of human growth hormone.[81]

Seeburg argued that, since he had made these reagents, they were rightfully his property, and he ought therefore to be able to take them with him to Genentech so that he could continue using them in his research there. Goodman took a very different view—as far as he was concerned, they were the property of UCSF. Because Goodman had banned Seeburg from entering the lab, it seemed that Goodman had won the day—but Seeburg had an ace up his sleeve.

Even molecular biologists hang up their lab coats and put down their pipettes for a momentary pause from work to celebrate on New Year's Eve. And so, figuring that the place would be deserted, Ullrich entered Goodman's lab at UCSF for the last time just before midnight on New Year's Eve, December 1978, with the intention of collecting his reagents from the freezer. With him was Seeburg, who had persuaded Ullrich to allow him to come along so that he could also collect his reagents that Goodman had locked away. Having procured all their reagents, they then left the lab and headed back across the UCSF campus, keeping an eye out for the campus police.[82]

At Genentech, they found a culture of 'work hard—play hard'. Swanson knew that his scientists, clad in their 'Clone or Die' T-shirts, were the company's most valuable asset.[83] In an address to Congress in 1988, he said that 'most of our technology walks out every night in tennis shoes'.[84] When Goeddel arrived for dinner at Perkins's mansion one evening, he was greeted with bewilderment by the butler who, on taking one look at Goeddel's attire, presumed he had come to the wrong address. As the dinner was being hosted for a delegation from the company Johnson & Johnson who had expressed an interest in buying Genentech, everyone else present was dressed in black tie. Having come straight from the lab— and being somewhat unaccustomed to black tie events anyway—Goeddel looked a little conspicuous in his jeans and T-shirt. But when he found Swanson and apologized to him for looking 'a complete slob', Swanson was delighted: 'No ... you look like a scientist, this is great'.[85]

Endearing though they were, such eccentricities could also be a headache. On one occasion, when a group of representatives from Lilly were visiting, Swanson spotted Heynecker walking around oblivious to a piece of paper hanging from his back that read 'Danger – Mad Scientist at work'. It had been stuck there as a prank by Goeddel and Kleid,[86] and was hardly the kind of behaviour likely to leave the Lilly executives confident that Swanson and his team were a serious operation.

For Kleid, having to hit the benchmarks set by Lilly was an entirely new way of working and he wasn't very happy about it. For if the benchmark targets were not met, then according to the terms of the deal, Lilly was free to take Genentech's insulin-producing bacteria without paying them anything. When Kleid protested to Swanson, saying that the task was impossible, he was told: 'Dennis ... what are you talking about? It's not impossible. I don't want to hear that word, "impossible." Just tell me what you need to accomplish it'.[87]

Swanson's optimism was justified. Thanks to the expertise of Giuseppe Miozarri, a new recruit to the lab, Kleid was able to snip out the piece of bacterial DNA that acted as a switch to turn on expression of the insulin protein and replace it with a new promoter region. This one regulated the bacterial genes involved in synthesis of the amino acid tryptophan, and it turbocharged the production of insulin, giving Kleid the fiftyfold increase he needed.[88] Kleid also observed that the bacteria were now producing so much insulin that it was accumulating inside their cells as a precipitate, which made its purification much more straightforward.

Increasing the yield brought other challenges, too, and although these were non-scientific, they were no less problematic. Under the regulations issued by the NIH on recombinant DNA research, all growth of bacteria carrying recombinant DNA was restricted to a maximum fermentation volume of ten litres. Because Genentech was a private company, it was actually exempt from having to comply with the NIH guidelines, but Swanson and Boyer felt that it would be wise to be seen as playing by the rules. With Boyer already under fire from certain academic colleagues about his commercial involvement, at the time of the press conference on somatostatin, Swanson had emphasized that Genentech set the example.

A year later, things had changed. The success of cloning insulin and the deal with Lilly meant that there was now an imperative to scale up production—even if it meant going way beyond the ten-litre limit. In June 1979, *The New York Times* reported that Swanson had been quite open about the fact that Genentech had already been carrying out fermentations in the order of sixty litres for some time.[89] Swanson said that he had informed the NIH about this and invoked the medical need to produce human insulin as his justification for taking this action.

His timing was perfect. Only a few years earlier, in July 1977, Walter Gilbert and 137 co-signatories who had attended the Gordon Conference on Nucleic Acids had written an open letter to Congress expressing their concerns that moves towards legislation on recombinant DNA might 'inhibit severely the further development of this field of research'.[90] Gilbert and his co-signatories felt that the motivation for such legislation was rooted in 'exaggerations of the hypothetical hazards of recombinant DNA research that go far beyond any reasoned assessment'.[91] However, for hostile commentators in the media, such as columnist Charles McCabe, such hazards were far from hypothetical:

> Those lovely people who gave us the atom bomb have another treat in store for us. Now they can create new forms of life, by jiggling with genes. I don't pretend to understand much about recombinant DNA, as the whole thing is called ... But I do know what the act of creation is. When it is placed in the hands of men who by definition do not know what they are doing, it scares the daylights out of me. Jiggling with genes may cure cancer. Then again, it may cause outbreaks of new forms of cancer ... Why, in the name of all that is sacred, can't we learn to let well enough alone? Why do we diddle ceaselessly with nature? Why will scientists persist in playing God?[92]

Yet, by 1980, fears that the NIH guidelines would be crystallized into restrictive legislation by Congress were receding—as were concerns among the public and media. This seismic shift had been driven by several developments over the previous couple of years. One of these was the development of an enfeebled strain of K12 *E. coli* that could not survive outside a lab. This was more than just a technical development, for the work had been done at the University of Alabama medical school by the microbiologist Roy Curtiss III, who had initially been convinced that recombinant DNA research posed a real threat. He had named the new strain phi1776 in honour of both his college fraternity and the year in which American colonists declared independence from Britain, and like the officer Benedict Arnold in the Revolutionary War of 1776, Curtiss had now made a dramatic switch of camps.

In a letter to NIH Director Donald Frederickson, Curtiss said he was now 'extremely concerned that, based on fear, ignorance and misinformation, we are about to embark on over-regulation' of recombinant DNA research and went on to explain why he had changed his position.[93] Over fourteen pages, he went into meticulous technical detail to show why he felt that if recombinant DNA experiments were carried out in this particular enfeebled bacterium, they would not pose any threat so long as NIH guidelines were followed.[94] Curtiss's letter was widely circulated in Congress where, along with a second letter written by Sherwood L. Gorbach, chief of the infectious disease unit at Tufts University School of Medicine, it had a significant impact.[95] Gorbach's letter summarized the outcome of a meeting held in June 1977 at Falmouth, Massachusetts, that had sought to address criticism that, as the NIH guidelines had largely been drawn up by molecular biologists, this was rather a case of putting the fox out to guard the chicken coop. In response, the Falmouth meeting brought together scientists from a wider range of disciplines in the life sciences who had expertise in infectious diseases, such as virologists and bacteriologists, to assess the risks from recombinant DNA. Although agreement was not unanimous, the general feeling was that the risks were largely speculative.

Another pivotal development was the publication of research by Cohen and S. N. Chang in which they claimed that recombinant DNA molecules could form naturally through the exchange of mammalian and bacterial genetic material. Cohen had begun these experiments in response to critics such as Chargaff and Robert Sinsheimer, who claimed that the construction of hybrid molecules containing both prokaryotic and eukaryotic DNA violated an evolutionary barrier that ought to be respected.[96] With Cohen's and Chang's demonstration that such exchanges of genetic material could occur naturally, Chargaff's and Sinsheimer's fears seemed to be built on a false premise.

Having met with a number of scientists and being particularly impressed by Cohen's and Chang's results, Senator Kennedy concluded that recombinant DNA research was not as hazardous as its opponents had claimed and consequently withdrew his bill.[97] But perhaps the most important factor in shifting the attitude of politicians was not scientific data, but economic realism. By the end of the 1970s, fears that the United States had lost its technological and industrial edge, coupled with economic recession, were giving politicians cause for concern. However, in June 1980, an article in *The New York Times* described how, in recent years, a small number of new companies had sprung up built on exciting novel technology that it declared to be 'An Elixir for America's Flagging Industry'.[98] Two sectors showed particular promise. The first of these was the electronics and microcomputer industry, for which *The New York Times* chose a small business called Apple Computer Inc. as its example of a leading light in the field. The other was recombinant DNA technology, for which the article named Genentech as the most promising example.

Sharing the conviction of *The New York Times* that 'There's Gold in Them Thar Recombinant Genetic Bits',[99] Congress now sought to adopt measures that would create a fertile environment for what was hailed as 'The Industry of Life'.[100] Such proposals included tax cuts, deregulation, and building stronger links between universities and industry—all of which it was hoped would restore the nation's economic vigour.[101] Nor did this political momentum stall when Ronald Reagan became president in January 1981. For with its recognition of the importance of science-based industries and a stark ideological commitment to deregulation, low taxes and a free market, the new administration sought to create favourable conditions so that others might follow in Boyer's and Swanson's lead.[102]

Despite the changing mood, there were still voices of opposition. A paper from one critic suggested that, if children living in the area around Lilly's production facility became infected with recombinant *E. coli* that produced human insulin, they might suffer hypoglycaemia and die.[103] But the political climate had changed, and in January 1980, Frederickson granted approval to special requests made by Genentech to exceed the ten-litre limit. Later that year, Lilly was granted permission to scale up the fermentation to 2000 litres and production began at facilities in Indianapolis and Speke, Liverpool, UK.

According to Irving Johnson, the rationale for choosing a production site in the UK was made not so much for economic or scientific reasons, but was shaped more by the Cold War:

> It was curious when we decided to go into production. Eventually we decided we would built a plant in England for it. One of the reasons for this was, we were still concerned about Communism taking over the world, and what if some catastrophic accident happened and destroyed our plant, like an atomic bomb or something, and what would the diabetics do? So we agreed that we would make a facility in England as well as in the United States.[104]

Cold War tensions rose in the late 1970s and early 1980s. The same year that Lilly built their production facility at Speke, the UK Government released 'Protect and Survive'—a leaflet containing advice on civil defence measures to be taken in the event of a nuclear attack. Though Johnson's concerns for diabetic patients were no doubt well meant, he appears not to have recognised that, after such a catastrophe, the procurement of a reliable supply of insulin would in all likelihood have been the least of their worries.[105]

Genentech and Lilly could quite easily have just ignored the NIH regulations. As private companies, they were not bound to comply with the regulations and could have forged ahead with large-scale production of recombinant insulin, but Frederickson's approval gave them legitimacy—and good PR. However, in the spring of 1980, Boyer and Swanson waited anxiously for the outcome of a Supreme Court concerning an application made in 1972 by the company General Electric to file a patent on a bacterium that had been deliberately modified by Ananda Chakrabarty, an industrial biochemist. Using methods that did not involve recombinant DNA technology, Chakrabarty had altered the bacterium so that it could degrade crude oil. General Electric felt that this was a sufficiently novel innovation to warrant protection with a patent, but the US Patent and Trademark office felt very differently and rejected the application on the grounds that forms of life were not eligible to be patented.

Until the Supreme Court reached a decision, Genentech and other new biotechnology start-up companies faced serious complications and delays over the intellectual property rights of their own recombinant bacteria—for example, those containing the DNA sequence for human insulin. Should the Supreme Court rule in favour of the decision made against General Electric, any hope that Genentech had of defending its own patents would be ruined, and the whole biotechnology industry might have been killed at birth.[106]

The Chakrabarty case has been described as 'one of the single most important legal-economic events of the 20th century' and Genentech made their voice heard, thanks to lawyer Tom Kiley.[107] At law firm Lyon & Lyon, Kiley had a reputation for taking on 'unconventional' clients, for example, Miss Nude American, a defendant in a law suit filed by Miss Universe Inc. over alleged trademark infringement.[108] His seniors felt that Genentech's situation fitted Kiley perfectly.[109] According to Kiley, when Swanson had first contacted Lyon & Lyon asking for someone to negotiate research contracts between Genentech and the university labs at UCSF and City of Hope the company's response had been: 'Let's see. Kiley represents lots of these weirdos ... send Swanson to him'.[110]

Kiley did not disappoint. He wrote a brief to the Supreme Court stressing how innovation through new technology companies like Genentech was essential for revitalizing the health of the domestic economy and that such fledgeling companies could only survive if their intellectual property were protected.[111] Swanson and Boyer breathed a sigh of relief when, in June 1980, swayed by Kiley's arguments, the Supreme Court voted by a margin of 5 to 4 in favour of General Electric. Chief Justice Burger ruled that 'Dr. Chakrabarty's invention is the result of human ingenuity and research', before adding that, while neither Newton nor Einstein could have patented their work, Chakrabarty's bacterium was 'not nature's handiwork, but his own; accordingly it is patentable subject matter'.[112]

With this ruling, the US Patent Office could now begin to process the backlog of over 100 patent applications it had amassed while waiting for the case decision.[113] One of these was the application first filed in 1974 by Neils Reimer on Cohen's and Boyer's techniques, and on December 2, 1980, this patent was finally granted.[114] In 1981, Stanford made licenses available on Cohen's and Boyer's technology, with companies paying an initial $10,000 fee for a license; by February 1982 Stanford had earned $1,420,000 as income from licenses.[115]

In the summer of 1981, the very first trials of recombinant insulin were carried out at Guy's Hospital, London under the supervision of the eminent diabetologist Professor Harry Keen. The trial was only preliminary and carried out on a very small scale using seventeen healthy volunteers. The aim of these tests was to compare recombinant human insulin with porcine insulin to show that it was effective in lowering blood sugar levels, that it did not elicit an immune response, and—crucially—that it was safe. Writing in *The Lancet*, Keen concluded that 'genetically synthesized human insulin seems to be safe and effective in man', but he also took care to acknowledge that this had only been a very limited preliminary trial and was 'but the first step in a longer process of investigation of its actions, validation of clinical efficacy in diabetics, and continuing vigilant surveillance for unexpected adverse effects'.[116]

Any sense of relief that Boyer and Swanson might have felt over the trial's success was short lived. Ullrich and Seeburg were both now working at Genentech using reagents they had procured from their New Year's Eve 'midnight raid' on Goodman's lab at UCSF, and this was not without its consequences. Although Ullrich and Seeburg considered these reagents to be their rightful property and that their action in taking them was therefore justifiable, the authorities at UCSF took a very different view. In February 1980, Rutter and Goodman requested that aggressive action be taken against Genentech on the grounds that its employees had unlawfully acquired reagents protected by patents filed by UCSF.

The situation was tense—venture capitalists like Perkins were not motivated by altruism: they expected a return on investment. Usually, if the start-up showed sufficient promise, it was purchased by a major pharmaceutical company. Despite the black-tie dinner at his mansion, Perkins had been unable to win the interest of Johnson & Johnson.

This left one other option—an initial public offering (IPO) of stock on Wall Street. However, prospective investors were hardly going to invest in a company embroiled in a legal battle with a major university. Genentech needed to resolve the situation quickly, and once again, Kiley saved the day. After negotiations with lawyers at UCSF, Kiley reached an agreement whereby Genentech would pay a one-off sum of $350,000 to UCSF as compensation. Although the legal disputes between Genentech and UCSF continued over the following years, Kiley's initial deal made UCSF back down in the months running up to Genentech's Wall Street debut.

Swanson breathed a sigh of relief. Never a man to do anything by halves, he now threw his heart and soul into promoting Genentech to prospective investors. When he and his wife Judy were married in Florida in September 1980, he immediately whisked her off to Paris—not only for a honeymoon, but as the first stop on a European-wide tour to drum up support from prospective investors:

> We did two cities a day. From Tuesday through Friday we did Paris, Geneva, Zurich, Edinburgh, Glasgow, London. We'd arrive and we'd tell my wife, 'We'll be back in two hours and we'll go to the next place. Have fun looking around'. Of course, we were always late. She was very understanding. That weekend there was the old Leeds Castle just south of London, which was dedicated by the Whitney family for medical conferences and things like that. So we were able to stay in the castle that weekend, along with the rest of this contingent ... left my poor wife on her own at Heathrow [Airport, London] with a rented car and said, 'Drive into the countryside. You're going to love the Lake District. I'll be back in a week'.[117]

It must have been true love. When people asked Swanson's new bride what she was doing wandering the Lake District alone, she replied, 'Well, I'm on my honeymoon'.[118]

Swanson was adamant everyone who worked at Genentech—from the company president to the lowliest assistant—would have a stake in the company.[119] While Swanson was confident that the IPO would be a resounding success, for many of the scientists, the world of stocks and shares was an alien environment. When Swanson called a meeting to explain to employees that they would all be receiving shares, one member of staff raised his hand and asked whether he could simply have cash in lieu of the shares. Swanson replied: 'Sure, that's okay ... Because you'll need that money to have your head examined'. Goeddel said it was the only time that he had ever seen Swanson put down an employee in front of others.[120]

Within only a few years of Swanson holding this meeting, UK pop duo The Pet Shop Boys captured the spirit and energy of the new decade with their single 'Opportunities', which hit number 10 in the US charts. The song told the story of two entrepreneurial types, one endowed with brains and the other with looks who had their sights set on making a fortune. Although the song is said to have been a satirical swipe at the values of the decade, the chorus could easily have been a hymn to Boyer's and Swanson's partnership. On the morning of October 14, 1980, as the bell rang to open the day's trading on Wall Street, all those investors sceptical of the initial offering of $35 as too high for a tiny company were proven spectacularly wrong.[121]

According to *The Washington Post*, Genentech was possibly 'the hottest new offering in Wall Street history', while the *Financial Times* described the mood on the trading floor as one of 'Euphoria'.[122,123] In the weeks leading up to the flotation, investors had scrambled to place their orders for shares. One securities analyst who had been following Genentech's research called her regular broker well in advance hoping to secure a bargain, only to be told 'Just because it's you, I'm doubling your allotment – two shares instead of one'.[124] Within only minutes of the start of trading, the price of shares in Genentech skyrocketed from $35 to $80.[125] Twenty minutes later, they had reached $89.[126,127] Watching in astonishment—and delight—one officer at investment company Merrill Lynch, Pierce, Fenner & Smith exclaimed, 'I have been with the firm 22 years ... I have never seen anything like this'.[128] As the price oscillated over the course of the day, Robert Antolini, a trader with Drexel Burnham Lambert, Inc. recalled how he resorted to popping aspirins and shouting expletives as a temporary computer failure froze the price for a few minutes—the tension no doubt aggravated by wondering whether his boss had been serious that morning when he had warned his team, 'No rally in Genentech, no job!'[129]

The Wall Street Journal hailed Genentech's debut as 'the most striking price explosion on a new stock within memory of most stockbrokers' and by the close of the day's trading, the shares had settled to a modest $71, driven down perhaps in part by calls for caution from John Whitehead, senior partner at Goldman Sachs & Co.[130] Whitehead was concerned that the stock market feeding frenzy might start an investment bubble that could 'attract issues of lesser quality'.[131] But despite Whitehead's call for sobriety, the closing price was still enough to make Boyer, the son of a Pittsburgh railwayman, a multimillionaire to the tune of $70.2 million.

One person who was perhaps left a little disappointed by Genentech's meteoric performance was Axel Ullrich, who had not shared Swanson's confidence about the forthcoming stock market flotation. Desperate to replace his old second-hand car that kept breaking down, Ullrich had sold a fifth of his 4000 shares ahead of the flotation in return for $8000 dollars with which he bought a second-hand VW Rabbit. It was, unsurprisingly, a decision that he quickly came to regret:

> … after we had gone public, the stock price went up and up and up. At some point, these eight hundred shares were worth more than a million dollars, and I bought a used Rabbit for that, a million-dollar Rabbit. Oh god![132]

Boyer, meanwhile, rushed out at the end of that first day and treated himself to a Porsche Targa.[133]

Nor were Genentech employees and Wall Street traders the only ones celebrating. Gilbert may have lost out to Boyer in the race to clone human insulin, but on the very same day that Genentech made its spectacular Wall Street debut, Gilbert received the news that he had been awarded the Nobel Prize in Chemistry for his work on developing methods of sequencing DNA.[134]

Sharing the award with Gilbert was Fred Sanger who, after first receiving this award in 1958 for his work on insulin, had now become the only British scientist ever to receive two Nobel Prizes. But for diabetes specialist Professor Sir George Alberti, such glittering accolades were not without their problems:

> I personally believe that such prizes and awards do more harm than good and should be abolished. Many a scientist has gone to their grave feeling deeply aggrieved because they were not awarded a Nobel Prize (it is actually rather comforting to be a bit stupid and not in the frame at all!).[135]

Speaking about the controversy over the award of the Prize to Banting and Macleod in1923, Alberti said, 'The real problem was the Nobel Prize. The difficulties could and would have been resolved much more rapidly if the Nobel Prize had not existed'.[136] And such controversies were showing no signs of disappearing any time soon. Over the course of the years following its discovery, insulin was to be intimately bound up with the Nobel Prize. It had brought a Nobel Prize to Sanger and Yalow, while in the case of Archer Martin, Richard Synge, and Dorothy Hodgkin, the work that earned them a Nobel Prize had proven also to be crucial in unravelling the structure of insulin.

But in some quarters, there was a feeling that one name was conspicuous by its absence from this list. When she was interviewed in 1994, Boyer's technician Mary Betlach, who later went on to earn a PhD and carve out a career in molecular biology, felt that a grave injustice had been done to her former boss by this omission:

> Every year when Nobel Prize time comes around, I feel upset for him. I think now he's gotten over a lot of this because he's doing so many different things in

his life now. About three months ago I finally sat down and told him how I felt, that I felt like it wasn't right, that I felt like he should have the Nobel Prize … I think for recombinant DNA and cloning, Herb deserves the Nobel Prize.[137]

James Watson agreed. For along with Gilbert and Sanger, the third co-recipient of the 1980 Nobel Prize in Chemistry was Paul Berg for his work on recombinant DNA. Writing in 2003, Watson said he found it inexplicable that 'neither Stanley Cohen nor Herb Boyer has been so honored'.[138] Boyer seems to have been more philosophical about this turn of events. Perhaps recalling just how much happiness the award of the Nobel prize had brought to Fred Banting back in 1923, he may have concluded that he was far better off with his shiny new Porsche.

13

'Don't You Want Cheap Insulin?'

When the March 9th edition of *Time* magazine hit the newsstands in 1981, Herb Boyer's beaming face took up the entire front cover, squeezing the news of the marriage of Prince Charles and Lady Diana Spencer that summer into the top right corner.[1] Although Genentech's share price had drifted back down to a more modest $45, Boyer still had good reason to be smiling from ear to ear.[2] By founding Genentech with Swanson, he had not only blazed a trail into exciting new territory, but had shown fellow academics a tantalizing glimpse of the glory that awaited them, were they to follow his lead. The years that followed saw a boom in small biotechnology companies springing up out of academic research from university labs. By 1983, eleven more biotechnology companies emerged and were floated on the stock market.[3] When Ronald Reagan ran for re-election the following year with the campaign slogan, 'It's morning in America again', his words rang true for patent lawyers specializing in the new field of biotechnology.[4]

In August 1982, two years after it was first tested in healthy volunteers, *The Lancet* published the results of a trial of the efficacy and safety of recombinant human insulin in diabetic patients.[5] The authors concluded that the new drug 'appears to be a safe alternative to porcine or bovine insulin' and that the method of producing it using recombinant DNA technology 'should overcome any problems in the supply of insulin for the treatment of diabetes'.[6] Later that year, the Food and Drugs Administration (FDA) granted approval for Lilly to launch its recombinant human insulin onto the market under the brand name 'Humulin'—and in response, Genentech's share price began once more to rise.[7]

However, investors hoping for a repeat of the Wall Street reaction two years earlier were disappointed. Having climbed from a modest $33 to $50, Genentech's shares closed at a sober $48[8]—not quite the same astronomical rise as its debut, but still enough to make a small fortune for those who had missed out first time round. But while investors still eagerly poured their dollars into Genentech shares, doctors were more cautious about the new drug. At a projected cost of $0.55 per dose, it might actually be twice as expensive as the cheapest bovine insulin.

David M. Nathan, director of the diabetes clinic at Massachusetts General Hospital, urged caution and questioned whether recombinant human insulin really would revolutionize the treatment of diabetes. He told *The Wall Street Journal*:

'It won't change our practice at all … I don't see it as the drug of choice; I don't see that it has any demonstrated clinical advantages over highly purified pork insulin'.[9]

But despite the general verdict that Humulin was 'beneficial, not revolutionary', clinicians did recognize that it might have one key advantage over animal insulins.[10,11] Writing in *The Lancet* in 1925, R. D. Lawrence had expressed concern over what he described as 'definite local effects and reactions at the site of injection' caused by insulin:

> These reactions vary with the individual, the site of injection, and the brand and batch of insulin used. The variations are not generally recognized and have been the cause of much worry to patients and physicians who have had, perhaps, only a few insulin cases to treat. They have been the cause of many complaints to the manufacturers about their insulin, and in some cases have led doctors to abandon the treatment, even where it seemed essential.[12]

By the time that Humulin was released onto the market, these reactions were becoming much less of a problem. In his work to identify proinsulin, the precursor of insulin, Don Steiner had shown that three distinct fractions could be separated from preparations of animal insulin—the proinsulin precursor, some other pancreatic proteins, and the active insulin.[13] These substances often co-purified with insulin, but work by Jorgen Schlichtkrull and his colleagues at Copenhagen's Novo research institute, and independently by Mary Root at Lilly, however, showed that if they were removed, then the immune reaction was significantly reduced.[14] As a result, Novo released this newly purified pork insulin onto the market in 1973 under the brand name Monocompetent, or MC, insulin, with Nordisk soon following suit.[15]

Despite this improvement, a small number of patients still continued to experience problems. For even though the amino acid sequence of porcine insulin differed from that of the human protein by only one single amino acid change, this was enough to elicit an immune response in some patients. For those using bovine insulin, which differed in its sequence from human insulin by three amino acids, this effect could be even more pronounced.

As a result, the Danish companies followed Lilly and turned their attention to the production of human insulin. They achieved this not by using genetic engineering but—initially at least—by a far less-controversial method. Even before US labs started their stampede to be the first to clone and express recombinant human insulin in bacteria, Michael Ruttenberg, a researcher in the Department of Chemistry at the University of California at San Diego had hit upon a piece of chemical alchemy that he believed might be of 'major utility in providing a practical source of human insulin'.[16]

Ruttenberg reasoned that, since human and porcine insulin differed only by one single amino acid at the very last position on the B chain, it might be possible to take pork insulin and convert it into its human analogue by simple chemical manipulation. His plan was straightforward. Using chemical and enzymatic treatment, he clipped off the last eight amino acids from the B chain of pork insulin and substituted in their place an artificially synthesized peptide chain that contained the first eight amino acids in the human protein.

Workers in Japan simplified Ruttenberg's method even further. Using an enzymatic reaction, they simply replaced the one amino acid that was different in porcine insulin, with that found in its human counterpart.[17] Using this improved method known as transpeptidation, researchers at Novo in collaboration with scientists at the University of York in the UK began producing what came to be known as semisynthetic, or 'humanized' insulin and which was shown in clinical trials to be as effective and safe as its porcine analogue.[18,19]

Both Eli Lilly and Novo were confident that patients would instinctively rather treat themselves with human insulin as opposed to animal forms—regardless of whether it was produced using recombinant DNA or semi-synthetically. Both companies directed their resources away from the manufacture of animal insulins to focus instead on the production of human insulin, with Novo also eventually adopting the recombinant DNA instead of the semisynthetic method but using genetically altered yeast instead of *E. coli* bacteria.[20]

The marketing strategy by Novo and Lilly stressed reduced immunogenicity and argued that, as this insulin was identical to that made by the body, it was the logical choice.[21] The investment in the extensive advertising campaigns for these products certainly proved to be money well spent. In the four years from 1984 to 1988, more than eighty per cent of patients in the UK had been switched from using animal to human insulins.[22]

The medical community was divided about the use of human insulin. While some doctors were sceptical about its benefits, others, like Dr. John Pickup of Guy's Hospital, London, argued that the new recombinant human insulin was to be welcomed:

> Without doubt the advent of new, biotechnological processes for insulin production should be welcomed for their potential value ... biosynthetic insulin may lead to much lower costs and increased supplies unlimited by the availability of animal pancreases. Whether the latter consideration will ever limit the provision of insulin remains arguable ... Nevertheless, the hope that one day large-scale production will be possible of cheap and pure human insulin must be relevant to diabetic patients everywhere.[23]

But in 1987 a paper appeared that seemed to cast doubt on the claim by certain manufacturers that 'human insulin is for all'.[24] Writing in *The Lancet*, two Swiss researchers, Arthur Teuscher and Willi G. Berger, presented data which they claimed showed that human insulin might have a very serious side effect. Type 1 diabetic patients are usually alerted to a drop in the concentration of their blood sugar by several warning signs, such as tremors, sweating, palpitations, and irregular heartbeat. Thanks to this physiological early warning system, the patient can recognize that they are entering hypoglycaemia and respond by ingesting some sugar quickly before the condition becomes serious. Teuscher and Berger, however, presented the case histories of three diabetic patients who claimed that their ability to detect the onset of a 'hypo' had been impaired after using human insulin as a replacement for animal insulins. This prompted the researchers to interview 176 type 1 patients, of whom thirty-six per cent claimed that moving from animal to human insulin had affected their ability to detect the onset of hypoglycaemia. Instead of the usual symptoms of tremors and sweating, they now experienced anxiety, light headedness, and an inability to concentrate.[25] Defending their work in the pages of *The Lancet*, Teuscher and Berger went on to say that when twelve of those sixty-six patients were transferred back from human to porcine insulin, they recovered their ability to detect hypoglycaemia.[26] From this analysis, Teuscher and Berger said that they had become 'concerned about the apparent marketing effort of manufacturers to influence physicians and patients to switch from animal to human insulin'.[27]

Unsurprisingly, this claim proved to be controversial. When clinical trials of recombinant human insulin had been conducted in 1982, three of the ninety-four patients had been withdrawn due to hypoglycaemia following the switch from animal to what was now being called Biosynthetic Human Insulin (BHI). The authors noted that this was 'a trend not found in the majority of patients' and concluded that BHI was 'a potent insulin preparation with no obvious side-effects that appear within 6 weeks of treatment'.[28]

But some patients remained unconvinced. An inquest into the death of Stephanie Wallace, a 30-year-old psychiatrist from Kettering, recorded that she had died in her sleep from suffocation. As a diabetic patient, she had taken her normal dose of insulin before going to bed, but a post-mortem revealed that her blood sugar levels had been 'immeasurably low'.[29] Speaking at the inquest, Dr. Patrick Toseland, a consultant pathologist and head of clinical chemistry at Guy's Hospital, London, said that he was becoming concerned about the apparent rise in sudden deaths among young diabetic patients.[30] In 1985, Toseland said that he had investigated two such cases; three years later this had risen to nine and by 1989, it was seventeen, of which Stephanie Wallace was the most recent.[31] About half of these cases were patients who had switched from animal to human insulin and as the story hit the newspapers with headlines, the finger of suspicion began to point at the new drug.[32]

Understandably, patients were worried by these headlines. In response, one diabetic patient wrote to describe her own experience of using human insulin:

> Within the last few years, insulin-dependent diabetics have been encouraged or, more to the point, told to change their insulin from the type that is extracted from animal pancreases, to 'human' insulin, a man-made insulin which has the same molecular structure as insulin produced in the human body. Since my doctor changed me to this new insulin a few years ago, I have experienced the same problems referred to in your news report, i.e. no warning signs usually

experienced when one's sugar level drops too low, quickened heartbeat, profuse sweating, numbness of the limbs, etc. etc. On various occasions my blood sugar levels dropped so low that I only realised what was happening when I was on the verge of passing out. During the night recently I was awakened by my boyfriend who was spooning honey into my mouth and talking to me, in an effort to bring me round ... I have mentioned these experiences to my doctor and to chemists both of whom have told me that other diabetics have complained of the same thing.

[...]

I have now insisted on changing back to animal insulin. Many of the insulin manufacturers have ceased making animal insulin having convinced themselves that 'human' insulin is the thing of the future, despite the opinions and wishes of diabetics themselves. I feel that the use of this new insulin should be questioned before more needless deaths occur amongst diabetics.[33]

Two other readers wrote in to say that they were very concerned that diabetic patients and their families were not being warned of the risk that human insulin might be associated with sudden hypoglycaemia. Having lost their brother in this way, they made a plea that this should change.[34]

Reporting on the cases in the UK, *The Wall Street Journal* pointed out that none of the patients who had died had been using human insulin made by Eli Lilly, but that the FDA would nevertheless begin conducting their own investigation.[35] However, for over two years, Eli Lilly had been warning that some patients had suffered a loss of physiological warning signs of impending hypoglycaemia when using human insulin. At the time, it was standard practice for pharmacists to remove the accompanying information leaflet from packs of insulin before passing them on to patients.[36] The leaflets were deemed to be written more for specialist clinicians than patients, but in August—within only weeks of the death of Stephanie Wallace and after several years of campaigning, Dr. Robert Tattersall, chair of the British Diabetic Association (BDA), announced that all manufacturers of human insulin would now be including a safety warning about the possibility of impaired awareness of hypoglycaemia.[37]

Both the BDA and the Committee for the Safety of Medicines expressed their concern about the situation and Novo Nordisk began an investigation into the safety of their human insulin.[38] But for some patients and their relatives, this was not enough. As two more papers from Teuscher's team appeared in the *British Medical Journal* claiming to show that human insulin—whether of recombinant or semisynthetic origin—was associated with an increased risk of hypoglycaemia and impaired warning of its onset, seven hundred patients and their families began to pursue legal action against Novo Nordisk.[39,40]

Some of these legal claims were actively encouraged by lawyers who placed advertisements in local newspapers advising diabetic patients that they might well be eligible for compensation. Clinicians were left distinctly unimpressed.[41] Writing in *The Observer*, consultant Dr. Anthony Barnett at East Birmingham Hospital described this conduct as unethical and stressed that there was still no clear evidence to show that switching from animal to human insulin resulted in impaired awareness of hypoglycaemia.[42]

In his letter, Dr. Barnett went on to point out that there were other important factors that might well be significant. Firstly, he claimed that, regardless of the source of their insulin, around one-third of diabetic patients lose their awareness of hypoglycaemia within ten years of diagnosis. Secondly, he pointed out that media coverage of the controversy had focused on only a handful of anecdotal accounts. Perhaps most significantly of all, the introduction of human insulin in the 1980s had coincided with a number of other significant clinical and technological developments in the management of type 1 diabetes.

One of these was the improvement in blood glucose monitoring systems that allowed patients to measure their own blood sugar levels by pricking their finger and drawing blood onto a disposable test strip inserted into a meter that gave a reading. The 1980s has been described as a 'bright start' for this technology, for at the same time as test strips were developed that required significantly smaller volumes of blood, blood glucose meters became more portable, easier to use, cheaper, and able to handle more data—all of which made the process of monitoring blood sugar levels easier for the patient.[43] These technological improvements allowed patients to monitor their own blood sugar levels more closely and the advice from clinicians at the time was to keep them as low as possible.[44]

This advice was given in order to avoid the long term complications of elevated blood sugar levels, but not every clinician felt that this was a wise strategy, for it risked tipping the patient into hypoglycaemia. In response to the controversy about possible deaths caused by impaired awareness of hypoglycaemia, Dr. Simon Wolff, a lecturer in Toxicology at University College and Middlesex School of Medicine, warned that:

> ... when diabetic patients keep their blood sugar very tightly regulated this alone makes them less aware of dangerously low blood sugar. In effect they appear to be exposed to risk and thus potentially damaged by their therapy. It is unfortunate that widespread adoption of an inappropriate policy of strict control of blood sugar coincided with the introduction of human insulin. It would be doubly unfortunate if human insulin became the scapegoat on which all adverse effects of the ill-judged policy were blamed.[45]

He also pointed out that the interpretation of blood sugar levels post-mortem was far from straightforward. This very same issue was raised by Robert Tattersall who, shortly after the death of Stephanie Wallace, explained in *The Guardian* how the tissues continue to metabolize glucose for some time after death, with the result that while blood sugar levels may well be hypoglycaemic at the time of autopsy, they may well have been normal at the actual time of death. To determine that a death had occurred by due to nocturnal hypoglycaemia was, therefore, notoriously difficult.[46]

In 1991, Tattersall and his colleague G. V. Gill published the results of their investigation into the cases of fifty type 1 diabetic patients who had died suddenly and unexpectedly. Having accounted for death by suicide, poisoning, and ketoacidosis, there remained twenty-four patients, most of whom who had been in good health with well-managed diabetes but who had gone to bed and been found dead the next morning. All of these patients had been switched from animal to human insulin from six months to two years earlier, and although all were found to have died as a result of hypoglycaemia or a hypoglycaemia-associated condition, the study concluded that 'there was nothing to implicate the species of insulin as a factor in these deaths'.[47]

The reputation of recombinant human insulin emerged untarnished from this episode, and, in a relatively short space of time, it had gone on to replace animal insulins as the drug of choice for managing type 1 diabetes. However, one question refused to go away: had all the effort and energy to create human insulin truly been in response to an impending shortage of bovine and porcine material or might there have been other motives?

Some doctors were indeed adamant that 'The real advantage [of human insulin] is that it represents an unlimited supply of the protein'.[48] And, as Irving Johnson of Eli Lilly pointed out, bacteria could do the job of producing human insulin far more efficiently than the body's own pancreatic cells:

> The significant advantage of this approach over producing insulin from animal pancreases was that only about 1% of the pancreas glands are islets of Langerhans, and only a fraction of the islet cells are β-cells, whereas in a bacterial fermentation of the rDNA organism, 100% of the cells are producing the desired protein.[49]

Moreover, when speaking at the Genentech press conference at City of Hope in 1978 at which the production of recombinant human insulin had first been announced to the world, Dr. Rachmiel Levine, director of research at City of Hope made the bold prediction that the new drug would solve the problem of supply of bovine and porcine material. When asked again seven years later, Levine acknowledged that the problem of supply had not yet reached the stage where it needed to be alleviated—although he cautioned that this might well change in the future.[50]

Others, however, expressed outright scepticism about the predicted looming shortage of bovine and porcine material. According to Paul Haycock, director of research at Squibb-Novo, 'The whole thing was rubbish ... There never was a shortage of pig pancreases, and there never will be'.[51] He claimed that the projected shortages were the result of a miscalculation by an official preparing figures for the Food and Drug Administration and such predictions have since been described as being 'remote and disputed'.[52]

Nevertheless, in 2019, the pharmaceutical company Wockhardt placed a full-page advertisement in *Balance*, the magazine of Diabetes UK, stating that 'Due to global bovine insulin raw unavailability', they had no choice but to discontinue producing bovine insulin.[53] Wockhardt was one of a handful of companies that had continued to manufacture animal insulins, but they were in a minority. Regardless of whether human insulin was actually any more clinically effective than the older animal insulins, the production of human insulin had become a holy grail for pharmaceutical companies.[54]

So what was really driving the race to be the first to produce human insulin? Historian Nicholas Rasmussen has suggested that it had very little to do with science. For Rasmussen, the allure of recombinant human insulin was not its therapeutic benefits, but its potency as a political weapon. In *Gene Jockeys*, his history of the biotechnology industry, Rasmussen argued that, had the scientists involved in recombinant DNA work genuinely wanted to answer questions about how genes are expressed, they could easily have chosen any number of plant or animal genes as their subject of study—yet they went for insulin. According to Rasmussen, this was no accident. Faced with formidable opposition from the media and public, the molecular biologists who pioneered this work, 'wanted the path smoothed for more such biology, choosing insulin as their emblem and opening move'.[55] It is a view shared by the writer Stephen Hall who, in his coverage of Mayor Al Vellucci's inquisition into recombinant DNA work at Cambridge, MA, identifies the testimony of Walter Gilbert as a pivotal moment:

> Gilbert's comments rendered very explicit the idea that medical applications justified the use of recombinant DNA research without delay. Insulin had been invoked as a potential benefit of recombinant DNA work.[56]

Rasmussen acknowledged that scientists like Gilbert may well have genuinely pursued this work to explore fundamental questions about how genes work, but also suggests that molecular biologists deliberately sought to ward 'off the challenge to their beneficent image by invoking human insulin as the prime example of the medical benefits recombinant DNA would bring'.[57] Certainly, when Paul Berg testified before the Senate subcommittee on Science, Technology and Space at the end of 1977, he told its members that 'the isolation of the insulin gene is a promising start' for the development of new medicines and industry.[58]

Erwin Chargaff would probably have agreed with Rasmussen that the creation of recombinant human insulin was done to win the trust of a fearful public and a sceptical press, but he also feared that it was something else, too. It was the symptom of a much deeper moral and widespread philosophical decay that had led science astray:

> it is the real purpose of science to teach us true things about nature, to reveal to us the reality of the world, the consequence of such teaching ought to be increased wisdom, a greater love of nature ... By confronting us with something incommensurably greater than ourselves, science should serve to push back the confines of the misery of human existence. These are the effects it may have had on men like Kepler or Pascal. But science, owing to the operation of forces that nobody, I

believe, can disentangle, has not persisted in this direction. From an undertaking designed to understand nature, it has changed into one attempting to explain, and then to improve on, nature ... Our kind of science is a disease of the Western mind.[59]

He stated earlier that this attempt to improve upon nature was not a change he welcomed:[60]

The black magic of our days—these mass media concerned with both the production and the distribution of so-called news; these forever titillating and nauseating intimacies, splashing all over us from newspapers and magazines, from radio and television; this bubbling and babbling emptiness of deadened imagination—has taken hold of science ... This has given rise to a popularization, but also to an enormous vulgarization, of science. Its achievements have begun to take on the form of a spectator sport, and young scientists start like race horses.[61]

For Chargaff, the race to clone and express human insulin symbolized a deeper moral and cultural rot at the heart not only of science, but of Western culture in general: 'The orderly, loving and careful study of life ... had been replaced by a frantic and noisy search for stunts and "breakthroughs"'.[62]

Chargaff derided scientists such as Gilbert or Howard Goodman as 'the modern version of King Midas ... whatever he touches turns into a publicity release'.[63] As Boyer and his team at Genentech worked at breakneck speed to clone human insulin, Chargaff lamented that science was 'falling more and more into the hands of entrepreneurs who, although there are exceptions, are on the whole a mindless, money-grabbing lot.[64]

Chargaff's problem with science was not simply that it had become about making a profit or courting public attention—it went much deeper. At the heart of the issue was one single question—what is the purpose of science? In her Presidential address to the British Association for the Advancement of Science in 1978, Dorothy Hodgkin used insulin to answer this question:

The history of research on insulin illustrates very well many of the complications of scientific discovery ... Insulin has a special place in protein history as the first protein of which the chemical structure was found—by Fred Sanger, twenty years ago ... In the case of penicillins, organic chemists can compete with micro-organisms in their efficiency in chemical synthesis—though in practice, fermentations are still largely used in the isolation of the nucleus compound to be modified. With insulin, chemical synthesis has also been achieved but is a far more difficult and lengthy process. For the present we have to depend on its synthesis in the pancreas of mammals for our main source of supply for medical use. Calculations show that with the improvement and extension of medical treatment of diabetics there will be a world shortage of insulin unless new methods of synthesis, chemical or biochemical, are found. Hence the practical interest in the experiments on bacteria now taking place.[65]

Having held up insulin as an example, she turned to the philosopher Baruch Spinoza (1632–1677) to clinch her argument:

> I wish to direct all Sciences in one direction or to one end, namely to obtain the greatest possible human perfection and thus everything in the sciences that does not promote this endeavour must be rejected as useless that is in a word all our endeavours and thoughts must be directed to this one end.[66]

Fellow Nobel Laureate Richard Synge was in hearty agreement with this sentiment. In a 1953 magazine interview that described him as one of the 'New Elizabethans', he was clear that science was there to provide solutions to practical problems so that 'human troubles should be overcome by human ingenuity rather than endured while blaming or accepting the will of God'.[67]

Chargaff took a radically different view. He felt that to relegate science from being an attempt to understand the universe to an activity the purpose of which was solely to serve practical and material ends was a denigration of a noble pursuit. To illustrate his point, he imagined Aristotle visiting a modern research laboratory and asking a scientist, 'What is the purpose of your actions?' to which came back the simple reply, 'Cheaper chicken'.[68] Having devoted much of his later career to research into nutrition at the Rowett Institute, Synge would have had no problem at all with this reply, so long as it did not conflict from a scientist's main public duty which he maintained was 'to know, just a little more clearly than others, just where the boundaries of human knowledge lie'.[69]

Synge firmly believed that, once education became accessible for all, people would 'develop all kinds of constructive and absorbing interests that will cause their only regret to be the speed at which a human lifetime runs out. And one direction which such interests will take will increasingly be devotion to natural science'.[70]

But Chargaff was much more pessimistic. For him, science was no longer the noble quest to understand the universe and elevate our minds that it had once been. 'Our present natural sciences have nothing to do with nature', he despaired.[71] Instead, science had become degraded to being little more than the servant of a consumer culture. Like some kind of overworked genie, it existed only to conjure up on demand whatever satisfied the appetites of society—be these new medicines, or new brands of soap.

This left Chargaff deeply pessimistic about the future: 'Most of the great scientists of the past could not have arisen ... most sciences could not have been founded ... if the present utility-drunk and goal-directed attitude had prevailed'.[72] He feared that scientists would now be forced onto a treadmill where they forever had to justify their research solely on the grounds of its perceived utility by the public. In the case of molecular biology, every grant application written would have to promise the cure for some disease.

Perhaps the stunning success of insulin was partly to blame for this. Maybe it helped to foster a lazy view among the public that, no matter how severe the disease, the men and women in white coats could always be relied upon to come up with a new pill. But for Chargaff there was another culprit: 'Our biology, no less than our technology, is a product of capitalism, governed by unwritten rules of supply and demand'.[73]

However, Chargaff seems to have overlooked one glaring—and inconvenient—fact. Scientific research conducted with clear utilitarian goals and fundamental studies into the nature of the universe were not necessarily mutually exclusive activities, and his own work was the very proof of this point. For Martin and Synge's development of partition chromatography was not driven by a desire to uncover the secrets of the universe, but rather, by the capitalist need to generate profit and create new markets for the textile industry, yet this had not prevented them, and later Chargaff, from helping to unlock mysteries that shed fundamental light on biology.

But whether Chargaff could—or would—admit this is debatable. As he watched molecular biology pick up pace, he reflected sadly that 'when I look at what is now called a scientist, I begin to wonder if I ever was one'.[74,75,76] At the same time, he was also painfully aware that his scepticism about the direction in which science was heading left him in a minority that had lost the moral high ground. For now that recombinant DNA technology seemed set to promise an easy supply of human insulin, Chargaff, and those like him, would be cast as anti-progressive reactionaries standing in the way of medical progress and the alleviation of human suffering. Furthermore, thanks to the publicity generated around the race to clone human insulin, Chargaff knew that every criticism he raised about recombinant DNA would now be shouted down:

> Without doubting the purity of their motives, I must say that nobody has, to my knowledge, set out clearly how he plans to go about curing everything from alkaptonuria to Zenker's degeneration, let alone replacing or repairing our genes. But screams and empty promises fill the air. 'Don't you want cheap insulin?'[77]

Chargaff could have responded that recombinant human insulin was not necessarily any cheaper, nor were the analogues that eventually replaced it. For by the 1990s, even human insulin was itself starting to look old fashioned as it in turn gave way to a successor.

Ever since the 1920s, clinicians had sought the holy grail of an insulin preparation that mimicked the action of the pancreas in being able to adjust its activity in response to changes in blood glucose levels. By maintaining a constant level of blood sugar that avoided violent oscillations between hypo- and hyperglycaemia, it was hoped that many long-term complications of diabetes might be overcome. The production of insulin preparations such as the protamine-insulin developed by Hagedorn at Nordisk was a major step forward in this direction. Furthermore, inspired by Hagedorn's findings, in 1936 Canadian chemist David Scott (1892–1971) from the Connaught laboratories in Toronto discovered that the solubility of insulin could be reduced simply by the addition of zinc.[78] This new preparation, known as Protamine-Zinc-Insulin (PZI) was effective for seventy-two hours after injection and could abolish peaks of blood sugar that were often seen after break-fast. Writing in the *British Medical Journal*, R. D. Lawrence welcomed these new developments:

> Insulin treatment has never been as simple as giving a bottle of medicine, and in the past it usually involved two or even three injections a day—no easy matter for many patients ... But I hope to show that treatment with zinc-protamine-insulin is easier and simpler for the doctor and his patients and that it ought to extend the benefit of insulin to thousands of patients who, although they needed it, did not get it.[79]

But despite PZI being hailed by Lawrence as 'the *practitioner's insulin* for choice', it was clear that long-acting, single-use insulins were still not without their problems.[80] This became dramatically evident to clinician Russell Wilder when he travelled to a conference in Kansas City with his colleagues Lydia Nelson, and R. G. Sprague, both of whom had type 1 diabetes:

> So far as anybody knew about us, everything went smoothly. However, on the second evening Miss Nelson, in great alarm, reported to me that Dr. Sprague was lost. We found him later, wandering in the streets distractedly in a delayed reaction from protamine insulin. No further trouble was encountered in Kansas City, but on the drive home Miss Nelson, too long without food, was unable to write her name in the registry of the hotel where we were stopping for the night. This promised to be scandalous, but a glass of orange juice re-established the proprieties and suspicion soon was laid—not, however, without great respect for our own pronouncement ... that although protamine insulin, in many cases, makes possible effective management of diabetes with only one administration of insulin a day, and with less insistence on rigid control of the diet, its careless use or disregard of the diet is attended with danger.[81]

As another article in the *British Medical Journal* pointed out, 'The very characteristics that constitute the advantages of the protamine insulins also have their drawbacks'.[82] Among these, the most serious was:

> the severity and nature of the hypoglycaemic attacks when they do occur. As a result of the prolonged action hypoglycaemic attacks when untreated last for a considerable time, and administration of sugar may need to be repeated many times.
> [...]
> A further unpleasant feature is the suddenness with which severe attacks burst on the patient. The fall of blood sugar produced by the new insulins appears to be so gentle that the warning symptoms of hypoglycaemia are not evoked and the blood sugar slides on unchecked to lower levels, where a severe attack is inevitable.[83]

Another problem was agents (e.g. protamine, globin, or histone) added to prolong the action of insulin by lowering its solubility, could also cause adverse allergic reactions in patients. In the early 1950s, however, researchers at Novo discovered a very simple solution. If insulin was prepared in a simple acetate buffer, it could be made insoluble at physiological pH without the inclusion of any other additives to give a long-lasting effect.[84] An added benefit was that this required only tiny amounts of zinc, which in the previous types of long-acting preparation had also been shown to cause adverse reactions.

When these 'Lente' insulins (as they became known) were first tested in clinical trials in the UK, most doctors were impressed—particularly with the fact that they caused no allergic reactions.[85] But history has been less kind. When assessing the benefit of these insulins in his 2019 book 'The Pissing Evil', Robert Tattersall notes that their use was coupled with a vogue among clinicians for relaxing dietary restrictions on diabetic patients, and the result was far from impressive:

> It seems that, except in a few clinics, the once-daily insulins introduction in the late 1930s was eagerly seized on by patients and physicians. Coupled with the 'free diet' movement, this led to three decades during which poor control of blood glucose was the rule rather than the exception.[86]

The recollections of Charles Fletcher, both a clinical specialist in diabetes and someone with type 1 diabetes, serve to illustrate the kind of problems that patients faced. Although his complaints about how insulin injections were an inconvenience on theatre trips and visits abroad were not likely to be a burning issue for all his fellow patients, his description of the challenges of trying to establish an insulin regime that prevented oscillations in blood sugar would no doubt strike a chord:

> At first I took long-acting insulin (protamine zinc), but I found this socially intolerable. It demands an evening meal at a fixed time which is often impracticable, especially after going to a theatre or in foreign countries where dinner may be very late. Twice-daily soluble insulin led to frequent late morning hypoglycaemia. At my wife's suggestion I started doing what the normal pancreas does and went over to three injections of soluble insulin daily before my main meals, supplementing the evening dose with a little isophane [a long-acting insulin] to cover the next early morning.[87]

Mrs. Fletcher had the right idea—the future of managing type 1 diabetes lay with using fast- and long-acting insulins in combination to emulate the action of the pancreas. And the advent of recombinant DNA technology, coupled with the discoveries of Sanger and Hodgkin, promised a means which this might be better achieved. The strategy was twofold. Firstly, patients needed what became known as a prandial insulin. This was an insulin that could be taken with meals, and which would be sufficiently fast acting to prevent the spike in blood sugar levels that usually occurred immediately after food. The second type of insulin, known as a basal insulin, was one which acted more slowly to maintain a constant low background level of activity.

Armed with recombinant DNA technology and the X-ray studies of the insulin structure, it now seemed possible that both these goals could be achieved. Hodgkin and her colleagues' work had shown that insulin was an innately sociable protein. A single insulin molecule, called a monomer, will naturally pair up with another to form a unit called a dimer. Three of these will then in turn gather around ions of zinc to form a larger unit composed of six individual molecules, known as a hexamer.[88,89]

It is this innate sociability of the insulin monomer that can make a diabetic patient an awkward dinner guest. This is because, for insulin to become physiologically active, these hexamers need to dissociate to release their constituent monomer units. And this takes time, which means that diabetic patients needed to remember to inject their insulin well in advance of actually sitting down to eat. Unsurprisingly, this could bring considerable logistical complications for diabetic patients at meal times.

Understanding the physical chemistry that drove the dissociation of insulin offered little consolation for countless dinners gone cold due to mistimed injections. However, the discovery of the three-dimensional structure of insulin by X-ray analysis did suggest how this might be avoided in future. It revealed the precise position of certain amino acids in the chains of insulin that were crucial in mediating the pairing up of monomers to form dimers and hexamers. Recombinant DNA technology now provided the tools with which to alter them.[90]

By deliberately altering the DNA sequence of insulin, the particular amino acids involved in dimer formation could be substituted in a variety of ways that made this process more difficult. Examples included the introduction of bulky amino acids that interfered with dimer formation due to their size, removing the particular amino acids involved in binding zinc, or inserting negatively charged amino acids so that when these came into close proximity with similarly charges on other insulin monomers, the resulting repulsion made association more difficult (Figure 59).[91,92]

The first of these analogues was not a success. For as well as being essential to stimulating the uptake of glucose by cells, insulin is also a growth factor and in one early insulin analogue, (Asp B10) a very small change had a very serious effect. Merely substituting the amino acid Histidine with Aspartic acid at position 10 in the B chain of insulin sent its power to stimulate cell division into overdrive to the point that it was shown to cause mammary tumours in mice. Unsurprisingly, this particular insulin analogue was abandoned, but in 1996, Eli Lilly gained approval for the release of a new form of modified human insulin called lispro (branded as Humalog) onto the market.[93] Four years later, Novo Nordisk followed suit with the release of their own fast-acting prandial insulin, Novolog. Despite containing slightly different modifications in their amino acid sequence, what both these new forms of insulin had in common was that they were absorbed more quickly and reached the peak of their activity faster, so there was no longer any need for a lengthy delay between a patient taking their injection and

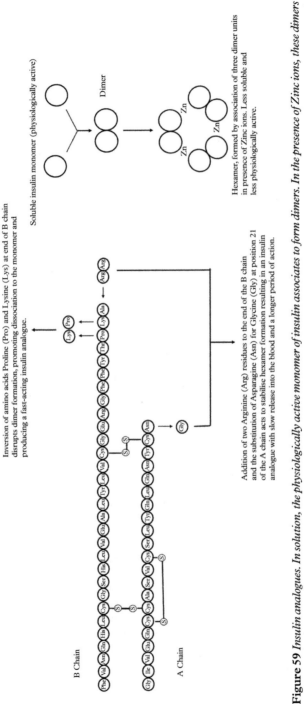

Figure 59 *Insulin analogues. In solution, the physiologically active monomer of insulin associates to form dimers. In the presence of Zinc ions, these dimers can then associate to form hexamers with low solubility. For insulin to be physiologically active, these hexamers need to dissociate into the monomeric form. By deliberately altering specific amino acids in the sequence of insulin which are involved in mediating this transition, analogues can be engineered which are faster or slower acting in the hope of allowing patients better control of their blood glucose levels. Diagram by K.T. Hall.*

sitting down to eat.[94] Because it also resulted in a flattened peak of glucose after a meal, it had the additional benefit of reducing the risk of hypoglycaemia between meals.[95]

Recent research has raised the possibility that the future of prandial insulin analogues may lie with *Conus geographus*, a species of humble sea snail that has evolved to use insulin as a chemical weapon. Hunting by night, these snails stalk schools of small fish in the crevices of underwater reefs and as they approach their prey, they use a net strategy, releasing a flood of venom from a false distended mouth, instead of a harpoon. Many of the components of this venom are neurotoxins that impair the sensory capacity of the prey, but one of the most potent compounds in this lethal mixture has recently been shown to be insulin, which incapacitates the prey by plunging it into hypoglycaemic shock, rendering them inactive, and providing an easy meal.[96]

In terms of molecular size, the insulin of the cone sea snail is the smallest yet found—and as well as being more potent, it has one other key advantage over human insulin—as it is monomeric, it is incredibly fast acting. There would be no complications of having to ensure that insulin was injected thirty minutes before a meal for diabetic patients injecting themselves with snail venom. With this in mind, a team of researchers reported in 2020 that, inspired by the structural principles of the cone sea snail insulin, they had engineered a monomeric form of human insulin with similar properties that they hoped might offer 'a new platform for therapeutic development of prandial insulins'.[97]

Structural studies of insulin offered an insight into how the opposite effect could be achieved to create a basal insulin. Armed with knowledge of the precise amino acid sequence of insulin, researchers at Sanofi used recombinant methods to alter the coding DNA sequence of human insulin so that an extra two amino acids were added to one end of the B chain, while a substitution was made at position 21 of the A chain.[98] The alteration of a mere three amino acids might seem trifling, but the effect was dramatic. Monomers of this modified insulin, which became known as 'Glargine' were soluble at acidic pH but at a more neutral pH such as that in the blood and tissues, they tended to associate into hexamers. As a result, when injected by a patient, instead of being quickly absorbed into the blood, the neutral pH of the body tissues caused the glargine to form hexamers, which dissociate very slowly to release physiologically active single monomers of insulin into the blood. In this way, glargine could be used a basal level of background insulin over the next 24 hours.

Not that recombinant DNA methods were the only means by which this could be achieved. In 2004, a slow-acting basal insulin analogue called Detemir was produced in which a hydrocarbon chain was artificially attached to a specific location on the insulin molecule. With this attachment, the insulin could then be bound in a complex to the protein albumin, which stabilized the formation of hexamers and slowed the activity of the insulin.[99]

Regardless of how these new analogues were produced, however, some doctors remained unconvinced that they were actually any more beneficial than the human—or even the animal—insulin that had preceded them. The authors of one paper published in a clinical diabetes journal in 2007 made their position quite clear with the title 'Nice Insulins, Pity About the Evidence'.[100] In their paper, the authors cited a report by the independent German body Institute for Quality and Efficiency in Healthcare (IQWiG) that they claimed showed that 'the benefits of short-acting analogues in type 1 diabetes were marginal at best'.[101] The IQWiG had been formed two years earlier in order to evaluate the clinical benefits of new drugs and advise German health insurance companies on their efficacy, but their claims that analogue insulins were no better than the human form quickly brought them into conflict with the pharmaceutical industry. Their head, Professor Peter Sawicki came under particularly heavy fire. Events took a new twist when it emerged that one of the most vocal journalists who was directing their fire at Sawicki for supposedly denying thousands of diabetic patients access to these new drugs had allegedly received 1.2 million Euros from the pharmaceutical industry.[102]

Despite these doubts about their benefits, analogue insulins were steadily replacing the human form. One study showed that in 2000, 86.3% of insulin had been human, where by 2008, that figure was down to 23.2%, with use of analogue insulins having risen in the same period from 10.7% to 76.1%.[103] Critics claimed that this increase in use was not because of any improved therapeutic benefit, but simply because of a marketing campaign that encouraged clinicians to prescribe them. Moreover, because insulin production was controlled by only three companies, this oligopoly had considerable power to shape the international market for insulin—first by promoting the replacement of animal insulins by the recombinant human form, and now by its analogues.[104]

Today, millions of patients manage their condition using a combination of prandial and basal insulin analogues. Analogues have also been developed in which the active insulin molecule consists not of a separate A and B chain joined together (as in the physiological form), but a single chain. It is hoped that single-chain insulins might be more stable and so be stored more easily in countries where refrigeration may be a challenge. Another strategy has been the development of analogues that act specifically to increase glucose uptake by the liver in the hope of achieving a better balance of blood sugar levels.[105] But despite these advances, the holy grail of using pharmaceutical intervention to mimic the action of the pancreas in being able to alter insulin activity in response to changes in blood glucose levels remains elusive. This may, however, yet change thanks to a new generation of insulin replacement therapies (IRTs)—or 'smart insulins'—of which, broadly speaking there are three main types.[106] The first of these is an engineering-based solution that couples continuous blood glucose monitoring technology with insulin pumps and employs algorithms to control the release of insulin in response to changes in blood glucose levels. The other two approaches also involve engineering, but at a molecular level. One of these involves sequestering insulin with a matrix, such as a gel, that is sensitive to glucose levels. Detection of glucose levels in these systems is by glucose binding proteins such as lectins, the inorganic compound phenyl boronic acid which binds reversibly to glucose, or the enzyme glucose oxidase. In the latter case, the enzyme acts on glucose to generate gluconic acid and the resulting drop in pH causes the matrix to release insulin into the blood.

The final type of system involves engineering the insulin molecule itself to be able to detect, and respond to, changes in the levels of blood glucose. Not actually a new idea, it was first suggested by Michael Brownlee and Anthony Cerami at Rockefeller University, New York in 1979.[107] Brownlee and Cerami synthesized a derivative form of insulin that contained a sugar portion attached to the protein chain. This allowed the insulin derivative to become bound in a complex with concanavalin A (conA), a member of a group of proteins called lectins, which bind sugars. When glucose was added to this complex, the free glucose molecules competed with the insulin for binding to the conA and, in the process, released bound insulin from the complex. In this way, Brownlee and Cerami proposed that the physiological activity of injected insulin could be fine-tuned to respond to blood glucose levels. Although this approach has since been pursued further, other strategies have also been developed, including the fusion of insulin with a glucose-sensitive enzyme like as glucose oxidase, or the attachment of inorganic glucose binding groups, for example, phenyl boronic acid.[108] More recently, X-ray crystallographic and cryo-electron microscopy studies of insulin bound to its receptor protein on the surface of cells has shown that the three-dimensional conformation of insulin can alternate between an active (open) or inactive (closed) formation.[109] This has raised the possibility that, by attaching a glucose sensitive group to the molecule, an insulin could be engineered that is induced into an active conformation when blood glucose levels are high, but shuts down into an inactive conformation as levels drop.

Quite evidently there is much confidence in these systems. When Novo Nordisk bought Ziylo, a small biotech company set up at the University of Bristol, UK by Harry Destecroix and his former PhD supervisor Professor Anthony Davis that was involved in the development of 'smart insulins', it was reported that the deal might eventually be worth $800 million.[110] Nearly four decades after Genentech's spectacular Wall Street debut, it seemed that insulin was once again proving to be profitable.

But although science might be able to deliver new improved insulins, the distribution of their benefits quickly becomes a question of politics and economics. As analogue insulins came to steadily replace human insulin in industrialized countries, some diabetes clinicians became concerned that this trend would soon be mirrored in the developing world.[111] Studies had already shown that the availability of insulin—or more accurately the lack thereof—was a big problem for diabetic patients in some of these countries.[112] The mortality rate for children born with type 1 diabetes in certain African countries was estimated to be similar to that in the West before the discovery of insulin.[113,114] As insulin analogues were significantly more expensive than human insulin, the prospect of them replacing human insulin in the developing world seemed unlikely to improve this situation.[115,116]

Nor were concerns about the availability of insulin confined only to the developing world. In a letter to *The Lancet* in 2012, a number of clinicians drew attention to the plight of patients in Greece after the financial crisis of 2008, and to those patients in the US who did not have health insurance. Even for those with health insurance, the situation wasn't much different—the price of a 10-ml vial of insulin in the US had risen from $20 to $275 since the 1990s.[117,118] For Lija Greenseid, a 46-year-old educational consultant from Minnesota and mother of a 13-year-old daughter with type 1 diabetes, this was unacceptable. Deciding to take action, Mrs. Greenseid and a small convoy of fellow 'insulin outlaws' drove over the border into Canada where they paid $1200 for insulin that would have cost $12,000 in the US.[119]

As US law allows a three-month supply of medicines to be imported for personal use, their action was not strictly illegal, but stories of such cross-border insulin convoys drew attention to the plight of diabetic patients in the US. According to a survey by the charity T1 International, founded to support type 1 patients, the monthly cost of insulin and related supplies in the US was the highest out of ninety countries, with a quarter of patients surveyed admitting that they sometimes rationed their insulin.[120]

For the diabetes specialists John Yudkin and D. Beran, this was nothing short of a 'double scandal':

> We are living in a world where resources are being wasted on products that show only small benefits, while people are not benefiting from a product that was first used some 90 years ago. With the increasing economic and health burdens of diabetes this should be an issue of concern to us all.[121]

Part of the problem was that the market price of insulin did always not reflect its production costs but was instead inflated by the costs of marketing and promotional campaigns.[122] Furthermore, the fact that control of production was in the hands of only three companies meant that there was little competition to drive down prices or bring generic alternatives onto the market.[123]

In 2012, to address these issues, a group of diabetes clinicians launched the '100 Campaign'. This aimed to achieve 100 per cent access to insulin by the 100th anniversary of its discovery in 2022 and, in the words of the campaign's founders, 'to ensure that, in 10 years' time, the leading cause of death in a child with diabetes is not lack of access to insulin'.[124] The campaign called for the World Health Organization to reject a proposal that analogue insulins be included on its Model List of Essential Medicines and to persuade insulin manufacturers to provide cheaper animal and human insulins to the developing world.[125,126]

It is a noble and laudable aim, but the campaigners also acknowledged that the challenge of type 1 diabetes in the developing world was not simply an issue of making insulin more available. One particular study on insulin availability in Zambia and Mozambique, where Novo-Nordisk were already operating a philanthropic scheme to provide insulin to their governments at a price not exceeding twenty per cent of its average price in North America, had a blunt but important lesson for the management of diabetes in both the developing and industrialized world:[127] 'Insulin is necessary, but not sufficient, for the survival of a patient with insulin-requiring diabetes'.[128]

It was a conclusion that would have been all too familiar to clinicians like Elliott Joslin who were the first to start using insulin in therapy. In addition to insulin, the successful management of type 1 diabetes required logistical considerations like the provision of needles, syringes, methods for testing of ketones and glucose in blood and urine, training of healthcare staff, and, crucially, the education of patients about their condition.

Yet, thanks to its prominence and miraculous aura in media stories and the public consciousness as a wonder drug, it had perhaps become a victim of its own success. This brought problems long before the advent of recombinant DNA technology. In his book *Bittersweet: Diabetes, Insulin and the Transformation of Illness*, Chris Feudtner observes, that all too often, the history of diabetes has been written solely as the story of the discovery of insulin. This approach, described by Feudtner as a 'technological ethos', holds a particular grip on our imagination and has come to dominate our thinking about medicine—particularly in the case of insulin:[129]

> Perhaps no story of medical progress, though, has been more influenced by this technology ethos than the history of diabetes. Stories of insulin have served various needs while reinforcing deeply held beliefs of twentieth-century Americans. A parable of salvation, the tale of diabetic deliverance has spoken to the imagination of doctors and lay people alike, serving as a potent and often cited symbol of scientific progress and the prospect of human mastery over disease ... [130]

As a diabetes specialist himself, Feudtner was not suggesting that the discovery of insulin was not to be welcomed—it had, after all, spared countless diabetic patients from a slow and inevitable death. But as Feudtner points out, all advances in medicine come with a cost—even if, as was the case with insulin, that cost is not always immediately apparent.[131] Diabetes had certainly been transformed from a fatal to a chronic condition and as the title of Feudtner's book reflects, this transformation was bittersweet. While insulin had spared patients from certain death, their new life would now be fraught with daily challenges. Not only was there the immediate danger of hypoglycaemia from an incorrect dose of insulin, but, as patients began to live longer, they also began to manifest the multiple severe complications that could arise due to elevated blood sugar levels. As a result, while patients' lives were extended by insulin, they were also now blighted by renal failure, neuropathy, blindness, stroke, heart disease, and amputation.[132] To live with diabetes was to walk a tightrope with a chasm on either side.

Reports like that in *The New York Times* in May 1923 proclaiming that 'Diabetes, Dread Disease, Yields to New Gland Cure' did nothing to help. In Feudtner's words, such overblown statements served only to 'conceal more than they reveal'.[133,134] The main problem with such mythologized and triumphalist accounts was that, by glamourizing the technology, they distracted attention from the daily struggles faced by the patients who used it.[135] As Feudtner was well aware, this was a problem that had only got worse with time:

> … portraits of miraculous therapeutic success present a modern yet mythic account of diabetes history, accentuating the potency of insulin as a heroic drug to rescue patients, vanquish disease, banish suffering, and finally secure an implied but unexamined 'happily-ever-after' ending. Mythical storytelling elements such as these permeate much of our current appreciation of other medical technologies. When pharmaceutical companies launch promotional advertising campaigns showing pictures of bald yet smiling cancer survivors; or when proponents of the human genome project speculate how gene therapy will eliminate certain inborn diseases; … or even when the biotechnology industry shows film clips on television of children spared from blindness due to rice supplemented with vitamin A, these examples of scientific achievement are all presented in the mythical aura of an idealistic quest for a better world. As they tap into our fears and desires, these stories about medicine reflect a broad technology ethos in our culture, the American propensity to embrace more technology as the best solution to our problems.[136]

Feudtner spoke candidly about how, in the course of his own career as a specialist in diabetes, he had experienced a personal epiphany about the philosophical shortcomings of medicine as it was currently being practiced:

> These days, medical understanding of disease prizes the reductive precision of scientific knowledge yet shies away from comprehending illness at a more individual and subjective level. As I came to appreciate after countless hours in darkened lecture halls, the biomedical view of disease looks piercingly through a patient toward some essential objective, solid reality of biology–and yet in doing so it loses, like an X-ray, almost any sense of the flesh of the person.[137]

As a result, Feudtner gathered the personal stories of patients in order to write a history of diabetes that sought 'to put the flesh back upon particular women and men and painting a more sensitive portrait that restores human detail'.[138] Diabetologists like Robert Tattersall were also very aware that, as a serious chronic health condition, diabetes is perhaps unique for the degree of involvement and responsibility that is placed upon the patient to manage their condition.[139] Technology like insulin and blood glucose monitoring systems are essential, but if not used in tandem with appropriate behaviour, they are of little benefit—|this was evident as far back as the early 1920s. Even as Elliott Joslin sang the praises of insulin, likening it to the vision of Ezekiel, he was also very aware that there was much more to the management of diabetes than regular injections of insulin:

> Insulin is a remedy which is primarily for the wise and not for the foolish, be they patients or doctors. Everyone knows it requires brains to live long with diabetes, but to use insulin successfully requires more brains.[140]

Joslin recognized that the successful management of diabetes required more than just technology in the form of injections with insulin—it also took discipline, determination, and thought on the part of the patient. Furthermore, although recombinant DNA technology might be able to coax bacteria into producing human insulin, it remains highly doubtful whether it could ever bestow upon us these essential character traits that are required to live with diabetes on a daily basis.

Echoing Joslin's concerns at the time, the President of the British Medical Association, Sir Thomas Horder reminded clinicians that insulin should be thought of, not as a cure for diabetes, but rather a 'walking stick for a lame pancreas'.[141] Leaning on this crooked timber for support, a diabetic patient might well be able to walk the path of a normal life - but only if they also played their part in using it properly. Since I first began writing this book, Joslin's and Horder's insights have become relevant to us all. According to a recent article in *The Guardian* newspaper a number of clinicians now suspect that that, in addition to its many other unwelcome long-term effects, Covid-19 may cause diabetes.[142] Only time and more research will tell. But regardless of whether or not there turns out to be a connection between Covid-19 and diabetes, the story of insulin may shed some light on what we can – and cannot – realistically expect in tackling the pandemic.

When Professor David Nabarro, the World Health Organization Special Envoy on Covid-19, was asked in an interview on BBC Radio 4's 'Today' programme about the use of tracing apps in trying to contain the spread of Covid-19, he cautioned that such measures were:

> not going to be the total answer. The total answer is basically us, just coming to terms with the fact that we're going to have to change our behaviour wherever possible. Physical distancing, face protection, protecting those who are most vulnerable, looking after the people who are in the front line who are getting infected all the time, like bus drivers or people in food processing plants.[143]

In the time since Professor Nabarro gave this interview, several effective vaccines have been developed against SARS-CoV-2. This is of course a welcome development and one that is playing a crucial role in bringing the pandemic under control, but it does not mean that we should become complacent. In *Vaxxers,* their account of the development of the Oxford Astra Zeneca vaccine, Professor Sarah Gilbert and Dr. Catherine Green stress the need to be vigilant for the emergence of new variants of SARS-CoV-2 that show increased transmissibility, or which can evade the immune response elicited by the vaccine.[144] Other experts in virology meanwhile have stressed that for vaccines to have the greatest protective effect and reduce the chances of new variants emerging, we need to vaccinate a large number of people quickly whilst at the same time keeping the number of infections in the general population low.[145]

Like insulin before them, these first vaccines against Covid-19 - powerful as they are - would perhaps be better thought of not as a silver bullet, but as a very welcome walking stick thanks to which societies can take cautious steps out of lockdowns towards something resembling normality. And for vaccines to have the maximum benefit in bringing the virus under control they, just like insulin, need to go hand in hand with appropriate behaviour.[146] For as Professor Nabarro made clear in his interview, and Elliott Joslin recognized nearly a century before him, all too often there are no technological quick fixes for complex problems – only crooked timbers. This is as true for the use of insulin in controlling Type 1 diabetes as it is for the management of Covid-19, or indeed the prevention of future pandemics – which, as Sarah Gilbert and Cath Green warn in the closing pages of their book, will happen.[147] And it will doubtless also prove to be the case for the other challenges that we will face in the future - such as those of climate change, the rise of artificial intelligence, or the spread of antibiotic resistant bacteria. In facing up to these challenges and thinking about how best to tackle them, the story of insulin has important lessons for us all – regardless of whether or not we happen to be injecting ourselves with it.

Bibliography

Abel, J. J. "Crystalline Insulin." *Proceedings of the National Academy of Sciences* 12 (1926): 132–36.

Abel, J. J., E. M. K. Geiling, G. Alles, and A. Raymond. "Researches on Insulin I. Is Insulin an Unstable Sulfur Compound." *Journal of Pharmacology and Experimental Therapeutics* 25 (1925): 423–45.

Abel, J. J., E. M. K. Geiling, C. A. Rouiller, F. K. Bell, and O. Wintersteiner. "Crystalline Insulin." *Journal of Pharmacology and Experimental Therapeutics* 31 (1927): 65–85.

Abir-Am, P. "Synergy or Clash: Disciplinary and Marital Strategies in the Career of Mathematical Biologist Dorothy Wrinch." In *Uneasy Careers and Intimate Lives*, edited by P. Abir-Am and D. Outram, 239–80. New Brunswick: Rutgers University Press, 1987.

Adams, V., and T. Ferrell. "The Transfer of Genes That Make Insulin." The New York Times, May 29, 1977.

Adlard, E. R. "Obituary: Archer Martin; Nobel Prize Winning Biochemist." The Independent, August 7, 2002.

"Advance in Insulin Study." *The Times,* August 15, 1969.

Alberti, G. "Lessons from the History of Insulin." *Diabetes Voice* 46 (2001): 33–34.

Alberti, G., and P. Lefebvre. "Paulesco: Science and Political Views." *The Lancet* 362 (2003): 2120.

Alexander, H. E. and G. Leidy, "Induction of Streptomycin Resistance in Sensitive Hemophilus Influenzae by Extracts Containing Desoxyribonucleic Acid from Resistant Hemophilus Influenzae." *Journal of Experimental Medicine* 97 (1953): 17–31.

Allan, F. N. "The Writings of Thomas Willis, MD: Diabetes Three Hundred Years Ago." *Diabetes* 2 (1953): 74–78.

Allan, F. N. "Diabetes before and after Insulin." *Medical History* 16 (1972): 266–73.

Allen, F. M. *Studies Concerning Glycosuria and Diabetes.* Cambridge, Harvard University Press, 1913.

Allen, F. M, E. Stillman, and R. Fitz. *Monographs of The Rockefeller Institute for Medical Research: Total Dietary Regulation in the Treatment of Diabetes.* New York: The Rockefeller Institute for Medical Research, 1919.

Allen, R. S., and J. Murlin, "Biuret-Free Insulin." *American Journal of Physiology* 75 (1925): 131–39.

Altwegg, K., et al "Prebiotic Chemicals-Amino Acid and Phosphorus in the Coma of Comet 67P/Churyumov-Gerasimenko." *Science Advances* 2, no. 5 (2016): 1–6. https://doi.org/10.1126/sciadv.1600285.

Ancel, J. *The History of the Holocaust: Romania Vol. 1.* Jerusalem: Yad Vashem Press, 2011.

Anderson, L. *The Story of WIRA: 70 Years of Wool Textile Research and Services.* Leeds, WIRA Technology Group Ltd, 1988.

Andreopoulos, S. "Gene Cloning by Press Conference." *New England Journal of Medicine* 302 (1980): 743–45.

Arber, W, and D. Dussoix. "Host Specificity of DNA Produced by Escherichia Coli I. Host Controlled Modification of Bacteriophage." *Journal of Molecular Biology* 5 (1962): 18–36.

Astbury, W. T. "The Evolution and Interpretation of Synthetic Fibres." In *Symposium on Coal, Petroleum and their Newer Derivatives*, 99–108. London, Royal Society of Chemistry, 1948.

Astbury, W. T. "In Praise of Wool." *Proceedings of the International Wool Textiles Research Conference* B (1955): 220–43.

Astbury, W. T. "Recent Advances in the X-Ray Study of Protein Fibres." *Journal of the Textile Institute* 36 (1936): P282–296.

Astbury, W. T. "Relation between 'Fibrous' and 'Globular' Proteins." *Nature* 140 (1937): 968–69.

Astbury, W. T. "X-Rays and the Stoichiometry of the Proteins." *Advances in Enzymology* 3 (1943): 63–108.

Astbury, W. T. "X-Ray Studies of Nucleic Acids." *Symposium of the Society for Experimental Biology* 66 (1947): 66–76.

Astbury, W. T. "X-Ray Studies of Protein Structure." *Cold Spring Harbor Symposia on Quantitative Biology* 2 (1934): 15–27.

Astbury, W. T., and F. O. Bell. "X-Ray Study of Thymonucleic Acid." *Nature* 141 (1938): 747–48.

Astbury, W. T., and A. Street. "X-Ray Studies of the Structure of Hair, Wool and Related Fibres. I. General." *Philosophical Transactions of the Royal Society, A.* 230 (1931) 75–101.

Astbury, W. T., and H. J. Woods. "4-The Molecular Structure of Textile Fibres." *The Journal of the Textile Institute* 33 (1932): T17–34.

Astbury, W. T., and H. J. Woods. "X-Ray Studies of the Structure of Hair, Wool and Related Fibres. II. The Molecular Structure and Elastic Properties of Hair Keratin." *Philosophical Transactions of the Royal Society, A.* 232 (1933): 333–94.

Avery, O. T., C. M. Macleod, and M. McCarty. "Studies on the Chemical Nature of the Substance Inducing Transformation of Pneumococcal Types: Induction of Transformation by a Desoxyribonucleic Acid Fraction Isolated from Pneumococcus Type III." *Journal of Experimental Medicine* 79 (1944): 137–58.

Bakh, N. A., A. B. Cortinas, M. A. Weiss, R. S. Langer, D. G. Anderson, Z. Gu, S. Dutta, and M. S. Strano. "Glucose-Responsive Insulin by Molecular and Physical Design." *Nature Chemistry* 9 (2017): 937–43.

Ballantyne, A. "Drug Test Hope for Alzheimer Victims." *The Guardian,* December 5, 1988.

Ballantyne, A. "Insulin Warning After Doctor Dies." *The Guardian,* July 29, 1989.

Banting, F. G. "Discussion on Diabetes and Insulin in Proceedings of Sections at the Annual Meeting, 1923." *British Medical Journal* 2 (1923): 445–51.

Banting, F. G. "Early Work on Insulin." *Science* 85 (1937): 594–96.

Banting, F. G. "This History of Insulin." *Edinburgh Medical Journal* 36 (1929): 1–18.

Banting, F. G. *The Story of Insulin.* Unpublished manuscript. Banting papers. University of Toronto, 1940.

Banting, F. G., and C. H. Best. "Pancreatic Extracts." *The Journal of Laboratory and Clinical Medicine* 7 (1922a): 464–72.

Banting, F. G., and C. H. Best. "The Internal Secretion of the Pancreas." *Journal of Laboratory and Clinical Medicine* 7 (1922b): 251–66.

Banting, F. G., C. H. Best, J. B. Collip, W. R. Campbell, and A. A. Fletcher. "Pancreatic Extracts in the Treatment of Diabetes Mellitus." *The Canadian Medical Association Journal* 2 (1922): 141–46.

Banting, F. G., C. H. Best, J. B. Collip, W. R. Campbell, A. A. Fletcher, J. J. R. Macleod, and E. C. Noble."The Effect Produced on Diabetes by Extracts of Pancreas." *Transactions of the Association of American Physicians* (1922): 1–11.

Banting, F. G., C. H. Best, J. B. Collip, J. J. R. Macleod, and E. C. Noble. "The Effect of Insulin on Normal Rabbits and on Rabbits Rendered Hyperglycaemic in Various Ways: 1. The Effect of Insulin on Normal Rabbits." *Transactions of the Royal Society of Canada* 5 (1922): 31–34.

Banting, F. G., W. R. Franks, and S. Gairns. "Physiological Studies in Metrazole Shocks. VII Anti-Insulin Activity of Insulin Treated Patients." *American Journal of Psychology* 95 (1938): 562–64.

Barr, M. L., and R. J. Rossiter. "James Bertram Collip, 1892–1965." *Biographical Memoirs of Fellows of the Royal Society* 19 (1973): 234–67.

Barron, M. "The Relation of the Islets of Langerhans to Diabetes with Special Reference to Cases of Pancreatic Lithiasis." *Surgery, Gynecology and Obstetrics* 31 (1920): 437–47.

Bayliss, W. M. "Research on Diabetes." The Times, 1922.

BBC Radio 4. "Interview with Professor David Nabarro, World Health Organisation," *Today*, May 15, 2020.

Bearn, A. G. "Oswald T. Avery and the Copley Medal of the Royal Society." *Perspectives in Biology and Medicine* 39 (1996): 550–54.

Beek, C. van. "Leonid V. Sobolev, 1876–1919." *Diabetes* 7 (1958): 245–48.

Bell, F. O. "*X-Ray and Related Studies of the Structure of the Proteins and Nucleic Acids.*" PhD thesis, University of Leeds, 1939. Thesis EBL S Bel MS2229. Online at: https://explore.library.leeds.ac.uk/special-collections-explore/650413.

Bell, G. I., R. L. Pictet, W. J. Rutter, B. Cordell, E. Tischer, and H. M. Goodman. "Sequence of the Human Insulin Gene." *Nature* 284 (1980): 26–32.

Benson, A. A., and M. Calvin. "The Dark Reductions of Photosynthesis." *Science* 105 (1947): 648–49.

Benson, A. A., and M. Calvin. "The Path of Carbon in Photosynthesis." *Science* 107 (1948a): 476–80.

Benson, A. A., and M. Calvin. "The Path of Carbon in Photosynthesis." *Cold Spring Harbor Symposia on Quantitative Biology* 13 (1948b): 6–10.

Beran, D., M. Basey, V. Wirtz, W. Kaplan, A. Atkinson, and J. S. Yudkin. "On the Road to the Insulin Centenary." *The Lancet* 380 (2012): 1648.

Beran, D., and J. S. Yudkin. "Diabetes in Sub-Saharan Africa." *The Lancet* 368 (2006): 1689–95.

Beran, D., and J. S. Yudkin. "The Double Scandal of Insulin." *Journal of the Royal College of Physicians of Edinburgh* 43 (2013): 194–96.

Beran, D., J. S. Yudkin, and M. de Courten. "Access to Care for Patients with Insulin-Requiring Diabetes in Developing Countries." *Diabetes Care* 28 (2005): 2136–40.

Berezkin, V. G. *Chromatographic Adsorption Analysis: Selected Works.* Translated by Mary Masson. New York, E. Horwood, 1990.

Berg, P., D. Baltimore, H. Boyer, S. Cohen, R. Davis, D. Hogness, D. Nathans, R. Roblin, J. D. Watson, S. Weissman, and N. D. Zinder. "Potential Biohazards of Recombinant DNA Molecules." *Science* 185 (1974): 303.

Berg, P., and J. Mertz. "Personal Reflections on the Origins and Emergence of Recombinant DNA Technology." *Genetics* 184 (2010): 9–17.

Berg, P. "A Stanford Professor's Career in Biochemistry, Science Politics, and the Biotechnology Industry." An Oral History Conducted in 1997 by Sally Smith Hughes. University of California, Berkeley, Regional Oral History Office, The Bancroft Library, 2000.

Berg, P., D. Baltimore, S. Brenner, R. O. Roblin 3rd, and M. F. Singer. "Asilomar Conference on Recombinant DNA Molecules." *Science* 188, no. 4192 (1975): 991–94. https://doi.org/10.1126/science.1056638.

Bergmann, M. "Complex Salts of Amino Acid and Peptides: II. Determination of l-Proline with the Aid of Rhodanilic Acid. The Structure of Gelatin." *Journal of Biological Chemistry* 110 (1935): 471–79.

Bergmann, M. "Proteins and Proteolytic Enzymes." *Harvey Lectures Series* 31 (1936): 37–56.

Bergmann, M. "Some Biological Aspects of Protein Chemistry,." *Journal of the Mount Sinai Hospital* 6 (1939): 169–86.

Bergmann, M., and C. Niemann. "On Blood Fibrin: A Contribution to the Problem of Protein Structure." *Journal of Biological Chemistry* 115 (1936): 77–85.

Bernal, J. D. "Dialectical Materialism and Modern Science." *Science and Society* 2 (1937): 58.

Bernal, J. D. "Vector Maps and the Cyclol Hypothesis." *Nature* 143 (1939): 74–75.

Berson, S. A., R. S. Yalow, M. A. Rothschild, and K. Newerly. "Insulin-I131 Metabolism in Human Subjects: Demonstration of Insulin Binding Globulin in the Circulation of Insulin Treated Subjects." *Journal of Clinical Investigation* 35 (1956): 170–90.

Best, C. H. "'A Canadian Trail of Medical Research,' in Proceedings of the Society for Endocrinology." *The Journal of Endocrinology* 19 (1959): 1–17.

Best, C. H. "'A Quarter of a Century in Medical Research' in Proceedings of the New York Diabetes Association, January 1946." *American Journal of Digestive Diseases* 13 (1946a): 148–59.

Best, C. H. "Charles Best – 'The Discovery of Insulin." *Proceedings of the American Diabetes Association* 6 (1946b): 87–93.

Best, C. H. "Frederick Grant Banting. 1891–1941." *Obituary Notices of Fellows of the Royal Society* 4 (1942a): 20–26.

Best, C. H. "Insulin: The Banting Memorial Lecture 1952." *Diabetes* 1 (1952): 257–67.

Best, C. H. "Nineteen Hundred Twenty-One in Toronto." *Diabetes* 21 (1972): 385–95.

Best, C. H. "Recent Work on Insulin." *Endocrinology* 1 (1924): 617–29.

Best, C. H. "Reminiscences of the Discovery of Insulin: The First Clinical Use of Insulin." *Diabetes* 5 (1956): 64–68.

Best, C. H. "Reminiscences of the Researchers Which Led to the Discovery of Insulin." *Canadian Medical Association Journal* 47 (1942b): 398–400.

Best, C. H., and J. J. R. Macleod. "Some Chemical Reactions of Insulin." *Journal of Biological Chemistry* 55 (1923): xxix–xxx.

Best, C. H., and N. B. Taylor. *The Physiological Basis of Medical Practice, University of Toronto Text in Applied Physiology*. 4th ed. London, Bailliere, Tindall and Cox, 1945.

Best, H. B. M. *Margaret and Charley: The Personal Story of Dr. Charles Best, the Co-Discoverer of Insulin*. Toronto, The Dundurn Group, 2003.

Betlach, M. C. "Early Cloning and Recombinant DNA Technology at Herbert W. Boyer's UCSF Laboratory in the 1970s." An Oral History Conducted in 1994 by Sally Smith Hughes. University of California, Berkeley, Regional Oral History Office, The Bancroft Library, 2002.

"Biologists Acclaim Victory over Diabetes." *The Mail and Empire*, December 29, 1922. https://insulin.library.utoronto.ca/islandora/object/insulin%3AC10056.

Bliss, M. *Banting: A Biography*. Toronto, McClelland and Stewart, 1984.

Bliss, M. *The Discovery of Insulin*. Toronto, University of Toronto Press, 1982b.

Bliss, M. "Rewriting Medical History: Charles Best and the Banting and Best Myth." *Journal of the History of Medicine and Allied Sciences* 48 (1993): 253–74.

Bliss, M. "Texts and Documents: Banting, Best's and Collip's Accounts of the Discovery of Insulin, With an Introduction by Michael Bliss." *Bulletin of the History of Medicine* 56 (1982a): 554–68.

Blum, F. "Uber Nebennierendiabetes." *Deutsches Archiv Der Klinische Medezin* 71 (1901): 146–47.

Blundell, T., G. Dodson, D. Hodgkin, and D. Mercola. "Insulin: The Structure of the Crystal and its Reflection in Chemistry and Biology." *Advances in Protein Chemistry* 26 (1972): 279–402.

Boivin, A. "Directed Mutation in Colon Bacilli, by an Inducing Principle of Desoxyribonucleic Nature: Its Meaning for the General Biochemistry of Heredity." *Cold Spring Harbor Symposia on Quantitative Biology* 12 (1947): 7–17.

Boivin, A., and R. Vendrely. "Sur le role possible des deux acides nucleiques dans la cellule vivante." *Experientia* 3 (1947): 32–34.

Borell, M. "Brown-Sequard's Organotherapy and Its Appearance in America at the End of the Nineteenth Century." *Bulletin of the History of Medicine* 50 (1976): 309–20.

Boseley, S. "Diabetes Killing African Children: Survey Highlights "Shocking" Shortage of Insulin." *The Guardian*, November 17, 2003.

Bourne, H. "The Insulin Myth." *The Lancet* 265 (1953): 964–68.

Boyer, H. W. "Recombinant DNA Research at UCSF and Commercial Application at Genentech." An Oral History Conducted in 1994 by Sally Smith Hughes. University of California, Berkeley, Regional Oral History Office, The Bancroft Library, 2001.

Brachet, J. "La localisation des acides pentose nucleiques dans les tissus animaux et les oeufs d'amphibiens en voie de developpement." *Archives Biologie* 53 (1942): 207–57.

Bragg, W. H. *Concerning the Nature of Things*. London, G. Bell and Sons, 1925.

Bragg, W. L. "Patterson Diagrams in Crystal Analysis." *Nature* 143 (1939): 73.

Bragg, W. L. B. "The Growing Power of X-Ray Analysis." In *Fifty Years of X-Ray Diffraction*, edited by P. P. Ewald, 131. Utrecht, International Union of Crystallography by N.V.A. Oosthoek's Uitgeversmaatschappj, 1962.

Brandenburg, D., H. G. Gattner, L. Herbertz, G. Krail, M. Weinert, and H. Zahn. "Semi-Synthetic Insulin Analogues." *The Biochemical Journal* 125 (1971): 51–52P.

Brange, J., U. Ribel, J. F. Hansen, G. Dodson, M. T. Hansen, S. Havelund, S. G. Melberg, F. Norris, K. Norris, L. Snel, A. R. Sørensen, and H. O. Voigt. "Monomeric Insulins

Obtained by Protein Engineering and Their Medical Implications." *Nature* 333 (1988): 679–82.

Brenner, S., F. Jacob, and M. Meselson. "An Unstable Intermediate Carrying Information from Genes to Ribosomes for Protein Synthesis." *Nature* 190 (1961): 576–81.

Brett, R. L. *Barclay Fox's Journal*. London: Bell & Hyman, 1979.

Brownlee, G. G. *Fred Sanger: Double Nobel Laureate - a Biography*. Cambridge, Cambridge University Press, 2014.

Brownlee, M., and A. Cerami. "A Glucose-Controlled Insulin-Delivery System: Semisynthetic Insulin Bound to Lectin." *Science* 206 (1979): 1190–91.

Brown-Sequard, C. E. in Compte Rendus de l'Academie des Sciences 1887; p.105; 1888; p.106; cited in Olmsted (1946); p.206.

Brown-Sequard, C. E. "On a New Therapeutic Method Consisting in the Use of Organic Liquids Extracted from Glands and Other Organs." *British Medical Journal* 1 (1893): 1145–47.

Burt, S., and K. Grady. *The Illustrated History of Leeds*. Derby, Breedon Books, 1994.

Calvin, M., and A. A. Benson. "The Path of Carbon in Photosynthesis IV: The Identity and Sequence of the Intermediates in Sucrose Synthesis." *Science* 109 (1949): 140–42.

"Cambridge Council Bids Harvard Delay Its Gene Research." *The New York Times*, July 8, 1976.

Cammidge, P. J. "Insulin and Diabetes." *British Medical Journal* 2 (1922): 997–98.

"Canadian Diabetes Cure: Encouraging Results." *The Times*, August 19, 1922.

Caroe, G. *William Henry Bragg 1862–1942, Man and Scientist*. Cambridge, Cambridge University Press, 1978.

Caspersson, T. "The Relations between Nucleic Acid and Protein Synthesis." *Symposium of the Society for Experimental Biology* 1 (1947): 127–51.

Chance, R., and B. Frank. "Research, Development Production and Safety of Biosynthetic Human Insulin." *Diabetes Care* 16 (1993): 133–42.

Chang, S., and S. N. Cohen. "In Vivo Site-Specific Genetic Recombination Promoted by the EcoRI Restriction Endonuclease." *Proceedings of the National Academy of Sciences of the United States of America* 74, no. 11(1977): 4811–15. https://doi.org/10.1073/pnas.74.11.4811.

Chargaff, E. "Amphisbaena." *In Essays on Nucleic Acids*. New York: Elsevier, 1963.

Chargaff, E. "Chemical Specificity of Nucleic Acids and Mechanism of Their Enzymatic Degradation." *Experientia* 6 (1950): 201–40.

Chargaff, E. *Heraclitean Fire: Sketches from a Life Before Nature*. New York, The Rockefeller University Press, 1978.

Chargaff, E. "On the Dangers of Genetic Meddling." *Science* 192 (1976): 939–40.

Chargaff, E. "Preface to a Grammar of Biology." *Science* 172 (1971): 637–42.

Chargaff, E. *Voices in the Labyrinth: Nature, Man and Science*. Edited by R. N. Anshen. New York: The Seabury Press, 1977.

Chargaff, E., R. Lipshitz, C. Green, and M. E. Hodes. "The Composition of the Desoxyribonucleic Acid of Salmon Sperm." *Journal of Biological Chemistry* 192 (1951): 223–30.

Chargaff, E., E. Vischer, R. Doniger, C. Green, and F. Misani. "The Composition of the Desoxypentose Nucleic Acids of Thymus and Spleen." *Journal of Biological Chemistry* 177 (1949): 405–16.

Chase, M. "Genentech's Insulin Excites Doctors Less Than It Did Brokers." *The Wall Street Journal,* November 2, 1982.

Chedd, G. "Threat to US Genetic Engineering." *New Scientist* 71 (1976): 14–15.

Chibnall, A. C. "Amino-Acid Analysis and the Structure of Proteins." *Proceedings of the Royal Society of London, Series B: Biological Sciences* 131 (1942): 136–60.

"Claims Priority in Cure for Diabetes in Animals." *Toronto Daily Star,* November 17, 1922. https://insulin.library.utoronto.ca/islandora/object/insulin%3AC10031.

Clark, A. E. "An Address on Commercial Influences in Therapeutics." *The Lancet* 202 (1923): 1067–69.

Clark, A. J. L., G. Knight, P. G. Wiles, and J. B. Scotton. "Biosynthetic Human Insulin In the Treatment of Diabetes: A Double-Blind Crossover Trial in Established Diabetic Patients." *The Lancet* 320 (1982): 354–57.

Clarke, S. F., and J. R. Foster. "A History of Blood Glucose Meters and Their Role Inself-Monitoring of Diabetes Mellitus." *British Journal of Biomedical Science* 69 (2012): 83–93.

Clowes Jr., G. H. A. "George Henry Alexander Clowes, PhD, Dsc, LLD (1877–1958): A Man of Science for All Seasons." *Journal of Surgical Oncology* 18 (1981): 197–217.

Cobb, M. *Life's Greatest Secret: The Race to Crack the Genetic Code.* London, Profile Books, 2015a.

Cobb, M. "Who Discovered Messenger RNA?" *Current Biology* 25 (2015b): R523–548.

Cobb, M. "A Speculative History of DNA: What If Oswald Avery Had Died in 1934?" *PLOS Biology* 14, no. 11 (2016): e2001197. doi:10.1371/journal.pbio.2001197.

Cobb, M. "60 Years Ago, Francis Crick Changed the Logic of Biology." *PLOS Biology* 15 (2017). https://doi.org/10.1371/journal.pbio.2003243.

Cohen, S. N. "Recombinant DNA: Fact and Fiction." *Science* 195, no. 4279(1977): 654–57. https://doi.org/10.1126/science.265099.

Cohen, S. N., A. C. Y. Chang, H. W. Boyer, and R. B. Helling. "Construction of Biologically Functional Plasmids in Vitro." *Proceedings of the National Academy of Sciences of the United States of America* 70 (1973): 3240–44.

Cohen, S. N. "Science, Biotechnology, and Recombinant DNA: A Personal History." An Oral History Conducted by Sally Smith Hughes in 1995. University of California, Berkeley, Regional Oral History Office, The Bancroft Library, 2009."

Collins, J. "Diabetes, Dread Disease, Yields to New Gland Cure." *The New York Times,* May 6, 1923.

Collip, J. B. "Frederick Grant Banting, Discoverer of Insulin." *The Scientific Monthly* 52 (1941): 472–74.

Collip, J. B. "John James Rickard Macleod, 1876-1935." *Biochemistry* 29 (1935): 1253–56.

Collip, J. B. "The Original Method as Used for the Isolation of Insulin in Semipure Form for the Treatment of the First Clinical Cases." *Journal of Biological Chemistry* 55 (1922a): xl.

Collip, J. B. "The Preparation of the Extracts as Used in the First Clinical Cases." *Transactions of the Royal Society of Canada* 5 (1922b): 27–29.

Collip, J. B. "Recollections of Sir Frederick Banting." *The Canadian Medical Association Journal* 47 (1942): 401–3.

Consden, R., A. H. Gordon, and A. J. P. Martin. "Partition Chromatography of Amino Acids." *The Biochemical Journal* 38 (1944a): ix.

Consden, R., A. H. Gordon, and A. J. P. Martin. "Qualitative Analysis of Proteins: A Partition Chromatographic Method Using Paper." *The Biochemical Journal* 38 (1944b): 224–32.

Consden, R., A. H. Gordon, A. J. P. Martin, and R. L. M. Synge. "Gramicidin S: The Sequence of the Amino-Acid Residues." *The Biochemical Journal* 41 (1947): 596–602.

Cookson, C. "Novo Nordisk Buys Bristol Lab Spin-Off in Deal to Pursue Diabetes Drugs." *Financial Times Weekend,* August 18–19, 2018.

"Cost of Insulin: British and American Prices." *The Times,* July 11, 1923g.

"Cost of Insulin: Cheaper Supplies from Strasbourg." *The Times,* July 18, 1923h.

"Cost of Insulin: To the Editor of The Times." *The Times,* July 10, 1923f.

Crea, R., A. Kraszewski, T. Hirose, and K. Itakura. "Chemical Synthesis of Genes for Human Insulin." *Proceedings of the National Academy of Sciences of the United States of America* 75, no. 12 (1978): 5765–69. https://doi.org/10.1073/pnas.75.12.5765.

Crea, R. "DNA Chemistry at the Dawn of Commercial Biotechnology." An Oral History Conducted in 2002 by Sally Smith Hughes, Ph.D. University of California, Berkeley, Regional Oral History Office, The Bancroft Library, 2004.

Crick, F. H. C. "On Protein Synthesis." *Symposium of the Society for Experimental Biology* 12 (1958): 138–63.

Crick, F. H. C. "Towards the Genetic Code." *Scientific American* 207 (1962): 8–16.

Crick, F. H. C. *What Mad Pursuit: A Personal View of Scientific Discovery.* London: Weidenfeld and Nicholson, 1988.

Crick, F. H. C., and L. E. Orgel. "Directed Panspermia." *Icarus* 19 (1973): 341–46.

"Crystalline Insulin." *British Medical Journal* (1926): 666.

Crowfoot, D. "X-Ray Single Crystal Photographs of Insulin." *Nature* 135 (1935): 591.

Crowfoot, D. M. "The Crystal Structure of Insulin. I. The Investigation of Air-Dried Insulin Crystals." *Proceedings of the Royal Society of London. Series A, Mathematical and Physical Sciences* 64 (1938): 580–602.

Culliton, B. "Recombinant DNA Bills Derailed: Congress Still Trying to Pass a Law." *Science* 199 (1978): 274–77.

Culliton, B. "Recombinant DNA: Cambridge City Council Votes Moratorium." *Science* 193 (1976): 300–301.

Daemmrich, A. A., N. R. Gray, and L. Shaper. *Reflections from the Frontiers, Explorations for the Future: Gordon Research Conferences, 1931–2006.* Philadelphia, Chemical Heritage Foundation, 2006.

Dahm, R. "Discovering DNA: Friedrich Miescher and the Early Years of Nucleic Acid Research." *Human Genetics* 122 (2008): 565–81.

Dale, H. H. "The Dedication of the Charles H. Best Institute: Special Convocation." *Diabetes* 3 (1954): 28–46.

Dale, H. H. "Insulin." *British Medical Journal* 2 (1922): 1241.

Dale, H. H. "Sir Frederick Banting, K.B.E., M.C., F.R.S., L.L.D., D.Sc.,F.R.C.P., F.R.C.S." *British Medical Journal* 1 (1941): 383.

Danna, K. J., G. H. Sack, and D. Nathans. "Studies of Simian Virus 40 DNA: VII. A Cleavage Map of the SV40 Genome." *Journal of Molecular Biology* 78 (1973): 363–76.

Davis, B. "Evolution, Epidemiology, and Recombinant DNA." *Science* 193, no. 4252 (1976): 442–442. https://doi.org/10.1126/science.11643326.

"Declares Best Shares Honor: Dr. Banting Pays Tribute to Partner in Discovery." *The Globe.* September 9, 1922. https://insulin.library.utoronto.ca/islandora/object /insulin%3AC10036.

Delhoume, L. *De Claude Bernard a d'Arsonval* Paris: Bailere & Fils, 1939.

Demerec, M. "Statement on Visa Action." *Science* 110 (1949): 335.

de Meyer, J. "Contribution a l'étude de la pathogénie de diabète pancréatique', Archive Internationale de Physiologie (1909): 121–80.

"Diabetics Warned After Rise in Deaths." *The Daily Telegraph,* October 13, 1989.

Dixon, G. H., and A. C.Wardlaw. "Regeneration of Insulin Activity from the Separated and Inactive A and B Chains." *Nature* 188 (1960): 721–24.

Dobson, M. "Experiments and Observations on the Urine in Diabetes." *Medical Observations and Inquiries* 5 (1776): 298–316.

Dodson, G. "Dorothy Hodgkin, Protein Crystallography and Insulin." *Current Science* 72 (1997): 466–68.

Dodson, G. "Dorothy Mary Crowfoot Hodgkin, O.M. 12 May 1910–29 July 1994." *Biographical Memoirs of Fellows of the Royal Society* 48 (2002): 179–219.

Donohue, J. "Fragments of Chargaff." *Nature* 276 (1978): 133–35.

Doroshaw, D. B. "Performing a Cure for Schizophrenia: Insulin Coma Therapy on the Wards." *Journal of the History of Medicine and Allied Sciences* 62 (2006): 213–43.

Dounce, A. "Duplicating Mechanism for Peptide Chain and Nucleic Acid Synthesis." *Enzymologia* 15 (1952): 251–58.

Dounce, A. "Nucleic Acid Template Hypothesis." *Nature* 172 (1953): 541.

Downing, T. *1983: The World at the Brink*. London, Abacus, 2019.

"Dr. Georg L. Zuelzer, Heart Specialist, 79." *The New York Times*, October 20, 1949.

Drügemöller, P., and L. Norpoth."Right and Wrong Avenues of Exploration in German Insulin Research." In *Diabetes: Its Medical and Cultural History*, edited by Dietrich von Engelhardt, 427–36. Berlin, Springer-Verlag, 1989.

Du, Y-C., Y. S. Zhang, Z. X. Lu, and C. L. Tsou. "Conditions for Successful Resynthesis of Insulin from Its Glycyl and Phenylalanyl Chains." *Scientia Sinica* 14 (1965): 229–36.

Du, Y-C., Y. S. Zhang, Z. X. Lu, and C. L. Tsou. "Resynthesis of Insulin from Its Glycyl and Phenylalanyl Chains." *Scientia Sinica* 10 (1961): 85–104.

Dubos, R. "Oswald Theodore Avery 1877–1955." *Biographical Memoirs of Fellows of the Royal Society* 2 (1956): 35–48.

Dubos, R. J. *The Professor, The Institute, and DNA*. New York: The Rockefeller University Press, 1976.

Dudley, H. W. "The Purification of Insulin and Some of its Properties." *The Biochemical Journal* 17 (1923): 376–90.

Duncan, G. G. "Frederick Madison Allen 1879–1964." *Diabetes* 1 (1964): 318–19.

Dunitz, J. D. "Dorothy Crowfoot Hodgkin—An Introduction to Her Work and Personality." *Current Science* 72 (1997): 447–50.

Dussoix, D., and W. Arber. "Host Specificity of DNA Produced by Escherichia Coli: II. Control over Acceptance of DNA from Infecting Phage λ." *Journal of Molecular Biology* 5, no. 1 (1962): 37–49. https://doi.org/10.1016/S0022-2836(62)80059-X.

Dyson, F. J. "Costs and Benefits of Recombinant DNA Research [1]." *Science* 193, no. 4247 (1976): 6. https://doi.org/10.1126/science.11643319.

Eastwood, J. D. "Insulin and Independence." *British Medical Journal* 293 (1986): 1659.

Edgerton, D. S., M. C. Moore, J. J. Winnick, M. Scott, B. Farmer, H. Naver, C. B. Jeppesen, P. Madsen, T. B. Kjeldsen, E. Nishimura, C. L. Brand, and A. D. Cherrington. "Changes in Glucose and Fat Metabolism in Response to the Administration of a Hepato-Preferential Insulin Analog." *Diabetes* 63 (2014): 3946–54.

Efstratiadis, A., F. C. Kafatos, A. M. Maxam, and T. Maniatis. "Enzymatic in Vitro Synthesis of Globin Genes." *Cell* 7 (1976): 279–88.

Efstratiadis, A., T. Maniatis, F. C., Kafatos, A Jeffrey, and J. N. Vournakis. "Full Length and Discrete Partial Reverse Transcripts of Globin and Chorion MRNAs." *Cell* 4 (1975): 367–78.

Egger, M., G. D. Smith, H. Imhoof, and A. Teuscher. "Risk of Severe Hypoglycaemia in Insulin Treated Diabetic Patients Transferred to Human Insulin: A Case Control Study." *British Medical Journal* 303 (1991): 617.

Egger, M., G. D. Smith, A. U. Teuscher, and A. Teuscher. "Influence of Human Insulin on Symptoms and Awareness of Hypoglycaemia: A Randomised Double Blind Crossover Trial." *British Medical Journal* 303 (1991): 622.

Elsden, S. R. "Obituary: Richard Synge." *The Independent,* August 24, 1994.

Elsden, S. R., and R. L. M. Synge. "Starch as a Medium for Partition Chromatography." *The Biochemical Journal* 38 (1944): proc ix.

Ettre, L. S. "The Birth of Partition Chromatography." *Liquid Chromatography Gas Chromatography (LCGC) Europe* 19 (2001): 506–12.

Ettre, L. S. and A. Zlatkis, eds. "A. J. P. Martin." In *75 Years of Chromatography: A Historical Dialogue,* 285–96. Journal of Chromatography Library 17. New York, Elsevier Science, 1979.

Ettre, L. S. and A. Zlatkis, eds. "Hendrik Boer." In *75 Years of Chromatography: A Historical Dialogue,* 11–19. Journal of Chromatography Library 17. New York, Elsevier Science, 1979.

Fara, P. "Pictures of Dorothy Hodgkin." *Endeavour* 27 (2002): 85–86.

Feasby, W. R. "The Discovery of Insulin." *Journal of the History of Medicine and Allied Sciences* 13 (1958): 68–74.

Ferriman, A. "Doctors Accuse Lawyers in Row on Diabetes." *The Observer,* September 29, 1991.

Ferry, G. *Dorothy Hodgkin: A Life.* 2nd ed. London, Bloomsbury, 2014.

Feudtner, C. *Bittersweet: Diabetes, Insulin and the Transformation of Illness.* Chapel Hill, University of North Carolina Press, 2003.

Fischbach, H., M. Mundell, and T. Eble. "Determination of Penicillin K by Partition Chromatography." *Science* 104 (1946): 84–85.

Fletcher, C. "One Way of Coping With Diabetes." *British Medical Journal* 280 (1980): 115–16.

Forgan, S. "Festivals of science and the two cultures: science, design and display in the Festival of Britain, 1951", *British Journal for the History of Science,* 31 (1998): 217–40.

Forschbach, J. "Versuche zur Behandlung des Diabetes mellitus mit dem Zuelzerschen Pankreashormon." *Deutschen Medizinisches Wochenschrift* 35 (1909): 2053–55.

Frazier, K. "Rise to Responsibility at Asilomar." *Science News* 107 (1975): 187.

Frederickson, D. S. *The Recombinant DNA Controversy: A Memoir—Science, Politics, and the Public Interest 1974-1981.* Washington D.C., ASM Press, 2001.

"Free Insulin." *The Lancet* 202, no. 5218 (1923): 471.

Friedman, J. "Discovery, Interrupted: How World War I Delayed a Treatment for Diabetes and Derailed One Man's Chance at Immortality." Harper's Magazine, November 2018.

Friling, T., R. Ioanid, and M. E. Ionescu. "Final Report: International Commission on the Holocaust in Romania." Bucharest, International Commission on the Holocaust in Romania, 2004. https://www.ushmm.org/m/pdfs/20080226-romania-commission-holocaust-history.pdf.

"Future of Insulin: Problem of Continuous Administration." *The Times*, August 2, 1923k.

Gamow, G. "Possible Relation between Deoxyribonucleic Acid and Protein Structures." *Nature* 173 (1954): 318.

Gamow, G. "Information Transfer in the Living Cell." *Scientific American* 193 (1955): 70–78.

Gemmill, C. L. "The Greek Concept of Diabetes." *Bulletin of the New York Academy of Medicine* 48 (1972): 1033–36.

Gilbert, S., & Green, C. *Vaxxers: The Inside Story of the Oxford AstraZeneca Vaccine and the Race Against the Virus.* Hodder & Stoughton (2021).

Gilbert, W. "Recombinant DNA Research: Government Regulation." *Science* 197, no. 4300 (1977): 208.

Gill, G., J. S. Yudkin, S. Tesfaye, M. de Courten, E. Gale, A. Motala, K. Ramaiya, N. Unwin, and S. Wild. "Essential Medicines and Access to Insulin." *The Lancet Diabetes and Endocrinology* 5 (2017): 324–25.

Gill, G. V., J. S. Yudkin, H. Keen, and D. Beran. "The Insulin Dilemma in Resource-Limited Countries. A Way Forward?" *Diabetologia* 54 (2011): 19–24.

"Gives Dr. Banting Credit for 'Insulin,'" *Toronto Daily Star*, September 7, 1922. https://insulin.library.utoronto.ca/islandora/object/insulin%3AC10034.

Gley, E. "Action des extraits de pancréas sclérosé sur des chiens diabétiques (par extirpation du pancréas." *Societe de Biologie, Compte Rendus* 87, December (1922): 1322.

Glick, S. "Rosalyn Sussman Yalow (1921–2011)." *Nature* 474 (2011): 580.

Goeddel, D. V., D. G. Kleid, F. Bolivar, H. L. Heyneker, D. G. Yansura, R. Crea, T. Hirose, A. Kraszewski, K. Itakura, and A. D. Riggs. "Expression in *Escherichia Coli* of Chemically Synthesized Genes for Human Insulin." *Proceedings of the National Academy of Sciences of the United States of America* 76 (1979): 106–10.

Goeddel, D. V., and A. D. Levinson. "Robert A. Swanson (1947–99)." *Nature* 403 (2000): 264.

Goeddel, D. V. "Scientist at Genentech, CEO at Tularik." An Oral History Conducted in 2001 and 2002 by Sally Smith Hughes. University of California, Berkeley, Regional Oral History Office, The Bancroft Library, 2003.

Gordon, A. H. "The Beginnings of Chromatography: How Paper Chromatography Was Discovered." *Trends in Biochemical Sciences* 2 (1977): N243–44.

Gordon, A. H., A. J. P. Martin, and R. L. M. Synge. "A Study of the Partial Acid Hydrolysis of Some Proteins, with Special Reference to the Mode of Linkage of the Basic Amino-Acids." *The Biochemical Journal* 35 (1941): 1369–87.

Gordon, A. H., A. J. P. Martin, and R. L. M. Synge. "A Study of the Partial Acid Hydrolysis of Cow-Hide Gelatin." *Biochemical Journal* 37, no. 1 (1943): 92–102. https://doi.org/10.1042/bj0370092.

Gordon, A. H., A. J. P. Martin, and R. L. M. Synge. "Partition Chromatography in the Study of Protein Constituents." *The Biochemical Journal* 37 (1943a): 79–86.

Gordon, A. H., A. J. P. Martin, and R. L. M. Synge. "Partition Chromatography of Free Amino Acids and Peptides." *The Biochemical Journal: Proceedings of the 229th Meeting of the Biochemical Society* 37 (1943b): xiii.

Gordon, A. H., A. J. P. Martin, and R. L. M. Synge. "The Amino-Acid Composition of Gramicidin." *The Biochemical Journal* 37 (1943c): 86–92.

Gordon, H. "Richard Laurence Millington Synge, 28 October 1914–18 August 1994." *Biographical Memoirs of Fellows of the Royal Society* 42 (1996): 455–76.

Gordon, H. "Synge, Richard Laurence Millington." In *Oxford Dictionary of National Biography* [online], September 23, 2004. https://doi.org/10.1093/ref:odnb/55773.

Graham, C., and C. F. Harris. "The New Treatment of Diabetes by Insulin: A Statement from The Medical Research Council." *The Lancet* 200 (1922): 1086.

Greenhouse, L. "Science May Patent New Forms of Life, Justices Rule, 5 to 4." *The New York Times,* June 17, 1980.

Grill, M. "Gib dem Affen Zucker' ('Give the monkeys sugar')." *Stern Magazine,* June 13, 2006. Online at https://www.stern.de/wirtschaft/news/pharmalobby-gib-dem-affen-zucker-3600554.html.

Gros, F., H. Hiatt, W. Gilbert, C. G. Kurland, R. W. Riseborough, and J. D. Watson. "Unstable Ribonucleic Acid Revealed by Pulse Labelling of *Escherichia Coli*." *Nature* 190 (1961): 581–85.

Gross, O. W. "Georg Ludwig Zuelzer." *Deutsche Medizinische Wochenschrift* 75 (1950): 153–54.

Grover, N. "Doctors Suggest Covid-19 Could Cause Diabetes: More Than 350 Clinicians Report Suspicions Of Covid-Induced Diabetes, Both Type 1 and Type 2." *The Guardian,* March 19, 2021.

Grubaugh, N. D., Hodcroft, E. B., Fauver, J. R., Phelan, A. L., and Cevik, M. "Public health actions to control new SARS-CoV-2 variants." *Cell* 184 (2021): 1127–1132.

Hagedorn, H. C., B. N. Jensen, N. B. Krarup, and I. Wodstrup. "Protamine Insulinate." *Journal of the American Medical Association* 106 (1936): 177–80.

Hall, K. T. *The Man in the Monkeynut Coat: William Astbury and the Forgotten Road to the Double-Helix.* Oxford, Oxford University Press, 2014. Paperback edition 2022.

Hall, K. T. ""In Praise of Wool": The development of partition chromatography and its under-appreciated impact on molecular biology" *Endeavour* 43 Issue 4 (2019): https://doi.org/10.1016/j.endeavour.2020.100708.

Hall, K. T., and N. Sankaran. "DNA Translated: Friedrich Miescher's Discovery of Nuclein in Its Original Context." *The British Journal for the History of Science* 54, no. 1 (2021): 99–107. https://doi.org/https://doi.org/10.1017/S000708742000062X.

Hall, S. *Invisible Frontiers: The Race to Synthesize a Human Gene.* Cheltenham, Tempus, 1987.

Hallas-Møller, K., M. Jersild, K. Petersen, and J. Schlichtkrull. "Zinc Insulin Preparations for Single Daily Injectionclinical Studies of New Preparations with Prolonged Action." *Journal of the American Medical Association* 150 (1952): 1667–71.

Hallas-Møller, K., K. Petersen, and J. Schlichtkrull. "Crystalline and Amorphous Insulin-Zinc Compounds with Prolonged Action." *Science* 116 (1952): 394–98.

Harding, M. M. "Gramicidin S: Some Stages in the Determination of Its Crystal and Molecular Structure." In *Structural Studies on Molecules of Biological Interest: A Volume in*

Honour of Professor Dorothy Hodgkin, edited by G. Dodson, J. P. Glusker, and D. Sayre, Ch. 20. Oxford, Clarendon Press, 1981.

Hargreaves, I. "Big Demand for Genentech Offer." *Financial Times,* October 15, 1980.

Hargreaves, I. "Technological Wonderland." *Financial Times,* October 18, 1980.

Havelund, S., P. Plum, U. Ribel, I. Jonassen, A. Vølund, J. Markussen, and P. Kurtzhals. "The Mechanism of Protraction of Insulin Detemir, a Long-Acting, Acylated Analog of Human Insulin." *Pharmaceutical Research* 21 (2004): 1498–504.

Heaton, H. *The Yorkshire Woollen and Worsted Industries.* 2nd ed. Oxford, Clarendon Press, 1965.

Hedgepeth, J., H. M. Goodman, and H. W. Boyer. "DNA Nucleotide Sequence Restricted by the RI Endonuclease." *Proceedings of the National Academy of Sciences of the United States of America* 69 (1972): 3448–52.

Hedon, E. "Greffe sous cutanee du pancreas: ses resultats au point de vue de la theorie due diabete pancreatique." *Societe de Biologie, Compte Rendus* 44 (1892): 678–80.

Heller, S. R. "Robert Tattersall: A Diabetes Physician Ahead of His Time." *Diabetes Care* 42 (2019): 1005–8.

Hetzel, K. S. "The Diet During Insulin Treatment of Diabetes Mellitus." *British Medical Journal* 1 , no. 3293 (1924): 230–32.

Heynecker, H. L., J. Shine, H. Goodman, H. W. Boyer, J. Rosenberg, R. E. Dickerson, S. A. Narang, K. Itakura, S. Lin, and A. D. Riggs. "Synthetic Lac Operator DNA is Functional in Vivo." *Nature* 263 (1976): 748–52.

Heynecker, H. L. "Molecular Geneticist at UCSF and Genentech, Entrepreneur in Biotechnology." An Oral History Conducted in 2002 by Sally Smith Hughes, Ph.D. University of California, Berkeley, Regional Oral History Office, The Bancroft Library, 2004.

Himsworth, H. P. "Protamine Insulin and Zinc Protamine Insulin in the Treatment of Diabetes Mellitus." *British Medical Journal* 1 (1937): 541–46.

His, W. *Die histochemischen und physiologischen Arbeiten von Friedrich Miescher.* Leipzig: F. C. W. Vogel, 1897.

Hoagland, M. B., M. L. Stephenson, J. F. Scott, L. I. Hecht, and P. C. Zamecnik. "A Soluble Ribonucleic Acid Intermediate in Protein Synthesis." *Journal of Biological Chemistry* 213 (1958): 241–57.

Hoagland, M. B., P. C. Zamecnik, and M. L. Stephenson. "Intermediate Reactions in Protein Biosynthesis." *Biochemica et Biophysica Acta* 24 (1957): 215–16.

Hodgkin, D. M. C. "Chinese Work on Insulin." *Nature* 255 (1975): 103.

Hodgkin, D. M. C. "Dorothy Wrinch - Obituary." *Nature* 260 (1976): 564.

Hodgkin, D. M. C. "The X-Ray Analysis of Complicated Molecules." Nobel Lecture, December 11, 1964. https://www.nobelprize.org/prizes/chemistry/1964/hodgkin/lecture/.

Hodgkin, D. M. C. "X Rays and the Structure of Insulin." *British Medical Journal* 4 (1971): 447–51.

Hodgkinson, N. "New Drug Helps Nobel Winner Regain His Memory." *The Sunday Times,* December 4, 1988.

Hodgson, A. J. "Treatment of Diabetes Mellitus." *Journal of the American Medical Association* 57 (1911): 1187–92.

Holleman, F., and E. A. M. Gale. "Nice Insulins, Pity about the Evidence." *Diabetologia* 50 (2007): 1783–90.

Hopson, J. L. "Recombinant Lab for DNA and My 95 Days in It." *The Smithsonian* 8, no. 3 (1977): 55–62.

"House of Commons: Cost of Insulin." *The Times*, July 26, 1923j.

Houssay, B. A. "The Discovery of Pancreatic Diabetes: The Role of Oscar Minkowski." *Diabetes* 1 (1952): 112–16.

Howey, D. C., R. R. Bowsher, R. L. Brunelle, and J. R. Woodworth. "[Lys(B28), Pro(B29)]-Human Insulin: A Rapidly Absorbed Analogue of Human Insulin." *Diabetes* 43 (1994): 396–402.

Hubbard, R. "Recombinant DNA: Unknown Risks." *Science* 193 (1976): 834–36.

Hufbauer, K. "George Gamow 1904–1968." In *National Academy of Sciences Biographical Memoirs*, 73, 3–38. Washington, DC: The National Academies Press, 2009.

Hughes, T. J. *Thomas Willis 1621–1675: His Life and Work*. London, Royal Society of Medicine, 1991.

Hull, S. E., R. Karlsson, P. Main, M. M.Woolfson, and E. J. Dodson. "The Crystal Structure of a Hydrated Gramicidin S–Urea Complex." *Nature* 275 (1978): 206–7.

Ingram, V. M. 1957. "Gene Mutations in Human Haemoglobin: The Chemical Difference between Normal and Sickle-Cell Haemoglobin." *Nature* 180: 326–28.

Ingram, V. M. "A Specific Chemical Difference Between the Globins of Normal Human and Sickle-Cell Anaemia Haemoglobin." *Nature* 178 (1956): 792–94.

"Insulin." *British Medical Journal*, 1, no. 3243 (1923): 341.

"Insulin and Diabetes." *British Medical Journal*, 2, no. 3227 (1922): 882.

"Insulin Available in This Country: Conditions of Sale and Precautions to Be Observed. Statement By Medical Research Council." *British Medical Journal* 1, no. 3252 (1923): 695–96. http://www.jstor.org/stable/20423215.

The Insulin Committee, University of Toronto. "Insulin: Its Action, Its Therapeutic Value in Diabetes and Its Manufacture." *Journal of the American Medical Association* 80 (1923): 1847–51.

"Insulin' from Fish: Scientific Research and Practical Problems." *The Times*, January 2, 1923.

"Insulin in Advanced Diabetes." *British Medical Journal*, October 27, 1923: 765. (https://www.bmj.com/content/2/3278/767).

"Insulin' on a Large Scale: New Extraction Methods." *The Times*, January 15, 1923b.

"Insulin Research Raises Debate on DNA Guidelines." *The New York Times*, June 29, 1979.

"Insulin Supply: British Manufacturers Output." *The Times*, July 20, 1923i.

"Insulin Value Proven, Children Are Benefited." *Toronto Daily Star*, April 7, 1923b. https://insulin.library.utoronto.ca/islandora/object/insulin%3AC10086.

Ionescu-Tirgoviste, C. *The Re-Discovery of Insulin*. Bucharest, Editura Geneze, 1996.

Ionescu-Tirgoviste, C., C. Guja, and S. Ioacara. Documents Regarding the Discovery of Insulin and Its Clinical Utilisation. Bucharest, Romanian Academy Publishing House, 2005.

Irwin, D. "The Contribution of Sir Frederick Banting to Silicosis Research." *The Canadian Medical Association Journal* 47, no. 5 (1942): 403–5.

Itakura, K., T. Hirose, R. Crea, A. D. Riggs, H. L. Heyneker, F. Boliver, and H. W. Boyer. "Expression in Eschericia Coli of a Chemically Synthesized Gene for the Hormone Somatostatin." *Science* 198 (1977): 1056–63.

Itakura, K. "DNA Synthesis at City of Hope for Genentech." An Oral History Conducted in 2005 by Sally Smith Hughes. University of California, Berkeley, Regional Oral History Office, The Bancroft Library, 2006.

Jackson, D. A., R. H. Symonds, and P. Berg. "Biochemical Method for Inserting New Genetic Information into DNA of Simian Virus 40: Circular SV40 DNA Molecules Containing Lambda Phage Genes and the Galactose Operon of Escherichia Coli." *Proceedings of the National Academy of Sciences of the United States of America* 69 (1972): 2904–9.

Jacob, F. *The Statue Within*. London: Unwin Hyman, 1988.

Jacob, F., and J. Monod. "Genetic Regulatory Mechanisms in the Synthesis of Proteins." *Journal of Molecular Biology* 3 (1961): 318–56.

James, A. T., and A. J. P. Martin. "Gas-Liquid Partition Chromatography: The Separation and Micro-Estimation of Volatile Fatty Acids from Formic Acid to Dodecanoic Acid." *The Biochemical Journal* 50 (1952): 679–90.

Jarosinski, M. A., D. Balamurugan, R. Nischay, C. Deepak, and M. A. Weiss. "'Smart' Insulin-Delivery Technologies and Intrinsic Glucose-Responsive Insulin Analogues." *Diabetologia* 64 (2021): 1016–29.

Johnson, I. "Eli Lilly & the Rise of Biotechnology." An Oral History Conducted in 2004 by Sally Smith Hughes. University of California, Berkeley, Regional Oral History Office, The Bancroft Library, 2006.

Johnson, I. S. "The Trials and Tribulations of Producing the First Genetically Engineered Drug." *Nature Reviews: Drug Discovery* 2 (2003): 747–51.

Jones, K. "Insulin Coma Therapy in Schizophrenia." *Journal of the Royal Society of Medicine* 93 (2000): 147–49.

Jörgens, V. "The Discovery of Insulin in 1914: Georg Zülzer, from Berlin, and Camille Reuter, the Forgotten Chemist from Luxembourg." *Diabetes & Metabolism* (2020a). doi: 10.1016/j.diabet.2020.07.007. Online ahead of print.

Jörgens, V. "They Got Very Near the Goal: Zuelzer, Scott and Paulescu." *Unveiling Diabetes - Historical Milestones in Diabetology* 29 (2020b): 58–72.

Joslin, E. P. "Obituary: George Richards Minot (1885-1950)." *The New England Journal of Medicine* 242 (1950): 565.

Joslin, E. P. "Pancreatic Extract in the Treatment of Diabetes." *Boston Medical and Surgical Journal* 145 (1922): 654.

Joslin, E. P. "Reminiscences of the Discovery of Insulin: A Personal Impression." *Diabetes* 5 (1956): 64–68.

Joslin, E. P. "The Routine Treatment of Diabetes with Insulin." *Journal of the American Medical Association* 80 (1923): 1581–83.

Joslin, E. P., H. Gray, and H. F. Root, "Insulin in Hospital and Home." *Journal of Metabolic Research* 8 (1922): 651–99.

Judson, H. F. *The Eighth Day of Creation*. Cold Spring Harbor, CSHL Press, 1996.

Judson, H. F. "No Nobel Prize for Whining." *The New York Times*, October 2003.

Jurdjevic, M., and C. Tillman. "Texts and Documents: E. C. Noble in June 1921, and His Account of the Discovery of Insulin." *Bulletin of the History of Medicine* 78 (2004): 864–75.

Kaarsholm, S. L., et al "Engineering Glucose Responsiveness into Insulin." *Diabetes* 67 (2018): 299–308.

Katsoyannis, P. G. "The Synthesis of the Insulin Chains and Their Combination to Biologically Active Material." *Diabetes* 13 (1964): 339–48.

Katsoyannis, P. G., K. Fukuda, A. Tometsko, K. Suzuki, and M. Tilak. "Insulin Peptides: X- The Synthesis of the B-Chain of Insulin and Its Combination with Natural or Synthetic A-Chain to Generate Insulin Activity." *Journal of the American Chemical Society* 86 (1964): 930–32.

Katsoyannis, P. G., and K. Suzuki. "Insulin Peptides II: Synthesis of a Protected Pentapeptide Containing the C-Terminal Sequence of the A-Chain of Insulin." *Journal of the American Chemical Society* 83 (1961): 4057–59.

Katsoyannis, P. G., and K. Suzuki. "Insulin Peptides VII: The Synthesis of Two Decapeptide Derivatives Containing the C-Terminal Sequence of the B Chain of Insulin." *Journal of the American Chemical Society* 85 (1963): 2659–61.

Katsoyannis, P. G., K. Suzuki., and A. Tometsko. "Insulin Peptides IV: Synthesis of a Protected Decapeptide Containing the C-Terminal Sequence of the A-Chain of Insulin." *Journal of the American Chemical Society* 85 (1963):1139–41.

Katsoyannis, P. G., A. Tometsko, and K. Fukuda. "Insulin Peptides IX: The Synthesis of the A-Chain of Insulin and Its Combination with Natural B-Chain to Generate Insulin Activity." *Journal of the American Chemical Society* 85 (1963): 2863–65.

Keen, H., J. C. Pickup, R. W. Bilious, A. Glynne, and R. Marsden. "Human Insulin Produced by Recombinant DNA Technology: Safety and Hypoglycæmic Potency in Healthy Men." *The Lancet* 316 (1980): 398–401.

Kelly, T. J., and H. O. Smith. "A Restriction Enzyme from Hemophilus Influenzae: II. Base Sequence of the Recognition Site." *Journal of Molecular Biology* 51 (1970): 393–400.

Khorana, H. G. et al "Total Synthesis of the Structural Gene for the Precursor of a Tyrosine Suppressor Transfer RNA from *Escherichia Coli*: 1. General Introduction." *Journal of Biological Chemistry* 251 (1976): 565–70.

Khorana, H. G., et al "CIII. Total Synthesis of the Structural Gene for an Alanine Transfer Ribonucleic Acid from Yeast." *Journal of Molecular Biology* 72 (1972): 209–17.

Kiley, T. D. "Genentech Legal Counsel and Vice President, 1976–1988, and Entrepreneur." An Oral History Conducted in 2000 and 2001 by Sally Smith Hughes. University of California, Berkeley, Regional Oral History Office, The Bancroft Library, 2002.

"King's College Hospital: Clinic for Diabetic Outpatients." *The Times*, May 12, 1934.

"King's College Hospital: Tributes to Progress." *The Times*, November 1, 1933.

Kleid, D. "Scientist and Patent Agent at Genentech." An Oral History Conducted in 2001 and 2002 by Sally Smith Hughes. University of California, Berkeley, Regional Oral History Office, The Bancroft Library, 2002.

Kleiner, I. S. "The Action of Intravenous Injections of Pancreas Emulsions in Experimental Diabetes." *Journal of Biological Chemistry* 40 (1919): 153–70.

Kleiner, I. S. "Hypoglycaemic Agents – Past and Present." *Clinical Chemistry* 5 (1959): 79–99.

Kleiner, I. S., and S. J. Meltzer. "Retention in the Circulation of Dextrose in Normal and Depancreatized Animals, and the Effect of an Intravenous Injection of an Emulsion of Pancreas upon This Retention." *Proceedings of the National Academy of Sciences* 1 (1915): 338–41.

Konforti, B. "The Servant with the Scissors." *Nature Structural Biology* 7 (2000): 99–100.

Kramer, B., J. Marker, and J. R. Murlin. "Pancreatic Diabetes in the Dog. II: Is the Glucose Retained when Sodium Carbonate Administered to Depancreatised Dogs Administered as Glycogen?" *Journal of Biological Chemistry* 27 (1916): 499–515.

Kuhn, T. S. *The Structure of Scientific Revolutions*. Chicago, University of Chicago Press, 1962.

Kung, Y. T., Y. C. Du, W. T. Huang, C. C. Chen, and L. T. Ke. "Total Synthesis of Crystalline Insulin." *Scientia Sinica* 15 (1966): 544–61.

Lancereaux, E. "Le diabete maigre: ses symptomes, son evolution, son prognostic et son traitement." *L'Union Médicale, Paris* 20 (1880): 205–11.

"A Landmark in Genetics." *The Times*, August 23, 1957.

Langerhans, P. "Contributions to the Microscopic Anatomy of the pancreas." MD Thesis, Berlin Faculty, 1869.

Langmuir, I., and D. Wrinch. "Vector Maps and Crystal Analysis." *Nature* 142 (1938): 581–83.

Laron, Z. "Nicolae Paulescu – Scientist and Politician." *The Israel Medical Association Journal* 10 (2008): 491–93.

Laurence, W. L. "Britons Discover Structure of B-12: Solve Puzzle of the Vitamin That is Vital to Building of Red Blood Cells Synthesis is Started Compound That Keeps Alive Pernicious Anemia Victims is Highly Complicated." The New York Times, July 26, 1955.

Lawrence, R. D. "Local Insulin Reactions." *The Lancet* 1 (1925): 1125–26.

Lawrence, R. D. "Zinc-Protamine-Insulin In Diabetes: Treatment By One Daily Injection." *British Medical Journal* 1 (1939): 1077–80.

Lawrence, R. D., and W. Oakley. "A New Long-Acting Insulin: A Preliminary Trial Of 'Lente' Novo Insulin." *British Medical Journal* 1 (1953): 242–44.

Lear, J. *Recombinant DNA: The Untold Story*. New York, Crown Publishers Inc., 1978.

Lederberg, J. "The Dawning of Molecular Genetics." *Trends in Microbiology* 8 (2000): 194–95.

Leopold, E. J. "Aretaeus the Cappadocian." *Annals of Medical History* 2 (1930): 424–35.

Lesser, F. "Human Insulin Comes under Scrutiny as Number of Deaths Rises." *New Scientist,* August 18, 1989.

Leyton, O. "Some Armchair Reflections Upon Insulin." *British Medical Journal* 2, no. 3232 (1922): 1144.

Leyton, O. "Insulin and Diabetes Mellitus." *British Medical Journal* 1 (1923): 882–83.

Leyton, O. "Hypoglycaemia." *Proceedings of the Royal Society of Medicine* 19 (1926): 47–50.

Lezard, N. "The Crooked Timber of Humanity: Chapters in the History of Ideas - A Review." *The Guardian* July 23, 2013.

Lobban, P., and A. D. Kaiser. "Enzymatic End-to End Joining of DNA Molecules." *Journal of Molecular Biology* 78 (1973): 453–71.

Lovelock, J. "Archer John Porter Martin CBE, 1 March 1910–28 July 2002." *Biographical Memoirs of Fellows of the Royal Society* 50 (2004): 157–70.

Lowe, F. "Question on Diabetics' Use of Insulin." *The Guardian*, 1989.

Lucas, C. C. "Chemical Examination of Royal Jelly." *The Canadian Medical Association Journal* 47, no. 5 (1942): 406–9.

MacFarlane, G. *Howard Florey: The Making of a Great Scientist*. Oxford, Oxford University Press, 1979.

Maclean, H. "The Use of Insulin in General Practice." *The Lancet* 202 (1923): 829–33.

Macleod, J. J. R. *Physiology and Biochemistry in Modern Medicine.* 3rd ed. London, H. Kimpton, 1921.

Macleod, J. J. R. "Insulin and Diabetes: A General Statement of the Physiological and Therapeutic Effects of Insulin." *British Medical Journal* 2, no. 3227 (1922): 833.

Macleod, J. J. R. "The Physiology of Insulin and Its Source in the Animal Body." Nobel Lecture, May 26, 1925. https://www.nobelprize.org/prizes/medicine/1923/macleod/lecture/.

"Magistrates Apology." *The Times,* August 11, 1923.

Major, R. H. "The Papyrus Ebers." *Annals of Medical History* 2 (1930): 547–55.

Maniatis, T., S. Gek Kee, A. Efstratiadis, and F. C. Kafatos. "Amplification and Characterization of a β-Globin Gene Synthesized in Vitro." *Cell* 8 (1976): 163–82.

Mann, F. C., and T. B. Magath. "The Liver as a Regulator of the Glucose Concentration of the Blood." *American Journal of Physiology* 55 (1921): 285–86.

Marks, V., and C. Richardson. *Insulin Murders: True Life Cases.* London, Royal Society of Medicine, 2007.

Markussen, J., K. Jørgensen, L. Thim, U. Damgaard, E. Sørensen, G. Dodson, and F. Chawdhury. "Human Monocompetent Insulin: Chemistry and Characteristics of Human Insulin (Novo)." *Diabetologia* 21 (1981): 302.

Martin, A. J. P. "The Development of Partition Chromatography." Nobel Lecture, December 12, 1952. https://www.nobelprize.org/uploads/2018/06/martin-lecture.pdf.

Martin, A. J. P. "A New Approach to the Problem of Structure in Proteins: An Investigation of a Partial Hydrolysate of Wool." In Fibrous Proteins: Proceedings of a Symposium Held at the University of Leeds on 23rd, 24th & 25th May, 1946, 1. Leeds, Society of Dyers and Colourists, 1946.

Martin, A. J. P., and R. L. M. Synge. "A New Form of Chromatogram Employing Two Liquid Phases. 1. A Theory of Chromatography 2. Application to the Micro-Determination of the Higher Monoamino-Acids in Proteins." *The Biochemical Journal* 35 (1941a): 1358–68.

Martin, A. J. P., and R. L. M. Synge. "Separation of the Higher Monoamino-Acids by Counter-Current Liquid-Liquid Extraction: The Amino-Acid Composition of Wool." *The Biochemical Journal* 35 (1941b): 91–121.

Maynard, L. A. "James Batcheller Sumner 1887–1955." In *National Academy of Sciences Biographical Memoirs,* 376–77. Washington, DC: The National Academies Press, 1958.

McCabe, C. "Playing God." The San Francisco Chronicle, April 4, 1977.

McElheny, V. "Gene Transplants Seen Helping Farmers and Doctors." *The* New York Times, May 20, 1974.

McElheny, V. "Gene Experiments Panel Urges Stiffer Guidelines." The New York Times, December 9, 1975.

McGuigan, H. "Sugar Metabolism and Diabetes." *The Journal of Laboratory and Clinical Medicine* 3 (1918): 319–37.

Meienhofer, J. E. Schnabel, H. Bremer, O. Brinkhoff, R. Zabel, W. Sroka, H. Klostermeyer, D. Brandenburg, T. Okuda, and H. Zahn. "Synthese Der Insulinketten Und Ihre Kombination Zu Insulinaktiven Präparaten." Zeitschrift *f* ür Naturforschung B Notizen 18 (1963): 1120.

Mellinghoff, K. H. "Georg Ludwig Zuelzers Beitrag Zur Insulinforschung." *Düsseldorfer Arbeiten zur Geschichte der Medizin* 36 (1971): 1–58.

Menting, J. G., et al "How Insulin Engages its Primary Binding Site on the Insulin Receptor." *Nature* 493 (2013): 241–45.

Menting, J. G., et al "Protective Hinge in Insulin Opens to Enable Its Receptor Engagement." *Proceedings of the National Academy of Sciences of the United States of America* 111 (2014): E3395–3404.

Merton, R. K. *On the Shoulders of Giants: A Shandean Postscript.* Chicago: University of Chicago Press, 1965.

Mertz, J. E., and R. W. Davis. "Cleavage of DNA by RI Restriction Endonuclease Generates Cohesive Ends." *Proceedings of the National Academy of Sciences of the United States of America* 69 (1972): 3370–74.

Meselson, M., and R. Yuan. "DNA Restriction Enzyme from E. Coli." *Nature* 217 (1968): 1110–14.

Metz, T. "New Genentech Issue Trades Wildly as Investors Seek Latest High-Flier." *The Wall Street Journal,* October 15, 1980.

Meyer, K., and H. Mark. "Über den Aufbau des Seiden-Fibroins." *Berichte der deutschen chemischen Gesellschaft* 61 (1928): 1932–36.

Miescher, F. "Ueber die chemische Zusammensetzung der Eiterzellen." *Medicinisch-chemische Untersuchungen* 4 (1871), 441–460.

Mihill, C. "Doctors Suggest Human Insulin Gives Less Warning of Blackouts." *The Guardian,* September 13, 1991.

Minkowski, O. "Über die bisherigen Erfahrungen mit der Insulinbehandlung des Diabetes." *Verhandlungen der Deutschen Gesellschaft für Innere Medizin* 36 (1924): 91–108.

Minkowski, O. "Die Lehre vom Pankreas-Diabetes in ihrer geschichtlichen Entwicklung." *München medizinisches Wochenschrift* 76 (1929): 311–15.

Moore, S., and W. Stein. "Partition Chromatography of Amino Acids on Starch." *Annals of the New York Academy of Science* 49 (1948): 265–78.

Morihara, K., T. Oka, and H. Tsuzuki. "Semi-Synthesis of Human Insulin by Trypsin-Catalysed Replacement of Ala-B30 by Thr in Porcine Insulin." *Nature* 280 (1979): 412–13.

Morris, P. J. T. "Martin, Archer John Porter." In Oxford Dictionary of National Biography [online] 2009. https://doi.org/10.1093/ref:odnb/77176.

Morrow, J. F., S. N. Cohen, A. C. Y. Chang, H. W. Boyer, H. M. Goodman, and R. B. Helling. "Replication and Transcription of Eukaryotic DNA in *Escherichia Coli.*" *Proceedings of the National Academy of Sciences of the United States of America* 71 (1974): 1743–47.

Murlin, J. R., H. D. Clough, C. B. F. Gibbs, & A. M. Stokes. "Aqueous Extracts of Pancreas: I. Influence on the Carbohydrate Metabolism of Depancreatized Animals." *Journal of Biological Chemistry* 56 (1923): 253–96.

Murlin, J. R., and B. Kramer. "The Influence of Pancreatic and Duodenal Extracts on the Glycosuria and the Respiratory Metabolism of Depancreatised Dogs." *Proceedings of the Society for Experimental Biology and Medicine* 10 (1913): 171–73.

Murlin, J. R., and B. Kramer. "Pancreatic Diabetes in the Dog: I. The Influence of Alkali and Acid Upon the Glycosuria and Hyperglycaemia." *Journal of Biological Chemistry* 27 (1916a): 481–98.

Murlin, J. R., and B. Kramer. "Pancreatic Diabetes in the Dog. III: The Influence of Alkali on Respiratory Metabolism After Total and Partial Pancreatectomy." *Journal of Biological Chemistry* 27 (1916b): 517–38.

Murlin, J. R., and B. Kramer. "A Quest for the Anti-Diabetic Hormone." *Journal of the History of Medicine* 11 (1956): 288–98.

Murlin, J. R., B. Kramer, and J. E. Sweet. "Pancreatic Diabetes in the Dog. VI. The Influence of Pancreatic Extracts without the Aid of Alkali upon the Metabolism of the Depancreatised Animal." *Journal of Metabolic Research* 1 (1922): 19–27.

Murlin, W. R. "History of Insulin." *Annals of Internal Medicine* 76 (1972): 330.

Murnaghan, J. H., and P. Talalay. "John Jacob Abel and the Crystallization of Insulin." *Perspectives in Biology and Medicine* 10 (1967): 334–80.

Murray, G. R. "Note on the Treatment of Myxoedema by Hypodermic Injections of an Extract of the Thyroid Gland of a Sheep." *British Medical Journal* 2, no. 1606 (1891): 796–97.

"Insulin: Credit for Its Isolation." *British Medical Journal* 3 (1969a): 651–52.

Murray, I. "The Search for Insulin." *The Scottish Medical Journal* 14 (1969b): 186–293.

Murray, I., and R. B. Wilson. "The New Insulins: Lente, Ultralente, And Semilente." *British Medical Journal* 2 (1953): 1023–26.

Nabarro, J. D. N., and J. M. Stowers. "The Insulin Zinc Suspensions." *British Medical Journal* 2 (1953): 1027–30.

Nakagawa, S. "Insulin Treatment in Japan." *Diabetes Research and Clinical Practice* 24 (1994): S247–250.

Neel, J. "The Inheritance of Sickle-Cell Anaemia." *Science* 110 (1949): 64–66.

Nelson, B. *The Woollen Industry of Leeds*. Leeds: D. & J. Thornton, 1980.

"New Diabetes Treatment. Canada's Gift. Insulin Patent for Britain." *The Times*, November 17, 1922b.

Nicol, D. S. H. W., and L. F. Smith. "Amino Acid Sequence of Human Insulin." *Nature* 187 (1960): 483–85.

Niemann, C., and M. Bergmann. "On the Structure of Proteins: Cattle Hemoglobin, Egg Albumin, Cattle Fibrin, and Gelatin." *Journal of Biological Chemistry* 118 (1937): 301–14.

Niemann, C., and M. Bergmann. "On the Structure of Silk Fibroin." *Journal of Biological Chemistry* 122 (1938): 577–96.

Niu, C-I., Y. T. Kung, W. T. Huang, L. T. Ke, C. C. Chen, Y. C. Chen, Y. C. Du, R. Q. Jiang, C. L. Tsou, S. C. Hu, S. Q. Chu, and K. Z. Wang. "Synthesis of Peptide Fragments of the B-Chain of Insulin: IX. Synthesis of the B-Chain of Insulin and Its Reconstitution with Natural A-Chain to Regenerate Insulin Activity." *Scientia Sinica* 13 (1964): 1343–45.

Niu, Ching-I., Y. T. Kung, W. T. Huang, L. T. Ke, and C. C. Chen. "Synthesis of Crystalline Insulin from Its Natural A-Chain and the Synthetic B-Chain." *Scientia Sinica* 15 (1966): 231–44.

Nixon, J. A. "Diabetes and Insulin." *British Medical Journal* 2, no. 3289 (1924): 53.

Norman, C. "Genetic Manipulation: Guidelines Issued." *Nature* 262 (1976): 2–4.

Oakley, W. "'Lente' Insulin (Insulin Zinc Suspension): Further Studies." *British Medical Journal* 2 (1953): 1021–23.

Oakley, W. "R. D. Lawrence, M.D., F.R.C.P., 1892–1968." *Diabetes* 18 (1969): 54–56.

'Obituary: Sir Frederick Banting.' *The Times*, February 26, 1941.

Ogston. A. G. "On the Theory of the Periodic Structure of Proteins." *Transactions of the Faraday Society* 39 (1943): 151–58.

O'Hara, M. "The NHS is a Precious Thing. Try Being Ill in the US If You Don't Believe This." *The Guardian,* November 5, 2019.

Olby, R. *The Path to the Double Helix: The Discovery of DNA.* New York: Dover Publications, 1994.

Olmsted, J. M. D. *Charles-Edouard Brown-Sequard: A Nineteenth Century Neurologist and Endocrinologist.* Baltimore: The Johns Hopkins Press, 1946.

Olmsted, J. M. D. "Claude Bernard, 1813–1878." *Diabetes* 2 (1953): 162–64.

Osler, W. *The Principles and Practice of Medicine,* 7th ed. New York, D Appleton-Century Company Ltd, 1909.

Owens, B. "Smart Insulin: Redesign Could End Hypoglycaemia Risk." *Nature Biotechnology* 36 (2018): 911–12.

Pallot, P. "Medical Alert Over Mystery Diabetic Deaths." The Daily Telegraph, August 17, 1989.

Papaspyros, N. S. *The History of Diabetes Mellitus.* 2nd ed. Stuttgart, Georg Thieme Verlag, 1964.

Parisi, A. J. "Industry of Life: The Birth of the Gene Machine." *The New York Times,* June 29, 1980.

Paulescu, N. C. "Action de l'extrait pancréatique injecté dans le sang chez un animal diabétique." *Comptes Rendus Des Seance de La Societe de Biologie* 85 (1921a): 555–57.

Paulescu, N. C. "Action de l'extrait pancréatique injecté dans le sang chez un animal normal." *Comptes Rendus Des Seance de La Societe de Biologie* 85 (1921b): 559.

Paulescu, N. C. "Influence de la quantité de pancréas employée pour préparer l'extrait injecté dan le sang chez un animal diabétique." *Comptes Rendus Des Seance de La Societe de Biologie* 85 (1921c): 558–59.

Paulescu, N. C. "Influence due laps de temps écoulé depuis l'injection intraveineuse de l'extrait pancréatique chez un animal diabétique." *Comptes Rendus Des Seance de La Societe de Biologie* 85 (1921d): 558.

Paulescu, N. C. "Recherche sur le role du pancréas dans l'assimilation nutritive." *Archives Internationales de Physiologie* 17 (1921e): 85–103.

Pauling, L., H. A. Itano, S. J. Singer, and I. C. Wells. "Sickle Cell Anaemia, a Molecular Disease." *Science* 110 (1949): 543–48.

Pauling, L., and Niemann, C. "The Structure of Proteins." *Journal of the American Chemical Society* 61 (1939): 1860–67.

Pavel, I. *The Priority of N.C. Paulescu in the Discovery of Insulin.* Bucharest: Editura Academici Republicii Socialiste Romania, 1976.

Pavia, W. "Insulin Price Drives Mother to Make Border "Drug Run"" *The Times,* June 19, 2019.

Perutz, M. "Forty Years' Friendship with Dorothy." *Current Science* 72 (1997): 450–53.

Pickup, J. "Human Insulin." *British Medical Journal* 292 (1986): 155–57.

Pickup J. C., and G. Williams. *Textbook of Diabetes,* 3rd ed. Oxford, Blackwell, 2002.

Plath, S. *The Bell Jar.* London, Faber and Faber, 1966.

Portugal, F. H. "Oswald T. Avery: Nobel Laureate or Noble Luminary?" *Perspectives in Biology and Medicine* 53 (2010): 558–70.

Portugal, F. H. *The Least Likely Man: Marshall Nirenberg and the Discovery of the Genetic Code.* Cambridge, MIT Press, 2015.

Pratt, J. H. "A Reappraisal of Researches Leading to the Discovery of Insulin." *Journal of the History of Medicine* 9 (1954): 281–89.

Prentice, T. "Side Effects of Human Insulin: Drug Firm Starts Inquiry." The Times, October 16, 1989.

Preston, R. D. "William Thomas Astbury, F.R.S., - Fibrous Polymer Extraordinary." In *Structure of Fibrous Biopolymers: Proceedings of the Twenty-Sixth Symposium of the Colston Research Society*, edited by E. D. T. Atkins and A. Keller, 1–20. London, Butterworths, 1974.

"Privileged Panel Patients." *The Times*, June 7, 1923e.

Ramamoorthy, B., R. G. Lees, D. G. Kleid, and H. G. Khorana. "Total Synthesis of the Structural Gene for the Precursor of a Tyrosine Suppressor Transfer RNA from Eschericia Coli. 12. Synthesis of a DNA Duplex Corresponding to a Sequence of 23 Nucleotide Units Adjoining the C-C-A End." *Journal of Biological Chemistry* 251 (1976): 676–94.

Ramaseshan, S. "Dorothy Hodgkin and the Indian Connection." *Current Science* 72 (1997): 457–63.

Rasmussen, N. *Gene Jockeys: Life Science and the Rise of Biotech Enterprise*. Baltimore, Johns Hopkins University Press, 2014.

Reichard, P. "Oswald Avery and the Nobel Prize in Medicine." *Journal of Biological Chemistry* 277 (2002): 13355–62.

Reimers, N. "Stanford's Office of Technology Licensing and the Cohen/Boyer Cloning Patents." An Oral History Conducted in 1997 by Sally Smith Hughes, Ph.D. University of California, Berkeley, Regional Oral History Office, The Bancroft Library, 1998.

Reinhold, R. "There's Gold in Them Thar Recombinant Genetic Bits." *The New York Times*, June 22, 1980.

Rennie, J., and T. Fraser. "The Islets of Langerhans in Relation to Diabetes." *The Biochemical Journal* 2 (1907): 7–19.

Rensberger, B. "Scientists Construct Functional Gene." *The New York Times*, August 28, 1976a.

Rensberger, B. "Synthesis of Working Gene Hailed as a Major Advance." *The New York Times*, August 29, 1976b.

"Retrospect 1889." *British Medical Journal* 2 (1889): 1433–78.

Reuter, C. "Le sécrétion interne du pancréas et le traitement du diabète sucré." *Section Des Sciences Naturelles, Physiques et Mathématiques de l'Institut Grand-Ducal* 8 (1924): 87–100.

Richards, B. "Reports of Deaths Among U.K. Diabetics Using Human Insulin Stir Concern Here." *The Wall Street Journal*, October 30, 1989.

Richards, D. W. "The Effect of Pancreas Extract on Depancreatized Dogs: Ernest L. Scott's Thesis of 1911." *Perspectives in Biology and Medicine* 10 (1966): 84–95.

Riggs, A. D. "Making, Cloning, and the Expression of Human Genes in Bacteria: The Path to Humulin." *Endocrine Reviews* 20 (2021): 1–7.

Riggs, A. D. "City of Hope's Contribution to Early Genentech Research." An Oral History Conducted in 2005 by Sally Smith Hughes. University of California, Berkeley, Regional Oral History Office, The Bancroft Library, 2006.

Riley, D. "Oxford: The Early Years." In *Structural Studies on Molecules of Biological Interest: A Volume in Honour of Professor Dorothy Hodgkin*, edited by G. Dodson, J. P. Glusker, and D. Sayre, 17–25. Oxford, Clarendon Press, 1981.

"Obituary Notices: Robert Daniel Lawrence." *British Medical Journal*, 3, no. 5618 (1968): 621.

Roberts, F. "Insulin." *British Medical Journal* 2, no. 1193 (1922): 1193–94.

Robinson, R. "Richard Willstätter, 1872–1942." *Obituary Notices of Fellows of the Royal Society* 8, no. 22(1953): 609–34.

Rollo, J. *An Account of Two Cases of the Diabetes Mellitus, With Remarks as They Arose During the Progress of the Cure.* London: C. Dilly, 1797. https://archive.org/details /b21469179_0002.

Root, H. F., P. White, and A. Marble. "Clinical Experience with Protamine Insulinate." *Journal of the American Medical Association* 106 (1936): 180–83.

Root, M. A., R. E. Chance, and J. A. Galloway. "Immunogenicity of Insulin." *Diabetes* 21, Suppl (1972): 657–60.

Rowe, J. L. "Designer Genes Are Snapped Up." The Washington Post, October 15, 1980.

"Rules Created to Control DNA Research." *Science News* 108 (1975): 372–73.

Ruttenberg, M. A. "Human Insulin: Facile Synthesis by Modification of Porcine Insulin." *Science* 177 (1972): 623–26.

Rutter, W. "The Department of Biochemistry and the Molecular Approach to Biomedicine at the University of California, San Francisco." An Oral History Conducted in 1992 by Sally Smith Hughes, Ph.D. University of California, Berkeley, Regional Oral History Office, The Bancroft Library, 1998.

Ryle, A. P., F. Sanger, L. F. Smith, and R. Kitai. "The Disulphide Bonds of Insulin." *The Biochemical Journal* 60 (1955): 541–56.

Sack, G. H., and D. Nathans. "Studies of SV40 DNA." *Virology* 51, no. 2 (1973): 517–20. https://doi.org/10.1016/0042-6822(73)90455-8.

Safavi-Hemani, H., J. Gajewiak, S. Karanth, S. D. Robinson, B. Ueberheide, A. Douglass, A. Schlegel, J. S. Imperial, M. Watkins, P. K. Bandyopadhyay, M. Yandell, Q. Li, A. W. Purcell, R. S. Norton, L. Ellgaard, and B. M. Olivera. "Specialized Insulin Is Used for Chemical Warfare by Fish-Hunting Cone Snails." *Proceedings of the National Academy of Sciences of the United States of America* 112 (2015): 1743–48.

Sanger, F. "The Arrangement of Amino Acids in Proteins." *Advances in Protein Chemistry* 7 (1952): 1–67.

Sanger, F. "The Free Amino Group of Gramicidin S." *The Biochemical Journal* 40 (1946): 261–62.

Sanger, F. "The Free Amino Groups of Insulin." *The Biochemical Journal* 39 (1945): 507–15.

Sanger, F. "Sequences, Sequences, Sequences." *Annual Review of Biochemistry* 57 (1988): 1–29

Sanger, F. "Some Chemical Investigations on the Structure of Insulin." *Cold Spring Harbor Symposia on Quantitative Biology* 14 (1950): 153–60.

Sanger, F. "The Terminal Peptides of Insulin." *The Biochemical Journal* 45 (1949): 563–74.

Sanger, F., and E. O. P. Thompson. "The Amino-Acid Sequence in the Glycyl Chain of Insulin. 1. The Identification of Lower Peptides from Partial Hydrolysates." *The Biochemical Journal* 53 (1953a): 353–66.

Sanger, F., and E. O. P. Thompson. "The Amino-Acid Sequence in the Glycyl Chain of Insulin. 2. The Investigation of Peptides from Enzymic Hydrolysates." *The Biochemical Journal* 53 (1953b): 366–74.

Sanger, F., and H. Tuppy. "The Amino-Acid Sequence in the Phenylalanyl Chain of Insulin.1. The Identification of Lower Peptides from Partial Hydrolysates." *The Biochemical Journal* 49 (1951a): 463–81.

Sanger, F., and H. Tuppy. "The Amino-Acid Sequence in the Phenylalanyl Chain of Insulin.2. The Investigation of Peptides from Enzymic Hydrolysates." *The Biochemical Journal* 49 (1951b): 481–90.

Scapin, G., V. P. Dandey, Z. Zhang, W. Prosise, A. Hruza, K. Kelly, T. Mayhood, C. Strickland, C. S. Potter, and B. Carragher. "Structure of the Insulin Receptor–Insulin Complex by Single-Particle Cryo-EM Analysis." *Nature* 566 (2018): 122–25.

Schiller, J. "Claude Bernard and Brown-Sequard: The Chair of General Physiology and the Experimental Method." *Journal of the History of Medicine* 21 (1966): 260–70.

Schlichtkrull, J. "Antigenicity of Monocompetent Insulins." *The Lancet* 2 (1974): 1260–61.

Schmeck, H. "'Gene Splicing' Faces New Debate in Congress." *The New York Times*, December 15, 1977a.

Schmeck, H. "Scientists Report Using Bacteria to Produce the Gene for Insulin." *The New York Times*, May 24, 1977b.

Schmeck, H. "Scientists Seek to Influence Legislation on Gene Research." *The New York Times*, July 6, 1977c.

Schmeck, H. "Substance Usually Made in Brain Grown in Bacteria." *The New York Times*, November 3, 1977d.

Schmeck, H. "U.S. to Process 100 Applications for Patents on Living Organisms." *The New York Times*, June 18, 1980.

Scott, D. A., and A. M. Fisher. "Crystalline Insulin." *The Biochemical Journal* 29 (1935): 1048–54.

Scott, D. A., and A. M. Fisher. "Studies on Insulin with Protamine." *Journal of Pharmacological and Experimental Therapeutics* 58 (1936): 78–92.

Scott, E. L. "On the Action of Intravenous Injections of an Extract of the Pancreas on Experimental Pancreatic Diabetes." *American Journal of Physiology* 29 (1912): 306–10.

Scott, E. L. "Priority in Discovery of a Substance Derived From the Pancreas, Active in Carbohydrate Metabolism." *Journal of the American Medical Association* 81 (1923): 1303–4.

Senechal, M. *I Died for Beauty: Dorothy Wrinch and the Cultures of Science.* Oxford, Oxford University Press, 2013.

Serafini, A. *Linus Pauling: A Man and His Science.* New York, Simon and Schuster, 1989.

Sgaramella, V. "Enzymatic Oligomerization of Bacteriophage P22 DNA and of Linear Simian Virus 40 DNA." *Proceedings of the National Academy of Sciences of the United States of America* 69 (1972): 3389–93.

Sharp, P. A., B. Sugden, and J. Sambrook. "Detection of Two Restriction Endonuclease Activities in Haemophilus Parainfluenzae Using Analytical Agarose-Ethidium Bromide Electrophoresis." *Biochemistry* 12 (1973): 3055–63.

Sherborne, M., and C. Priest. *H. G. Wells: Another Kind of Life.* London, Peter Owen, 2012.

Shonle, H. A., and J. H. Waldo. "Some Chemical Reactions of the Substance Containing Insulin." *Journal of Biological Chemistry* 58 (1924): 731–36.

Siekevitz, P. "Recombinant DNA Research: A Faustian Bargain?" *Science* 194 (1976): 256–57.

Simring, F. R. "On the Dangers of Genetic Meddling." *Science* 192 (1976): 940.

Singer, M, and D. Soll. "Guidelines for DNA Hybrid Molecules." *Science* 181 (1973): 1114.

Singer, M. F., and P. Berg. "Recombinant DNA: NIH Guidelines." *Science* 193 (1976): 186–88.

Slama, G. "Nicolae Paulesco: An International Polemic." *The Lancet* 362 (2003): 1422.

Smith, D. C. *H.G. Wells: Desperately Mortal - a Biography.* New Haven, Yale University Press, 1986.

Smith, H. O., and K. W. Wilcox. "A Restriction Enzyme from Hemophilus Influenzae:I. Purification and General Properties." *Journal of Molecular Biology* 51, no. 2 (1970): 379–91. https://doi.org/10.1016/0022-2836(70)90149-X.

Smith-Hughes, S. *Genentech: The Beginnings of Biotech.* Paperback. Chicago, University of Chicago Press, 2013.

Smith-Hughes, S. "Making Dollars Out of DNA: The First Major Patent in Biotechnology and the Commercialization of Molecular Biology, 1974–1980." *Isis* 92 (2001): 541–75.

"Soldier Patients Laud Work of Insulin Clinics." *Toronto Daily Star*, January 1923a.

"Some Clinical Results of the Use of Insulin. A Report to the Medical Research Council." *British Medical Journal* 1, no. 3252 (1923): 737–39.

Stahl, G. A. "An Interview with A. J. P. Martin." *Journal of Chemical Education* 54 (1977): 80–83.

Stalvey, R. M. "A Chat with Dr. Charles Best." *Nutrition Today* 6 (1971): 5–7.

Starling, E. H. "Croonian Lecture: On the Chemical Correlation of the Functions of the Body." *The Lancet* 2 (1905): 339–41.

Stein, W. H. "Introduction." *Annals of the New York Academy of Science* 47 (1946): 59–62.

Steiner, D. F., D. Cunningham, L. Spiegelman, and B. Aten. "Insulin Biosynthesis: Evidence for a Precursor." *Science* 157 (1967): 697–700.

Steiner, D. F., and P. E. Oyer. "The Biosynthesis of Insulin and a Probable Precursor of Insulin by a Human Islet Cell Adenoma." *Proceedings of the National Academy of Sciences of the United States of America* 57 (1967): 473–80.

Stevenson, L. G. "J. J. R Macleod and the Discovery of Insulin." *Trends in Biochemical Sciences* 4 (1979): 158–60.

"Still His Mother's Boy Despite Honors and Glory." *Evening Telegram*, January 18, 1923.

Straus, E. *Nobel Laureate Rosalyn Yalow: Her Life and Work in Medicine.* Cambridge, Perseus Books, 1998.

"Success of New Insulin Cure." *The Times*, April 28, 1923d.

Sumner, J. B. "Enzyme Urease." *Journal of Biological Chemistry* 69 (1926a): 435–41. http://www.jbc.org/content/69/2/435.full.pdf+html.

Sumner, J. B. "The Isolation and Crystallization of the Enzyme Urease. Preliminary Paper." *Journal of Biological Chemistry* 69 (1926b): 435–41. https://doi.org/10.4159/harvard.9780674366701.c115.

Sumner, J. B. "The Story of Urease." *Journal of Chemical Education* 14, no. 6 (1937): 255–59. https://doi.org/10.1021/ed014p255.

"The Supply of Insulin: New Diabetes Cure in Use at Hospitals." *The Times*, February 21, 1923c.

Suter, F. "Prof. F. Miescher. Persönlichkeit und Lehrer." *Helvetica Physiologica et Pharmacologia Acta Supplementa* 2 (1944): 5–17.

Swanson, R. A. "Co-Founder, CEO, and Chairman of Genentech, Inc., 1976–1996." An Oral History Conducted in 1996 and 1997 by Sally Smith Hughes. University of California, Berkeley, Regional Oral History Office, The Bancroft Library, 2001.

"SWC to French Health Minister and Romanian Ambassador: Cancel Paris Hospital Tribute to Antisemitic Hatemonger." Press release, Los Angeles, California,

Simon Wiesenthal Center, August 22, 2003. http://www.avancesendiabetologia.org/gestor/upload/revistaAvances/26-6-15.pdf.

Synge, R. L. M. "Applications of Partition Chromatography." Nobel Lecture, December 12, 1952. https://www.nobelprize.org/prizes/chemistry/1952/synge/lecture/.

Synge, R. L. M. *Some New Methods in Amino-Acid Analysis: The Amino-Acid Composition of Wool.* PhD thesis, Cambridge University, 1940.

Synge, R. L. M. "Analysis of a Partial Hydrolysate of Gramicidin by Partition Chromatography with Starch." *The Biochemical Journal* 38 (1944): 285–94.

Synge, R. L. M. "Science For The Good of Your Soul." In *The Science of Science*, edited by M. Goldsmith and A. Mackay, 214–25. London, Penguin, 1964.

Szekeres, A., B. Worcester, M. S. Ascher, D. Tuxen, R. Heyendal, K. M. Walsh, and M. A. Charles. "Comparison of the Biologic Activity of Porcine and Semisynthetic Human Insulins Using the Glucose-Controlled Insulin Infusion System in Insulin-Dependent Diabetes." *Diabetes Care* 6 (1983): 191–93.

Tattersall, R. B. "Diabetes Deaths and Human Insulin." *The Guardian,* August 10, 1989.

Tattersall, R. B. "Charles-Edouard Brown-Sequard: Double Hyphenated Neurologist and Forgotten Father of Endocrinology." *Diabetic Medicine* 11 (1994): 728–31.

Tattersall, R. B. *Diabetes: The Biography.* Oxford, Oxford University Press, 2009.

Tattersall, R. B. "A Force of Magical Activity: The Introduction of Insulin Treatment in Britain 1922–1926." *Diabetic Medicine* 12 (1995): 739–55.

Tattersall, R. B. *The Pissing Evil: A Comprehensive History of Diabetes Mellitus.* East Ayrshire, Swan & Horn, 2019.

Tattersall, R. B., and G. V. Gill. "Unexplained Deaths of Type 1 Diabetic Patients." *Diabetic Medicine* 8 (1991): 49–58.

"Telegrams in Brief." *The Times,* June 5, 1924.

Teuscher, A., and W. G. Berger. "Awareness of Hypoglycaemia in Diabetes." *The Lancet* 330 (1987a): 919.

Teuscher, A., and W. G. Berger. "Hypoglycaemia Unawareness in Diabetics Transferred from Beef/Porcine Insulin to Human Insulin." *The Lancet* 329 (1987b): 382–85.

Thornton, D. *Leeds: The Story of a City.* Ayr, Fort Publishing, 2002.

Todd, A. P. "Structure of Vitamin B-12." *The New York Times,* August 13, 1955.

Torlone, E., C. Fanelli, A. M. Rambotti, G. Kassi, F. Modarelli, A. Di Vincenzo, L. Epifano, M. Ciofetta, S. Pampanelli, P. Brunetti, and G. B. Bolli. "Pharmacokinetics, Pharmacodynamics and Glucose Counterregulation Following Subcutaneous Injection of the Monomeric Insulin Analogue [Lys(B28),Pro(B29)] in IDDM." *Diabetologia* 37 (1994): 713–20.

"The Treatment of Diabetes by Insulin." *The Lancet* 200 (1922): 1081–82.

"The Treatment of Diabetes by Insulin." *The Lancet* 201 (1923): 391–92.

Tsou, C-L. "Chemical Synthesis of Crystalline Bovine Insulin: A Reminiscence." *Trends in Biochemical Sciences* 20 (1995): 289–92.

Tsvett, M. S. "Adsorption Analysis and the Chromatographic Technique. Application to the Chemistry of Chlorophyll." *Berichte der Deutschen Botanischen Gesellschaft* 24 (1906a): 316–26.

Tsvett, M. S. "Physico-Chemical Studies of Chlorophyll. Adsorption." *Berichte der Deutschen Botanischen Gesellschaft* 24 (1906b): 316–26.

Ullrich, A. "Molecular Biologist at UCSF and Genentech." An Oral History Conducted in 1994 and 2003 by Sally Hughes. University of California, Berkeley, Regional Oral History Office, The Bancroft Library, 2006.

Uzbekova, D. G. "Nicolai Kravkov's Pancreotoxine." *Journal of Medical Biography* 26 (2018): 189–193.

Vainshtein, B. K. "Meetings with Dorothy." *Current Science* 72 (1997): 455–56.

Valenstein, E. S. *Great and Desperate Cures: The Rise and Decline of Psychosurgery and Other Radical Treatments for Mental Illness.* New York, Basic Books, 1986.

Veigl, S., O. Harmann, and E. Lamm. "Friedrich Miescher's Discovery in the Historiography of Genetics: From Contamination to Confusion, from Nuclein to DNA." *Journal of the History of Science* 53 (2020): 451–84.

Villa-Komaroff, L., A. Efstratiadis, S. Broome, P. Lomedico, R. Tizard, S. P. Naber, W. L. Chick, and W. Gilbert. "A Bacterial Clone Synthesizing Proinsulin." *Proceedings of the National Academy of Sciences* 1978, no. 75 (1978): 3727–31.

Vischer, E., and E. Chargaff. "The Separation and Characterization of Purines in Minute Amounts of Nucleic Acid Hydrolysates." *Journal of Biological Chemistry* 168 (1947): 781–82.

Vischer, E., S. Zamenhof, and E. Chargaff. "Microbial Nucleic Acids: The Desoxypentose Nucleic Acids of Avian Tubercle Bacilli and Yeast." *Journal of Biological Chemistry* 177 (1949): 429–38.

Voet, D., and J. Voet. *Biochemistry.* 4th ed. New York, John Wiley & Sons, Inc., 2012.

Volovici, L. *Nationalist Ideology and Antisemitism: The Case of Romanian Intellectuals in the 1930s,* translated by C. Kornos. Oxford, Pergamon Press, 1991.

Wade, N. "Gene-Splicing: At Grass-Roots Level a Hundred Flowers Bloom." *Science* 195, no. 4278 (1977a): 558–60. https://doi.org/10.1126/science.11643359.

Wade, N. "Gene Splicing: Senate Bill Draws Charges of Lysenkoism." *Science* 197 (1977b): 348–49.

Wade, N. "Genetic Manipulation: Temporary Embargo Proposed on Research." *Science* 185 (1974): 332–34.

Wade, N. "Genetics: Conference Sets Strict Controls to Replace Moratorium." *Science* 187 (1975a):931–34.

Wade, N. "Microbiology: Hazardous Profession Faces New Uncertainties." *Science* 182 (1973): 566–67.

Wade, N. "Recombinant DNA: A Critic Questions the Right to Free Inquiry." *Science* 194, no. 4262 (1976a): 303–6. https://doi.org/10.1126/science.11643339.

Wade, N. "Recombinant DNA at the White House." *Science* 193 (1976b): 468.

Wade, N. "Recombinant DNA: NIH Group Stirs Storm by Drafting Laxer Rules." *Science* 190 (1975b): 767–69.

Wade, N. "Recombinant DNA: NIH Rules Broken in Insulin Gene Project." *Science* 197 (1977c): 1342–45.

Wade, N. "Recombinant DNA: The Last Look Before the Leap." *Science* 192 (1976c): 236–37.

Wald, G. "The Case Against Genetic Engineering." *The Sciences* 16 (1976): 6–8.

Waller, J. *Fabulous Science.* Oxford, Oxford University Press, 2002.

Wang, J-H. "The Insulin Connection: Dorothy Hodgkin and the Beijing Insulin Group." *Trends in Biochemical Sciences* 23 (1998): 497–500.

Warwick, O. H. "James Bertram Collip—1892–1965—An Appreciation." *The Canadian Medical Association Journal* 93 (1965): 425–26.

Watson, J. D. "An Imaginary Monster." *The Bulletin of Atomic Scientists* 33 (1977): 12–13.

Watson, J. D. *The Double-Helix: A Personal Account of the Discovery of the Structure of DNA.* 8th ed. New York, Penguin, 1986.

Watson, J. D. *Genes, Girls and Gamow.* Oxford, Oxford University Press, 2001.

Watson, J. D., and A. Berry. *DNA: The Secret of Life.* London, Arrow Books, 2003.

Watson, J. D., and F. H. C. Crick. "Molecular Structure of Nucleic Acids: A Structure for Deoxyribose Nucleic Acid." *Nature* 171 (1953): 737–38.

Watson, J. D., N. H. Hopkins, J. W. Roberts, J. A. Steitz, and A. M. Weiner. *Molecular Biology of the Gene.* 4th ed. Menlo Park, Benjamin/Cummings, 1987.

Watson, J. D., and J. Tooze. *The DNA Story: A Documentary History of Gene Cloning.* San Francisco: W. H. Freeman and Company, 1981.

Wauchope, G. M., W. Oakley, and A. Grunberg. "Insulin Zinc Suspensions." *British Medical Journal* 2 (1953): 1325–26.

Weill, N. "Paris manque d'honorer l'inventeur antisemite de l'insuline." Le Monde, August 26, 2003.

Weinberg, J. H. "Decision at Asilomar." *Science News* 107 (1975): 194–96.

Wells, H. G. "The Select Company of Diabetics: For the Benefit of Their Cult." *The Times,* April 19, 1933.

Wells, H. G. "Diabetics in Sympathy: An Association for Rich and Poor." *The Times,* February 15, 1934.

Wells, H. G. "Diabetic Children: Provision of Holidays." *The Times,* July 16, 1935.

Whyte, C. *More Than a Legend.* 3rd ed. London, Hamish Hamilton, 1961.

Wilder, R. "Recollections and Reflections on Education, Diabetes, Other Metabolic Diseases, and Nutrition in the Mayo Clinic and Associated Hospitals, 1919–50." *Perspectives in Biology and Medicine* 1 (1958): 237–77.

Williams, G. *A Monstrous Commotion.* London, Orion Books, 2015.

Williams, G. *Unravelling the Double Helix: The Lost Heroes of DNA.* London, Weidenfeld and Nicholson, 2019.

Williams, J. R. "An Evaluation of the Allen Method of Treatment of Diabetes Mellitus." *American Journal of the Medical Sciences* 162 (1921): 62–72.

Williams, T. I. *Howard Florey: Penicillin and After.* Oxford, Oxford University Press, 1984.

Wintersteiner, O., V. Du Vigneaud, and H. Jensen. "Studies on Crystalline Insulin V. The Distribution of Nitrogen in Crystalline Insulin." *The Journal of Pharmacology and Experimental Therapeutics* 32 (1928): 397–411.

Wise, J. "Covid-19: The E484K Mutation and the Risks It Poses." *British Medical Journal* 372 (2021): 359.

Wolff, S. P. "Human Insulin: The Gap Between Medical Opinion and Fact." *The Guardian,* September 2, 1991.

"Would Share Honors with His Colleagues." *The Mail and Empire,* August 27, 1923. https://insulin.library.utoronto.ca/islandora/object/insulin%3AC10099.

Wright Jr, J. R. "From Ugly Fish to Conquer Death: J. J. R. Macleod's Fish Insulin Research, 1922–1924." *The Lancet* 359 (2002): 1238–42.

Wrinch, D. "The Pattern of Proteins." *Nature* 137 (1936): 411–12.

Wrinch, D. M. "Is There a Protein Fabric?" *Cold Spring Harbor Symposia on Quantitative Biology* 6 (1938): 122–39.

Wrinch, D. M. "On the Pattern of Proteins." *Proceedings of the Royal Society of London. Series A, Mathematical and Physical Sciences* 160 (1937a): 59–86.

Wrinch, D. M. "On the Structure of Insulin." *Science* 85 (1937b): 566–67.

Xiong, X., et al "A Structurally Minimized Yet Fully Active Insulin Based on Cone-Snail Venom Insulin Principles." *Nature Structural and Molecular Biology* 27 (2020): 615–24.

Yalow, R. S. "Radioimmunoassay: A Probe for Fine Structure of Biological Systems." Nobel Lecture, December 8, 1977. https://www.nobelprize.org/uploads/2018/06/yalow-lecture.pdf.

Yalow, R. S. "The Role of Technology in Creative Biologic Research." *Perspectives in Biology and Medicine* 25 (1982): 573–82.

Yalow, R. S., and S. A. Berson. "Assay of Plasma Insulin in Human Subjects by Immunological Methods." *Nature* 184 (1959): 1648–49.

Yalow, R. S., and S. A. Berson. "Immunoassay of Endogenous Plasma Insulin in Man." *Journal of Clinical Investigation* 39 (1960): 1157–75.

Yudkin, J. S. "Insulin for the World's Poorest Countries." *The Lancet* 355 (2000): 919–21.

Yudkin, J. S., and D. Beran. "Prognosis of Diabetes in the Developing World." *The Lancet* 362 (2003): 1420–21.

Zahn, H. "My Journey from Wool Research to Insulin." *The Journal of Peptide Science* 6 (2000): 1–10.

Zahn, H., B. Gutte, E. F. Pfeiffer, and J. Ammon. "Resynthese von Insulin aus Präoxydierter A-Kette und Reduzierter B-Kette." *Liebigs Annalen der Chemie* 691 (1966): 225–31.

Zamenhoff, S., G. Brawerman, and E. Chargaff. "On the Desoxypentose Nucleic Acids from Several Microorganisms." *Biochemica et Biophysica Acta* 9 (1952): 402–5.

Zatman, L. J. "B. C. J. G. Knight, Obituary 1904–1981." *The Journal of General Microbiology* 129 (1983): 1261–68.

Zaykov, A. N., J. P. Mayer, and R. D. DiMarchi. "Pursuit of a Perfect Insulin." *Nature Reviews: Drug Discovery* 15 (2016): 425–39.

Ziff, E. "Benefits and Hazards of Manipulating DNA." *New Scientist* 260 (1973): 274–75.

Zuelzer, G. "Das Scharlachprobleme." Typewritten manuscript by Zuelzer held in the papers of the Geheimes Staatsarchiv, Preussischer Kulturbesitz (1931) UI 8774/31.

Zuelzer, G. "Diskussionsbeitrag zur Diabetesbehandlung mit Insulin." *Verhandlungen Der Deutschen Gesellschaft Für Innere Medizin* 36 (1924): 135–36.

Zuelzer, G. "Experimentelle Untersuchungen über den Diabetes." *Berliner Klinisches Wochenschrift* 44 (1907): 474–75.

Zuelzer, G. "Über Acomatol, das deutsche Insulin." *Medizinische Klinik* 19 (1923): 1551–52.

Zuelzer, G. "Über Versuche einer specifischen Fermentherapie des Diabetes." *Zeitschrift Fur Experimentelle Pathologie Und Therapie* 5 (1908): 307–18.

Zuelzer, G., M. Dohrn, and A. Marxer. "Neuere untersuchungen über den experimentellen Diabetes." *Deutsche Medizinische Wochenschrift* 32 (1908): 1380–85.

Zuelzer, G. L. "Die Geschichte meiner Entdeckung des Acomatols, des deutschen Insulins." *Wir Zuckerkranken – publication of the German Diabetic Association (Deutschen Diabetikerbund)*, 136–38. Held in the papers of the Geheimes Staatsarchiv, Preussischer Kulturbesitz (n.d.) UI 8774/31.

Notes

Abbreviations used in Endnotes

IC - Insulin Collection.

TFRBL UT. - Insulin Collection Thomas Fisher Rare Book Library, University of Toronto.

IC UT BG IC ARMS UT - Insulin Collection, University of Toronto, Board of Governors, Insulin Committee, Archives and Records Management Services, University of Toronto.

IC SALP BNL SA - Insulin Collection, Sanofi Aventis Limited (formerly Connaught) Papers, Balmer Neilly Library, Sanofi Aventis Ltd.

IC UT OP ARMS UT - Insulin Collection, University of Toronto, Office of the President, Archives and Records Management Services, University of Toronto.

PCRLMS TCL UC - The papers and correspondence of Richard Laurence Millington Synge, Trinity College Library, University of Cambridge. GBR/0016/SYNG.

DHPBL UO - Dorothy Hodgkin Papers, Bodleian Library, University of Oxford

ULSC BL - University of Leeds Special Collections, Brotherton Library

Preface

1 P. Hennessy, 'Theresa May: I Have Diabetes; Theresa May, the Home Secretary, is Suffering from Type 1 Diabetes, She Revealed on Saturday Night', *The Daily Telegraph*, July 27, 2013.

2 D. McRae, 'Henry Slade: "There Was Never Any Thought I Wouldn't Keep Playing with Diabetes"', *The Guardian*, July 14, 2020.

3 L. Clark, 'Diabetes News: James Norton Has Revealed an Unusual Way of Coping with Type 1', *The Daily Express*, December 20, 2017.

4 M. W. Nirenberg, 'Man's Power to Shape His Own Biologic Destiny. Will Society Be Prepared to Use it Wisely?', *Research Corporation Quarterly Bulletin* (1967): p1, 4. http:// profiles.nlm.nih.gov/ps/access/JJBCBG.pdf. Cited in Portugal 2015: 139.

5 G. Kolata, S-L. Wee, and P. Belluck, 'Chinese Scientist Claims to Use CRISPR to Make First Genetically Edited Babies', *The New York Times*, November 26, 2018.

6 A. N. Whitehead, 'The Organisation of Thought', address given to *Section A. Mathematical and Physical Sciences*, Meeting of the British Association for the Advancement of Science, September 6, 1916.

7 S. Brenner, 'The Rough and the Smooth', *Nature* 317 (1985): 209–10; 209.

8 Some of the sources of the 'shoulders of giants' quotation have now quite reliably traced at least as far back to Bernard of Chartres (12th century) who apparently

likened us to dwarfs who can only see further by being carried by giants (See Merton, R.K. "On the Shoulders of Giants: a Shandean Postscript" (1965) New York, The Free Press, Collier-Macmillan Limited, London). This echoes the legend of Orion who, in Greek mythology, having been blinded as a punishment, was guided by the dwarf Cedalion, who sat on his shoulders.

9 N. Lezard, 'The Crooked Timber of Humanity: Chapters in the History of Ideas—A Review', *The Guardian* July 23, 2013.

Introduction

1 Tattersall 2009, 1.
2 Synge 1964, 221.

Chapter 1

1 In acknowledgement of Dr. Robert Tattersall, who also used this memorable title for his own book on the history of diabetes; Robert Tattersall, *The Pissing Evil: A Comprehensive History of Diabetes Mellitus*, Swan, 2019.
2 Major 1930, 548.
3 Major 1930, 552.
4 Gemmill 1972.
5 Leopold 1930.
6 Leopold 1930, 433.
7 Allan 1953, 74.
8 Hughes 1991, 25.
9 Pickup and Williams 2002, 1.2–1.3.
10 Tattersall 2009, 10.
11 McGuigan 1918, 319.
12 Papaspyros 1964, 4.
13 Dobson 1776, 299–300.
14 Dobson 1776, 300–301.
15 Dobson 1776, 305.
16 Home, cited in Tattersall 2019, 6.
17 Pickup and Williams 2002, 1.3.
18 Rollo 1797, 14; Vol. 1. London, MDCCXCVII. Eighteenth Century Collections Online. Gale. University of Leeds. March 23, 2021. Online at http://find.gale.com/ecco/infomark.do?&source=gale&prodId=ECCO&userGroupName=leedsuni&tabID=T001&docId=CW3307091832&type=multipage&contentSet=ECCO Articles&version=1.0&docLevel=FASCIMILE.
19 Rollo 1797, 87; Vol. 1. London, MDCCXCVII. Eighteenth Century Collections Online. Gale. University of Leeds. March 23, 2021. Online at http://find.gale.com/ecco/infomark.do?&source=gale&prodId=ECCO&userGroupName=leedsuni&

tabID=T001&docId=CW3307091855&type=multipage&contentSet=ECCO
Articles&version=1.0&docLevel=FASCIMILE.

20 Diabetes UK 2014, 6.
21 United Nations General Assembly, Resolution Adopted by the General Assembly—
61/225. World Diabetes Day, New York: UN General Assembly, 2006. Cited in
Tattersall 2009, 1.
22 Tattersall 2019, 22.
23 Tattersall 2019, 23.
24 Olmsted 1953, 162.
25 Macleod 1921, 706.
26 Houssay 1952, 113.
27 'Retrospect' 1889, 1440.
28 Johann Conrad Brunner (1653–1727) was one such physician. Brunner had
observed that dogs whose pancreas had been removed 'drank immoderately from the
brook flowing through the town' and 'ran into the courtyard to urinate and watered a
considerable area of ground'. Another was Thomas Cawley, ex-chief surgeon of the
British army hospital in Jamaica who, in 1778, described abnormalities in the pancreas of a patient who had died from diabetes. Later, in 1833, Dr. Richard Bright
wrote about a case in which he suspected diabetes had been caused by a disease of
the pancreas. See Tattersall 2019, 27; 33.
29 Tattersall 2019, 29.
30 Hedon 1892, cited in Tattersall 2019, 29; Bliss 1982b.
31 Schiller 1966, 264.
32 Starling 1905, 339.
33 Brown-Séquard 1893, 1145.
34 Borell 1976, 311.
35 Tattersall 1994, 729.
36 Brown-Séquard 1887, 105; Brown-Séquard 1888, 106; both cited in Olmsted 1946,
206.
37 Olmsted 1946, 207.
38 'The Pentacle of Rejuvenescence', *British Medical Journal*, 1889.
39 Olmsted 1946, 208.
40 Tattersall 1994, 729.
41 'Pentacle', 1416.
42 'Dr. Brown-Sequard's Experiments', *British Medical Journal*, 1889, 347.
43 'Dr. Brown-Séquard's Hypodermic Fluid', *British Medical Journal*, 1889, 29.
44 Murray 1891.
45 'Animal Extracts as Therapeutic Agents', *British Medical Journal*, 1893.
46 Delhoume 1939, 384; cited in Olmsted 1946, 216.
47 Brown-Séquard 1893, 1146–1147, cited in Olmsted 1946.
48 Borell 1976, 318.
49 Olmsted 1946, 217.
50 Delhoume 1939, 393.
51 Brown-Séquard 1893, 1146.

52 Tattersall 2009, 36.

53 Langerhans 1869, cited in Bliss 1982b, 25.

54 The name 'Islets of Langerhans' was coined by physiologist Gustave Laguesse.

55 Van Beek 1958.

56 Tattersall 2009, 41–42.

57 Bliss 1982b, 27.

58 Meyer 1904, cited in Pickup and Williams 2002, 1.1.

59 Some diabetes specialists have argued that there are forms of diabetes which share characteristics of both Type 1 and Type 2 and should therefore be considered as 'Type 1.5' or 'Type 3'.

60 In 1880, French physician Etienne Lancereaux (1829–1910) used the obesity of patients to distinguish two different forms of diabetes which he described as 'diabète maigre' (skinny diabetes) and 'diabète gras' (fat diabetes). Lancereaux 1880, cited in Pickup and Williams 2002, 1.1.2. It may well be this distinction by Lancereaux that evolved into the relatively recent modern classification of Type 1 and Type 2, although at the time Lancereaux was accused by the English clinician George Harley (1829–1896) of having plagiarized this classification from his own work in 1866. Tattersall, 2019, 16.

61 Bliss 1982b, 23.

62 Cited in Tattersall 2009, 20.

63 Allen, Stillman, and Fitz 1919, 29.

64 Tattersall 2009, 19.

65 Tattersall 2009, 22.

66 Cited in Allen, Stillman, and Fitz 1919, 25.

67 Bliss 1982b, 23.

68 Tattersall 2009, 20.

69 Bliss 1982b, 23.

70 Allen, Stillman, and Fitz 1919, 27.

71 Duncan 1964, 319.

72 Bliss 1982b, 39.

73 Bliss 1982b, 37.

74 Tattersall 2009, 50.

75 Williams 1921, 65.

76 Williams 1921, 70–71.

77 Hodgson 1911, 1187.

78 Hodgson 1911, 1190; 1187.

79 Tattersall 2009, 49.

80 Bliss 1982b, 39.

81 Tattersall 2009, 50.

82 Bliss 1982c, 39.

83 Tattersall 2009, 51.

84 Allen 1913, 813, 815, 816; cited in Bliss 1982b, 33.

Chapter 2

1 Banting 1940, 14e; 'The story of the discovery of insulin', MS COLL 76 (Banting) Box 1, Folders 9–13; IC, F. G. Banting (Frederick Grant, Sir) Papers, TFRBL UT. Online at https://insulin.library.utoronto.ca/islandora/object/insulin%3AW10027_ 0258.

2 Bliss 1982c, 230.

3 Banting 1940, 15e. Online at https://insulin.library.utoronto.ca/islandora/object /insulin%3AW10027_0259.

4 Banting 1940, 23. Online at https://insulin.library.utoronto.ca/islandora/object /insulin%3AW10027_0045; Bliss 1984, 34.

5 F. Banting to Isabel Knight, May 6, 1917, cited in Bliss 1984, 37.

6 Banting 1940, 25. Online at https://insulin.library.utoronto.ca/islandora/object /insulin%3AW10027_0049.

7 Banting 1940, 27. Online at https://insulin.library.utoronto.ca/islandora/object /insulin%3AW10027_0053.

8 Banting 1940, 29; Letter to Mother, October 15, 1918; MS COLL 76 (Banting) Scrapbook 1, Box 3, Page 179; IC, F. G. Banting (Frederick Grant, Sir) Papers, TFRBL UT. Online at https://insulin.library.utoronto.ca/islandora/object/insulin %3AL10062.

9 Lt. Col. W. H. K. Anderson to Captain Banting, August 26, 1918; MS COLL 76 (Banting) Scrapbook 1, Box 3, Page 204; IC, F. G. Banting (Frederick Grant, Sir) Papers, TFRBL UT. Online at https://insulin.library.utoronto.ca/islandora/ object/insulin%3AL10063.

10 Banting 1940, 30. Online at https://insulin.library.utoronto.ca/islandora/object /insulin%3AW10027_0059.

11 Banting 1940, 46–47. Online at https://insulin.library.utoronto.ca/islandora/object /insulin%3AW10027_0291.

12 'Still His Mother's Boy Despite Honors and Glory', Evening Telegram 1923; MS COLL 76 (Banting) Scrapbook 1, Box 1, Page 32; IC, F. G. Banting (Frederick Grant, Sir) Papers, TFRBL UT. Online at https://insulin.library.utoronto.ca/ islandora/object/insulin%3AC10059.

13 'Letter to dearest mother', September 29, 1918; MS COLL 76 (Banting) Scrapbook 1, Box 3, Page 179; IC, F. G. Banting (Frederick Grant, Sir) Papers, TFRBL UT. Online at https://insulin.library.utoronto.ca/islandora/object/insulin%3AL10059.

14 'Letter to dearest mother'.

15 F. G. Banting's certificate of service in the Canadian Expeditionary Force, June 17, 1920; MS COLL 76 (Banting) Scrapbook 1, Box 3, Page 180; IC, F. G. Banting (Frederick Grant, Sir) Papers, TFRBL UT. Online at https://insulin. library.utoronto.ca/islandora/object/insulin%3AE10011.

16 Banting 1940, 38. Online at https://insulin.library.utoronto.ca/islandora/object /insulin%3AW10027_0075.

17 Bliss 1984, 49; 'First page of laboratory notebook entitled Daily Accounts 1920–1921', July 29, 1920; MS COLL 76 (Banting), Box 26, Folder 3; IC,

F. G. Banting (Frederick Grant, Sir) Papers, TFRBL UT. Online at https://insulin.library.utoronto.ca/islandora/object/insulin%3AN10021

18 Banting 1940, 35. Online at https://insulin.library.utoronto.ca/islandora/object/insulin%3AW10027_0069.

19 Banting 1940, 38. Online at https://insulin.library.utoronto.ca/islandora/object/insulin%3AW10027_0075.

20 Banting 1940, 39. Online at https://insulin.library.utoronto.ca/islandora/object/insulin%3AW10027_0077.

21 Banting 1940, 40–41. Online at https://insulin.library.utoronto.ca/islandora/object/insulin%3AW10027_0079.

22 Banting 1940, 42. Online at https://insulin.library.utoronto.ca/islandora/object/insulin%3AW10027_0083.

23 Banting 1940, 43. Online at https://insulin.library.utoronto.ca/islandora/object/insulin%3AW10027_0083.

24 Banting 1940, 43. Online at https://insulin.library.utoronto.ca/islandora/object/insulin%3AW10027_0085.

25 Banting 1940, 44. Online at https://insulin.library.utoronto.ca/islandora/object/insulin%3AW10027_0087.

26 Banting 1940, 52. Online at https://insulin.library.utoronto.ca/islandora/object/insulin%3AW10027_0103.

27 Banting 1940.

28 Banting 1940, 44. Online at https://insulin.library.utoronto.ca/islandora/object/insulin%3AW10027_0087.

29 Barron 1920.

30 Banting 1940, 46–47. Online at https://insulin.library.utoronto.ca/islandora/object/insulin%3AW10027_0091.

31 Banting 1940, 47.

32 Stevenson 1979, 158.

33 Bliss 1982a, 558; Banting 1940, 61. Online at https://insulin.library.utoronto.ca/islandora/object/insulin%3AW10027_0121

34 Richards 1966.

35 Scott 1912.

36 Cited in Tattersall 2009, 44.

37 Banting 1940, 62. Online at https://insulin.library.utoronto.ca/islandora/object/insulin%3AW10027_0123.

38 Stevenson 1979, 299.

39 Stevenson 1978, 301.

40 F. G. Banting: Account of the Discovery of Insulin, September 1922; MS COLL 76 (Banting), Box 37, Folder 2; IC, F. G. Banting (Frederick Grant, Sir) Papers, TFRBL UT. Online at https://insulin.library.utoronto.ca/islandora/object/insulin%3AW10015

41 'Banting: Account of the Discovery of Insulin'.

42 J. J. R. Macleod to F. G. Banting, March 11, 1921; Academy of Medicine, 123 (Banting), Folder 8; IC, Academy of Medicine Collection, TFRBL UT. Online at https://insulin.library.utoronto.ca/islandora/object/insulin%3AL10054.

43 Bliss 1984, 60.

44 Stevenson 1979, 302.

45 Jurdjevic and Tillman 2004, 865.

46 Dale 1941, 383.

47 Banting 1940, 66–67. Online at https://insulin.library.utoronto.ca/islandora/object /insulin%3AW10027_0131.

48 F. G. Banting: account of the discovery of insulin, September 1922; MS COLL 76 (Banting), Box 37, Folder 2; IC, F. G. Banting (Frederick Grant, Sir) Papers, TFRBL UT. Online at https://insulin.library.utoronto.ca/islandora/object /insulin%3AW10015.

49 Dale 1954a, 32.

50 Best's departure for the training camp must have been some time between 17th June—28th July 1921, as a lab notebook containing entries by Noble, Best, and Banting shows that during this period, measurements of blood sugar levels are written in Noble's hand. Up until this point, Noble had been making measure-ments of blood sugar levels in turtles, but during this period he was clearly assisting Banting in testing the antidiabetic effects of pancreatic extracts. Jurdje-vic and Tillman 2004. Online at https://insulin.library.utoronto.ca/islandora/object/ insulin%3AN10014_0060.

51 Banting 1940, 70. Online at https://insulin.library.utoronto.ca/islandora/object /insulin%3AW10027_0139.

52 Banting 1940, 71–73. Online at https://insulin.library.utoronto.ca/islandora/object /insulin%3AW10027_0141.

53 Stevenson 1979, 302.

54 C. H. Best to J. J. R. Macleod August 9, 1921; MS COLL 76 (Banting), Box 62, Folder 3; IC, F. G. Banting (Frederick Grant, Sir) Papers, TFRBL UT. Online at https://insulin.library.utoronto.ca/islandora/object/insulin%3AL10027.

55 Bliss 1982b, 67.

56 Best 1942b, 400.

57 F. G. Banting to J. J. R. Macleod, August 9, 1921. MS COLL 76 (Banting), Box 62, Folder 2A; IC, F. G. Banting (Frederick Grant, Sir) Papers, TFRBL UT. Online at https://insulin.library.utoronto.ca/islandora/object/insulin:L10026.

58 Bliss 1982b, 67.

59 Entry for 27th July; Laboratory notebook 1 21/01/–10/08/1921; p.76; MS COLL 76 (Banting), Box 6A, Folder 3; IC, F. G. Banting (Frederick Grant, Sir) Papers, TFRBL UT. Online at https://insulin.library.utoronto.ca/islandora/object/ insulin%3AN10014_0083.

60 Laboratory notebook 1 21/01/–10/08/1921, 30th July, 82. Online at https://insulin .library.utoronto.ca/islandora/object/insulin%3AN10014_0089.

61 Bliss 1982b, 72–73; Banting 1940, 33b. Online at https://insulin.library.utoronto .ca/islandora/object/insulin%3AW10027_0192 .

62 Banting 1940, 17b–18b. Online at https://insulin.library.utoronto.ca/islandora/ object/insulin%3AW10027_0176.

63 27th July Lab book 1 Jan–Aug 1921; 4th August, 40; MS COLL 76 (Banting), Box 6A, Folder 3; IC, F. G. Banting (Frederick Grant, Sir) Papers, TFRBL UT. Online at https://insulin.library.utoronto.ca/islandora/object/insulin%3AN10014_0044.

64 Lab book 1 Jan–Aug 1921, 5th August, 43. Online at https://insulin.library.utoronto .ca/islandora/object/insulin%3AN10014_0047.

65 Lab book 1 Jan–Aug 1921, 5th August, 44. Online at https://insulin.library.utoronto .ca/islandora/object/insulin%3AN10014_0049.

66 Lab book 1 Jan–Aug 1921, 7th August, 48. Online at https://insulin.library.utoronto .ca/islandora/object/insulin%3AN10014_0054.

67 Lab book 1 Jan–Aug 1921, 49. Online at https://insulin.library.utoronto.ca /islandora/object/insulin%3AN10014_0055.

68 F. G. Banting to J. J. R. Macleod August 9, 1921; MS COLL 76 (Banting), Box 62, Folder 2A; IC, F. G. Banting (Frederick Grant, Sir) Papers, TFRBL UT. Online at https://insulin.library.utoronto.ca/islandora/object/insulin%3AL10026.

69 J. J. R. Macleod to F. G. Banting, August 23, 1921. MS COLL 76 (Banting), Box 62, Folder 4; IC, F. G. Banting (Frederick Grant, Sir) Papers, TFRBL UT. Online at https://insulin.library.utoronto.ca/islandora/object/insulin:L10028.

70 Best 1972, 388.

71 Best 1972, 388.

72 Kleiner and Meltzer 1915.

73 Anonymous, 'Find Diabetes Cause; Now Seek a Remedy: Rockefeller Institute Doctors Say Disease is Due to Failure of the Pancreas', *The New York Times* (1857–1922), August 12, 1915, 18. Online at https://www.nytimes.com/1915/08/12/ archives/find-diabetes-cause-now-seek-a-remedy-rockefeller-institute-doctors. html.

74 'Find Diabetes Cause'.

75 'Find Diabetes Cause'.

76 Friedman 2018.

77 Friedman 2018.

78 Kleiner 1919, 169.

79 S. Flexner to S. J. Meltzer, November 30, 1918; Thanks to Professor Jeffrey Friedmann, Rockefeller Institute.

80 S. J. Meltzer to S. Flexner, April 4, 1918; Thanks to Professor Jeffrey Friedmann, Rockefeller Institute.

81 William Shakespeare, *Henry V*, 3.1.

82 Friedman 2018.

83 Kleiner 1959, 83.

84 Banting and Best 1922b, 256.

85 Banting 1940, 30b/31b. Online at https://insulin.library.utoronto.ca/islandora/object /insulin%3AW10027_0189.

86 Banting, 3b. Online at https://insulin.library.utoronto.ca/islandora/object/insulin %3AW10027_0162.

87   C. H. Best transcribed by W. Feasby, March 20, 1956, 3, National Film Board file, MS COLL 235 Box 5, Folder 15; IC, Feasby (William R.) Papers, TFRBL UT.

88   Banting 1940, 8–9. Online at https://insulin.library.utoronto.ca/islandora/object/insulin%3AW10027_0176.

89   Banting, 5b. Online at https://insulin.library.utoronto.ca/islandora/object/insulin%3AW10027_0164.

90   MacLeod, cited in Stevenson 1979, 304.

91   Banting, cited in Bliss 1982a, 560.

92   Banting 1940, 10b. Online at https://insulin.library.utoronto.ca/islandora/object/insulin%3AW10027_0169.

93   Macleod, cited in Stevenson 1979, 299.

94   Stevenson 1979, 304.

95   Banting and Best 1922b, 265.

96   Scott 1912.

97   Banting and Best 1922a.

98   Banting and Best 1922a.

99   E. P. Joslin to J. J. R. Macleod, November 19, 1921; MS COLL 241 (Best) Box 51, Folder 8; IC, C. H. Best (Charles Herbert) Papers, TFRBL UT. Online at https://insulin.library.utoronto.ca/islandora/object/insulin%3AL10039.

100   J. J. R. Macleod to E. P. Joslin, November 21, 1921; MS COLL 241 (Best) Box 51, Folder 8; IC, C.H. Best (Charles Herbert) Papers, TFRBL UT. Online at https://insulin.library.utoronto.ca/islandora/object/insulin%3AL10040.

101   J. J. R. Macleod to E. P. Joslin, November 21, 1921; MS COLL 241 (Best) Box 51, Folder 8; IC, C. H. Best (Charles Herbert) Papers, TFRBL UT. Online at https://insulin.library.utoronto.ca/islandora/object/insulin%3AL10040.

102   Banting 1940, 7d. Online at https://insulin.library.utoronto.ca/islandora/object/insulin%3AW10027_0200.

103   Macleod, cited in Stevenson 1979, 305.

104   Banting, cited in Bliss 1982a, 561.

105   Bliss, 561.

106   Banting 1940, 8d. Online at https://insulin.library.utoronto.ca/islandora/object/ insulin%3AW10027_0201.

107   Clowes Jr. 1981, 208.

108   Macleod, cited in Stevenson 1979, 306.

109   Banting 1940, 18d. Online at https://insulin.library.utoronto.ca/islandora/object/insulin%3AW10027_0212.

110   Note card recording the first clinical use of extract on Joe Gilchrist on December 20, 1921. MS COLL 76 (Banting), Box 62, Folder 7; IC, F. G. Banting (Frederick Grant, Sir) Papers, TFRBL UT. Online at https://insulin.library.utoronto.ca/islandora/object/insulin%3AW10011.

111   'Use of extract on Joe Gilchrist'.

112   (Banting, Best, Collip, Campbell, and Fletcher 1922, 144.

113   Bliss 1982b, 112; Allan 1972, 267.

114 Bliss, 112; Tattersall 2019, 93, ascribes this description to Ed Jeffrey, the house doctor who was working with Campbell at the time.

115 Best ground up adult bovine pancreas in alcohol, then filtered the resulting solution to remove solid residue before evaporating off most of the alcohol and dissolving what remained in water.

116 Dec 1921–Jan 1922. Patient records for Leonard Thompson. MS COLL 76 (Banting), Box 8B, Folder 17B; Insulin Collection, F. G. Banting (Frederick Grant, Sir) Papers, TFRBL UT. Online at https://insulin.library.utoronto.ca/islandora/object/insulin%3AM10015.

117 Banting, Best, Collip, Campbell, and Fletcher 1922, 144.

118 Banting, Best, Collip, Campbell, and Fletcher 1922, 144.

119 Barr and Rossiter 1973.

120 Tattersall 2019, 92.

121 Collip 1922a; Collip 1922b.

122 J. B. Collip to H. M. Tory, January 25, 1922; MS COLL 269 (Collip), Box 37, Folder 2; IC, Collip (James Bertram) Papers, TFRBL UT. Online at https://insulin.library.utoronto.ca/islandora/object/insulin%3AL10002.

123 Warwick 1965a, 425.

124 Banting 1940, 6d. Online at https://insulin.library.utoronto.ca/islandora/object/insulin%3AW10027_0199.

125 'Work on Diabetes Shows Progress Against Disease', *Toronto Star Weekly*, January 14, 1922: MS COLL 76 (Banting) Scrapbook 1, Box 1, Page 1; IC, F. G. Banting (Frederick Grant, Sir) Papers, TFRBL UT. Online at https://insulin.library.utoronto.ca/islandora/object/insulin:C10024.

126 'Work on Diabetes Shows Progress Against Disease'.

127 'Medical Discovery Proves Epoch Making', *Toronto Star Weekly*, March 26, 1922; MS COLL 76 (Banting) Scrapbook 1, Box 1, Page 1; IC, F. G. Banting (Frederick Grant, Sir) Papers, TFRBL UT. Online at https://insulin.library.utoronto.ca/islandora/object/insulin%3AC10021.

128 'Have They Robbed Diabetes of its Terrors?' *Toronto Star Weekly*, March 26, 1922; MS COLL 76 (Banting) Scrapbook 1, Box 1, Page 3; IC, F. G. Banting (Frederick Grant, Sir) Papers, TFRBL UT. Online at https://insulin.library.utoronto.ca/islandora/object/insulin%3AC10025.

129 'Toronto Doctors on Track of Diabetes Cure', *Toronto Daily Star*, March 22, 1922. MS COLL 76 (Banting) Scrapbook 1, Box 1, Page 3; IC, F. G. Banting (Frederick Grant, Sir) Papers, TFRBL UT. Online at https://insulin.library.utoronto.ca/islandora/object/insulin%3AC10026.

130 'Medical Men Gather to Discuss Recent Discovery', February 14, 1922. MS COLL 76 (Banting) Scrapbook 1, Box 1, Page 1; IC, F. G. Banting (Frederick Grant, Sir) Papers, TFRBL UT. Online at https://insulin.library.utoronto.ca/islandora/object/insulin%3AC10023.

131 'Biology Club Held Interesting Meeting', February 14, 1922. MS COLL 76 (Banting) Scrapbook 1, Box 1, Page 1; IC, F. G. Banting (Frederick Grant, Sir)

Papers, TFRBL UT. Online at https://insulin.library.utoronto.ca/islandora/object/insulin%3AC10057.

132 Banting 1940, 17d. Online at https://insulin.library.utoronto.ca/islandora/object/insulin%3AW10027_0211.

133 C. H. Best to Sir Henry Dale, February 22, 1954. Dale Papers, The Royal Society Archive, MS HD/24/1/19

134 Best to Dale, February 22, 1954.

135 'Memorandum in reference to the co-operation of the Connaught Antitoxin Laboratories', January 25, 1922. MS COLL 76 (Banting) Scrapbook 2, Box 1, Page 40; IC, F. G. Banting (Frederick Grant, Sir) Papers, TFRBL UT. Online at https://insulin.library.utoronto.ca/islandora/object/insulin%3AW10022.

136 Banting 1940, 18d–19d. Online at https://insulin.library.utoronto.ca/islandora/object/insulin%3AW10027_0212.

137 Banting 1940, 20–21d. Online at https://insulin.library.utoronto.ca/islandora/object/insulin%3AW10027_0215.

138 Banting 1940, 21d.

139 Banting, Best, Collip, Campbell, Fletcher, Macleod, and Noble 1922, 8; This paper is recognized as being the official public announcement of the discovery of insulin.

140 Joslin 1922.

141 Joslin 1922.

142 Woodyatt, cited in Banting, Best, Collip, Campbell, Fletcher, et al. 1922, 10.

143 Banting, Best, Collip, Campbell, Fletcher, et al. 1922, 4.

144 In an article written later that year (Macleod 1922), Macleod credited the English physiologist Sir Edward A. Sharpey-Schafer with coining the term 'insulin'. Yet the very first suggestion of this name for the anti-diabetic agent in the pancreatic extract is thought to have actually been by the Belgian clinician and physiologist Jean de Meyer in 1909.

145 Bliss 1982b, 128; 268.

146 A. Hughes to F. G. Banting, July 3, 1922. MS COLL 76 (Banting), Box 8A, Folder 26A; IC, F. G. Banting (Frederick Grant, Sir) Papers, TFRBL UT. Online at https://insulin.library.utoronto.ca/islandora/object/insulin%3AL10016.

147 'Canadian Diabetes Cure: Encouraging Results', *The Times*, August 19, 1922a.

Chapter 3

1 J. J. R. Macleod to W. B. Cannon, April 29, 1922; MS COLL 241 (Best) Box 51, Folder 14; IC, C.H. Best (Charles Herbert) Papers, TFRBL UT. Online at https://insulin.library.utoronto.ca/islandora/object/insulin%3AL10041.

2 Macleod to Cannon, April 29, 1922.

3 These included Best noting that the water pressure in the vacuum pumps that were used to evaporate off the alcohol showed significant variations which were causing fluctuations in temperature and distillation time. Macleod meanwhile discovered

that these elevated temperatures could cause the insulin to lose activity and Collip suggested that this problem could be overcome by using acetone as the main extractant instead of alcohol, allowing the temperature to be kept down. Bliss 1982b, 133–134.

4   Bliss 1982b, 134.

5   J. J. R. Macleod to G. H. A Clowes, April 3, 1922; Aventis Pasteur Limited Archives 95-025-01; IC SALP BNL SA. Online at https://insulin.library. utoronto.ca/islandora/object/insulin%3AL10338.

6   G. H. A Clowes to J. J. R. Macleod, May 11, 1922; Aventis Pasteur Limited Archives 95-025-01; IC SALP BNL SA. Online at https://insulin.library. utoronto.ca/islandora/object/insulin%3AL10340.

7   Telegram from G. H. A Clowes to J. J. R. Macleod, May 25, 1922; Aventis Pasteur Limited Archives 95-025-01; IC SALP BNL SA. Online at https://insulin.library. utoronto.ca/islandora/object/insulin%3AL10342.

8   G. H. A. Clowes to J. J. R. Macleod, June 28, 1922; Aventis Pasteur Limited Archives 95-025-01; IC SALP BNL SA. Online at https://insulin.library. utoronto.ca/islandora/object/insulin%3AL10374.

9   J. J. R. Macleod to G. H. A Clowes, April 3, 1922; Aventis Pasteur Limited Archives 95-025-01; IC SALP BNL SA. Online at https://insulin.library. utoronto.ca/islandora/object/insulin%3AL10338.

10 Noble, cited in Jurdjevic and Tillman 2004, 871.

11 Jurdjevic and Tillman 2004, 871.

12 Banting, Best, Collip, Campbell, Fletcher, et al. 1922, 6.

13 Banting, Best, Collip, Macleod, et al. 1922, 32.

14 Bliss 1982, 109.

15 A year earlier, two researchers, F.C. Mann and T.B. Magath, had observed that removal of the liver causes a dramatic fall in the levels of blood glucose which resulted in coma and death. Following the injection of a solution of glucose however, the animal was shown to recover; Mann and Magath 1921.

16 Some patients do suffer loss of hypo awareness in which the physiological early warning system that would normally alert them to the onset of hypoglycaemia fails to do so. Why this happens is poorly understood.

17 Leyton, 1926, 47.

18 Leyton, 1926, 47.

19 Leyton, 1926, 47.

20 Fletcher 1980, 1116.

21 Tattersall 2019, 139.

22 Jones 2000, 147.

23 Valenstein 1986, 56, cited in Doroshaw 2006, 214.

24 J. D. Ratcliff, 'Minds that Come Back', *Collier's Magazine*, February 12, 1938, 38, cited in Doroshaw 2006, 214.

25 Jones 2000, 147.

26 Plath 1966, 184.

27 Banting, Franks, and Gairns 1938.

28 Doroshaw 2006, 213.

29 Bourne 1953.

30 Bourne 1953, 968.

31 Doroshaw 2006, 217.

32 Marks and Richardson 2007.

33 J. J. R. Macleod to G. H. A. Clowes, October 20, 1922; Aventis Pasteur Limited Archives 95-025-01; IC SALP BNL SA. Online at https://insulin.library.utoronto.ca/islandora/object/insulin%3AL10376.

34 G. H. A. Clowes to J. J. R. Macleod, August 12, 1922; Aventis Pasteur Limited Archives 95-025-01; IC SALP BNL SA. Online at https://insulin.library.utoronto.ca/islandora/object/insulin%3AL10346.

35 J. J. R. Macleod to G. H. A Clowes, August 15, 1922; Aventis Pasteur Limited Archives 95-025-01; IC SALP BNL SA. Online at https://insulin.library.utoronto.ca/islandora/object/insulin%3AL10344.

36 J. J. R. Macleod to J. G. Fitzgerald, November 11, 1922; Aventis Pasteur Limited Archives 95-025-01; IC SALP BNL SA. Online at https://insulin.library.utoronto.ca/islandora/object/insulin%3AL10369.

37 J. J. R. Macleod to G. H. A Clowes, September 23, 1922; Aventis Pasteur Limited Archives 95-025-01; IC SALP BNL SA. Online at https://insulin.library.utoronto.ca/islandora/object/insulin%3AL10349

38 J. J. R. Macleod to G. H. A. Clowes, October 10, 1922; Aventis Pasteur Limited Archives 95-025-01; IC SALP BNL SA. Online at https://insulin.library.utoronto.ca/islandora/object/insulin%3AL10458.

39 Macleod to Clowes, October 20, 1922.

40 Clowes to Macleod, May 11, 1922.

41 G. H. A. Clowes to J. J. R. Macleod, June 20, 1922; Aventis Pasteur Limited Archives 95-025-01; IC SALP BNL SA. Online at https://insulin.library.utoronto.ca/islandora/object/insulin%3AL10343.

42 G. H. A. Clowes to J. J. R Macleod, September 23, 1922; Aventis Pasteur Limited Archives 95-025-01; IC SALP BNL SA. Online at https://insulin.library.utoronto.ca/islandora/object/insulin%3AL10350.

43 The Insulin Committee 1923, 1849; 1851.

44 The Insulin Committee 1923, 1850.

45 The Insulin Committee 1923, 1850.

46 'Never Received Cent Dr. Banting Asserts', *Toronto Star*, September 9, 1937; MS COLL 76 (Banting) Scrapbook 2, Box 2, Page 101; IC, F. G. Banting (Frederick Grant, Sir) Papers, TFRBL UT. Online at https://insulin.library.utoronto.ca/islandora/object/insulin%3AC10144.

47 J. J. R. Macleod to F. G. Banting, August 23, 1921; MS COLL 76 (Banting), Box 62, Folder 4; IC, F. G. Banting (Frederick Grant, Sir) Papers, TFRBL UT. Online at https://insulin.library.utoronto.ca/islandora/object/insulin:L10028_0003.

48 Murlin and Kramer 1913; Murlin and Kramer 1916; Kramer, Marker, and Murlin 1916; Murlin and Kramer 1916.

49 Murlin and Kramer 1956.

50 'Claims Priority in Cure for Diabetes in Animals', *Toronto Star*, November 17, 1922; MS COLL 76 (Banting) Scrapbook 1, Box 1, Page 6; IC, F. G. Banting (Frederick Grant, Sir) Papers, TFRBL UT. Online at https://insulin. library.utoronto.ca/islandora/object/insulin%3AC10031.

51 Murlin and Kramer 1916, 482.

52 Murlin and Kramer 1913, 172; Murlin and Kramer 1916.

53 Murlin and Kramer 1956, 291.

54 Murlin 1972.

55 Murlin, Kramer, and Sweet 1922, 19.

56 Murlin, Kramer, and Sweet 1922, 21–22.

57 Murlin, Kramer, and Sweet 1922, 23.

58 'Claims Priority in Cure for Diabetes in Animals'.

59 'Canadian Diabetes Cure: Encouraging Results', *The Times*, August 19, 1922.

60 'Canadian Diabetes Cure'.

61 'Canadian Diabetes Cure'.

62 F. M. Allen to F. G. Banting, October 12, 1922; MS. COLL. 76 (Banting), Box 1, Folder 22A; IC, F.G. Banting (Frederick Grant, Sir) Papers, TFRBL UT. Online at https://insulin.library.utoronto.ca/islandora/object/insulin%3AL10310.

63 J. R. Murlin to F. G. Banting, August 9, 1922; University of Toronto Archives. A1982-0001, Box 13, Folder 10; IC UT BG IC ARMS UT Online at https:// insulin.library.utoronto.ca/islandora/object/insulin%3AL10251.

64 J. R. Murlin to F. G. Banting, November 11, 1922; University of Toronto Archives. A1982-0001, Box 24, Folder 11; IC UT BG IC ARMS UT. Online at https://insulin. library.utoronto.ca/islandora/object/insulin%3AL10295.

65 'Discovery of Extract That Has Power to Restore Capacity Lost in Diabetes is Made Public by Dr. John R. Murlin', *Democrat and Chronicle* (Rochester, NY), November 11, 1922. MS COLL 76 (Banting) Scrapbook 1, Box 1, Page 6; IC, F.G. Banting (Frederick Grant, Sir) Papers, TFRBL UT. Online at https://insulin. library.utoronto.ca/islandora/object/insulin%3AC10032.

66 G. H. A. Clowes to J. J. R. Macleod, October 5, 1922; Aventis Pasteur Limited Archives 95-025-01; IC SALP BNL SA. Online at https://insulin. library.utoronto.ca/islandora/object/insulin%3AL10352.

67 J. R. Murlin to C. H. Riches, October 7, 1922; University of Toronto Archives. A1982-0001, Box 13, Folder 10; IC UT BG IC ARMS UT. Online at https://insulin. library.utoronto.ca/islandora/object/insulin%3AL10252.

68 Murlin to Riches, October 7, 1922.

69 J. R. Murlin to J. J. R. Macleod, September 25, 1922; University of Toronto Archives. A1982-0001, Box 13; IC UT BG IC ARMS UT. Online at https://insulin. library.utoronto.ca/islandora/object/insulin%3AL10124.

70 Murlin to Macleod, September 25, 1922.

71 Insulin Committee Minutes 17/08/1922 to 29/09/1925; 28th–30th September 1922; University of Toronto Archives. A1982 - 0001, Box 044; IC UT BG IC ARMS UT. Online at https://insulin.library.utoronto.ca/islandora/object/ insulin%3AW10028.

72 J. R. Murlin to C. H. Riches, October 7, 1922; University of Toronto Archives. A1982-0001, Box 13, Folder 10; IC UT BG IC ARMS UT. Online at https://insulin. library.utoronto.ca/islandora/object/insulin%3AL10252.

73 G. H. A. Clowes to J. J. R. Macleod, November 1922; Aventis Pasteur Limited Archives 95-025-01; IC SALP BNL SA. Online at https://insulin. library.utoronto.ca/islandora/object/insulin%3AL10382.

74 J. R. Williams to G. H. A. Clowes, December 19, 1922; Aventis Pasteur Limited Archives 95-025-01; IC SALP BNL SA. Online at https://insulin. library.utoronto.ca/islandora/object/insulin%3AL10358.

75 Charles Riches to U of T Governors February 22, 1923; University of Toronto Archives. A1967 - 0007, Box 81; IC UT BG IC ARMS UT. Online at https:// insulin.library.utoronto.ca/islandora/object/insulin%3AL10275

76 Riches to U of T Governors February 22, 1923; Insulin Committee Minutes 17/08/1922 to 29/09/1925.

77 G. B. Schley to Eli Lilly, December 8, 1922; University of Toronto Archives. A1982-0001, Box 18, Folder 4; IC UT BG IC ARMS UT. Online at https:// insulin.library.utoronto.ca/islandora/object/insulin%3AL10465.

78 J. J. R. Macleod to F. M. Allen, November 29, 1922; University of Toronto Archives. A1982-0001, Box 12; IC UT BG IC ARMS UT. Online at https:// insulin.library.utoronto.ca/islandora/object/insulin%3AL10245.

79 F. G. Banting to Sir Robert Falconer, January 27, 1923; University of Toronto Archives. A1982-0001, Box 62, Folder 1; IC UT BG IC ARMS UT. Online at https://insulin.library.utoronto.ca/islandora/object/insulin%3AL10256.

80 F. G. Banting to C. E. Hughes, US Secretary of State, November 21, 1922; MS COLL 76 (Banting), Box 1, Folder 23; IC, F.G. Banting (Frederick Grant, Sir) Papers, TFRBL UT. Online at https://insulin.library.utoronto.ca/ islandora/object/insulin%3AL10317.

81 C. E. Hughes, US Secretary of State, to F. G. Banting, November 25, 1922; MS COLL 76 (Banting), Box 1, Folder 23; IC, F. G. Banting (Frederick Grant, Sir) Papers, IC TFRBL UT. Online at https://insulin.library.utoronto.ca/ islandora/object/insulin%3AL10318.

82 J. J. R. Macleod to G. H. A. Clowes, December 21, 1922; Aventis Pasteur Limited Archives 95-025-01. Online at https://insulin.library.utoronto.ca/ islandora/object/insulin%3AL10360.

83 'Biologists Acclaim Victory Over Diabetes: Prof. Murlin Humorously Admits Dr. Banting Forestalled Him. Applaud Discoverer', *Mail and Empire*, December 29, 1922; MS COLL 76 (Banting) Scrapbook 1, Box 1, Page 30. IC TFRBL UT. Online at https://insulin.library.utoronto.ca/islandora/object/insulin%3AC10056.

84 Biologists Acclaim Victory.

85 Biologists Acclaim Victory.

86 Biologists Acclaim Victory.

87 'Banting Given an Ovation by an Assembly of Scientists', MS COLL 76 (Banting) Scrapbook 1, Box 1, Page 54; IC, F. G. Banting (Frederick Grant, Sir)

Papers, TFRBL UT. Online at https://insulin.library.utoronto.ca/islandora/object/ insulin%3AC10076.

88 Gley 1922; Bliss 1982c, 170; Tattersall 2019, 80.

89 J. R. Williams to J. J. R Macleod, May 29, 1923; University of Toronto Archives. A1982-0001, Box 12, Folder 5; IC UT BG IC ARMS UT. Online at https://insulin. library.utoronto.ca/islandora/object/insulin%3AL10244.

90 Williams to Macleod, May 29, 1923.

91 'Research on Diabetes', W. M. Bayliss, *The Times*, August 24, 1922.

92 'Research on Diabetes'.

93 J. J. R. Macleod to W. M. Bayliss, September 8, 1922; MS COLL 241 (Best) Box 51, Folder 19; IC, C. H. Best (Charles Herbert) Papers, TFRBL UT. Online at https://insulin.library.utoronto.ca/islandora/object/insulin%3AL10046.

94 'Gives Dr. Banting Credit for "Insulin"', *Toronto Daily Star*, September 7, 1922; MS COLL 76 (Banting) Scrapbook 1, Box 1, Page 10; IC, F. G. Banting (Frederick Grant, Sir) Papers, TFRBL UT. Online at https://insulin.library. utoronto.ca/islandora/object/insulin%3AC10034.

95 'Gives Dr. Banting Credit'.

96 W. M. Bayliss to J. J. R. Macleod, September 29, 1922; MS COLL 241 (Best) Box 51, Folder 19; IC, C. H. Best (Charles Herbert) Papers, TFRBL UT. Online at https://insulin.library.utoronto.ca/islandora/object/insulin%3AL10175.

97 J. J. R. Macleod to A. B. MacCallum, September 14, 1922; MS COLL 241 (Best) Box 51, Folder 19; IC, C. H. Best (Charles Herbert) Papers, TFRBL UT. Online at https://insulin.library.utoronto.ca/islandora/object/insulin%3AL10047.

98 In one of the four patients (Case III), the extract was administered by hypodermic injection. Rennie and Fraser 1907, 13.

99 Rennie and Fraser 1907, 8.

100 Rennie and Fraser 1907, 8.

101 Tattersall 2019, 81.

102 Rennie and Fraser 1907, 13.

103 Wright Jr 2002, 1240.

104 'Insulin From Fish: Scientific Research and Practical Problems', *The Times*, January 2, 1923a.

105 Wright Jr 2002, 1240.

106 Wright Jr 2002, 1240.

107 This lack of interest makes an interesting contrast with the situation in Japan where, despite unsuccessful attempts by Professor Kumagai of Tohoku University to make pancreatic extracts for the treatment of diabetes at around the same time as Banting and Best, Japan was so successful in the industrial preparation of insulin from fish that, until about 1956, all its insulin requirements were met in this way; (Nakagawa 1994)

108 J. J. R. Macleod to H. H. Dale, December 7, 1923; MS COLL 241 (Best) Box 51, Folder 55; IC, C. H. Best (Charles Herbert) Papers, TFRBL UT. Online at https://insulin.library.utoronto.ca/islandora/object/insulin%3AL10134.

109 Macleod to MacCallum, September 14, 1922.

110 F. G. Banting to J. G. Fitzgerald, October 5, 1922; MS COLL 76 (Banting), Box 1, Folder 22A; IC, F. G. Banting (Frederick Grant, Sir) Papers, TFRBL UT. Online at https://insulin.library.utoronto.ca/islandora/object/insulin%3AL10309.

111 'Declares Best Shares Honor: Dr. Banting Pays Tribute to Partner in Discovery', *The Globe*, September 9, 1922; MS COLL 76 (Banting) Scrapbook 1, Box 1, Page 10; IC, F. G. Banting (Frederick Grant, Sir) Papers, TFRBL UT. Online at https://insulin.library.utoronto.ca/islandora/object/insulin%3AC10036.

112 A. Gooderham to F. G. Banting, September 16, 1922; MS COLL 76 (Banting), Box 62, Folder 21; IC, F. G. Banting (Frederick Grant, Sir) Papers, TFRBL UT. Online at https://insulin.library.utoronto.ca/islandora/object/insulin%3AL10030.

113 Gooderham to Banting, September 16, 1922.

114 J. J. R. Macleod to J. B.Collip, September 18, 1922; MS COLL 269 (Collip), Item 1; IC, Collip (James Bertram) Papers, TFRBL UT. Online at https://insulin.library.utoronto.ca/islandora/object/insulin%3AL10001.

115 J. B. Collip to J. J. R. Macleod, September 1922; MS COLL 269 (Collip), Box 37, Folder 4; IC, Collip (James Bertram) Papers, TFRBL UT. Online at https://insulin.library.utoronto.ca/islandora/object/insulin%3AL10004.

116 'New Diabetes Treatment. Canada's Gift. Insulin Patent for Britain', *The Times*, November 17, 1922; University of Toronto Archives. A1982-0001, Box 24, Folder 11; IC UT BG IC ARMS UT. Online at https://insulin.library.utoronto.ca/islandora/object/insulin%3AC10188.

117 Roberts 1922.

118 Roberts 1922, 1194.

119 Dale 1922, 1241.

120 The original application had been on the process of making insulin and therefore included only the names of Best and Collip. On the advice of patent lawyers, this draft was then amended to include both product and process and Banting's name was added. J.J.R. Macleod to F. G. Banting, October 31, 1922; MS COLL 76 (Banting), Box 1, Folder 22A. TFRBL UT. Online at https://insulin.library.utoronto.ca/islandora/object/insulin%3AL10314; Documents submitted with application to the Canadian Patent Office for a patent; MS COLL 76 (Banting), Box 10, Folder 11; Insulin Collection, F. G. Banting (Frederick Grant, Sir) Papers, TFRBL UT. Online at https://insulin.library.utoronto.ca/islandora/object/insulin%3AW10017.

121 'Assignment to the Governors of the University of Toronto, January 15, 1923; MS COLL 76 (Banting), Box 10, Folder 10; IC, F. G. Banting (Frederick Grant, Sir) Papers, TFRBL UT. Online at https://insulin.library.utoronto.ca/islandora/object/insulin%3AQ10013.

122 Macleod 1922, 835.

123 Joslin 1923, 1581.

124 Envelope addressed to the 'Dr. Who Cures Diabetis'; MS COLL 76 (Banting) Scrapbook 1, Box 1, Page 16. TFRBL UT. Online at https://insulin.library.utoronto.ca/islandora/object/insulin%3AE10007; 'Letter to Dr. Fred' November

10, 1922; MS COLL 76 (Banting) Scrapbook 1, Box 1, Front Cover. Online at https://insulin.library.utoronto.ca/islandora/object/insulin%3AL10055; 'Monsieur le professeur, University of Medicine, Who Has Found a Means of Curing Diabetis', November 15, 1922; MS COLL 76 (Banting) Scrapbook 1, Box 1, Page 18; IC, F. G. Banting (Frederick Grant, Sir) Papers, TFRBL UT. Online at https://insulin.library.utoronto.ca/islandora/object/insulin%3AE10008.

125 Bliss 1982b, 143.

126 'Soldier Patients Laud Work of Insulin Clinics'. (The Toronto Daily Star 1923a); MS. COLL 76 (Banting) Scrapbook 1, Box 1, Page 66; Insulin Collection, F.G. Banting (Frederick Grant, Sir) Papers, Thomas Fisher Rare Book Library, University of Toronto; https://insulin.library.utoronto.ca/islandora/object/insulin%3AC10089

127 'Soldier Patients Laud Work of Insulin Clinics'.

128 'Insulin value proven, children are benefited',(The Toronto Daily Star 1923b) 7th April 1923; MS. COLL 76 (Banting) Scrapbook 1, Box 1, Page 64; Insulin Collection, F.G. Banting (Frederick Grant, Sir) Papers, Thomas Fisher Rare Book Library, University of Toronto; https://insulin.library.utoronto.ca/islandora/object/insulin%3AC10086

129 'Soldier Patients Laud Work of Insulin Clinics'.

130 'Soldier Patients Laud Work of Insulin Clinics'.

131 'Soldier Patients Laud Work of Insulin Clinics'.

132 E. P. Joslin to G. W. Ross, March 20, 1923; MS COLL 76 (Banting), Box 1, Folder 27; IC, F. G. Banting (Frederick Grant, Sir) Papers, TFRBL UT. Online at https://insulin.library.utoronto.ca/islandora/object/insulin%3AL10321.

133 'Europe Discovers a Boston Doctor', *Boston Globe*, October 8, 1934; MS COLL 76 (Banting) Scrapbook 2, Box 2 Page 88; IC, F. G. Banting (Frederick Grant, Sir) Papers, TFRBL UT. Online at https://insulin.library.utoronto.ca/islandora/object/insulin%3AC10139.

134 Joslin 1950, 565.

135 Joslin, Gray, and Root 1922, 672.

136 Joslin 1956, 68.

137 Joslin 1923, 1581.

138 'A Cure for Diabetes', *The Times*, October 18, 1922..

139 Joslin, Gray, and Root 1922, 654.

140 Joslin, Gray, and Root 1922, 652-653.

141 'New diabetes treatment. Canada's gift. Insulin patent for Britain', *The Times*, November 17, 1922b.

142 'Insulin and Diabetes', *British Medical Journal*, November 4, 1922, 882.

143 'The Treatment of Diabetes by Insulin', *The Lancet*, November 18, 1922a, 1081.

144 'Insulin and Diabetes', 882.

145 'Insulin and Diabetes',

146 Graham and Harris 1922, 1086.

147 'Insulin', British Medical Journal, February 24, 1923, 341.

148 Dudley 1923)

149 'The Treatment of Diabetes by Insulin', 391–92.

150 'Insulin', 341.

151 'The Treatment of Diabetes by Insulin', 391.

152 'Insulin', 341.

153 'The Supply of Insulin: New Diabetes Cure in Use at Hospitals', *The Times*, February 21, 1923c.

154 'Some Clinical Results of the Use of Insulin', *British Medical Journal*, April 28, 1923, 737.

155 'Success of New Insulin Cure', *The Times*, April 28, 1923c.

156 Leyton 1922.

157 'Insulin Available in This Country: Conditions of Sale and Precautions to Be Observed', *British Medical Journal*, April 21, 1923, 695.

158 Leyton 1922)

159 'Insulin Available in This Country', 695.

160 'Insulin Available in This Country'.

161 'Insulin Available in This Country', 696.

162 'The Cost of Insulin: To the Editor of The Times', *The Times*, July 10, 1923c.

163 'Cost of Insulin: British and American Prices', *The Times*, July 11, 1923d.

164 Nixon 1924, 55.

165 Leyton 1923, 882.

166 'Privileged Panel Patients', *The Times*, June 7, 1923b.

167 Tattersall 1995, 745.

168 Tattersall 1995, 745.

169 Osler 1909, cited in Tattersall 1995, 746.

170 'Cost of Insulin: Cheaper Supplies from Strasbourg', *The Times*, July 18, 1923b.

171 Clark 1923, 1067.

172 'Cost of Insulin'.

173 'Insulin Supply: British Manufacturers Output', *The Times*, 20th July 1923c.

174 'House of Commons: Cost of Insulin', *The Times*, 26th July 1923d.

175 'House of Commons: Cost of Insulin'

176 'Insulin Available in This Country', 696.

177 Hetzel 1924, 231.

178 'Free Insulin', *The Lancet*, September 1, 1923, 471.

179 'Free Insulin'.

180 'Insulin Available in This Country', 696.

181 'The Future of Insulin: Problem of Continuous Administration', *The Times*, August 2, 1923b.

182 Maclean 1923, 829.

183 Maclean 1923, 829.

184 'Insulin in Advanced Diabetes', *British Medical Journal*, October 27, 1923, 765.

185 Banting's reply in Banting 1923, 'Discussion on Diabetes and Insulin', Proceedings of Sections at the Annual Meeting, 1923, *British Medical Journal*, September 15, 1923, 451.

186 'The Future of Insulin: Problem of Continuous Administration', *The Times*, August 2, 1923k.

187 'Insulin', *The Times*, August 7, 1923b.

188 Insulin: 'A Patient's Point of View', *The Times*, August 7, 1923b.

189 Tattersall 2019, 110–111.

190 'Insulin: A Patient's Point of View,' *The Times London*, August 7, 1923b.

191 Joslin 1923, 1581.

192 Eastwood 1986.

193 Eastwood 1986, 1659.

194 Eastwood 1986, 1660.

195 Joslin, Gray, and Root 1922, 695.

196 'Robert Daniel Lawrence', *British Medical Journal*, 7th September, 1968: 621.

197 Oakley 1969, 54.

198 Smith 1986, 582, n. 20.

199 'The Select Company of Diabetics: For the Benefit of Their Cult', *The Times*, April 19, 1933.

200 'King's College Hospital: Tributes to Progress', *The Times*, November 1, 1933.

201 'King's College Hospital: Clinic for Diabetic Outpatients', *The Times*, May 12, 1934.

202 'Diabetics in Sympathy: An Association for Rich and Poor', *The Times*, February 15, 1934.

203 'Diabetics in Sympathy'.

204 Sherborne and Priest 2012, 304.

205 'Diabetic Children: Provision of Holidays', *The Times*, July 19, 1935.

206 Wells 1933.

207 F. M. Allen to G. W. Ross, March 19, 1923; MS COLL 76 (Banting), Box 1, Folder 27; IC, F.G. Banting (Frederick Grant, Sir) Papers, TFRBL UT. https://insulin.library.utoronto.ca/islandora/object/insulin%3AL10320

208 'Plan Chair of Medicine to Honor Dr. Banting'; MS COLL 76 (Banting) Scrapbook 1, Box 2, Page 78; IC, F. G. Banting (Frederick Grant, Sir) Papers, TFRBL UT. Online at https://insulin.library.utoronto.ca/islandora/object/insulin%3AC10105; A Chair of Medicine in the University of Toronto 'Professorship for Dr. Banting ", *The Times*, April 4, 1923: 9.

209 G. W. Ross to Prime Minister Mackenzie King, May 8, 1923; MS COLL 76 (Banting), Box 1, Folder 29; IC, F. G. Banting (Frederick Grant, Sir) Papers, TFRBL UT. Online at https://insulin.library.utoronto.ca/islandora/object/insulin%3AL10324.

210 Prime Minister Mackenzie King to F. G.Banting, July 23, 1923; MS COLL 76 (Banting), Box 62, Folder 25; IC, F. G. Banting (Frederick Grant, Sir) Papers, TFRBL UT. Online at https://insulin.library.utoronto.ca/islandora/object/insulin%3AL10032.

211 'Banting Opens CNE', *The Times*, June 30, 1923, 11.

212 J. G. Kent to F. G. Banting, June 27, 1923; MS COLL 76 (Banting), Box 1, Folder 30; IC, F. G. Banting (Frederick Grant, Sir) Papers, TFRBL UT. Online at https://insulin.library.utoronto.ca/islandora/object/insulin%3AL10327.

213 Kent to Banting, June 27, 1923.

214 'Toronto Scientists Honored at Ottawa', *Toronto Daily Star*, February 14, 1923; MS COLL 76 (Banting) Scrapbook 1, Box 1, Page 29; IC, F. G. Banting (Frederick Grant, Sir) Papers, TFRBL UT. https://insulin.library.utoronto.ca/islandora/object/insulin%3AC10158.

215 'Would Share Honors with His Colleagues', *The Mail and Empire*, August 27, 1923; MS COLL 76 (Banting) Scrapbook 1, Box 2, Page 74; IC, F. G. Banting (Frederick Grant, Sir) Papers, TFRBL UT. Online at https://insulin.library.utoronto.ca/islandora/object/insulin%3AC10099; 'Banting and Best in Seats of Honor at Fair Luncheon', MS COLL 76 (Banting) Scrapbook 1, Box 1, Page 47; IC, F. G. Banting (Frederick Grant, Sir) Papers, TFRBL UT. Online at https://insulin.library.utoronto.ca/islandora/object/insulin%3AC10072.

216 Banting 1940, 1l; 'The Story of the Discovery of Insulin', MS COLL 76 (Banting) Box 1, Folders 9–13; IC, F. G. Banting (Frederick Grant, Sir) Papers, TFRBL UT. Online at https://insulin.library.utoronto.ca/islandora/object/insulin%3AW10027_0245.

217 The Australian scientist Howard Florey—who, thanks to his later crucial work on penicillin, would himself become no stranger to disputes between scientists over priority claims—was present at the meeting and is alleged to have said that, 'Banting of insulin fame was there—a most poisonous looking fellow'. Florey cited in MacFarlane, G. 1979. "Howard Florey: The Making of a Great Scientist." In. Oxford University Press: p.77.

218 Banting 1923, 445; 447.

219 'Insulin on a Large Scale', *The Times*, July 15, 1923b.

220 'Magistrates Apology', *The Times*, August 11, 1923.

221 "Telegrams in Brief", *The Times*, June 5, 1924.

222 Banting 1940, 4l. Online at https://insulin.library.utoronto.ca/islandora/object/insulin%3AW10027_0248

223 Banting 1940), 5l. Online at https://insulin.library.utoronto.ca/islandora/object/insulin%3AW10027_0249.

224 Banting 1940), 1l. Online at https://insulin.library.utoronto.ca/islandora/object/insulin%3AW10027_0245.

225 Draft of letter from F. G. Banting to E. P. Joslin, May 22, 1923; MS COLL 241 (Best) Box 26, Folder 5; IC, C. H. Best (Charles Herbert) Papers, TFRBL UT. Online at https://insulin.library.utoronto.ca/islandora/object/insulin%3AL10065.

226 Banting 1940, 14l. Online at https://insulin.library.utoronto.ca/islandora/object/insulin%3AW10027_0258.

227 'Toronto Doctors Honored: Nobel Prize to Banting', October 26, 1923; MS COLL 76 (Banting) Scrapbook 1, Box 2, Page 81; IC, F. G. Banting (Frederick Grant, Sir) Papers, TFRBL UT. Online at https://insulin.library.utoronto.ca/islandora/object/insulin%3AC10111.

Chapter 4

1 Speech given at the Banting–Macleod banquet 26/11/1923 by Sir Robert Falconer; University of Toronto Archives. A1967–0007, Box 82, Banting-Macleod Banquet Folder; IC UT OP ARMS UT. Online at https://insulin.library.utoronto.ca/islandora/object/insulin%3AW10038.

2 Banting-Macleod Nobel Prize banquet; University of Toronto Archives. A1967–0007, Box 76a, Banting Folder; IC UT OP ARMS UT. Online at https://insulin.library.utoronto.ca/islandora/object/insulin%3AE10017.

3 T. W. Todd to J. J. R Macleod, October 29, 1923; MS COLL 241 (Best) Box 51, Folder 53; IC, C. H. Best (Charles Herbert) Papers, TFRBL UT. Online at https://insulin.library.utoronto.ca/islandora/object/insulin%3AL10126.

4 H. R. Geyelin to F. G. Banting, November 7, 1923; MS COLL 76 (Banting), Box 1, Folder 38; IC, F. G. Banting (Frederick Grant, Sir) Papers, TFRBL UT. Online at https://insulin.library.utoronto.ca/islandora/object/insulin%3AL10335.

5 Zuelzer, G. of Berlin, Germany, Assignor to Chemische Fabrik Auf Actien (Vorm. E. Schering), of Berlin, Germany, 'Pancreas Preparation Suitable for the Treatment of Diabetes'. United States Patent Office, Serial No. 431,226. Application filed May 6, 1908. Published on May 28, 1912.

6 'Zuelzer patent application'.

7 Zuelzer would doubtless also feel aggrieved that November 14, Banting's date of birth, was declared World Diabetes Day by the World Diabetes Foundation.

8 Blum 1901.

9 Zuelzer 1924, 136.

10 Tattersall 2019, 30.

11 Zuelzer 1908, 307–308; Zuelzer, Dohrn, and Marxer 1908, 1380–1381.

12 Drügemöller and Norpoth 1989, 432; Mellinghoff 1971, 43.

13 Banting and Best were using the Lewis–Benedict method of blood sugar determination, which required only 0.2ml of blood; Tattersall 2019, 91.

14 Zuelzer 1907, 475; 'Um ein nicht zu toxisch wirkendes Pankreassekret zu erhalten, müssen sämtliche Eiweisskörper des Pankreas entfernt werden'.

15 Zuelzer, Dohrn, and Marxer 1908, 6; 1384.

16 Gross 1950.

17 Gross 1950; Mellinghoff 1971, 20.

18 Mellinghoff 1971, 21.

19 Zuelzer 1908, 310.

20 Jörgens 2020b, 61.

21 Forschbach 1909, 2055.

22 Forschbach 1909, 2055.

23 Mellinghoff 1971, 44; Zuelzer "Die Geschichte meiner Entdeckung des Acomatols, des deutschen Insulins." ('The story of my discovery of Acomatol, the German Insulin').

24 Zuelzer "Die Geschichte meiner Entdeckung des Acomatols, des deutschen Insulins.", 136.

25 Zuelzer "Die Geschichte meiner Entdeckung des Acomatols, des deutschen Insulins.", 136; Mellinghoff 1971, 44; Drügemoller and Norpoth 1989, 431.

26 Zuelzer "Die Geschichte meiner Entdeckung des Acomatols, des deutschen Insulins.", 137.

27 Zuelzer "Die Geschichte meiner Entdeckung des Acomatols, des deutschen Insulins.", 137.

28 Zuelzer "Die Geschichte meiner Entdeckung des Acomatols, des deutschen Insulins.", 137.

29 Jörgens 2020b, 62; Reuter 1924, 97.

30 Zuelzer "Die Geschichte meiner Entdeckung des Acomatols, des deutschen Insulins.", 137.

31 Zuelzer "Die Geschichte meiner Entdeckung des Acomatols, des deutschen Insulins.", 137.

32 Zuelzer "Die Geschichte meiner Entdeckung des Acomatols, des deutschen Insulins.", 137.

33 Reuter 1924, 92.

34 Zuelzer "Die Geschichte meiner Entdeckung des Acomatols, des deutschen Insulins.", 137; Jörgens 2020a; Jörgens 2020b, 62.

35 Zuelzer "Die Geschichte meiner Entdeckung des Acomatols, des deutschen Insulins.", 138.

36 Minkowski 1924, 107–108; Minkowski's reference to the use of alcohol in the purification of insulin having now been abandoned is probably a reference to the fact that by now the large- scale production of insulin used the new isoelectric method developed by George Walden at Lilly.

37 Zuelzer 1924, 135–136. Translation by K. Hall.

38 G. Zuelzer, 'The Overcoming of Diabetes: New Facts about Insulin', October 16, 1923; MS COLL 76 (Banting) Box 62, Folder 26; IC, F. G. Banting (Frederick Grant, Sir) Papers, TFRBL UT. Online at https://insulin.library.utoronto.ca/islandora/object/insulin%3AW10041.

39 Zuelzer, 'The Overcoming of Diabetes', 4.

40 Zuelzer 1923, 1552.

41 Zuelzer, 'The Overcoming of Diabetes'.

42 Zuelzer, 'The Overcoming of Diabetes'.

43 Zuelzer, 'The Overcoming of Diabetes'.

44 Zuelzer, 'The Overcoming of Diabetes'.

45 Zuelzer wrote two letters to Professor Goran Liljestrand, Secretary of the Nobel Committee on December 22, 1923, and August 18, 1924; see Bliss 1982, 281, n. 44. When contacted during the writing of this book, the Nobel Institute were unable to locate the letters.

46 Reuter 1924, 95; Zuelzer n.d.; Jörgens 2020a; Jörgens 2020b, 62.

47 H. F. Lorne, January 14, 1925; 'Report to the Insulin Committee on F. Lorne Hutchison's mission to Europe in the summer of 1924', 14; University of Toronto Archives. A1982–0001, Box 06, Folder 1; IC UT BG IC ARMS UT. Online at https://insulin.library.utoronto.ca/islandora/object/insulin%3AW10030.

48 Jörgens 2020a, 1.

49 F. L. Hutchinson to C. H Riches, October 17, 1925; University of Toronto Archives. A1982–0001, Box 15, Folder 5; IC UT BG IC ARMS UT. Online at https://insulin. library.utoronto.ca/islandora/object/insulin%3AL10239.

50 C. H. Riches to the University of Toronto Insulin Committee, January 8, 1926; University of Toronto Archives. A1982–0001, Box 15, Folder 5; p.3; IC UT BG IC ARMS UT. Online at https://insulin.library.utoronto.ca/islandora/object/ insulin%3AL10240.

51 'Application for United States patent no. 1,027,790: Pancreas Preparation Suitable for the Treatment of Diabetes', by G. Zuelzer—with letters from patent examiner and subsequent amendments. The application was originally filed May 6, 1908 and was amended five times. The parts of the application filed May 6, 1908 are the petition, affidavit, and specification. Amendments to the application were made July 8, 1909, July 14, 1909, August 3, 1910, September 29, 1911, and November 4, 1922. Samples were filed with the July 14, 1909 amendment. University of Toronto Archives. A1982–0001, Box 15, Folder 5; IC UT BG IC ARMS UT. Online at https://insulin.library.utoronto.ca/islandora/object/ insulin%3AQ10002

52 Riches to the University of Toronto Insulin Committee, January 8, 1926.

53 'Now Hoped That Insulin Will Be Permanent Cure', *Toronto Daily Star*, January 30, 1923; MS COLL 76 (Banting) Scrapbook 1, Box 1, Page 36; Insulin Collection, F. G. Banting (Frederick Grant, Sir) Papers, TFRBL UT. Online at https://insulin. library.utoronto.ca/islandora/object/insulin%3AC10061.

54 'Now Hoped That Insulin Will Be Permanent Cure'.

55 Macleod 1925, 4.

56 G. Zuelzer to J. J. R. Macleod, October 7, 1927; MS COLL 241 (Best) Box 52, Folder 30; IC, C. H. Best (Charles Herbert) Papers, TFRBL UT. Online at https://insulin.library.utoronto.ca/islandora/object/insulin%3AL10051.

57 Zuelzer to Macleod, October 7, 1927.

58 G. Zuelzer to J. J. R. Macleod, November 5, 1927; MS COLL. 241 (Best) Box 52, Folder 30; IC, C. H. Best (Charles Herbert) Papers, TFRBL UT. Online at https://insulin.library.utoronto.ca/islandora/object/insulin%3AL10201.

59 Minkowski 1929, 315. Translated K. Hall.

60 Bliss 1982b, 165; Tattersall 2019, 113.

61 O. Minkowski to J. J. R. Macleod, May 8, 1924; MS COLL 241 (Best) Box 52, Folder 6; IC, C. H. Best (Charles Herbert) Papers, TFRBL UT. Online at https:// insulin.library.utoronto.ca/islandora/object/insulin%3AL10194.

62 J. J. R. Macleod to O. Minkowski, April 6, 1923; University of Toronto Archives A1982–0001 Box 5, Folder 11; IC UT BG IC ARMS UT. Online at https://insulin. library.utoronto.ca/islandora/object/insulin%3AL10269.

63 Drugemoller and Norpoth 1989, 427.

64 Cammidge 1922, 997; Banting 1929, 7–8.

65 Zuelzer 1931; Gross 1950, 154; Mellinghoff 1971, 17; *The New York Times* 1934.

66 Jörgens 2020b, 63.

67 Mellinghoff 1971, 17.

68 'Dr. Georg L. Zuelzer, Heart Specialist, 79', *The New York Times*, October 20, 1949.

69 'Dr. Georg L. Zuelzer, Heart Specialist, 79'; Gross 1950, 154; In a letter dated April 24, 1947 to his daughter Hertha Louise Ernst, Zuelzer claims to have developed the first hormonal treatment for weight loss; cited in Mellinghoff 1971, 17.

70 Jörgens 2020b, 63.

Chapter 5

1 Scott 1923, 1304.

2 Scott 1923, 1304.

3 Scott 1923, 1304.

4 Tattersall 2019, 84.

5 Murlin and Kramer 1956, 297.

6 Murlin 1972.

7 Murlin, J.R., Kramer 1956, 288.

8 Bliss 1993.

9 C. H. Best to F. G. Banting, June 28, 1923; MS COLL 76 (Banting), Box 1, Folder 30; IC, F. G. Banting (Frederick Grant, Sir) Papers, TFRBL UT. Online at https://insulin.library.utoronto.ca/islandora/object/insulin%3AL10328.

10 F. G. Banting to F. Hipwell, July 15, 1923; MS COLL 76 (Banting), Box 1, Folder 31; IC, F. G. Banting (Frederick Grant, Sir) Papers, TFRBL UT. Online at https://insulin.library.utoronto.ca/islandora/object/insulin%3AL10330.

11 F. G. Banting to H. R. Geyelin, November 10, 1923; MS COLL 76 (Banting), Box 1, Folder 38; IC, F. G. Banting (Frederick Grant, Sir) Papers, TFRBL UT. Online at https://insulin.library.utoronto.ca/islandora/object/insulin%3AL10336.

12 F. G. Banting to C. H. Best, Telegram October 26, 1923, Banting papers, University of Toronto, MS COLL 76 (Banting) Scrapbook 1, Box 3, Page 172; IC, F. G. Banting (Frederick Grant, Sir) Papers, TFRBL UT. Online at https://insulin.library.utoronto.ca/islandora/object/insulin%3AL10058.

13 'Banting Shares Prize with Fellow-Worker', November 1923; MS COLL 76 (Banting) Scrapbook 1, Box 2, Page 103; IC, F. G. Banting (Frederick Grant, Sir) Papers, TFRBL UT. Online at https://insulin.library.utoronto.ca/islandora/object/insulin%3AC10118.

14 Banting also used half of his prize money to establish a medical fund. 'Banting to Use Nobel Award in Aiding Research', *New-York Tribune*, November 5, 1923; MS COLL 76 (Banting) Scrapbook 1, Box 2, Page 76; IC, F. G. Banting (Frederick Grant, Sir) Papers, TFRBL UT. Online at https://insulin.library.utoronto.ca/islandora/object/insulin%3AC10104.

15 'Banting Will Use Nobel Prize Fund for Research: Will Share his Portion of Award with C. Best', *The Evening Telegram*, October 27, 1923; MS COLL 76 (Banting) Scrapbook 1, Box 1, Page 57; IC, F. G. Banting (Frederick Grant, Sir) Papers, TFRBL UT. Online at https://insulin.library.utoronto.ca/islandora/object/insulin%3AC10079.

16 Dr. H. H. Best to F. G. Banting, November 29, 1923; MS COLL. 76 (Banting), Box 1, Folder 38; IC, F. G. Banting (Frederick Grant, Sir) Papers, TFRBL UT. Online at https://insulin.library.utoronto.ca/islandora/object/insulin%3AL10333. As a result of his later research, Best actually went on to be nominated for the Nobel Prize in Physiology or Medicine, 14 times between 1950 and 1954. One of these nominations was by his old friend and mentor Sir Henry Dale while in 1950, Professor Ulf von Euler who went on to become chairman of the board of the Nobel Foundation, reviewed Best's role in the discovery of insulin and concluded that he should have been awarded the Nobel Prize along with Banting. (Erling Norrby, 'Nobel Prizes and Life Sciences', pp. 161–173).

17 Sir H. H. Dale to J. J. R. Macleod, April 16, 1923; MS COLL. 241 (Best) Box 51, Folder 45; IC, C. H. Best (Charles Herbert) Papers, TFRBL UT. Online at https://insulin.library.utoronto.ca/islandora/object/insulin%3AL10152.

18 'Dr. Banting as Dog Stealer', *The Abolitionist*, February 1, 1924; MS COLL 76 (Banting) Scrapbook 1, Box 2, Page 114; IC, F. G. Banting (Frederick Grant, Sir) Papers, TFRBL UT. Online at https://insulin.library.utoronto.ca/islandora/object/insulin%3AC10119.

19 Banting to Best, July 15, 1923. Best papers, cited in Bliss 1982b, 223.

20 'Research Institute New Toronto Boast', *The Mail and Empire*, September 5, 1923; MS COLL 76 (Banting) Scrapbook 1, Box 3, Page 201; IC, F. G. Banting (Frederick Grant, Sir) Papers, TFRBL UT. Online at https://insulin.library.utoronto.ca/islandora/object/insulin%3AC10132.

21 'Saunders and Banting Knighted in King's Birthday Honors', *The Mail and Empire*, June 4, 1934; MS COLL 76 (Banting) Scrapbook 2, Box 2, Page 79; IC, F. G. Banting (Frederick Grant, Sir) Papers, TFRBL UT. Online at https://insulin.library.utoronto.ca/islandora/object/insulin%3AC10138.

22 Certificate granting F. G. Banting the title of K. B. E., June 4, 1934; MS COLL 76 (Banting) Scrapbook 2, Box 1, Page 34; IC, F.G. Banting (Frederick Grant, Sir) Papers, TFRBL UT. Online at https://insulin.library.utoronto.ca/islandora/object/insulin%3AA10011.

23 'Noted for Science But a Devotee of Art', *Toronto Star Weekly*, June 18, 1938; MS COLL 76 (Banting) Scrapbook 2, Box 2, Page 101; IC, F. G. Banting (Frederick Grant, Sir) Papers, TFRBL UT. Online at https://insulin.library.utoronto.ca/islandora/object/insulin%3AC10145.

24 'Vision of Further Conquests Forms Keynote of Ceremonies When Banting Institute Opens', *The Mail and Empire*, September 17, 1930; MS COLL 76 (Banting) Scrapbook 2, Box 2, Page 58; IC, F. G. Banting (Frederick Grant, Sir) Papers, TFRBL UT. Online at https://insulin.library.utoronto.ca/islandora/object/insulin%3AC10135.

25 'Dr. Banting's Hard Work Won Nobel Prize', *New York Herald*, November 4, 1923; MS COLL 76 (Banting) Scrapbook 1, Box 1, Page 57; IC, F. G. Banting (Frederick Grant, Sir) Papers, TFRBL UT. Online at https://insulin.library.utoronto.ca/islandora/object/insulin%3AC10080.

26 'Charles H. Best Named to Chair of Physiology', *The Mail and Empire*, January 11, 1929; MS COLL 76 (Banting) Scrapbook 1, Box 3, Page 191; IC, F. G. Banting (Frederick Grant, Sir) Papers, TFRBL UT. Online at https://insulin.library.utoronto.ca/islandora/object/insulin%3AC10130.

27 'Banting Institute Supreme Among Medical Buildings', *The Evening Telegram*, September 1930; MS COLL 76 (Banting) Scrapbook 1, Box 3, Page 201; IC, F. G. Banting (Frederick Grant, Sir) Papers, TFRBL UT. Online at https://insulin.library.utoronto.ca/islandora/object/insulin%3AC10131.

28 'Charles H. Best Named to Chair of Physiology'.

29 Banting, cited in Bliss 1984, 290.

30 Bliss 1984, 295.

31 Bliss 1984, 238.

32 Rous was eventually awarded the Nobel Prize in Physiology or Medicine for this work in 1966.

33 Bliss 1984, 157.

34 Bliss 1984, 211.

35 Irwin 1942; Bliss 1984, 184–186; Lucas 1942.

36 Bliss 1984, 243.

37 Bliss 1984, 243.

38 'Obituary: Sir Frederick Banting', *The Times*, February 26, 1941.

39 Bliss 1984, 259–260.

40 Bliss 1984, 284–285.

41 Interview with Dr. L. W. Billingsley, November 30, 1983. Transcript in Bliss insulin papers, MS COLL 232: Box 6, Folder 6; TFRBL UT.

42 'Banting Lived Brief Time Companions Died Instantly in Newfoundland Crash', February 25, 1941; MS COLL 76 (Banting) Scrapbook 2, Box 2, Page 113; IC, F. G. Banting (Frederick Grant, Sir) Papers, TFRBL UT. Online at https://insulin.library.utoronto.ca/islandora/object/insulin%3AC10153.

43 Bliss, 1984, 305.

44 'Banting Pilot Writes Own Story of Crash', *Toronto Star*, March 13, 1941; MS COLL 76 (Banting) Mapcase; IC, F. G. Banting (Frederick Grant, Sir) Papers, TFRBL UT. Online at https://insulin.library.utoronto.ca/islandora/object/insulin%3AC10007.

45 'Banting Pilot Writes Own Story of Crash'.

46 'Banting Pilot Writes Own Story of Crash'.

47 'Banting Pilot Writes Own Story of Crash'.

48 'Photo Spread Covering F. G. Banting's Funeral', *Toronto Daily Star*, March 5, 1941; MS COLL 76 (Banting) Scrapbook 2, Box 2, Page 116; IC, F. G. Banting (Frederick Grant, Sir) Papers, TFRBL UT. Online at https://insulin.library.utoronto.ca/islandora/object/insulin%3AC10156.

49 'Death of Sir Frederick Tragic but Triumphant', March 4, 1941; MS COLL 76 (Banting) Mapcase; IC, F. G. Banting (Frederick Grant, Sir) Papers, TFRBL UT. Online at https://insulin.library.utoronto.ca/islandora/object/insulin%3AC10005.

50 Bliss 1984, 308–309.

51 Recollection of Dr. Keith MacDonald, October 23, 1981, cited in Bliss 1982b, 234.

52 Collip 1935, 1256.

53 'Asks Time to Consider Prize Award Disposal', *Toronto Daily Star*, November 2, 1923; MS COLL 76 (Banting) Scrapbook 1, Box 1, Page 55; IC, F. G. Banting (Frederick Grant, Sir) Papers, TFRBL UT. Online at https://insulin.library.utoronto.ca/islandora/object/insulin%3AC10077.

54 'Macleod Awards Collip Half of His Nobel Prize', *Toronto Daily Star*, November 7, 1923; MS COLL 269 (Collip), Item 9; IC, Collip (James Bertram) Papers, TFRBL UT. Online at https://insulin.library.utoronto.ca/islandora/object/insulin%3AC10012.

55 'Four Men Will Share in the Nobel Prize: Prof. J. J. R. Macleod Divides his $20,000 with Dr. J.B. Collip', November 1923; MS COLL 76 (Banting) Scrapbook 3, Page 6; IC, F. G. Banting (Frederick Grant, Sir) Papers, TFRBL UT. Online at https://insulin.library.utoronto.ca/islandora/object/insulin%3AC10167.

56 J. B. Collip to J. J. R. Macleod, January 10, 1924; MS COLL 241 (Best) Box 52, Folder 2A; IC, C. H. Best (Charles Herbert) Papers, TFRBL UT. Online at https://insulin.library.utoronto.ca/islandora/object/insulin%3AL10049.

57 J. J. R. Macleod to J. B. Collip, February 28, 1923; MS COLL 241 (Best) Box 51, Folder 40; IC, C. H. Best (Charles Herbert) Papers, TFRBL UT. Online at https://insulin.library.utoronto.ca/islandora/object/insulin%3AL10144.

58 J. J. R. Macleod to B. P. Watson, January 3, 1924; MS COLL 241 (Best) Box 52, Folder 2B; IC, C. H. Best (Charles Herbert) Papers, TFRBL UT. Online at https://insulin.library.utoronto.ca/islandora/object/insulin%3AL10050.

59 J. J. R. Macleod to Dr. Pritchard, Chair of the Carnegie Corporation, June 22, 1922; J. J. R. Macleod to Sir R. Falconer, June 23, 1922; MS COLL 241 (Best) Box 51, Folder 16; IC, C. H. Best (Charles Herbert) Papers, TFRBL UT. Online at https://insulin.library.utoronto.ca/islandora/object/insulin%3AL10044.

60 J. J. R. Macleod to J. B. Collip, June 26, 1922; MS COLL 241 (Best) Box 51, Folder 16; IC, C. H. Best (Charles Herbert) Papers, TFRBL UT. Online at https://insulin.library.utoronto.ca/islandora/object/insulin%3AL10045.

61 J. B. Collip to J. J. R. Macleod, April 21, 1923; MS COLL 241 (Best) Box 51, Folder 46; IC, C. H. Best (Charles Herbert) Papers, TFRBL UT. Online at https://insulin.library.utoronto.ca/islandora/object/insulin%3AL10168.

62 Collip to J. J. R. Macleod, April 21, 1923.

63 A. B. MacCallum to J. J. R. Macleod, November 28, 1923; MS COLL 241 (Best) Box 51, Folder 54; IC, C. H. Best (Charles Herbert) Papers, TFRBL UT. Online at https://insulin.library.utoronto.ca/islandora/object/insulin%3AL10127.

64 A. B. MacCallum to J. J. R. Macleod, December 3, 1923; MS COLL 241 (Best) Box 51, Folder 55; IC, C. H. Best (Charles Herbert) Papers, TFRBL UT. Online at https://insulin.library.utoronto.ca/islandora/object/insulin%3AL10133.

65 'Co-Discoverer of Insulin, Dr. J. B. Collip, Honored by Citizens of Alberta', *Edmonton Journal*, May 25, 1923; MS COLL 269 (Collip), Item 9; IC, Collip (James Bertram) Papers, TFRBL UT. Online at https://insulin.library.utoronto.ca/islandora/object/insulin%3AC10010.

66 'Co-Discoverer of Insulin, Dr. J. B. Collip'.

67 A. B. MacCallum to J. J. R. Macleod, December 7, 1923; MS COLL 241 (Best) Box 51, Folder 55; IC, C. H. Best (Charles Herbert) Papers, TFRBL UT. Online at https://insulin.library.utoronto.ca/islandora/object/insulin%3AL10170.

68 J. J. R. Macleod to A. B. MacCallum, December 5, 1923; IC, C. H. Best (Charles Herbert) Papers, TFRBL UT. Online at https://insulin.library.utoronto.ca/islandora/object/insulin%3AL10169.

69 Macleod to MacCallum, December 5, 1923.

70 J. J. R. Macleod to J. B. Collip, December 18, 1923; MS COLL 241 (Best) Box 51, Folder 55; IC, C. H. Best (Charles Herbert) Papers, TFRBL UT. Online at https://insulin.library.utoronto.ca/islandora/object/insulin%3AL10171.

71 'Collip Claims the Credit', November 1923; MS COLL 76 (Banting) Scrapbook 1, Box 1, Page 59; IC, F. G. Banting (Frederick Grant, Sir) Papers, TFRBL UT. Online at https://insulin.library.utoronto.ca/islandora/object/insulin%3AC10083.

72 'Collip Claims the Credit'.

73 F. G. Banting to J. B. Collip, June 30, 1925; MS COLL 76 (Banting), Box 62, Folder 31; IC, F. G. Banting (Frederick Grant, Sir) Papers, TFRBL UT. Online at https://insulin.library.utoronto.ca/islandora/object/insulin%3AL10033.

74 J. B. Collip to F. G. Banting, July 7, 1923; MS COLL 76 (Banting), Box 62, Folder 21; IC, F. G. Banting (Frederick Grant, Sir) Papers, TFRBL UT. Online at https://insulin.library.utoronto.ca/islandora/object/insulin%3AL10034.

75 Collip 1942, 401.

76 Collip 1941, 474.

77 Collip 1941, 474.

78 Warwick 1965, 426.

79 Barr and Rossiter 1973b, 248.

80 Best 1942a 25.

81 J. B. Collip to J. J. R. Macleod, June 15, 1922; MS COLL 241 (Best) Box 51, Folder 16; IC, C. H. Best (Charles Herbert) Papers, TFRBL UT. Online at https://insulin.library.utoronto.ca/islandora/object/insulin%3AL10042.

82 Barr and Rossiter 1973b, 254, 255.

83 Warwick 1965.

84 J. J. R. Macleod to Sir Robert Falconer, December 12, 1928; University of Toronto Archives. A1967–0007, Box 115, Macleod Folder; IC UT OP ARMS UT. Online at https://insulin.library.utoronto.ca/islandora/object/insulin%3AL10290.

85 J. J. R. Macleod to C. H. Best, January 26, 1929; MS COLL 241 (Best) Box 52, Folder 35; IC, C. H. Best (Charles Herbert) Papers, TFRBL UT. Online at https://insulin.library.utoronto.ca/islandora/object/insulin%3AL10053.

86 C. H. Best to J. J. R. Macleod, February 14, 1935; MS COLL 241 (Best) Box 52, Folder 38; IC, C. H. Best (Charles Herbert) Papers, TFRBL UT. Online at https://insulin.library.utoronto.ca/islandora/object/insulin%3AL10136.

87 Kuhn 1962, 1.

88 Best and Taylor 1945, 575.

89 Best 2003 153.

90 Voet and Voet 2012, 1102.

91 Best 1942a, 23.

92 Dr. Edward Tolstoi in Best 1946a, 148.

93 Best 1946a, 149.

94 Best 1946a, 150.

95 Best 1946b, 91.

96 Best 1946b, 87.

97 Best 1946b, 87.

98 Best 1952, 257.

99 Best 1952, 262.

100 Best 1952, 258.

101 Bliss 1982b, 95.

102 Dale 1954b, 32.

103 H. H. Dale to C. H. Best, November 5, 1953, MS 93HD, Dale Papers, The Royal Society.

104 Dale to Best, November 5, 1953; Dale 1954, 33.

105 H. H. Dale to C. H. Best, February 16, 1954, MS 93HD, Dale Papers, The Royal Society.

106 C. H. Best to H. H. Dale, February 22, 1954, MS 93HD, Dale Papers, The Royal Society.

107 Best to Dale, February 22, 1954.

108 Best to Dale, February 22, 1954.

109 C. H. Best to W. Feasby, 'Comments of Dr. MacFarlane's Script of "The Seekers."', March 1, 1956, 1. MS COLL 235 Box 5, Folder 16; Feasby (William R.) Papers, TFRBL UT.

110 C. H. Best to L. Macfarlane, August 17, 1954, 3. Transcribed by W. Feasby: MS COLL 232 Box 5, Folder 5; Feasby (William R.) Papers, TFRBL UT.

111 Best to Macfarlane, August 17, 1954, 3.

112 Best to Feasby, 'Comments of Dr. MacFarlane's Script', 1.

113 Best to Feasby, 'Comments of Dr. MacFarlane's Script', 1.

114 Best to Feasby, 'Comments of Dr. MacFarlane's Script', 6.

115 Best to Macfarlane, August 17, 1954, 5.

116 C. H. Best to L. MacFarlane, August 17, 1954, 5.

117 J. B. Collip to W. Feasby, July 17, 1956. MS COLL 235 Box 5, Folder 1; Feasby (William R.) Papers, TFRBL UT.

118 C. H. Best dictation, April 27, 1956, 1. Transcribed by W. Feasby. MS COLL 235 Box 5, Folder 15; Page 1. Feasby (William R.) Papers, TFRBL UT. p.3.

119 C. H. Best dictation, April 27, 1956.

120 W. Feasby cited in C. H. Best dictation, April 27, 1956, 2.

121 W. Feasby cited in C. H. Best dictation, April 27, 1956, 1.

122 W. Feasby cited in C. H. Best dictation, April 27, 1956, 1–2.

123 C. H. Best to W. Feasby, 'Comments of Dr. MacFarlane's Script of "The Seekers."', March 1 1956; 1. MS COLL 235 Box 5, Folder 16; Feasby (William R.) Papers, TFRBL UT.

124 C. H. Best dictation, April 27, 1956, 3.

125 Leslie Macfarlane to Nick Balls, August 2, 1956, 'The Quest' file, NFB archives Montreal, cited in Bliss 1993, 262.

126 Pratt 1954, 289.

127 Pratt 1954, 289.

128 Best 1956, 66.

129 Best 1956, 66.

130 Best 1956, 66.

131 Best 1959, 7.

132 Bliss 1982b, 130.

133 Best 1956), 66–67.

134 Feasby 1958, 82.

135 C. H. Best dictation, January 17, 1957; 1957 file, Feasby papers, cited in Bliss 1993, 264.

136 Best, Oslerian Lecture 1957, cited in Bliss 1993, 267.

Chapter 6

1 Murray 1969a.

2 Murray 1969b.

3 I. Murray to I. Pavel, November 17, 1968, cited in Ionescu-Tirgoviste 1996, 146; Murray 1969b.

4 Ionescu-Tirgoviste 1996, 49.

5 Paulescu 1921a in Pavel 1976, 39; English translation, 45.

6 Paulescu 1921d in Pavel 1976, 42; English translation, 48.

7 Paulescu 1921c in Pavel 1976, 43; English translation, 49.

8 Paulescu 1921b in Pavel 1976, 44; English translation, 50.

9 Paulescu 1921e, in Pavel 1976, 53; English translation in Ionescu-Tirgoviste, Guja, and Ioacara 2005, 39–56.

10 Reproduced in Ionescu-Tirgoviste 1996, 83–84; Paulescu 1921e, cited in Ionescu-Tirgoviste 1996, 85.

11 N. C. Paulescu, November 6, 1923, reprinted in Ionescu-Tirgoviste 1996, 131–132.

12 Ionescu-Tirgoviste 1996, 33.

13 Murlin, Clough, Gibbs, and Stokes 1923, 255.

14 N. C. Paulescu to F. G. Banting, February 5, 1923; University of Toronto Archives. A1982–0001, Box 62, Folder 1; IC, F. G. Banting (Frederick Grant, Sir) Papers, TFRBL UT. Online at https://insulin.library.utoronto.ca/islandora/object/insulin%3AL10259.

15 C. H. Best, Notes on an article by Nicolae Paulescu; MS COLL 76 (Banting), Box 62, Folder 6; IC, F. G. Banting (Frederick Grant, Sir) Papers, TFRBL UT. Online at https://insulin.library.utoronto.ca/islandora/object/insulin%3AW10010.

16 C.H. Best, Notes on an article by Nicolae Paulescu; MS. COLL. 76 (Banting), Box 62, Folder 6; IC, F.G. Banting (Frederick Grant, Sir) Papers, TFRBL UT. https://insulin.library.utoronto.ca/islandora/object/insulin%3AW10010

17 Banting and Best 1922b, 252–253.

18 Pavel 1976, 107.
19 Professor C. Achard to N. C. Paulescu, February 19 1924, cited in Ionescu-Tirgoviste 1996, 134.
20 Paulescu, cited in Ionescu-Tirgoviste 1996, 136–137.
21 C. H. Best to I. Pavel, October 15, 1969, cited in Pavel 1976, 109.
22 Pavel 1976, 108, 221.
23 Professor A. Tiselius to Professor I. Pavel, December 29, 1969, cited in Pavel 1976, 119.
24 Pavel 1976, 148.
25 Pavel 1976, 150.
26 Pavel 1976, 220.
27 Pavel 1976, 164.
28 Pavel 1976, 208.
29 Uzbekova 2018, 191.
30 Uzbekova 2018, 191.
31 Uzbekova 2018, 191.
32 Uzbekova 2018, 193.
33 Pavel 1976, 192.
34 Pavel 1976, 190.
35 Alberti 2001.
36 Slama 2003.
37 Weill 2003.
38 'SWC to French Health Minister and Romanian Ambassador: Cancel Paris Hospital Tribute to Antisemitic Hatemonger', Press release, Los Angeles, California, Simon Wiesenthal Center, August 22, 2003.
39 'SWC'; Laron 2008.
40 Friling, Ioanid, and Ionescu 2004, 33.
41 Friling, Ioanid, and Ionescu 2004, 33.
42 Friling, Ioanid, and Ionescu 2004, 35.
43 Paulescu, N.C., Fiziologia filozofica. Talmudl, Cahalul, Franc-Masoneria. Bucharest 1913, 11; cited in Volovici 1991, 29.
44 Volovici 1991, 35.
45 Volovici 1991, 22, 35.
46 Ancel 2011, 128.
47 Volovici 1991, 61.
48 Ancel 2011, 119.
49 Volovici 1991, 49; Ancel 2011, 132, 157, 159, 163, 122,159.
50 Alberti and Lefebvre 2003, 2120.
51 Excerpts from minutes of the IDF board meetings, 2004, 2005; Professor Pierre Lefebvre, personal communication.
52 Professor Pierre Lefebvre, personal communication.
53 Waller 2002, 243.
54 Best, interviewed in Stalvey 1971, 21.
55 Bliss 1993, 274.

56 Best 2003, 390.
57 Best 2003, 390.
58 Best 2003, 102, 193.
59 Best 1956, 67.
60 Bliss 1982b, 157.

Chapter 7

1 Altwegg *et al.* 2016.
2 Murnaghan and Talalay 1967, 344.
3 'Crystalline Insulin', 1926, 666.
4 Murnaghan and Talalay 1967, 339. Pure adrenaline was first isolated by the Japanese chemist Takamine.
5 Murnaghan and Talalay 1967, 340.
6 Abel, Geiling, Alles, and Raymond 1925.
7 Murnaghan and Talalay 1967, 342.
8 Murnaghan and Talalay 1967, 343.
9 J. J. Abel to J. J. R. Macleod, January 20, 1926; MS COLL 241 (Best) Box 52, Folder 17; IC, C. H. Best (Charles Herbert) Papers, TFRBL UT. Online at https://insulin.library.utoronto.ca/islandora/object/insulin%3AL10163.
10 J. J. Abel to J. J. R. Macleod, February 6, 1926; MS COLL 241 (Best) Box 52, Folder 18; IC, C. H. Best (Charles Herbert) Papers, TFRBL UT. Online at https://insulin.library.utoronto.ca/islandora/object/insulin%3AL10206.
11 Abel 1926, 135.
12 J. J. Abel to J. J. R. Macleod, March 12, 1926; MS COLL 241 (Best) Box 52, Folder 19; IC, C. H. Best (Charles Herbert) Papers, TFRBL UT. Online at https://insulin.library.utoronto.ca/islandora/object/insulin%3AL10207.
13 Shonle and Waldo 1924, 731.
14 Shonle and Waldo 1924, 736.
15 Murlin, Kramer, and Sweet 1922, 27.
16 Allen and Murlin 1925.
17 Allen and Murlin 1925, 139.
18 Best 1924, 629.
19 Best and Macleod 1923.
20 Wintersteiner, Du Vigneaud, and Jensen 1928, 397.
21 Robinson 1953.
22 Abel, Geiling,Rouiller, Bell, and Wintersteiner 1927, 79.
23 Scott and Fisher 1935.
24 Maynard 1958, 377.
25 Maynard 1958, 379.
26 Sumner 1926a; Sumner 1926b.
27 Sumner 1937, 257.
28 Sumner 1937, 257.

29 Sumner 1937, 258.

30 Murnaghan and Talalay 1967, 353.

31 Meyer and Mark 1928.

32 Astbury and Woods 1932.

33 Astbury 1936.

34 Astbury and Street 1931; Astbury and Woods 1933, 344.

35 Bergmann 1939, 169.

36 Tattersall 2019, 129.

37 Bergmann 1936, 40–41.

38 Bergmann 1939, 169.

39 Astbury 1934, 22.

40 Preston 1974, 2.

41 W. Weaver to W. T. Astbury, May 27, 1948. MS 419 E153. Astbury papers, ULSC BL

42 Bergmann 1935, 475.

43 Bergmann and Niemann 1936, 79.

44 Niemann and Bergmann 1937, 301; Niemann and Bergmann 1938, 582.

45 Niemann 1937, 306, 307; Bergmann and Niemann 1936, 80; Astbury 1934.

46 Astbury 1937, 969.

47 W. Astbury to A. C. Chibnall, June 12, 1942. MS 419 E.29. Astbury papers, ULSC BL

48 Astbury 1948, 99, cited in Olby 1994, 46.

49 Gordon 1996, 455.

50 Gordon 1996, 456.

51 Thanks to Dr. Charlotte Synge for kindly sharing this recollection of her father.

52 Gordon 1996, 456.

53 Gordon 1996, 457.

54 Elsden 'Obituary: Richard Synge', The Independent, 24th August 1994.

55 Gordon 1996, 458; Synge 1952, 374.

56 Synge 1940.

57 In 1968 the IWS had a technical centre built on Valley Drive, Ilkley, West Yorkshire and the original building can still be seen today.

58 Synge 1940, 11.

59 This is mentioned in a collection of handwritten reminiscences recorded in a desk diary for 1990 about the various people with whom Synge had worked with during his life: *Anecdotes – Was Nicht in Dem Annalen Steht* (*What is not to be found in the Annals*). R. Synge, 'Archer Martin' in 'Anecdotes'. A14–15. Entry recorded on page for March 1, 1990. PCRLMS TCL UC.

60 Ettre and Zlatkis 1979, 14.

61 Martin 1952, 360.

62 Stahl 1977, 80.

63 Stahl 1977, 80.

64 Stahl 1977, 80.

65 R. Synge, 'Archer Martin' in 'Anecdotes'. A14–15. Entry recorded on page for March 1, 1990. PCRLMS TCL UC

66 Martin 1979, 286.

67 Tsvett 1906a, reprinted in Berezkin 1990, 21–26.

68 Tsvett 1906b, reprinted in Berezkin 1990, 27–34.

69 Martin 1952, 360.

70 Martin 1979, 287.

71 Martin 1979, 287.

72 Stahl 1977, 81.

73 Ettre 2001.

74 Anderson 1988, 19. See also Hall, 'The Man in the Monkeynut Coat' (2014/2022): pp. 30–32; pp. 136–138.

75 Anderson 1988, 37.

76 Nelson 1980, 14–15; Thornton 2002, 32; Heaton 1965, 1–2, cited in Burt and Grady 1994, 19.

77 Nelson 1980, 4–5.

78 Burt and Grady 1994, 35–36.

79 Defoe, D. 1724–26. *A Tour Through the Whole Island of Great Britain.* New Haven, Yale University Press, 1991. Abridged and Edited by P. N. Furbank and W. R. Owens. Picture research by A. J. Coulson.

80 Olby 1994, 42.

81 Anderson 1988, 23.

82 A. Martin to R. Synge, June 2, 1944. J186–188. PCRLMS TCL UC.

83 Caroe 1978, 29.

84 Brett 1979, 110.

85 Caroe 1978, 52.

86 R. Synge, 'Archer Martin' in 'Anecdotes'. A14–15. Entry recorded on page for March 3, 1990. PCRLMS TCL UC.

87 Stahl 1977, 80–83; R. Synge, 'Archer Martin' in 'Anecdotes'. A14-15. Entry recorded on page for March 2, 1990. PCRLMS TCL UC.

88 H. Marston to R. L. M. Synge, November 9, 1939. J178–185. PCRLMS TCL UC.

89 J.R.R. Tolkien, 'The Clerke's Compleinte', *The Gryphon*, N.S 4, no. 3 (December 1922), 95. Translated by Dr. Alaric Hall, School of English, University of Leeds.

90 Martin 1979, 288.

91 R. Synge, 'Anecdotes'. A14–15. Entries recorded on pages for March 7–9, 1990. PCRLMS TCL UC.

92 R. Synge, 'Anecdotes'. A14–15.

93 R. Synge, 'Anecdotes'. A14–15.

94 R. Synge, 'Anecdotes'. A14–15.

95 R. Synge, 'Anecdotes'. A14–15.

96 R. Consden to R. L. M. Synge, September 3, 1943. J70. PCRLMS TCL UC.

97 R. Synge, 'Anecdotes'. A14–15. PCRLMS TCL UC.

98 R. Synge, 'Anecdotes'. A14–15.

99 Archer Martin and Judith Bagenal were married in the Leeds Register Office on January 11, 1943. General Register Office, England and Wales Civil Registration Marriage Index, 1916–2005, London, England: Registration district: Leeds; Volume 9b; Page 506.

100 R. Synge, 'Anecdotes'. A14–15.—PCRLMS TCL UC.

101 Stahl 1977, 71.

102 The term 'partition chromatography' was actually suggested by Dr. E. Lester Smith to distinguish it from classical chromatography. Gordon, Martin, and Synge 1943, 79. In his Nobel speech, Martin pointed out that around the same time at least two other researchers, Sverre Stene in Norway and Van Dyck in Holland had also conceived of a similar idea of 'partition chromatography' but he maintained that, since they had no immediate practical problem upon which to apply this technique, their work went unnoticed. Martin 1952, 365.

103 Martin 1952, 367.

104 *WIRA Bulletin*, Vol. 15. January 1953, 11–12. A.194. PCRLMS TCL UC.

105 Stahl 1977, 81.

106 Recollection of Dr. Maryon Dougill, a former employee at WIRA who knew Martin and Synge. Personal communication.

107 Gordon, Martin, and Synge 1943b.

108 Consden, Gordon, and Martin 1944a.

109 Consden, Gordon, and Martin 1944b.

110 Stahl 1977, 81.

111 Stahl 1977, 81.

112 W. T. Astbury to Sir Charles Martin, February 17, 1941. MS 419 E115. Astbury Papers, ULSC BL.

113 Sir Charles Martin to R. Synge, July 10, 1941. J189. PCRLMS TCL UC.

114 W. T. Astbury to Sir Charles Martin, February 17, 1941. MS 419 E115. Astbury Papers, ULSC BL.

115 Hall 2014.

116 R. Synge, 'Archer Martin' in 'Anecdotes'. A14–15. PCRLMS TCL UC.

117 R. Synge to R. Olby, July 1, 1980. A209. PCRLMS TCL UC.

118 W. T. Astbury to K. Bailey, November 1, 1941. MS 419, Box E6. Astbury Papers, ULSC BL.

119 Ogston 1943.

120 Chibnall 1942, 159.

121 Martin and Synge 1941a.

122 Martin and Synge 1941b.

123 Gordon, Martin, and Synge 1941, 1369.

124 Gordon, Martin, and Synge 1941, 1369.

125 Gordon, Martin, and Synge 1941, 1369. Bergmann had himself actually pointed out that the sure way to test his theory was to break a protein chain down into dipeptides or larger fragments but although he said that work was underway, it does not appear that he carried it out to the extent of the WIRA workers; Bergmann 1935, 475.

126 Martin 1946, 1.

127 Gordon, Martin, and Synge 1941, 1385.

128 Two years later, Gordon, Martin and Synge published another paper showing how partition chromatography on silica-gel could also be used to separate amino acids and short peptide fragments; Gordon, Martin, and Synge 1943.

129 Astbury 1943, 82.

130 Martin 1946, 2.

131 Astbury, cited in Martin 1946, 3.

132 Thanks to Professor Steven French, School of Philosophy, Religion and History of Science for first pointing out the comic relationship between Sheldon and Leonard as an example of the tension between theoretical and experimental scientists in his lecture 'The Air Pump and the Laws of Nature', given in 'History and Philosophy of Science in 20 Objects', a public lecture series at the University of Leeds in 2018: https://www.youtube.com/watch?v=QFoKDYTcgjs&list=PL6vDA9mrYRYTPTM YB5Me0gjW_NtgVN1Dg&index=6.

133 Gordon, Martin, and Synge 1943, 92.

134 Astbury 1955, 220.

135 Adlard 2002.

136 Astbury 1955, 235.

137 Best 1952, 264.

138 Martin 1946, 1.

Chapter 8

1 R. L. M. Synge to Research Manager, ICI Ltd, October 23, 1941. A159. PCRLMS TCL UC.

2 R. L. M. Synge to Research Manager, ICI Ltd, October 23, 1941.

3 M. P. Applebey to R. L. M. Synge, October 29, 1941. A159. PCRLMS TCL UC.

4 R. L. M. Synge to R. Dubos, November 25, 1941. J87. PCRLMS TCL UC.

5 Zatman 1983.

6 R. L. M. Synge to R. Dubos, March 31, 1942. J87. PCRLMS TCL UC.

7 R. L. M. Synge to R. Hotchkiss, July 11, 1943. J137–138. PCRLMS TCL UC.

8 Gordon, Martin, and Synge 1943b, 88.

9 R. L. M. Synge to G. Gauze, May 3, 1944. J107. PCRLMS TCL UC.

10 Report Sent to Moscow on Studies of Gramicidin by D. Crowfoot, G. M. J. Schmidt, and R. L. M. Synge, July 26, 1944. E67. PCRLMS TCL UC.

11 Williams 1984, 188–189.

12 R. L. M. Synge to D. Hodgkin, June 2, 1944, B472. DHPBL UO, 5606. While partition chromatography was used in the analysis of Gramicidin and Gramicidin S, it was also used in the preparation of another antibiotic, Penicillin K; Fischbach, Mundell, and Eble 1946.

13 R. L. M. Synge to Dr. N. A. Lapteva, May 21, 1945. E67. PCRLMS TCL UC.

14 N. A. Lapteva to R. L. M. Synge, May 28, 1945. E67. PCRLMS TCL UC.

15 Elsden and Synge 1944.

16 Synge 1944; Moore and Stein 1948.

17 Stein 1946.

18 R. L. M. Synge to D. Hodgkin, September 7, 1945, B.472. DHPBL UO, 5606.

19 R. L. M. Synge to Professor P. Sergiev, May 21, 1945. E67. PCRLMS TCL UC.

20 R. L. M. Synge to Professor G. Gauze, October 12, 1945. E67. PCRLMS TCL UC.

21 Martin 1952, 368.

22 Gordon 1977, 244.

23 A. J. P. Martin to R. L. M. Synge, n.d. J186–188. PCRLMS TCL UC.

24 Gordon 1977, N244.

25 A. J. P. Martin to R. L. M. Synge, n.d. J186–188.

26 A. J. P. Martin to R. L. M. Synge, n.d. J186–188.

27 A. J. P. Martin to R. L. M. Synge, n.d. J186–188.

28 A. J. P. Martin to R. L. M. Synge, n.d. J186–188.

29 A. J. P. Martin to R. L. M. Synge, n.d. J186–188.

30 R. Synge, 'Dorothy Hodgkin' in 'Anecdotes'. A.14–15. Entry recorded on page for April 5, 1990. PCRLMS TCL UC.

31 Ferry 2014, 125; also in 'Chemistry in Oxford 1928–1960', March 3, 1989. DHPBL UO

32 R. Synge, 'Dorothy Hodgkin' in 'Anecdotes'. A.14–15.

33 Perutz 1997, 450.

34 Ferry 2014, 29

35 Ferry 2014, 2.

36 Bragg 1925, 6.

37 Ferry 2014, 227.

38 Ferry 2014, 228.

39 D. Hodgkin to R. L. M. Synge, October 6, 1943. J129–133. PCRLMS TCL UC.

40 Ferry 2014, 256.

41 W. T. Astbury to D. Hodgkin, February 12, 1945; DHPBL UO MS Eng. c. 5599/1; NCUACS 47/3/94 B.339; cited in Ferry 2014, 258.

42 Laurence 1955.

43 Todd 1955.

44 Ferry 2014, 320.

45 Bragg 1962, 131.

46 *The Observer* (colour supplement)' December 13, 1964; With thanks to Georgina Ferry.

47 Fara 2002, 85.

48 Fara 2002, 86.

49 H. H. Dale, 'Lectures on Certain Aspects of Biochemistry', University of London Press, 1926; cited in Ferry 2014, 129.

50 Scott and Fisher 1935.

51 Ferry 2014, 130.

52 Dodson 2002, 204.

53 Crowfoot 1935.

54 Tattersall, 2009, 79.

55 Krogh's wife had been diagnosed with what was then known as 'Maturity onset diabetes' but today is called type 2 diabetes. Nevertheless, in the later stages of her illness it had become so serious that she needed to treat it with insulin; Tattersall 2019, 102.

56 A. Krogh to J. J. R. Macleod, October 23, 1922; University of Toronto Archives A1982-0001 Box 13, Folder 2; IC UT BG IC ARMS UT. Online at https://insulin.library.utoronto.ca/islandora/object/insulin%3AL10246.

57 A. Krogh to J. J. R. Macleod, November 1922; MS COLL 241 (Best) Box 51, Folder 26; IC, C. H. Best (Charles Herbert) Papers, TFRBL UT. Online at https://insulin.library.utoronto.ca/islandora/object/insulin%3AL10116.

58 Nomination Archive. Nobel Media AB 2021. Wed. 14 Apr 2021. Online at https://www.nobelprize.org/nomination/archive/show.php?id=13377.

59 A. Krogh to J. J. R. Macleod, September 20, 1923; MS COLL 241 (Best) Box 51, Folder 52; IC, C. H. Best (Charles Herbert) Papers, TFRBL UT. Online at https://insulin.library.utoronto.ca/islandora/object/insulin%3AL10160.

60 J. J. R. Macleod to C. H. Best; MS COLL 241 (Best) Box 51, Folder 25; IC, C. H. Best (Charles Herbert) Papers, TFRBL UT. Online at https://insulin.library.utoronto.ca/islandora/object/insulin%3AL10159.

61 Hagedorn, Jensen, Krarup, and Wodstrup 1936, 177.

62 Tattersall 2009, 80.

63 Hagedorn's formulation was later further developed by Nordisk under the name Neutral Protamine Hagedorn (NPH) insulin.

64 Banting 1937, 596.

65 Root, White, and Marble 1936, 183.

66 M. B. Funder, Nordisk Insulinlaboratorium to D. M. C. Hodgkin, August 9, 1938; A. T. Jensen to D. M. C. Hodgkin, July 20, 1938. DHPBL UO. MS Eng. c. 5599/1; NCUACS 47/3/94. B.106.

67 Ferry 2014, 122.

68 Abir-Am 1987, 251.

69 Abir-Am 1987, 252.

70 Senechal 2013, 101.

71 Abir-Am 1987, 258.

72 Wrinch 1936, 411.

73 Wrinch 1936, 411.

74 Wrinch 1936, 412.

75 Wrinch 1937b.

76 Wrinch 1937b, 567.

77 Wrinch 1938, 123.

78 Wrinch 1937a.

79 Crowfoot 1938, 597.

80 Langmuir and Wrinch 1938.

81 Riley 1981, 24.

82 D. Hodgkin to D. Wrinch, April 23, 1940; Correspondence with Hodgkin, Dorothy Crowfoot, 1938–47, 1950–53, 1969, undated; Box 4, Folder 38; Wrinch Papers, Sophia Smith Collection of Women's History, Smith College.

83 D. Hodgkin to D. Wrinch, August 17, 1941; Correspondence with Hodgkin, Dorothy Crowfoot, 1938–47, 1950–53, 1969, undated; Box 4, Folder 38; Wrinch Papers, Sophia Smith Collection of Women's History, Smith College.

84 Abir Am 1987, 264–265.

85 Bragg 1939, 73.

86 Bernal 1939, 74–75.

87 Serafini 1989, 55.

88 D. Wrinch to L. Pauling, February 9, 1937. Special Collections & Archives Research Center, Oregon State University Libraries. Online at http://scarc.library. oregonstate.edu/coll/pauling/proteins/corr/corr438.12-wrinch-lp-19370209.html.

89 W. Weaver to L. Pauling, February 11, 1937. Special Collections & Archives Research Center, Oregon State University Libraries. Online at http://scarc.library. oregonstate.edu/coll/pauling/proteins/corr/sci14.038.5-weaver-lp-19370211.html.

90 L. Pauling to W. Weaver, March 6, 1937. Special Collections & Archives Research Center, Oregon State University Libraries. Online at http://scarc.library .oregonstate.edu/coll/pauling/proteins/corr/sci14.038.5-lp-weaver-19370306.html.

91 L. Pauling to W. Weaver, February 23, 1938. Special Collections & Archives Research Center, Oregon State University Libraries. Online at http://scarc.library. oregonstate.edu/coll/pauling/proteins/corr/sci14.037.8-lp-weaver-19380223.html.

92 Abir-Am 1987, 268.

93 Pauling and Niemann 1939, 1860.

94 Pauling and Niemann 1939, 1860.

95 Abir-Am 1987, 273.

96 D. Wrinch to E. H. Neville, October 1940. Dorothy Wrinch Papers, Sophia Smith Collection, Smith College, Northampton, Massachusetts; cited in Abir-Am 1987, 268.

97 P. Wrinch to L. Pauling, undated but believed to be 1939 or 1940. Dorothy Wrinch Papers, Sophia Smith Collection, Smith College, Northampton, Massachusetts; cited in Senechal 2013, 171.

98 D. Wrinch to O. Glaser, May 3, 1941. Dorothy Wrinch Papers, Sophia Smith Collection, Smith College, Northampton, Massachusetts; cited in Abir-Am 1987, 271.

99 D. Wrinch to D. Hodgkin, June 2, 1939. H.255. DHPBL UO, 5606.

100 D. Wrinch to D. Hodgkin, July 17, 1944. H.255. DHPBL UO, 5606.

101 D. Wrinch to D. Hodgkin, February 25, 1951. H.255. DHPBL UO, 5606.

102 It would seem that concerns about Wrinch procuring data were not confined to insulin. Synge recalled that when he was still working at Cambridge before moving to Leeds, Norman Pirie warned him to stay on guard in the laboratory in case Wrinch, who was visiting Bernal, 'tried to steal TMV [Tobacco Mosaic Virus]' on which Pirie was working. R. Synge, 'N. W. Pirie' in 'Anecdotes'. A14–15. PCRLMS TCL UC.

103 D. Hodgkin to D. Wrinch, March 4, 1951: H.255. DHPBL UO, 5606.

104 Abir-Am 1987, 275.

105 D. Wrinch to D. Hodgkin, August 25, 1951: H.255. DHPBL UO, 5606.

106 Hodgkin 1976, 564.

107 D. Hodgkin to Dr. C. Cohen, Brandeis University, February 11, 1977. H.256. DHPBL UO, 5606.

108 W. Astbury to D. Crowfoot, November 6, 1942. DHPBL UO, 5606.

109 W. Astbury to D. Crowfoot, June 26, 1941. DHPBL UO. 5606.

110 Astbury 1943, 99

111 Abir-Am 1987, 280.

112 R. L. M. Synge to E. Chargaff, March 3, 1959. J.51. PCRLMS TCL UC.

113 Harding 1981, 198.

114 R. L. M. Synge, 'Proposals for further work on Gramicidin and Gramicidin S', B.475. DHPBL UO, 5606.

115 Hull, Karlsson, Main, Woolfson, and Dodson 1978.

116 Forgan 1998; p.228.

117 'Barbara Low: Biochemist Who Played an Important Role in the Development of Penicillin as an antibiotic But Was Once Mistaken for a Tea Lady', *The Times (London)*, March 28, 2019.

118 Perutz 1997, 451.

Chapter 9

1 Stahl 1977, 83.

2 R. L. M. Synge, 'Dorothy Crowfoot' in 'Anecdotes'. A14–15; Entry recorded on page for April 5, 1990 diary. PCRLMS TCL UC.

3 Baroness Thatcher, interviewed by Georgina Ferry, cited in Ferry 2014, 459.

4 Ferry 2014, 438.

5 Ferry 2014, 447.

6 Pugwash was a group of scientists who, in 1957 at Pugwash, Nova Scotia, held the first of a series of conferences devoted to the abolition of nuclear weapons.

7 Ferry 2014, 457–458.

8 Baroness Thatcher, interviewed by Georgina Ferry, cited in Ferry 2014, 459.

9 Vainshtein 1997, 456.

10 R. Synge to the secretary of PM M. Thatcher. A437. PCRLMS TCL UC. See also, letter from Sir John Cornforth to Margaret Thatcher, March 19, 1989. A.437. PCRLMS TCL UC.

11 R. Synge, diary entry, November 22, 1990. Pocket diaries. A59. PCRLMS TCL UC.

12 R. L. M. Synge to Dr. J. Legget, Director of Science, Greenpeace. June 23, 1989. A378–419. PCRLMS TCL UC.

13 R. L. M. Synge to Dr. J. Legget, Director of Science, Greenpeace.

14 Autumn 1983 had proven to be a particularly close call. In September, Stanislav Petrov, a Lieutenant Colonel working as the watch officer for the USSR's missile detection system received an automated warning that a satellite had detected what appeared to be five incoming US intercontinental ballistic missiles. Asking himself why the US would fire only five missiles, Petrov ignored the warning and chose not to pass the information up the chain of command, which would certainly have resulted in retaliation. His level-headed judgment of the situation proved to be fortunate for

the rest of humanity. It later transpired that the satellite had simply misinterpreted sunlight reflected from low-lying clouds as an ICBM launch; Downing 2019, 195.

15 Stahl 1977, 83.

16 Stahl 1977, 82.

17 Stahl 1977, 83.

18 Morris 2009.

19 R. L. M. Synge to Professor G. Gauze, October 12, 1945. E67. PCRLMS TCL UC.

20 M. Demerec to R. L. M. Synge, January 21, 1949. G3-4. PCRLMS TCL UC.

21 The term 'molecular biology' is said to have been coined by Warren Weaver, director of the Rockefeller Institute, but it was popularized by William Astbury, who described it as being 'simple and expressive of everything that we want to do'. W. T. Astbury to W. Weaver, January 11, 1948. Astbury Papers MS419 E153, ULSC BL. Synge, however, took a very different view. In one letter he complained 'If only that tautologous expression [molecular biology] had never been adopted'. R. Synge to J. Fruton, May 22, 1992. J106. PCRLMS TCL UC.

22 R. L. M. Synge to M. Demerec, June 9, 1949. G3-4. PCRLMS TCL UC.

23 R. L. M. Synge to M. Demerec, June 9, 1949. PCRLMS TCL UC.

24 R. L. M. Synge to M. Demerec, June 9, 1949. PCRLMS TCL UC.

25 R. L. M. Synge to M. Demerec, June 9, 1949. PCRLMS TCL UC.

26 H. B. Day, American Consul to R. L. M. Synge, May 3, 1949. G3-4. PCRLMS TCL UC.

27 R. L. M. Synge to M. Demerec, June 9, 1949. PCRLMS TCL UC.

28 R. L. M. Synge to M. Demerec, May 4, 1949. G3-4. PCRLMS TCL UC.

29 R. L. M. Synge to M. Demerec, June 9, 1949. PCRLMS TCL UC.

30 Demerec 1949, 335.

31 W. Stein to R. L. M. Synge, June 18, 1949. G3–4· PCRLMS TCL UC.

32 Only a few years later, in 1953, Hodgkin was invited to attend a major conference on protein structure in Pasadena and, like Synge, she too was denied an entry visa as a result of her left-wing politics.

33 Letter signed by nine delegates, including Dorothy Crowfoot and Fred Sanger, to R. L. M. Synge, June 13, 1949. G3–4· PCRLMS TCL UC.

34 Richard Synge to M. Demerec, June 2, 1949. G3–4· PCRLMS TCL UC.

35 S. Moore to R. L. M. Synge, June 20, 1949. G3-4. PCRLMS TCL UC.

36 *The Simpsons*, Series 3, Episode 10, 'Flaming Moe's'. In this particular episode, gas chromatography is used to analyse the chemical composition of a new cocktail invented by Moe, the local bartender and one of Bart Simpson's classmates cites Archer Martin as being his favourite scientist for having invented this technique.

37 Martin and Synge 1941a, 1359.

38 Lovelock 2004, 159.

39 James and Martin 1952.

40 Lovelock 2004, 159.

41 Recollection of Desty, in Ettre and Zlakis 1979, 33.

42 Lovelock 2004, 159. Lovelock was referring to Carson, R. *Silent Spring*, Boston, Houghton Mifflin Company, 1962. It has been credited as being a major driving force behind the growth of the environmental movement.

43 Dr. Charlotte Synge, personal communication.

44 Whyte 1961, x.

45 Williams 2015, 117.

46 S. Moore to R. L. M. Synge, November 5, 1952. A167–194. PCRLMS TCL UC.

47 'Blobs Win Nobel Prize For Two Young Scientists', *News Chronicle*, November 7, 1952. A167–194. PCRLMS TCL UC.

48 S. Moore to R. L. M. Synge, November 5, 1952. PCRLMS TCL UC.

49 A. C. Chibnall to R. L. M. Synge, November 6, 1952. A167–194. PCRLMS TCL UC.

50 S. Fox to R. L. M. Synge, November 7, 1952. A167–194. PCRLMS TCL UC.

51 W. Astbury to R. L. M. Synge, November 7, 1952. A167–194. PCRLMS TCL UC.

52 Tiselius, cited in the *Manchester Guardian*, November 7, 1952. A167–194. PCRLMS TCL UC.

53 B. H. Wilsden to R. L. M. Synge, May 7, 1953. J331. PCRLMS TCL UC.

54 B. H. Wilsden to R. L. M. Synge, February 1, 1954. J331. PCRLMS TCL UC.

55 Benson and Calvin 1947, 1948a, 1948b.

56 Calvin and Benson 1949.

57 R. L. M. Synge to M. Calvin, February 7, 1953. A180. PCRLMS TCL UC.

58 M. Calvin to R. L. M. Synge, February 17, 1953. A180. PCRLMS TCL UC.

59 F. Sanger to R. L. M. Synge, November 8, 1952. A167–194. PCRLMS TCL UC.

60 Brownlee 2014, 66.

61 Brownlee 2014, 69.

62 Brownlee 2014, 69.

63 Brownlee 2014, 70.

64 Brownlee 2014, 69.

65 Brownlee 2014, 51.

66 Brownlee 2014, 68.

67 Brownlee 2014, 74.

68 Brownlee 2014, 75.

69 Brownlee 2014, 78.

70 Brownlee 2014, 82.

71 F. Sanger to R. L. M. Synge, May 31, 1945. J265. PCRLMS TCL UC.

72 F. Sanger to R. L. M. Synge, June 16, 1945. J265. PCRLMS TCL UC.

73 F. Sanger to R. L. M. Synge, June 29, 1945. J265. PCRLMS TCL UC.

74 F. Sanger to R. L. M. Synge, August 7, 1945. J265. PCRLMS TCL UC.

75 F. Sanger to R. L. M. Synge, December 12, 1945. J265. PCRLMS TCL UC.

76 F. Sanger to R. L. M. Synge, November 26, 1946. J265. PCRLMS TCL UC.

77 F. Sanger to R. L. M. Synge, December 12, 1945. PCRLMS TCL UC.

78 F. Sanger to R. L. M. Synge, March 17, 1947. J265. PCRLMS TCL UC.

79 Sanger 1946.

80 Sanger 1945.

81 Sanger 1949.

82 F. Sanger to R. L. M. Synge, March 11, 1949. J265. PCRLMS TCL UC.

83 F. Sanger to R. L. M. Synge, March 11, 1949. PCRLMS TCL UC.

84 Sanger 1950, 153.

85 Consden, Gordon, Martin, and Synge, 1947, 602.

86 F. Sanger, 'Report of Work for 1949–1950'; SA/BMF/A.2/231; Papers of Fred Sanger, Wellcome Archive.

87 The peptide fragments were separated on the basis of their overall electrical charge using a method that Sanger learned from Synge, while analysis of their constituent amino acids was done using paper chromatography.

88 Judson 1996, 89.

89 F. Sanger to R. L. M. Synge, May 23, 1951. J265. PCRLMS TCL UC.

90 Sanger and Tuppy 1951a, 463; Sanger and Tuppy 1951b.

91 Sanger and Thompson 1953a; Sanger and Thompson 1953b.

92 Sanger and Thompson 1953a; Sanger and Thompson 1953b.

93 Brownlee 2014, xv.

94 'Dr. Sanger Awarded Nobel Prize for Chemistry', *The Times*, October 29, 1958, 8.

95 Tattersall 2019, 211.

96 Sanger 1988, 9.

97 Sanger 1952, 2.

98 The impact of Sanger's work can also be understood in terms of ideas about information theory that have been crucial in the emergence of modern communication networks and computing. One of the pioneers of this field, the mathematician Claude Shannon (1916–2001) who worked at Bell Labs, proposed a mathematical definition of information content in which there was an inverse relationship between the information content of a message and its predictability. As an example, consider the following two strings, both of which contain forty-eight characters:

String   1:ABCABCABCABCABCABCABCABCABCABCABCABCABCAB-CABCABC

String 2: DNA was discovered by Friedrich Miescher in 1869

Both strings contain 48 characters, but according to Shannon's definition, their information content differs. This is true not only in the sense of their obvious difference in meaning, but also in the mathematical understanding of information as defined by Shannon. Because of its repetitive nature, the first string of characters can be written far more simply with the instruction (or algorithm) 'Write (ABC) 16 times', or '16*(ABC)'. It is therefore said to have a high degree of algorithmic compression and a low information content by Shannon's definition.

Although an algorithm could also be written to represent the second string, it would actually be bigger than the original string itself. So this string is said to have a low degree of algorithmic compression and therefore a high information content. Sanger's work on insulin showed that, according to Shannon's definition, protein sequences carry a high information content in contrast to the model proposed by Astbury, Bergmann, and Niemann in which, due to being made up of simple repeating patterns of amino acids, they had a low information content.

99 Hodgkin 1964, 88.

100 Ferry 2014, 372.

101 Wang 1998, 498.

102 Dodson 1997, 466.

103 F. Sanger to D. Hodgkin, June 1, 1951; MS Eng 5588/1-23 B.109; DHPBL UO.

104 D. Hodgkin to R. Corey, December 7, 1954; MS Eng 5588/1-23 B.90; DHPBL UO.

105 R. Corey to D. Hodgkin, January 18, 1955; MS Eng 5588/1-23 B.90; DHPBL UO.

106 Ryle, Sanger, Smith, and Kitai 1955, 556.

107 Cited in Perutz 1997, 452.

108 Ferry 2014, 389.

109 Ramaseshan 1997 459.

110 Hodgkin 1964, 90.

111 D. Hodgkin to Roseman, May 18, 1967. MS Eng 5588/1-23 B.98; DHPBL UO.

112 Ramaseshan 1997, 460.

113 Hodgkin 1971, 449.

114 Perutz 1997, 453.

115 Ramaseshan 1997, 460.

116 Ferry 2014, 396.

117 'Advance in Insulin Study', *The Times*, August 15, 1969.

118 Dunitz 1997, 449.

119 MS Eng 5592/25 B.204; DHPBL UO.

120 N. R. Lazarus to D. Hodgkin, July 14, 1972; DHPBL UO.

121 'Discoveries and their Uses', Presidential Address to the British Association for the Advancement of Science, September 4–8, 1978; MS Eng 5654/7 D.148; DHPBL UO.

122 Dixon 1960.

123 Dixon 1960.

124 Katsoyannis and Suzuki 1963, 2659.

125 Katsoyannis and Suzuki 1961.

126 Katsoyannis, Suzuki, and Tometsko 1963.

127 Katsoyannis and his team first synthesized the complete A chain of sheep insulin and combined it its recombination with the natural B chain, to generate a hybrid molecule with full biological activity; Katsoyannis, Tometsko, and Fukuda 1963.

128 Katsoyannis, Fukuda, Tometsko, Suzuki, and Tilak 1964.

129 'The First Man-Made Protein in History', *Life*, May 8, 1964.

130 Katsoyannis 1964, 339.

131 Meienhofer *et al.* 1963.

132 Zahn, Gutte, Pfeiffer, and Ammon 1966.

133 Brandenburg, Gattner, Herbertz, Krail, Weinert, and Zahn 1971.

134 Zahn 2000.

135 Tang, interviewed by Georgina Ferry, cited in Ferry 2014, 413.

136 Bernal 1937, 58.

137 Tang cited in Ferry 2014, 413.

138 Du, Zhang, Liu, and Tsou 1961.

139 Niu, Kung, Wang, and Chen 1966.

140 Niu *et al.* 1964.

141 Du, Zhang, Liu, and Tsou 1965.
142 Ferry 2014, 416.
143 Ferry 2014, 416.
144 Kung, Du, Huang, Chen, and He 1966, 544.
145 Tsou 1995, 291·
146 Kung, Du, Huang, Chen, and He 1966, 544.
147 Kung, Du, Huang, Chen, and He 1966, 545.
148 Kung, Du, Huang, Chen, and He 1966, 545.
149 Tsou 1995, 291.
150 Tsou 1995, 291.
151 Wang 1998, 499.
152 Hodgkin 1975, 103.
153 Ferry 2014, 421.
154 Katsoyannis 1964, 340.
155 Ruttenberg 1972.
156 Tattersall, 2009, 147.
157 Martin and Synge 1941a, 1358.

Chapter 10

1 Watson 1986, 155

2 Crick, cited in Olby 2009, 467, n. 26.

3 See the article by Professor Matthew Cobb, 'Happy 100th Birthday, Francis Crick (1916–2004)', in which he writes: 'Watson's own description of the discovery of the structure of DNA did not contain any striking new revelations, with one exception. He finally admitted that when he wrote in *The Double Helix* that Crick strode into the Eagle pub and proclaimed "We have discovered the secret of life," this was not true. Watson said he made it up, for dramatic effect. Crick always denied saying any such thing, and historians have long known that *The Double Helix* cannot be taken as an entirely reliable source'. Online at https://whyevolutionistrue.com/2016/06/08/happy-100th-birthday-francis-crick-1916-2004/.

4 Crick 1988, 77.

5 Williams 2019, 10.

6 Dahm 2008, 567.

7 Suter 1944, 10; cited in Veigl, Harmann, and Lamm 2020, 456.

8 Hall and Sankaran 2021, https://doi.org/10.1017/S000708742000062X.

9 Miescher 1871, 458; Translated in Hall and Sankaran, 2021, 35.

10 Miescher 1871, 459; translated in Hall and Sankaran, 2021, 37.

11 Miescher 1871, 460; translated in Hall and Sankaran, 2021, 39.

12 His 1897, 31, 2.

13 F. Miescher to Hoppe-Seyler, Letter XV, 1870 in His 1897, 56: 'Ich glaube, dass der Physiologe der Wissenschaft einen grösseren Dienst tut, wenn es ihm für präzise,

rein Chemische Probleme gelingt … Leider sind so wenige Physiker und Chemiker geneigt, sich aufzuopfern für die Bewältigung von Aufgaben, deren Bedeutung mehr auf dem Gebiet der angewandten, als auf dem der reinen Disziplin liegt'. Translated by K. Hall.

14 Judson 1996, 13.
15 Avery, Macleod, and McCarty 1944.
16 Lederberg 2000, 194.
17 Watson, Hopkins, Roberts, Steitz, and Weiner 1987, 69.
18 Boivin 1947.
19 Alexander and Leidy 1953.
20 Reichard 2002, 13362.
21 Portugal 2010, 567.
22 Bearn 1996, 553.
23 Dubos 1976, 168.
24 Dubos 1976, 159.
25 Portugal 2010, 559.
26 Reichard 2002, 13357.
27 Judson 1996, 72.
28 Portugal 2010, 567.
29 Reichard 2002, 13361.
30 Dubos 1976, 159.
31 Dubos 1956, 41.
32 A. Cobourn to J. Lederberg, September 28, 1965. Online at https://profiles. nlm.nih.gov/ps/retrieve/ResourceMetadata/CCAAIW.
33 Judson 2003; Tiselius' acknowledgement that Paulescu should have had a share in the 1923 Nobel Prize was made before Paulescu's extremist political activities and his anti-Semitism had come to light.
34 O.T. Avery to R. Avery, May 26, 1943; cited in Dubos 1976, 219.
35 W. T. Astbury to F. B. Hanson, October 19, 1944, Astbury Papers MS419 E152 ULSC BL.
36 W. T. Astbury to O. T. Avery, January 18, 1945, Astbury Papers MS419 E152 ULSC BL.
37 Bell 1939, 39.
38 Astbury and Bell 1938.
39 Astbury and Bell 1938.
40 Astbury 1947, 68.
41 'Woman Scientist Explains', *Yorkshire Evening News*, March 23, 1939, in Press Clippings Book, Astbury Papers MS419, Box. A1, ULSC BL.
42 'Aged 25, She Will Lecture to Scientists', in Press Clippings Book, Astbury Papers MS419, Box. A1, University of Leeds Special Collections, Brotherton Library.
43 The complete text of the plaque reads: 'The X-ray spectroscopy image of the DNA molecule taken by Franklin (1920–1958) and PhD student Ray Gosling at King's

College London in 1952 can claim to be one of the world's most important photographs. It demonstrates the helical structure of DNA and enabled James Watson and Francis Crick of Cambridge to build the first model of the molecule in 1953'.

44 Watson 1986, 132.

45 Watson and Crick also needed crucial X-ray data found in a report written by Franklin for the MRC, which told them that the DNA molecule was composed of two chains, and that these ran in opposite directions.

46 Brenda Maddox, *Rosalind Franklin: The Dark Lady of DNA*. New York, Harper Collins, 2002, 316.

47 Watson and Crick 1953, 737.

48 Crick, cited in Olby 1994, 386–387.

49 Olby 1994, 386–387.

50 Chargaff 1978, 37.

51 Portugal 2010, 563.

52 Chargaff 1971, 639.

53 Chargaff 1978, 86. It is not entirely clear to which of Avery's experiments Chargaff is referring here, as the original paper (Avery, Macleod, and McCarty 1944) contains no mention of whether a preparation of nucleic acid isolated from calf thymus could transform bacteria in a similar manner to that isolated from pneumococcus.

54 Chargaff *et al.* 1949, 406.

55 Chargaff 1978, 86.

56 Chargaff 1978, 89–90.

57 Consden, Gordon, and Martin 1944b, 231.

58 Olby 1994, 436.

59 Synge 1952, 379.

60 Vischer, cited in Olby 1994, 212.

61 Vischer and Chargaff 1947; Chargaff 1978, 90; Consden, Gordon, and Martin 1944b; Consden, Gordon, and Martin 1944a.

62 Chargaff 1950, 203.

63 Chargaff 1978, 101.

64 Chargaff, cited in Judson 1996, 119.

65 *OED*, 'A person who sells gadgets, novelties, etc., esp. on the street or in a fair. Now chiefly: a person who delivers a sales pitch, esp. on television; an advertiser, a promoter'.

66 Chargaff 1978, 101.

67 Chargaff 1978, 102.

68 Chargaff *et al.* 1949.

69 Vischer, Zamenhof, and Chargaff 1949.

70 Vischer, Zamenhof, and Chargaff 1949, 436.

71 Chargaff 1950, 206.

72 Chargaff 1950, 206.

73 Olby 1994, 388.

74 F. Crick to M. Crick, March 19, 1953. Wellcome archive, PP/CRI/D/4/3:Box 243.

75 Chargaff 1978, 101.

76 Chargaff 1978, 102.

77 Chargaff *et al.* 1951, 229.

78 Chargaff 1978, 101.

79 Chargaff 1978, 94.

80 According to Donohue, in one of Chargaff's papers (Zamenhoff, Brawerman, and Chargaff 1952), although the ratio of A/T was indeed always close to 1, the ratio of C/G ranged from 0.85 to 0.93.

81 Donohue 1978.

82 Cobb 2015a.

83 Watson and Crick 1953.

84 Cobb 2016, 4.

85 Chargaff, Vischer, Doniger, Green, and Misani 1949, 415; Vischer, Zamenhof, and Chargaff 1949, 429.

86 Chargaff 1950, 206.

87 Chargaff 1950, 209.

88 G. Gamow to F. H. C. Crick, Wellcome Library for the History and Understanding of Medicine. Francis Harry Compton Crick Papers. Box 29. Folder: PP/CRI/D/4/1.

89 Hufbauer 2009, 14.

90 Judson 1996, 264.

91 Watson 2001, 41.

92 F. H. C. Crick to J. D. Watson, February 10, 1955: Wellcome Library for the History and Understanding of Medicine. Francis Harry Compton Crick Papers. Box: 26. Folder: PP/CRI/D/2/45. Online at http://archives.wellcome.ac.uk/.

93 Watson 2001, 24.

94 G. Gamow to J. D. Watson and F. H. C. Crick, July 8, 1953. Wellcome Library, PP/CRI/D/4/1:Box 29.

95 Gamow 1954, 318.

96 Gamow 1955.

97 Boivin and Vendrely 1947.

98 Brachet 1942; Caspersson 1947.

99 Oxford University Biochemistry Finals Part I, 1991.

100 I speak from personal experience because I was one such desperate finalist.

101 Hoagland, Zamecnik, and Stephenson 1957; Hoagland *et al.* 1958.

102 Crick 1988, 119.

103 Jacob 1988, 312.

104 In 1952, the biochemist Alexander Dounce had already proposed a theoretical model for how an RNA intermediate might direct protein synthesis; Dounce 1952; Dounce 1953.

105 Jacob and Monod 1961, 350.

106 Jacob 1988, 313–314, cited in Cobb 2015a, 166.

107 Hall 1987, 32.

108 'Telegram from James D. Watson to Sydney Brenner', *CSHL Archives Repository*, Reference SB/1/1/757/9, accessed May 8, 2020. Online at http://libgallery.cshl.edu/items/show/66514.

109 Brenner, Jacob, and Meselson 1961; Gros *et al.* 1961.

110 Gros *et al.* 1961.

111 For a much more detailed account of the story of the discovery of mRNA, see Cobb 2015b.

112 Dounce 1952; Dounce 1953.

113 G. Gamow to J. D. Watson, May 26, 1954. Wellcome Library. PPCRI/D/4/1.

114 G. Gamow to J. D. Watson, November 5, 1954. Wellcome Library. PPCRI/D/4/1.

115 G. Gamow to J. D. Watson, December 17, 1954. Wellcome Library. PPCRI/D/4/1.

116 Judson 1996, 269.

117 F. H. C. Crick 'On degenerate templates and the adaptor hypothesis', Unpublished note, Wellcome Library. PPCRI/H/1/38.

118 Judson 1996, 299.

119 Neel 1949.

120 Pauling *et al.* 1949.

121 Ingram 1956, 793.

122 More recently we have become aware of another example of just how powerful such a single substitution mutation can be, with the emergence of the E484K variant of the Sars-Cov-2 virus. In this variant, an alteration in the genetic code of the virus results in substitution of the negatively charged amino acid glutamic acid by lysine at position 484 in the viral spike protein. Because lysine carries a positive charge, this results in a structural change in the folding of the protein chain, which may make the virus more resistant to binding by antibodies and so possibly reduce the effectiveness of vaccines. See Wise 2021.

123 Judson 1996, 330; Cobb 2017.

124 Ingram 1957, 328.

125 'A Landmark in Genetics', *The Times*, August 23, 1957.

126 Crick 1958, 143; It should be mentioned however, that the idea that DNA might direct the synthesis of proteins, acting via an RNA intermediate was first proposed by the microbiologists Boivin and Vendrely 1947, while five years before Crick's lecture, the biochemist Alexander Dounce had suggested that 'the specific arrangement of amino acid residues in a given peptide chain is derived from the specific arrangement of nucleotide residues in a corresponding specific nucleic acid molecule'. See Dounce 1952, 251.

127 Judson 1996, 330.

128 Crick 1958, 152.

129 Watson 2001 265.

130 Watson 2001 265.

131 Portugal 2015, xiii.

132 Portugal 2015, 88.

133 Judson 1996, 463.

134 Watson 2001, 265.

135 Watson 2001, 265.

136 Judson 1996, 464.

137 Olby 1994, 285.

138 Watson 2001, 265.

139 Judson 1996, 469.

140 Portugal 2015, 123.

141 Crick 1962, 16.

142 Crick and Orgel 1973.

143 Perhaps a more imaginative name for this theory might be the 'Battlestar Galactica' hypothesis, after the popular sci-fi space opera TV series from the late 1970s (given a dramatic and imaginative reboot in 2004), which began each episode began with the words, 'There are those who believe that life here began out there—far across the universe ...'. It would appear from his paper that Crick was one such person, as were the astronomers Sir Fred Hoyle (1915–2001) and Chandra Wickramasinghe (1939–).

144 Crick and Orgel 1973, 343.

145 Crick 1962, 16.

146 Judson 1996, 463.

147 Judson 1996, 637.

148 Judson 1996, 631.

149 Chargaff 1978, 55.

150 Chargaff 1977, 51.

151 Chargaff 1978, 55.

152 Chargaff 1977, 133.

153 Chargaff 1963, 176.

154 Chargaff 1977, 96.

155 Chargaff 1978, 177.

156 Chargaff 1977, 95.

157 Chargaff 1977, 102.

158 Chargaff 1976.

159 Chargaff 1977, 124.

160 Judson 1996, 612–613.

161 Judson 1996, 614.

162 Chargaff 1977, 152.

163 Chargaff 1977, 113.

164 Chargaff 1977, 112.

165 Chargaff 1978, 3.

166 Chargaff 1978, 136.

167 Chargaff 1978, 3.

168 Chargaff 1976, 938.

169 Chargaff 1977, 113.

170 Chargaff 1977, 113.

171 Judson 1996, 632.

172 Chargaff 1978, 199.

173 Chargaff 1978, 198.

174 Chargaff 1977, 30; Chargaff 1978, 176.

175 Chargaff 1978, 176, 164.

Chapter 11

1   Boyer and Smith-Hughes 2001, 45.
2   Cohen and Smith-Hughes 2009, 79.
3   Cohen and Smith-Hughes 2009, 79.
4   Lear 1978, 66.
5   Daemmrich, Gray, and Shaper 2006, 131.
6   Cohen and Smith-Hughes 2009, 48; Boyer and Smith-Hughes 2001, 31; (Smith-Hughes 2013.
7   Hopson 1977, 56.
8   Smith-Hughes 2013, 13.
9   Smith-Hughes 2013, 8.
10 Berg and Smith-Hughes 2000, 56.
11 Boyer and Smith-Hughes 2001, 18
12 Smith Hughes 2013, 6.
13 Smith Hughes 2013, 6.
14 Arber and Dussoix 1962.
15 Dussoix and Arber 1962.
16 Smith and Wilcox 1970; Kelly and Smith 1970; Meselson and Yuan 1968.
17 Sack and Nathans 1973; Danna, Sack, and Nathans 1973.
18 Cohen and Smith-Hughes 2009; Har Gobind Khorana had also proposed this idea but, unlike Lobban, had proposed adding only a single base.
19 Jackson, Symonds, and Berg 1972. In this paper the authors gave credit to PhD student Pete Lobban and A. D. Kaiser for achieving the same results with DNA from bacteriophage P22; see Lobban and Kaiser 1973.
20 Lear 1978, 1.
21 Wade 1973, 567.
22 Wade 1973, 567.
23 There is some controversy about priority here, with some arguing that cohesive ends were discovered first by V. Sgaramella (Sgaramella 1972).
24 Mertz and Davis 1972; Hedgepeth, Goodman, and Boyer 1972; Sgaramella 1972.
25 Berg and Smith-Hughes 2000, 70.
26 Smith Hughes 2013, 12.
27 Sharp, Sugden, and Sambrook 1973.
28 Boyer, cited in Smith Hughes 2013, 16.
29 Cohen *et al.* 1973.
30 Lear 1978, 70.
31 Lear 1978, 72.
32 Ziff 1973, 275.
33 Singer and Soll 1973, 1114.
34 Boyer and Smith-Hughes 2001, 109.
35 Morrow *et al.* 1974.
36 McElheny 1974.
37 McElheny 1974.

38 Smith-Hughes 2001, 550.

39 Reimers and Smith-Hughes 1998, 4. Whether Reimers was actually the first person to suggest that insulin might be the ideal candidate to be produced using recombinant DNA technology is a debatable point. Boyer's technician Mary Betlach has said that Boyer was talking about doing this 'long before anybody else' (Betlach and Smith-Hughes 2002, 11). According to a memorandum (Bill Carpenter to File S74–43, memo re: Dr. Herbert Boyer, September 18, 1974, S74–43, patent correspondence 1974–1980, Office of Technology and Licensing, Stanford) cited in Smith Hughes 2013, 26, Boyer also made this suggestion. Keichi Itakura, a chemist who later worked with Boyer, has also said that he was thinking about it too as early as 1974 (Itakura and Smith-Hughes 2006, 18–19). Also, in a news release by Stanford (News release, Stanford University News Service, May 20, 1974, patent correspondence S74–43, 1974–1980, Office of Technology and Licensing, Stanford University) cited in Smith Hughes, 2013, 20, the Nobel laureate Joshua Lederberg also suggested insulin.

40 Smith Hughes 2001, 550.

41 Smith Hughes 2001, 550. Cohen did actually reverse his decision some years later.

42 S. N. Cohen and H. W. Boyer, December 2, 1980, Process for Producing Biologically Functional Molecular Chimeras, United States Patent no. 4,237,224.

43 Berg and Mertz 2010, 14.

44 Berg *et al.* 1974.

45 Berg *et al.* 1974, 303.

46 Berg *et al.* 1974, 303.

47 Wade 1974)

48 Berg *et al.* 1974, 303.

49 'Halt in Genetic Work Urged', *The Washington Post*, July 18, 1974; 'Genetic Tests Renounced Over Possible Hazards', *The New York Times*, July 18, 1974; 'The Scientific Conscience', *The Washington Post*, July 18, 1974, cited in Watson and Tooze 1981, Document 1.6, 12.

50 Berg and Smith-Hughes 2000, 94.

51 Michael Rogers, 'The Pandora's Box Congress', Rolling Stone 189, June 19, 1975, cited in Watson and Tooze 1981); Document 2.1, p.28.

52 Berg *et al.* 1975, 991; The original programme for the Asilomar conference is at 'Appendix C: Program of the International Conference on Recombinant DNA Molecules'. The Maxine Singer Papers (Profiles in Science). US National Library of Medicine. Online at https://collections.nlm.nih.gov/catalog/nlm:nlmuid-101584644X97-doc. Other documents that in this collection include 'Appendix I: Summary Statement of the Asilomar Conference on Recombinant DNA Molecules' (https://collections.nlm.nih.gov/catalog/nlm:nlmuid-101584644X105-doc); 'Appendix D: Proposed Guidelines on Potential Biohazards Associated with Experiments Involving Genetically Altered Microorganisms' (https://collections.nlm.nih.gov/catalog/nlm:nlmuid-101584644X139-doc); 'Appendix E: Preparation and Use of Recombinant Molecules Involving Animal Virus Genomes'

(https://collections.nlm.nih.gov/catalog/nlm:nlmuid-101584644X140-doc); 'Appendix F: Cloned Eukaryotic DNA' (https://collections.nlm.nih.gov/catalog/nlm:nlmuid-101584644X141-doc); 'Appendix J: A Plausible Model for Monitoring Compliance with Containment Guidelines' (https://collections.nlm.nih.gov/catalog/nlm:nlmuid-101584644X143-doc); 'Appendix H: Provisional Statement of the Conference Proceedings' (https://collections.nlm.nih.gov/catalog/nlm:nlmuid-101584644X100-doc); and 'Report of the Organizing Committee for the Asilomar Conference on Recombinant DNA Molecules' (https://collections.nlm.nih.gov/catalog/nlm:nlmuid-101584644X138-doc).

53 Berg, in Watson and Tooze 1981, Document 2.1, 31.
54 Watson and Tooze 1981, Document 2.1, 31.
55 Chargaff 1976, 938.
56 Rogers, in Watson and Tooze 1981, Document 2.1, 36.
57 Watson and Tooze 1981, Document 2.1, 35.
58 J. Lederberg, February 28, 1975, 'Statement concerning Safety Hazards Associated with Research on Recombinant DNA Molecules', Joshua Lederberg Papers (Profiles in Science) US National Library of Medicine. Online at https://collections.nlm.nih.gov/catalog/nlm:nlmuid-101584906X13225-doc.
59 Wade 1975a.
60 Rogers, in Watson and Tooze 1981, Document 2.1, 34.
61 Watson and Tooze 1981, Document 2.1, 35.
62 Watson and Tooze 1981, Document 2.1, 35.
63 Berg *et al.* 1975, 991.
64 Novick, cited in Watson and Tooze 1981, Document 6.4,149.
65 Wade 1975b, 767.
66 For an example of the kind of confusion that this caused, and its consequences, see the case of Dr. Charles Thomas of Harvard University, which is the subject of discussion in 'Minutes of the Eleventh Meeting of the Executive Recombinant DNA Committee, May 25, 1978'. The Maxine Singer Papers (Profiles in Science). US National Library of Medicine. Online at https://collections.nlm.nih.gov/catalog/nlm:nlmuid-101584644X145-doc.
67 Wade 1975b, 769.
68 A draft of NIH guidelines on recombinant DNA research can be found at 'Summary of the Proposed Guidelines for Research on Recombinant DNA Molecules, February 27, 1976'. The Donald S. Frederickson Papers (Profiles in Science). US National Library of Medicine. Online at https://collections.nlm.nih.gov/catalog/nlm:nlmuid-101584939X145-doc; 'Rules Created to Control DNA Research', Science News 1975, 373.
69 McElheny 1975; Norman 1976.
70 Weinberg 1975, 194.
71 Frazier 1975, 187.
72 Watson 1977.
73 Watson, cited in Lubow 1977, 16; Watson and Tooze 1981, Document 5.11, 118.
74 Sinsheimer, cited in Wade 1976c, 236.

75 Sinsheimer cited in Wade 1976a, 303.

76 Wade 1976a, 303.

77 Siekevitz 1976, 256.

78 Siekevitz 1976, 257.

79 Simring 1976, 940.

80 Chargaff 1976, 940.

81 Singer and Berg 1976, 188.

82 Cohen 1977, 655.

83 Cohen 1977, 655.

84 Dyson 1976, 6.

85 F. Dyson to P. Berg, June 19, 1975. The Paul Berg Papers (Profiles in Science). US National Library of Medicine. Online at https://collections.nlm.nih.gov/catalog/nlm:nlmuid-101584580X97-doc.

86 F. Dyson to P. Berg, June 19, 1975.

87 Davis 1976, 442.

88 'Summary Statement by Donald S. Frederickson, M.D., Director National Institutes of Health on Recombinant DNA Technology Before the Subcommittee on Health and Scientific Research of the Senate Committee on Human Resources', April 6, 1977. The Donald S. Frederickson Papers (Profiles in Science). US National Library of Medicine. Online at https://collections.nlm.nih.gov/catalog/nlm:nlmuid-101584939X125-doc

89 Hall 1987, 23.

90 Hubbard 1976, 836.

91 Hall 1987, 26.

92 Chedd 1976, 14.

93 Lublow 1977, 51, cited in Watson and Tooze 1981, 121.

94 Culliton 1976, 300.

95 Chedd 1976, 14.

96 Hall 1987, 47.

97 Hall 1987, 51.

98 Hall 1987, 26.

99 Hall 1987, 47.

100 Arthur Kornberg, cited in Wald 1976; Watson and Tooze 1981, Document 5.9, 112.

101 Hall 1987, 38.

102 Hall 1987, 53.

103 Hall 1987, 53.

104 Johnson and Smith-Hughes 2006, 19.

105 R. Synge, 'Archer Martin (3)' in 'Anecdotes'.

106 Morris 2009, 4.

107 R. Synge, 'Archer Martin (3)' in 'Anecdotes'. A14–15. The papers and correspondence of Richard Laurence Millington Synge. Trinity College Library, Cambridge. GBR/0016/SYNG.

108 R. Synge, 'Archer Martin (3)' in 'Anecdotes'; also Dr. Charlotte Synge, personal communication.

109 Adlard 2002.

110 Johnson 2003, 748.

111 Thanks to my former boss Dr. Tony Balmforth for first suggesting this analogy in to illustrate the challenge of hunting for genes involved in cardiovascular disease.

112 Bell, Pictet, Rutter, Cordell, Tischer, and Goodman 1980.

113 Efstratiadis *et al.* 1975, 1976; Maniatis *et al.* 1976.

114 Hall 1987, 18.

115 Hall 1987, 92.

116 Rutter and Smith-Hughes 1998, 112.

117 Rutter and Smith-Hughes 1998, 116–117.

118 'Cambridge Council Bids Harvard Delay Its Gene Research', *The New York Times*, July 8, 1976.

119 A. Vellucci to P. Handler, May 16, 1977, cited in Watson and Tooze 1981, Document 7.1, 206.

120 Hall 1987, 36.

121 Rutter and Smith-Hughes 1998, 110.

122 Hall 1987, 109.

123 Hall 1987, 110.

124 Ullrich and Smith-Hughes 2006, 9.

125 Ullrich and Smith-Hughes 2006, 10.

126 Ullrich and Smith-Hughes 2006, 6, 9.

127 Rutter and Smith-Hughes 1998, 138.

128 Ullrich and Smith-Hughes 2006, 12.

129 Rutter and Smith-Hughes 1998, 142.

130 Rutter and Smith-Hughes 1998, 142; Stetten, cited in Rutter 1998.

131 Lederberg 2000, cited in Watson and Tooze 1981, Document 3.2, 59.

132 Wade 1977a.

133 Wade 1976b.

134 Wade 1976b.

135 Schmeck 1977c.

136 Wade 1977b)

137 E. Kennedy, September 22, 1977, 'Opening Statement to the Senate Subcommittee on Recombinant DNA Research and the NIH Guidelines', cited in Watson and Tooze 1981; Document 6.2, 144–145.

138 Watson and Tooze 1981; Document 6.2, 144–145.

139 Rutter and Smith-Hughes 1998, 174.

140 Hall 1987, 139.

141 Rutter and Smith-Hughes 1998, 143.

142 Hall 1987, 279.

143 Andreopoulos 1980, 743.

144 Schmeck 1977b.

145 Adams and Ferrell 1977.

146 Hall 1987, 166–167.

147 Wade 1977c.

148 Wade 1977c, 1343.

149 Wade 1977c, 1343.

150 Hopson 1977, 61.

151 Johnson and Smith-Hughes 2006, 19; Hopson 1977, 57.

152 Johnson 2003, 748.

153 'An Overview of the Role of NIH and Other Federal Agencies in the Conduct of Research with Recombinant DNA', US National Academy of Sciences, March 7–9, 1977. The Donald S. Frederickson Papers (Profiles in Science). US National Library of Medicine. Online at https://collections.nlm.nih.gov/catalog/nlm:nlmuid-101584939X127-doc.

154 An Overview of the Role of NIH.

155 Federal Register, July 28, 1978, cited in Watson and Tooze 1981, Document 12.4, 349.

156 Frederickson 2001, 136; Rasmussen 2014, 51.

157 Rutter and Smith-Hughes 1998, 154.

158 Rutter and Smith-Hughes 1998, 154.

159 Rutter and Smith-Hughes 1998, 155.

160 Rutter and Smith-Hughes 1998, 155.

161 Hall 1987, 171; Stevenson, cited in Frederickson 2001, 175—Memorandum from Joseph Hernandez, November 9, 1977, DSF Papers, Folder dna\dna77 \legisl\fed\November.

162 Hall 1987, 159.

163 Hall 1987, 160.

164 Hall 1987, 160–161.

165 Hall 1987, 161.

166 Hall 1987, 177.

167 Hall 1987, 178.

168 Hall 1987, 181.

169 Hall 1987, 189.

170 Hall 1987, 186.

171 Hall 1987, 182.

172 Villa-Komaroff *et al.* 1978.

173 Cited in Hall 1987, 235.

174 'Bacterium Makes Insulin', *Nature* 273 (1978): 485; 'Discoveries and their Uses' Presidential Address to the British Association for the Advancement of Science, September 4–8, 1978; MS Eng 5654/7 D.148; Dorothy Hodgkin Papers Bodleian Library, University of Oxford.

175 Hall 1987, 241–242.

176 Hall 1987, 255.

177 Hall 1987, 256–257.

178 Hall 1987, 260.

179 Hall 1987, 265.

180 Hall 1987, 206.

181 Lilly had actually approached Walter Gilbert, but as he was in the process of setting up a company called Biogen and wished to retain both his academic post at Harvard, along with his position at Biogen, Irving Johnson of Lilly felt that this might cause some difficulties, so no deal was reached; see Johnson and Smith-Hughes 2006, 26.

Chapter 12

1 Goeddel and Levinson 2000.
2 Boyer and Smith-Hughes 2001, 62.
3 Swanson and Smith-Hughes 2001, 13.
4 Swanson and Smith-Hughes 2001, 21.
5 Swanson and Smith-Hughes 2001, 16–17.
6 The brilliant but flawed advertising genius of the TV 1960s period piece drama *Mad Men*.
7 Swanson and Smith-Hughes 2001, 23.
8 Swanson and Smith-Hughes 2001,
9 Nicol and Smith 1960.
10 Khorana et al. 1972.
11 Khorana et al. 1976; Rensberger 1976b; Rensberger 1976a.
12 Heynecker *et al.* 1976.
13 Riggs 2021.
14 Riggs and Smith-Hughes 2006, 35.
15 Swanson and Smith-Hughes 2001, 48.
16 Heynecker and Smith-Hughes 2004, 29.
17 Heynecker and Smith-Hughes 2004, 22.
18 Heynecker and Smith-Hughes 2004, 19.
19 Swanson and Smith-Hughes 2001, 36.
20 Riggs and Smith-Hughes 2006, 41.
21 Riggs and Smith-Hughes 2006, 41; Swanson and Smith-Hughes 2001, 75.
22 Swanson and Smith-Hughes 2001, 36.
23 Riggs and Smith-Hughes 2006, 42–43.
24 Itakura *et al.* 1977.
25 Riggs and Smith-Hughes 2006. 47.
26 Schmeck 1977d; Smith-Hughes 2013, 67; Riggs 2006, 45.
27 Smith-Hughes 2001, 566.
28 Swanson and Smith-Hughes 2001, 43.
29 Ullrich and Smith-Hughes 2006, 27.
30 Ullrich and Smith-Hughes 2006, 28.
31 Ullrich and Smith-Hughes 2006, 29.
32 Boyer and Smith-Hughes 2001, 96.
33 Boyer and Smith-Hughes 2001, 98.
34 Boyer and Smith-Hughes 2001, 98.
35 Swanson and Smith-Hughes 2001, 56.

36 Boyer and Smith-Hughes 2001, 87.

37 Ramamoorthy *et al.* 1976.

38 Kleid and Smith-Hughes 2002, 180.

39 Kleid and Smith-Hughes 2002, 32.

40 Kleid and Smith-Hughes 2002, 25, 34.

41 Kleid and Smith-Hughes 2002, 35, 36.

42 Goeddel and Smith-Hughes 2003, 26.

43 Crea and Smith-Hughes 2004, 25.

44 Crea and Smith-Hughes 2004, 25.

45 Steiner and Oyer 1967; Steiner *et al.* 1967; in 1986, the method of synthesizing the A
    and B chains of human insulin separately in bacteria before reconstituting them was
    superseded when Lilly began using recombinant DNA methods to produce insulin
    from bacteria carrying the human proinsulin gene; see Chance and Frank 1993.

46 Crea and Smith-Hughes 2004, 27.

47 Kleid and Smith-Hughes 2002, 39.

48 Goeddel and Smith-Hughes 2003, 113.

49 Kleid and Smith-Hughes 2002, 42.

50 Heynecker and Smith-Hughes 2004, 73.

51 Goeddel and Smith-Hughes 2003, 113; Heynecker and Smith-Hughes 2004, 74.

52 Crea and Smith-Hughes 2004, 33.

53 Kleid and Smith-Hughes 2002, 46.

54 Kleid and Smith-Hughes 2002, 45; Goeddel and Smith-Hughes 2003, 25.

55 Goeddel and Smith-Hughes 2003, 26.

56 'A Madame Curie From the Bronx', *The New York Times Magazine*, April 9, 1978,
    cited in Straus 1998, 104.

57 Glick 2011.

58 Straus 1998, 135.

59 Berson *et al.* 1956, 172.

60 Yalow 1977, 448.

61 Straus 1998, 13–14.

62 Straus 1998, 14.

63 Straus 1998, 14.

64 Yalow and Berson 1959; Yalow and Berson 1960.

65 Straus, 1998, 11.

66 Yalow 1977, 453.

67 Straus, 1998, 152.

68 Yalow 1977, 465.

69 Yalow 1982, 581.

70 Goeddel and Smith-Hughes 2003, 122.

71 The work was published as two papers in *Proceedings of the National Academy of
    Sciences of the United States of America* at the end of 1978 and the start of 1979: Crea
    et al. 1978; Goeddel *et al.* 1979.

72 Goeddel and Smith-Hughes 2003, 120.

73 Kleid and Smith-Hughes 2002, 54.

74 Kleid and Smith-Hughes 2002, 54.
75 Goeddel and Smith-Hughes 2003, 51.
76 Kleid and Smith-Hughes 2002, 55.
77 Kleid and Smith-Hughes 2002, 52; Heynecker and Smith-Hughes 2004, 133.
78 Lilly used a range of methods including amino acid sequence analysis, immunological testing, X-ray crystallography, and nuclear magnetic resonance spectroscopy to confirm that insulin made using recombinant DNA (rDNA insulin) was identical to human insulin; Johnson 2003, 750.
79 Crea and Smith-Hughes 2004, 60.
80 Kleid and Smith-Hughes 2002, 55.
81 Ullrich and Smith-Hughes 2006, 32.
82 Ullrich and Smith-Hughes 2006, 32.
83 Swanson and Smith-Hughes 2001, 108.
84 Smith-Hughes 2013, 166.
85 Goeddel and Smith-Hughes 2003, 127.
86 Goeddel and Smith-Hughes 2003, 50.
87 Kleid and Smith-Hughes 2002, 56.
88 Kleid and Smith-Hughes 2002, 58.
89 'Insulin Research Raises Debate on DNA Guidelines', *The New York Times*, June 29, 1979.
90 Gilbert 1977.
91 Gilbert 1977.
92 McCabe 1977, cited in Watson and Tooze 1981, Document 6.11, 165.
93 R. Curtiss to D. Frederickson, April 12, 1977. Maxine Singer Papers (Profiles in Science). US National Library of Medicine. Online at https://collections.nlm.nih.gov/catalog/nlm:nlmuid-101584644X132-doc.
94 Culliton 1978, 275.
95 See also P. Berg's to D. Frederickson, December 17, 1979, 'Letter Expressing Confidence in the *E. coli* K12 Strain'. The Paul Berg Papers (Profiles in Science). US National Library of Medicine. Online at https://collections.nlm.nih.gov/catalog/nlm:nlmuid-101584580X197-doc.
96 Chang and Cohen 1977.
97 Culliton 1978, 277.
98 Parisi 1980.
99 Reinhold 1980.
100 Parisi 1980.
101 Smith-Hughes 2001, 544, 570.
102 Smith-Hughes 2001, 570; Smith-Hughes 2013, 124.
103 Johnson 2003, 750.
104 Johnson and Smith-Hughes 2006, 40.
105 'Protect and Survive', 1980 website of the Imperial War Museum. Online at https://www.iwm.org.uk/collections/item/object/1500124311.
106 Crea and Smith-Hughes 2004, 51.
107 Kiley and Smith-Hughes 2001, 8.

108 Kiley and Smith-Hughes 2001, 7.
109 Smith-Hughes 2013, 55–57.
110 Smith-Hughes 2013, 56; Kiley and Smith-Hughes 2001, 9.
111 Smith-Hughes 2013, 149.
112 Greenhouse 1980.
113 Schmeck 1980.
114 Smith-Hughes 2001, 569.
115 Smith-Hughes 2001, 570.
116 Keen *et al.* 1980, 398.
117 Swanson and Smith-Hughes 2001, 100–101.
118 Swanson and Smith-Hughes 2001, 101.
119 Swanson and Smith-Hughes 2001, 110.
120 Goeddel and Smith-Hughes 2003, 61.
121 Hargreaves 1980a.
122 Rowe 1980.
123 Hargreaves 1980b.
124 Metz 1980.
125 The Economist 1980a.
126 Metz 1980.
127 The Economist 1980b.
128 The Economist 1980b.
129 The Economist 1980b.
130 The Economist 1980b.
131 The Economist 1980b.
132 Ullrich and Smith-Hughes 2006, 38.
133 Smith-Hughes 2013, 158.
134 Smith-Hughes 2013, 159.
135 Alberti 2001, 34.
136 Alberti 2001, 34.
137 Betlach and Smith-Hughes 2002, 32–33.
138 Watson and Berry 2003, 104.

Chapter 13

1 The cover can be seen at https://content.time.com/time/covers/0,16641,1981 0309,00.html.
2 Smith-Hughes 2013, 161.
3 Smith-Hughes 2013, 161.
4 Crea and Smith-Hughes 2004, 51.
5 Clark *et al.* 1982.
6 Clark *et al.* 1982, 354.
7 Smith Hughes 2013, 125.
8 Chase 1982.

9 Chase 1982.

10 Chase 1982.

11 Chase 1982.

12 Lawrence 1925.

13 Steiner and Over 1967.

14 Schlichtkrull 1974; Root, Chance, and Galloway 1972.

15 Tattersall 2009, 143.

16 Ruttenberg 1972.

17 Morihara, Oka, and Tsuzuki 1979.

18 Markussen *et al.* 1981, 302.

19 Szekeres *et al.* 1983.

20 Szekeres *et al.* 1983.

21 Szekeres *et al.* 1983.

22 Tattersall 2009, 149.

23 Pickup 1986, 156.

24 Teuscher and Berger 1987b.

25 Teuscher and Berger 1987b.

26 Teuscher and Berger 1987a.

27 Teuscher and Berger 1987a, 385.

28 Clark, Knight, Wiles, and Scotton 1982.

29 Lesser 1989.

30 Ballantyne 1989.

31 Pallot 1989.

32 Pallot 1989; 'Diabetics Warned After Rise in Deaths' *The Daily Telegraph*, October 13, 1989.

33 Lowe 1989.

34 Lowe 1989.

35 Richards 1989.

36 Lesser 1989.

37 Tattersall 1989.

38 Prentice 1989.

39 Egger, Smith, Teuscher, *et al.* 1991; Egger, Smith, Imhoof, *et al.* 1991.

40 Mihill 1991.

41 Ferriman 1991.

42 Ferriman 1991.

43 Clarke and Foster 2012.

44 Lesser 1989.

45 Wolff 1991.

46 Tattersall 1989.

47 Tattersall and Gill 1991, 49.

48 Chase 1982.

49 Johnson 2003, 750.

50 Hall 1987, 304.

51 Haycock cited in Hall 1987, 304.

52 Hall 1987, 304; Rasmussen 2014, 40.

53 Diabetes UK, *Balance*, Summer Issue 2019, 62.

54 Tattersall 2009, 148.

55 Rasmussen 2014, 68.

56 Hall 1987, 53.

57 Rasmussen 2014, 52.

58 P. Berg, 'Testimony to the U.S. Subcommittee on Science, Technology and Space, November 2, 1977. The Paul Berg Papers (Profiles in Science), US National Library of Medicine. Online at https://collections.nlm.nih.gov/catalog/nlm:nlmuid-101584580X193-doc.

59 Chargaff 1978, 120, 206.

60 Chargaff 1977, 51.

61 Chargaff 1977, 27–28.

62 Chargaff 1978, 107.

63 Chargaff 1977, 14.

64 Chargaff 1977, 72–73.

65 D. Hodgkin, 'Discoveries and their Uses', Presidential Address to the British Association for the Advancement of Science, September 4–8, 1978; MS Eng 5654/7 D.148; DHPBL UO.

66 D. Hodgkin, 'Discoveries and their Uses'.

67 R. Synge, cited in 'New Elizabethan', *The Courier*, April 1953. A167–194. PCRLMS TCL UC.

68 Chargaff 1978, 122.

69 Synge 1964, 217.

70 Synge 1964, 220.

71 Chargaff 1978, 203.

72 Chargaff 1978, 55.

73 Chargaff 1977, 32.

74 Chargaff 1977, 165.

75 Chargaff 1978, 55.

76 Chargaff 1978, 204.

77 Chargaff 1976, 938.

78 Scott and Fisher 1936; the addition of zinc ions stabilizes the formation of insulin hexamers.

79 Lawrence 1939.

80 Lawrence 1939.

81 Wilder 1958, 266.

82 Himsworth 1937, 544.

83 Himsworth 1937, 544.

84 Hallas-Møller *et al.* 1952; Hallas-Møller, Petersen, and Schlichtkrull 1952.

85 'Lente' was actually a mixture of a form of long-acting crystalline zinc insulin called 'ultra-Lente' and amorphous zinc insulin, known as 'semi-Lente'. Lawrence and Oakley 1953; Wauchope, Oakley, and Grunberg 1953; Oakley 1953; Nabarro and Stowers 1953; Murray and Wilson 1953; Oakley 1953.

86 Tattersall 2019, 149.

87 Fletcher 1980, 1115.
88 Blundell *et al.* 1972.
89 Tattersall 2009, 175.
90 Blundell et al. 1972.
91 Brange *et al.* 1988.
92 Bakh *et al.* 2017.
93 Howey *et al.* 1994; Torlone et al. 1994.
94 The secret of Humalog's faster action was the inversion of the amino acids proline and lysine at positions 28 and 29, respectively, in the B chain of insulin, whereas in Novolog, proline at position 28 was substituted by the negatively charged amino acid aspartic acid. Both of these types of modification disrupted the association of insulin into dimers.
95 Tattersall 2009, 176.
96 Safavi-Hemani *et al.* 2015.
97 Xiong *et al.* 2020.
98 Two arginine residues were added to the C terminus of the B chain, while asparagine at position 21 of the A chain was substituted with glycine.
99 Havelund *et al.* 2004.
100 Holleman and Gale 2007.
101 Holleman and Gale 2007, 1783.
102 Grill 2006.
103 Gill *et al.* 2011, 21.
104 Gill *et al.* 2011, 21.
105 Zaykov, Mayer, and DiMarchi 2016; Edgerton *et al.* 2014.
106 Jarosinski *et al.* 2021.
107 Brownlee and Cerami 1979.
108 Kaarsholm *et al.* 2018; Bakh *et al.* 2017; Jarosinski *et al.* 2021.
109 Menting *et al.* 2013, Menting *et al.* 2014; Scapin *et al.* 2018.
110 Owens 2018; Cookson 2018.
111 Gill *et al.* 2011, 21.
112 Beran, Yudkin, and de Courten 2005; Beran and Yudkin 2006.
113 Yudkin, cited in Boseley 2003.
114 Yudkin and Beran 2003.
115 Gill et al., 2011, 20.
116 Gill et al. 2017, 324.
117 O'Hara 2019.
118 Pavia 2019.
119 Pavia 2019.
120 Pavia 2019.
121 Beran and Yudkin 2013, 195.
122 Yudkin 2000, 921.
123 Beran & Yudkin 2013, 195.
124 Beran *et al.* 2012.
125 Gill et al. 2017, 234.

126 Yudkin 2000, 921.

127 Beran & Yudkin 2006, 1691.

128 Beran, Yudkin, and de Courten 2005, 2136.

129 Feudtner 2003, 4.

130 Feudtner 2003, 9.

131 Feudtner 2003, 40.

132 Feudtner 2003, 39.

133 Feudtner 2003, 10.

134 Collins 1923.

135 Feudtner 2003.

136 Feudtner 2003, 9.

137 Feudtner 2003, xv.

138 Feudtner 2003, xvi.

139 Heller 2019, 1008.

140 Joslin, Gray, and Root 1922, 652.

141 Banting, F.G. (1923); 445.

142 'Doctors suggest Covid-19 could cause diabetes: More than 350 clinicians report suspicions of Covid-induced diabetes, both type 1 and type 2.' N. Grover, The Guardian, 19th March 2021.

143 Professor David Nabarro interviewed on BBC Radio 4 'Today', 15th May 2020

144 Gilbert and Green 2021, p.277.

145 See the lecture by Professor Paul Kellam, Professor of Virus Genomics, Imperial College London, April 2021, https://www.youtube.com/watch?v=dYSTfr9T_5c

146 Grubaugh, Hodcroft, Fauver, Phelan and Cevik 2021: 1130,1131.

147 Gilbert and Green 2021, p.276.

Index

Note: Figures are indicated by an italic *f* following the page number.

Numbers
100 campaign 346

A
Abel, John 137, 139, 140, 141, 143
acetone, use as solvent 53
Ackermann, Barbara 276–7
Acomatol 93, 94–8
adenine 228, 229*f*
alanine 138*f*
Alberti, George 326
alcohol, use as solvent 40–1, 47, 116, 120, 122, 124
Alexander, Hattie 223–4
alkaline injections, Murlin's research 59, 60
Allen, Frederick Madison 11, 12, 15, 57–8, 84
American Physiological Society, Banting's presentation, 1921 41–2
amino acids 137, 138*f*, 141
  chain formation 143, 144*f*
  sequences 162–5, 168 *see also* protein structure
analogue insulins *see* insulin analogues
antibiotic resistance 255
antibiotics 166–7
  gramicidin S 167, 189, 203, 205*f*
  penicillin, determination of structure 173–5
anti-Semitism 132–3
Anti-Vivisection Society, opposition to Banting's work 71–2
Arbers, Silvia 261
Arbers, Werner 257
Aretaeus of Cappadocia 4
arginine 138*f*
Arrhenius 247
artificial DNA 301–2
  Riggs and Itakura's work 302
Asilomar meeting 252–4, 268–71, 270*f*

Askonas, Brigitte 206
asparagine 138*f*
aspartic acid 138*f*
Association of American Physicians, Macleod's presentation, 1922 50–1
Astbury, William 143, 145–7, 145*f*, 160–1, 199
  model of protein structure 163*f*
  work on DNA 225
  work on insulin 188–9
availability of insulin 78–9, 345–7
  100 campaign 346
Avery, Oswald 223–5, 223*f*, 247
  Chargaff's support for 231

B
bacteriophages (phages) 239, 256, 257
  recombinant DNA research 258
Bagenal, Judith *see* Martin, Judith
Baltimore, David 268
Banting, Fred 2, 14*f*, 15, 27*f*, 56
  alcohol, use as solvent 40–1
  alcohol consumption 50
  APS presentation, 1921 41–2
  artistic abilities 21
  audience with King George V 86–7
  Barron's paper, impact of 22
  on Best's contribution 68, 85
  BMJ address 80
  Canadian Parliament annuity award 84–5
  cancer research 105
  Dale's defence of 70
  death and funeral 107–8, 109*f*

dogs, treatment of 29, 38
early life 17
early medical career 20–2
early research success 30
first human trial of pancreatic extract 44–5, 49
foetal tissue, use of 40
and insulin patent 59, 63–4
knighthood 103
lab notebook extracts 37*f*
letter from Paulescu 128
Macleod's *Toronto Star* interview 48
medical training 18–19, 20
Murlin's claims, reaction to 60–1
Nobel Prize 16–17, 87–8, 102–3
opposition to work 71–2
patients' testimonies 72
press, contempt for 87
Professorial Chair in Medicine 84
reaction to fame 103–4, 106
relationship with Best 26, 27–8, 102, 104
relationship with Collip 48–50, 111–12
relationship with Macleod 22–4, 38–40, 42–3, 49–50, 66–7, 68, 85
research plan 25*f*
research problems 28–9, 35
Roberts' criticism of research 69–70
tributes to 84–7
victory over Murlin 65
wartime work 18, 19–20, 106–7
Washington AAP presentation, absence from 51
whole pancreas extracts 36
working conditions 26–7, 31*f*

Banting Research
   Institute 103
Barlow murder case 57
Barnett, Anthony 332–3
Barron, Moses 22
basal insulin 340
   insulin analogues 343
Bayliss, William 65–6
Beijing University, work on
   insulin synthesis 215
Bell, Florence 226–7, 226*f*,
   227*f*
Beran, D. 346
Berg, Paul 255–6, 255*f*,
   267–8, 273, 335
   Asilomar meeting 268–71,
      270*f*
   Nobel Prize award 261,
      327
   recombinant DNA
      research 258–61, 259*f*
Berger, Willi G. 330
Bergmann, Max 143–4, 146,
   162
Bernal, John Desmond 171,
   184, 209
Bernard, Claude 6, 9
Berson, Solomon 313
   insulin antibody
      theory 314
   insulin
      radioimmunoassay 315
Best, Charles 26*f*, 27*f*, 68,
   85, 112, 140, 165
   AAP presentation, absence
      from 51
   accolades and honours 118
   account of discovery of
      insulin 116–17, 119,
      124, 134–5
   Chair of Physiology 104
   Dale's defence of 70
   depression 135
   on early production
      problems 123–4
   early research success 30
   and insulin patent 59
   involvement in proposed
      film 120–2
   lack of acknowledgment
      of 102, 104
   Macleod's *Toronto Star*
      interview 48
   *The Physiological Basis of
      Medical Practice* 113–16,
      115*f*
   relationship with
      Banting 27–8

relationship with
   Collip 112–13
research problems 28–9
Roberts' criticism of
   research 69–70
share of Nobel Prize
   money 102–3
translation of Paulescu's
   letter 128–9
working conditions 26–7,
   31*f*
Best, George 116
beta-lactam structure 173–4
Betlach, Mary 286, 288,
   326–7
biological weapons 107
biotechnology com-
   panies 328 *see*
   also Genentech
Bird, William 107
Bliss, Michael 117
blood sugar monitoring 333
Blum, Ferdinand 91
Blum, Leon 78
Blundell, Tom 211
Boivin, Andre 223–4, 237
Bolivar, Francisco 306*f*
Bolivar, Paco 286
Bouchardat, Apollinaire 11
Bourne, Harold 56–7
Boyer, Herb 257, 286, 290,
   298, 306*f*, 325, 326, 328
   appearance before Senate
      committee 292
   Asilomar meeting 252–4
   collaboration with
      Cohen 262–3, 264–5
   involvement with
      Genentech 307–8
   Nobel Prize, lack of 326–7
   partnership with Swan-
      son 299–300, 301*f see
      also* Genentech
   patenting of work 265–6
   recombinant DNA
      research 262–3
Bragg, Lawrence 170, 176,
   184
Bragg, William 154, 170–1
Brenner, Sydney 239–40,
   270*f*
   Asilomar meeting 270–1,
      270*f*
British Drug Houses (BDH),
   insulin production 76,
   78
Brownlee, Michael 345

Brown-Séquard,
   Charles-Edouard 7–9

C

California Institute of
   Technology (CalTech),
   Abel's work 139
Calvin, Melvin 199
   Nobel Prize award 200
Cambridge Experimentation
   Review Board 284
'Cambridge incident,'
   recombinant DNA
   technology 276–80
Cambridge University
   Bernal and Hodgkin's
      work 171
   Sanger's work 202–7
   Todd's work 175
   Watson and Crick's
      work 228–30, 232–4
Canadian National Exhibition
   (CNE) 85
Canadian Parliament, annuity
   award to Banting 84–5
cancer research, Banting 105
Carlisle, Harry 172–3
Carrasco-Forminguera,
   Rosendo 99
casting waters 5
Cerami, Anthony 345
Chain, Ernst 166–7, 173
Chakrabarty, Ananda 322–3
Chamberlain, Neville 78
Chang, S.N. 321
Charaka 5
Chargaff, Erwin 230–2, 230*f*,
   240, 321
   and base pairing 233–4
   disillusionment with
      science 247–51, 335–6,
      337–8
   meeting with Watson and
      Crick 232–3
   on recombinant DNA
      technology 269, 273,
      275
Chargaff's rule 234
chemical weapons 106–7
Chen-Lu Tsou 215, 216
Chevreul, Michel 6
Chibnall, Albert C. 161–2,
   199, 202
Chick, William 284
China
   Cultural
      Revolution 215–16
   Great Leap Forward 215

work on insulin
synthesis 214–17
Chirgwin, John 285
cholesterol, Hodgkin's
work 172–3
chromatography 152
gas–liquid 197 *see
also* partition
chromatography
Churchill, Winston 195
Clark, Alfred 77
cloning
Boyer and Cohen's
work 262–6 *see
also* recombinant DNA
technology
Clostridium welchii 166
Clowes, George 43, 53, 61,
63, 74
patenting of insulin 58–9
Coburn, Alvin 225
codons 246*f*
Codreanu, Corneliu
Zelea 133
Cohen, Stanley N. 253*f*,
274–5
Asilomar meeting 252–4
collaboration with
Boyer 262–3, 264–5
on natural genetic material
exchange 321
patenting of work 265–6
plasmid research 255
recombinant DNA
research 262–3
Cold Spring Harbor sympo-
sium, 1949 195, 203,
205
Cold War 195, 322
Collip, James Bertram 45–7,
46*f*, 68–9, 74, 86, 116
attempted election to Royal
Society 110–11
concerns over proposed
film 121
experiments on
hypoglycaemia 54–5
and insulin patent 59
insulin production
problems 52
lack of acknowledgment
of 116–18, 119, 120–1,
124
modesty 113
obituary for Macleod 108
purification of pancreatic
extract 47

relationship with
Banting 48–50, 111–12
relationship with
Best 112–13
share of Nobel Prize
money 109–10, 111
Connaught Anti-Toxin
laboratory 49, 52
Consden, Raphael 159
Conus geographicus (cone sea
snail) 343
Cornell University Medical
College
Murlin's research 59
Sumner's work 142
Cornforth, John 173, 174,
192
cost of insulin 346
early production 76–7, 78
Humulin 328
Covid-19 275, 349
Crea, Roberto 306*f*, 309,
310*f*, 316
DNA fragment
synthesis 311
Crick, Francis 206, 220,
228–30
on base pairing 233–4
collaboration with
Gamow 235–6
commemorative plaque,
Eagle pub 219*f*
Directed Panspermia
hypothesis 247
meeting with
Chargaff 232–3
Sequence Hypothesis 243
work on genetic
code 241–2
work on mRNA 239–40
crystallization of insulin 139,
141
Cultural Revolution,
China 215–16
Curtiss, Roy 293, 320–1
Cuza, Alexandru C. 132
cyclol theory of protein
structure 181–4, 182*f*,
187–8
Pauling's criticism 184–6
cysteine 138*f*
cytosine 228, 229*f*

D

DAFNE (Dose-Adjusted-
For-Normal-Eating) 82
Dale, Henry 69, 70–1, 103,
118–19, 224

Davis, Anthony 345
Davis, Bernard 275
de Meyer, Jean 10, 51
Department of Scientific
and Industrial Research
(DSIR) 153
Destecroix, Harry 345
detemir 343
diabetes 80
causes 10–11
complications 2, 5, 347
dietary treatments 5,
11–15, 44
discovery of cause 6–7
early descriptions of 4–5
effects on the body 2, 11
forms of 10
origins of the term 4
personal experiences 1–2,
80
type 2 313–14
Diabetic Association 83
dimer formation 341, 342*f*
Directed Panspermia
hypothesis 247
Directo, Leonore 306*f*
DNA (deoxyribonucleic
acid)
artificial synthesis 301–2
Astbury's work 225
Avery's work 223–4
base pairing 228–30, 229*f*,
233–4
Bell's work 226
Chargaff's work 231–3
discovery by
Miescher 220–2
Franklin's work 228
function, early ideas 222
structure, discovery
of 226, 228, 234 *see
also* genetic code
Dobson, Matthew 5
dogs, use by Banting and Best
25*f*, 27*f*, 28–31
Dubos, Rene 224, 225
Dudley, Harold Ward 75
Dunitz, Jack 211
Dussoix, Daisy 257
Dyson, Freeman 274

E

Eagle, The (Cambridge
pub) 219–20
commemorative
plaque 219*f*
Eastwood, Jack 81–2
Ebers papyrus 4

Ednam, Viscount 78
Efstrtiadis, Argiris 281, 293, 296–7
Eli Lilly 43, 53, 74, 298
  agreement with Genentech 316–17, 318, 319
  development of human insulin 281
  Speke production facility 322
Elsden, Sydney 167
enzymes 140–1
  as proteins 142, 143–4
  urease 142
Ephrussi, Boris 241
Escherichia coli (*E.coli*)
  enfeebled strains 293, 320–1
  PaJaMo experiments 238–9
  use in recombinant DNA research 258, 262, 263, 269, 271, 273, 275, 293
Escherichia coli Restriction Endonuclease I (EcoRI) 257, 261–2
ethical issues
  Cohen–Boyer patent 266–8
  recombinant DNA technology 253, 263–4, 266–8, 272–6

F

Falconer, Robert 72, 89
false hopes 73–4, 75, 76
Fara, Patricia 176
Feasby, W.R. 119, 120, 121, 122, 124
Feudtner, Chris 347–8
Feynmann, Richard 240
fibroin, X-ray studies 143
film proposal, NFB of Canada 119–22
First World War
  gas gangrene 166
  impact on Kleiner's work 33–4
  impact on Murlin's work 59
  impact on Paulescu's work 127
  impact on Zuelzer's work 94
Fischer, Emil 143
fish as source of insulin 67–8, 93

Fitzgerald, John 16, 111
Fletcher, Charles 55, 340
Flexner, Simon 34–5
Fliflet, A.T. 195–6
Florey, Howard 166–7
fluorodinitrobenzene (FDNB) 202, 203, 204*f*
foetal tissue, use by Banting and Best 40
Ford, Gerald 287
Forschbach, Joseph 93
Fourier, Jean-Baptiste Joseph 172
Fourier transforms 172, 174
Fox, Barclay 154
Fox, Sydney 199
Franklin, Rosalind 228
Fraser, Thomas 67
Frederickson, Donald 275, 322
Friends of the Earth 273, 291
Fuller, Forrest 279, 293–4

G

Gabriel, Vivian 78
Gamow, George
  background 235
  collaboration with Watson and Crick 235–6
  model of genetic code 236–7, 237*f*, 241
  RNA Tie Club 240–1
gas gangrene 166
gas–liquid chromatography (GC) 197
gelatine, structure of 145, 146
gel electrophoresis 262–3
Genentech 135
  agreement with Eli Lilly 316–17, 318, 319
  attempted recruitment of Ullrich 306–7
  company philosophy 308
  criticisms of 307–8
  culture 319
  foundation 300
  human insulin, choice as a product 300–1
  human insulin cloning strategy 303*f*
  human insulin DNA sequence synthesis 311–12
  human insulin production 319–20

  human insulin synthesis 316
  initial public offering 324–5
  intellectual property rights 323
  legal disputes with UCSF 324
  press conference announcement 317
  recruitment of Kleid and Goeddel 308–9
  recruitment of Ullrich and Seeburg 318
  San Francisco site 306
  share prices 325, 328
  somatostatin synthesis 304–6
gene regulation 238
genetic code 234, 246*f*
  Crick's Sequence Hypothesis 243
  Crick's work 241–2
  degenerate nature of 245, 247
  Gamow's model 236–7, 237*f*
  Nirenberg's work 244–5
  RNA Tie Club 240–1
  triplet nature of 245
  universality of 247
genetic engineering
  Chargraff's fears 249–51 *see also* recombinant DNA technology
George V, King 86–7
Geyelin, Henry Rawle 90
Gilbert, Walter 278, 279, 279–80, 279*f*, 335
  Nobel Prize award 326
  work at Porton Down 296–7
  work on human insulin synthesis 282*f*, 283, 284–5, 293–4
  work on mRNA 240
Gilchrist, Joe 44, 71
Gill, G.V. 334
glargine 343
Gley, Eugene 65
glucose 10
Glusker, Jenny (nee Pickworth) 175
glutamic acid 138*f*
glutamine 138*f*
glycine 137, 138*f*
glycogen 6

glycoproteins, Synge's work 149
Goeddel, David 308–9, 310*f*, 319
  human insulin synthesis 316
  insulin DNA sequence synthesis 311–12
Gooderham, Albert 16–17, 68–9
Goodman, Howard 283, 318
  work on human insulin synthesis 285
Gorbach, Sherwood L. 321
Gordon, Hugh 159
Gordon, Neil E. 254
Gosling, Raymond 228
Graham, Duncan 44
gramicidin S 167, 189
  amino acid sequence 205*f*
  Sanger's work 203
'Great Leap Forward,' China 214–15
Great War Veterans Association (GWVA), opposition to Banting's work 72
greenhouse effect 192–3
Greenseid, Lija 346
Griffiths, John 230
guanine 228, 229*f*
Gull, William 11

H

haemoglobin 143
  Ingram's work 242–3
Hagedorn, Hans Christian 178, 179, 338
Hambling, Maggi, portrait of Dorothy Hodgkin 176, 177*f*
Harvard, work on human insulin synthesis 283–5, 293–4
Haycock, Paul 334
Hédon, Édouard 7
Heidenhain, Rudolf 46–7
Henderson, Velvyn 39
Hetzel, K.S. 78–9
hexamer formation 341, 342*f*
  glargine 343
Heynecker, Herb 304, 306*f*, 312, 319
Hill, C.A. 78
Hirose, Tadaki 306*f*
histidine 138*f*

Hodgkin, Dorothy Crowfoot 169*f*, 207, 211*f*
  background 170–1
  on discovery of insulin 212
  and Margaret Thatcher 191–2
  Nobel Prize 176
  political stance 191–2
  portraits of 176, 177*f*
  on purpose of science 336–7
  relationship with Wrinch 186–8
  studies of cholesterol 172–3
  studies of gramicidin S 189
  studies of insulin 177–8, 181, 183, 207, 209–12
  studies of penicillin 173–4
  studies of pepsin 171
  studies of vitamin B12 175–6
  support for Synge 196
  visits to China 215, 217
  working conditions 170
Hofmeister, Franz 143
Hollerith machine 174
Home, Francis 5
Hopkins, Gowland 201–2
Hoppe-Seyler, Felix 220, 221
Hopson, Janet 290
Horder, Thomas 85
hormones 7–8, 140
Hubbard, Ruth 276–7
Hughes, Charles 64
Hughes, Elizabeth 51
human insulin
  advantages of 329–30, 334
  clinical trials 323–4
  cost 328
  FDA approval 328
  and impaired hypoglycaemia awareness 330–4
human insulin synthesis 3, 281
  beginning of mass production 322
  Genentech 316, 319–20
  Genentech's announcement of 317
  Genentech's strategy 303*f*

Gilbert's team 283, 284–5, 293–8
  media coverage of 288
  motivation for 334–5
  and NIH guidelines 319–21, 322
  patenting of UCSF work 288
  UCSF team's work 285–6, 288–90, 297–8
'humanized' insulin 329–30
Humphrey, John 148
Humulin 328
hypoglycaemia
  clinical uses 55–7
  early research 54–5
  effects 55
  human insulin safety concerns 330–4
  and protamine insulins 339
  Zuelzer's pancreatic extracts 94, 96

I

immune responses to insulin 329
industrialization of science 248–9, 335, 338
Ingram, Vernon 242–3
injection site reactions 329
institutionalization of science 248–9
insulin
  actions of 10
  administration, responsibility for 79
  as agent of murder 57
  determination of chemical nature 139–40
  determination of structure 177–8
  discovery of 71
  early availability 78–9
  early production problems 52–3, 123–4
  initial cost 76–7, 78
  naming of 10, 51
  patenting 58–9
  purification 74
  standardisation 74, 75
  supply problems 57–8, 71, 280, 334
  treatment challenges 79–80 *see also* human insulin

insulin analogues 342*f*
  availability 345–7
  basal insulins 343
  doubts over benefits 343–4
  prandial insulins 341, 343
  single-chain insulins 344
  uptake of 344
insulin antibodies, Yalow's
  work 314–15
insulin coma therapy
  (ICT) 55–7
Insulin Committee 57–8
  Murlin's letter to 63
insulin dimer and hexamer
  formation 341, 342*f*
  glargine 343
insulin DNA sequence
  synthesis 311–12
insulin gene sequencing 281
insulinoma cells 284
insulin radioimmunoas-
  say 315–16
insulin regimens 340
insulin replacement therapies
  (IRTs) 344–5
insulin structure 310
  Astbury's work 188–9
  Chibnall's work 202
  Hodgkin's work 207,
    209–12
  Sanger's work 202–7,
    204*f*, 208*f*
  Wrinch's work 181–6
insulin synthesis
  Chinese research 214–17
  clinical benefits 217
  Katsoyannis's work 213
  Zahn's work 213–14 *see
    also* human insulin
    synthesis
intellectual proper-
  ty rights 323 *see
    also* patents
International Diabetes
  Federation (IDF)
  Pavel's pleas on behalf of
    Paulescu 130–2
  reaction to Paulescu's
    anti-Semitism 133–4
introns 281
Iron Guard, The 133
Islets of Langerhans 9–10
isoelectric point 74
isoleucine 138*f*
Itakura, Keiichi 302, 306*f*,
  310*f*
  DNA fragment
    synthesis 311

  work on somatostatin
    synthesis 304–6

J
Jacob, Francois 238–40
James, Tony 197
Jeffrey, Ed 44
Jensen, Hans Friedrich 140,
  144
Jensen, Norman 179
Johnson, Irving 280, 281,
  290–2, 334
Joslin, Elliott P. 13–15, 13*f*,
  41, 57, 71, 72–3, 73–4,
  347, 348–9
  advocacy of
    self-management 81, 82

K
Kalckar, Hermann 224
Katsoyannis, Panayotis 213
Keen, Harry 323
Kendrew, John 209, 215
Kennedy, Edward 287, 321
keratin, X-ray studies 143
ketones 2, 11
Khorana, Har Gobind 245
  artificial DNA
    synthesis 301–2
Kiley, Tom 323, 324
Kleid, Dennis 317
  human insulin
    synthesis 319
  insulin DNA sequence
    synthesis 311–12
  recruitment by
    Genentech 308–9
Kleiner, Israel 32–5, 33*f*
Knight, B.C.J.G. 166–7
Kornberg, Arthur 255, 279
Kramer, Benjamin 59
Kravkov, Nikolai 131
Krogh, August 178–9
Kuhn, Thomas 113
Kuznetsov, Anatoly 131

L
Lancereaux, Etienne 126
Langerhans, Paul 9–10
Langmuir, Irving 180*f*, 183,
  184
Laron, Zvi 134
Lawrence, R.D. 82–3, 329,
  339
Lazarus, Norman 212
League of National Christian
  Defence (LNCD) 132
Leder, Philip 245

Lederberg, Joshua 269, 279,
  286–7
Leeds 154, 155
  Martin and Synge's
    laboratory 156*f*
  Martin and Synge's
    work 155–65
  Wool Industries Research
    Association 153–4
Leeds University, Astbury's
  work 145–7
Legion of the Archangel
  Michael, The ('The
  Legion') 133
Lente insulins 340
leucine 138*f*
Levine, Rachnmiel 334
Leyton, Otto 55, 76–7
Liebrich, Oscar 221
ligases 304
lispro 341
Lobban, Pete 258
Loch Ness monster 197–8
long-acting insulins 339–40
Lovelock, James 197
Low, Barbara 173, 174,
  189–90
lysine 138*f*
lysozyme 242

M
MacCallum, Archibald
  B. 110–11
Macfarlane, Leslie 120
Mackey, Joseph 107–8
Macleod, Colin 223
Macleod, J.J.R. 6–7, 17*f*, 140
  AAP presentation,
    1922 50–1
  APS presentation,
    1921 41–2
  correspondence with
    Zuelzer 98–9
  early publication,
    justification of 52
  final days 108
  Insulin Committee 57–8
  interview with *Toronto
    Star* 48
  monkfish extract
    preparations 67–8
  naming of insulin 51
  negotiations with Clowes
    and Eli Lilly 53–4
  Nobel Prize award 16–17
  Nobel Prize money,
    sharing of 109–10, 111

patenting of insulin 58–9, 71
relationship with Banting 22–4, 38–40, 42–3, 49–50, 66–7, 68–9
supervision of Banting & Best 30–2
support of Collip 110–11
William Bayliss's claims 65–6
Mao Zedong 214, 215
Mark, Hermann 143
Martin, Archer 2, 150*f*, 231, 280–1
  amino acid sequencing 162–5
  background 150–1
  depression 151–2
  domestic life 157–8, 168
  mention in *The Simpsons* 197
  move to Leeds 154
  Nobel Prize 198–9, 198*f*
  partition chromatography 158–60
  political stance 193–4
  problems with chromatography 168–9
  Sanger's praise for 202
  separating machines 152, 153
  wool protein structure model 163*f*
  work on chemical separation 152–3
  work on gramicidin S 167
  work on wool 155–62
Martin, Judith (nee Bagenal) 158, 168
Matthaei, Heinrich 245
Mc Carty, MacLyn 223
Mead, Margaret 292
Medical Research Council (MRC)
  control of insulin use 75–8
  license for insulin production 74–5
Meltzer, Samuel 32, 34–5
Mendel, Lafayette 34
Mertz, Janet 258, 261
  work on EcoRI 261–2
Meselson, Mat 240, 244, 257
messenger RNA (mRNA) 239–40, 244*f*, 283
  Nirenberg's work 245
  sources of 284, 285

metformin 10
methionine 138*f*
Meyer, Kurt 143
Miescher, Johann Friedrich 220–2
Miller, F.R. 22
Minkowski, Oskar 7, 93, 95, 99
Minot, George 73
Miozarri, Giuseppe 319
Mirsky, Alfred 223
Mirsky, I. Arthur 313–14
molecular biology
  Chargraff's disillusionment with 247–51
  Sequence Hypothesis 243
monkfish as source of insulin 67–8
Monocompetent (MC) insulin 329
Monod, Jacques 238–9
moon landings 210–11
Moore, Henry 176
Moore, Stanford 198–9
morphine dependency, insulin coma therapy 55
Morrow, John 264
murder, insulin as agent of 57
Murlin, John 65, 101–2, 128, 140
  alkali injections 59
  claims over pancreatic extract use 60–1
  connection to Wilson and Company 61, 63
  intention to produce insulin 61–3
  newspaper clipping 62*f*
  patenting challenge 63–4
Murphy, William P. 73
Murray, George 9
Murray, Ian 125
mustard gas antidote 106–7
myxoedema 9

N

Nabarro, David 349
Nathan, David M. 328
Nathans, Daniel 257
national health insurance, Panel system 77
National Institute for Medical Research (NIMR) 75
National Institutes for Health (NIH), rDNA technology guidelines 271–2,

291–2, 296, 319–21, 322
  breach of 289–90, 292
Neel, James 242
Nelson, Lydia 339
Niemann, Carl 146, 162
ninhydrin 159
Nirenberg, Marshall 244–5, 246*f*
Nixon, J.A. 76
Nobel Prizes
  Alberti's view 326
  Avery's omission 224–5
  Banting and Macleod 16–17, 87, 88*f*, 89–90, 94–7, 127–8
  Berg 261
  Boyer's omission 326–7
  Calvin 200
  Gilbert and Sanger (1980) 326
  Hodgkin 176
  Martin and Synge 198–9, 198*f*
  Nirenberg 245
  Perutz and Kendrew 209
  Sanger (1958) 206–7
  sharing of prize money 102–3, 109–10, 111
  Yalow 315
Noble, Clark 54
Northrop, John 142
Novo Nordisk 329
  legal action against 332
nuclein, Miescher's discovery of 220–2
nucleotide bases 228–30, 229*f*
  Chargaff's work 231–4
  genetic code 234–7

O

Ochoa, Severo 245
Opie, Eugene 10
Orgel, Leslie 247
  RNA Tie Club 240–1
Oughton, Beryl 189
Oxford University, Hodgkin's work 171–8, 209–12

P

PaJaMo experiments 238–9
pancreas
  duct ligation research 25*f*, 28–9
  early experiments on 7
  effect of duct blockage 22

pancreas (*Continued*)
  functions of 9–10
  Islets of Langerhans 9–10
pancreatic extracts
  acetone as solvent 53
  alcohol as solvent 40–1, 47
  bovine 40–1
  degenerated 24–5, 29, 114
  first human trial 44–5, 49
  foetal tissue 40
  whole pancreas extracts 36
pancreine 127, 131
pancreotoxine 131
Panel system 77
Pardee, Arthur 238–9
partition chromatog-
    raphy 158–60,
    160*f*
  Nobel Prize award 198–9,
    198*f*
  practical problems 168–9
  studies of nucleic
    acids 231–3
  use by Calvin 199–200
  use by Sanger 200, 202,
    205–6, 207
  use of potato starch 167
patents 58–9, 71
  Acomatol 97–8
  Chakrabarty case 322–3
  on cloning of coding
    sequence for rat
    insulin 288
  Cohen–Boyer
    patent 265–6, 323
  Murlin's challenge 63–4
  Zuelzer 91
patient education 82, 83 *see
    also* self-management
Patterson, Arthur Lindo 172
Paulescu, Nicolai 125, 126*f*,
    212, 225
  anti-Semitism 132–4
  background 126–7
  call for justice 129
  letter to Banting 128
  Pavel's support for 129–32
  publications 127
  reaction to Nobel Prize
    award 127–8
Pauling, Linus 242
  criticism of Wrinch's
    work 184–6
Pavel, Ion 125, 129–32, 212
Pavy, Frederick William 11
pBR322 plasmid 286, 289,
    292–3
pCR1 plasmid 288

penicillin, determination of
    structure 173–5
pepsin 142
  Bernal and Hodgkin's
    work 171
Perkins, Tom 311–12
pernicious anaemia 73
Perutz, Max 207, 209, 210
Pet Shop Boys,
    'Opportunities' 325
Pflüger, Eduard Friedrich
    Wilhelm 91
phase problem, X-ray
    crystallography 172,
    209
phenylalanine 138*f*
photosynthesis, Calvin's
    work 199–200
*Physiological Basis of Medical
    Practice, The*, Best and
    Taylor, 1937 113–16,
    115*f*
Pickup, John 330
Pickworth, Jenny 175
Piorry, Pierre 11–12
'Pissing Evil' 5
plasmids 254–5
  Cohen's work 262
  use in human insulin
    synthesis 285–6, 288,
    289, 292, 311
  validation process 286,
    288
Plath, Sylvia, *The Bell Jar* 56
pMB9 plasmid 288
political views
  Hodgkin 191–2
  Martin 193–4
  Synge 192–3, 194–6
Pollack, Robert 258–60
pork insulin,
    'humanized' 329–30
Porton Down 106
  Gilbert's work 296–7
prandial insulin 340
  insulin analogues 341,
    342*f*, 343
Pratt, Joseph H. 122
Pringsheim, Hans 142
proinsulin 310
proline 138*f*
protamine 179
Protamine-Zinc-Insulin
    (PZI) 338–9
proteins 137, 141
  enzymes as 142
  function, early ideas 222
  haemoglobin 143

insulin as 139–40
protein structure 143–5,
    144*f*
  amino acid
    sequencing 162–5, 168
  Astbury's and Martin's
    models 163*f*
  Astbury's work 145–7,
    188–9
  Bergmann and Niemann's
    work 146
  Synge's work on gramicidin
    S 167, 189
  Synge's work on
    wool 149–65
  Wrinch's cyclol the-
    ory 181–6, 182*f*,
    187–8
protein synthesis 244*f*
  Gamow's model 236–7,
    237*f*
  PaJaMo
    experiments 238–9
pSC101 265
psychiatric disorders, insulin
    coma therapy 56–7
Ptashne, Mark 276, 278
purification of insulin 74
  Abel's work 139, 140, 141
  Collip's work 47–9, 119,
    123
  use of alcohol 116, 120,
    122, 124
purified insulins 329

Q

Quest, The (film) 122

R

radioimmunoassay
    (RIA) 315
Ramaseshan, S. 210
Rasmussen, Nicholas 335
Reagan, Ronald 321–2, 328
Recombinant DNA Advisory
    Committee 271
recombinant DNA
    technology
  Asilomar meeting 252–4,
    268–71, 270*f*
  Berg letter 267–8
  Berg's work 258–61
  Boyer and Cohen's
    work 262–3, 264–5
  'Cambridge
    incident' 276–80
  Cohen–Boyer
    patent 265–6

concerns and opposition to 263–4, 272–6, 320, 322
EcoRI 261–2
guidelines 270, 271–2, 289–90, 291–2, 319–21, 322
intellectual property rights 323
legislation, threat of 286–7, 292, 320–1
moratorium on 279, 284
motivation for insulin research 334–5
natural genetic material exchange 321
potential benefits 274
public attitudes 290–1, 291*f*
safety precautions 295, 296
suggested uses 264–5
support from Congress 321–2
Yalow's view 315 *see also* human insulin synthesis
recombinant human insulin *see* human insulin
regulation of insulin production 75, 77–8
Reimer, Neils 265–6
Rennie, John 67
restriction endonucleases (restriction enzymes) 257, 260*f*
EcoRI 261–2
Reuter, Camille 94, 97
reverse transcriptase 281–2
ribosomal RNA (rRNA) 238
ribosomes 236
Riches, Charles 63, 97
Riggs, Art 302, 306*f*, 310*f*, 316
work on somatostatin synthesis 304–6
Riley, Denis 183
Rimmer, Beryl (nee Oughton) 189
Ringer, A.I. 57–8
RNA (ribonucleic acid) 238
messenger RNA 239–40, 244*f*, 283
RNA Tie Club 240–1
Roach, Edith 20, 24
Roberts, F. 69–70
Robinson, Robert 173, 177

Roche 97
funding of Zuelzer 93–4
Rockefeller Institute
Bergmann and Niemann's work 146
Kleiner's research 32–5
Rollo, John 5–6
Root, Mary 329
Rosetta probe 137
Rosovsky, Henry 277
Ross, G.W. 84
Rous, Peyton 105
Rous sarcoma 105
Rowe, Wallace 260
Royal Society, portrait of Dorothy Hodgkin 176
Royal Society of Canada, Collip's attempts at election 110–11
Russell, Bertrand 180
Ruttenberg, Michael 329–30
Rutter, Bill 283–4, 287, 297–8
appearance before Senate committee 292
violation of NIH rules 289–90
work on human insulin synthesis 285–6

S
Sakel, Manfred 55–6
Samuels, Shimon 132
Sanger, Fred 200*f*, 310
background 200–1
collaboration with Crick 241
Nobel Prize awards 206–7, 326
relationship with Synge 202–3
work on insulin structure 202–7, 204*f*, 208*f*
Sawicki, Peter 344
scarlet fever, Zuelzer's work 100
schizophrenia, insulin coma therapy 56–7
Schlichtkrull, Jorgen 209–10, 329
Schrödinger, Erwin 222
science, purpose of 335–8
Scott, David 141
Scott, Ernest Lyman 23, 41, 101, 128
Second World War, Banting's work 106–7

secretin 38
Seeburg, Pete 297
move to Genentech 318, 324
self-management
early days 80–1
Jack Eastwood 81–2
patient education 83
Sequence Hypothesis 243
serine 138*f*
'Servant With the Scissors' 261
Shanghai Institute of Biochemistry, work on insulin synthesis 214–15
sickle-cell anaemia 242–3
Siekevitz, Phillip 273
Simian Virus 40 (SV40) 256
effect of EcoRI 257
proposed recombinant DNA research 258, 259*f*
tumour risk from 259
Simpsons, The 197
Singer, Maxine 263–4, 270*f*, 273
single-chain insulins 344
Sinsheimer, Robert 272–3, 321
smart insulins 344–5
Smith, Hamilton 257
Snailham, Bill 107
Sobolev, Leonid 10
Soll, Dieter 264
somatostatin 302
artificial synthesis of 304–6
Spinoza, Baruch 337
Sprague, R.G. 339
Stahl, Frank 240
standardisation of insulin 74, 75
Stanley, Wendell Meredith 142
Starling, Ernest 8
Starr, Clarence 18
starvation diets 11–15, 44
Stein, William 167, 196
Steiner, Don 310, 329
Stetten, Hans DeWitt 286
Stevenson, Adlai 292
Sumner, James Batcheller 141–2
Sushruta 5
Svedberg, Theodor 143
Swanson, Bob 298, 299–300, 316, 319
attempted recruitment of Ullrich 306–7

Swanson, Bob (*Continued*)
  Genentech IPO 324–5
  partnership with
    Boyer 299–300,
    301*f see also* Genentech
Synge, Richard 2, 3, 147*f*,
    232
  amino acid
    sequencing 162–5
  background 147–9
  domestic life 157–8
  and the Loch Ness
    monster 197–8
  move to Leeds 154–5
  Nobel Prize 198–9, 198*f*
  partition
    chromatography 158–60
  political stance 192–3,
    194–6
  on purpose of science 337
  refusal of admission to
    USA 195–6, 200–1
  relationship with
    Sanger 202–3
  work on gramicidin S 167,
    189
  work on wool 149–50,
    155–62
syringes and needles, early
    problems 80–1

T

Tang Youqi 214–15
Tattersall, Robert 2, 332,
    333–4, 340, 348
Taylor, Norman, *The
    Physiological Basis of
    Medical Practice* 113–16,
    115*f*
technology ethos 347–9
Teller, Edward 240
Temin, Howard 279
testicular extracts,
    Brown-Séquard's
    experiments 8–9
Teuscher, Arthur 330
Thatcher, Margaret 191–2
Thompson, Leonard 44–5,
    45*f*
  proposed depiction in
    film 120–1, 122
Thompson, Ted 206
threonine 138*f*
thymine 228, 229*f*
thyroid extracts, use in
    myxoedema 9
Tiselius, Arne 129–30, 199,
    216, 225

Todd, Alexander 175
Todd, Thomas Wingate 89
Tolkein, J.R.R. 155
Tolstoi, Edward 116
Toronto Star, Macleod's
    interviews 48, 66
Toseland, Patrick 331
transfer RNA (tRNA) 238
Trueman, A.W. 121
trypsin 36
trypsinogen 46–7
tryptophan 138*f*
Tsvett, Mikhail 152
Tuppy, Hans 206
type 1 diabetes 10–11
type 2 diabetes 10, 313–14
typhus, Zuelzer's work 99

U

Ullrich, Axel 283, 297–8,
    326
  move to Genentech 318,
    324
  recruitment attempt by
    Genentech 306–7
  work on human insulin
    synthesis 285–6, 287–90
ultracentrifuge 143
United States, availability of
    insulin 346
University of Aberdeen,
    Rennie's work 67
University of California at
    San Francisco (UCSF)
  Boyer's work 257
  breach of NIH guide-
    lines 289–90,
    292–3
  legal disputes with
    Genentech 324
  work on human insulin
    synthesis 285–6, 287–9,
    297–8
University of Cambridge
  Martin's work 151–3
  Synge's work 149–50
University of Chicago, Scott's
    work 101
University of Leeds 153
  Astbury's work 143,
    145–7, 160–1
University of Pittsburgh,
    Katsoyannis's work 213
University of Rochester,
    Murlin's research 59–64
University of Toronto

Banting and Best's
    research 24–32, 35–41,
    43–5
Banting Research
    Institute 103
Banting's Professorial
    Chair 84
Collip's work 45–7
Insulin Committee 57–8
patenting of insulin 58–9,
    63–4, 71
reactions to Nobel Prize
    award 89–90
uracil, Nirenberg's work 245
urease, Sumner's work 142
urine, sweetness of 5, 6

V

vaccines, risks versus
    benefits 274–5
valine 138*f*
Variot, M. 9
Vellucci, Alfred 275–6,
    277–9, 284
Vendrely, Roger 237
Vijayan, M. 210
Villa-Komaroff, Lydia 294–6
Vischer, Ernst 231–2
vitamin B$_{12}$
  deficiency of 73
  determination of
    structure 175
Vokhminzeva, Lyobov
    (Rho) 235
von Mering, Joseph 7
von Noorden, Carl H. 14

W

Wade, Nicholas 289–90,
    292–3
Wald, George 276
Walden, George 74
Wallace, Stephanie 331
Watson, James 220, 228,
    234, 270*f*, 327
  collaboration with
    Gamow 235–6
  commemorative plaque,
    Eagle pub 219*f*
  meeting with
    Chargaff 232–3
  on recombinant DNA
    technology 269, 272
  RNA Tie Club 240–1
  work on mRNA 240
Wells, H.G. 83–4
Whipple, George Hoyt 73
Wilder, Russell 339

Williams, John 65
Williams, John R. 12
Willis, Thomas 5
Willstätter, Richard 141, 142
Wilsden, B.H. 199
Wilson and Company,
　Murlin's connection
　to 61–3
Winterstein, A. 152
Wirsungian (pancreatic)
　duct 9, 10
　effect of blockage 22, 24–5
Wockhardt 334–5
Wolff, Simon 333
wool, Synge's early
　work 149–50
Wool Industries Research
　Association
　(WIRA) 153–4
　Martin and Synge's
　work 155–65
Wrinch, Dorothy 180*f*
　background 179–81
　cyclol theory of protein
　structure 181–4, 182*f*,
　187–8
　Pauling's criticism 184–6
　relationship with
　Hodgkin 186–8

X

Xenopus DNA research 264
Xi, Lily 306*f*
X-ray crystallography 170–1
　Astbury's work 225
　Bell's work 226–7
　Franklin's work 228
　Hodgkin's work 171–8
　phase problem 172, 209
　studies of
　　cholesterol 172–3
　studies of DNA 225–8
　studies of insulin 177–8,
　　181, 183, 210
　studies of penicillin 173–4
　studies of vitamin
　　B12 174–6
X-ray studies of proteins 143
　Astbury's work 145–7

Y

Yale University, APS meeting,
　1921 41–2
Yalow, Rosalyn
　affectionate treatment of
　　guinea pigs 314–15
　background 312–13, 313*f*
　insulin antibody
　　theory 314

insulin
　radioimmunoassay 315
　Nobel Prize award 315
Yashamori, Robert 257
Yuan, Robert 257
Yudkin, John 346

Z

Zahn, Helmut 213–14
Ziff, Edward 263
zinc atoms, presence in
　insulin 210
Zinder, Norton 270*f*
Zuelzer, Georg 90*f*, 122
　background 91
　collaboration with
　　Reuter 94
　correspondence with
　　Macleod 98–9
　funding by Roche 93–4
　later work 99–100
　military service 94
　obstacles to work 93
　patenting of work 91, 97–8
　publication of work 92–3
　reaction to Nobel
　　Prize 94–7
　research on pancreatic
　　extracts 91–2
　trials in human patients 92